Handbook of Immunoassay Technologies

Handbook of Immunoassay Technologies

Approaches, Performances, and Applications

Edited by

Sandeep K. Vashist

John H.T. Luong

ACADEMIC PRESS

An imprint of Elsevier

Academic Press is an imprint of Elsevier
125 London Wall, London EC2Y 5AS, United Kingdom
525 B Street, Suite 1800, San Diego, CA 92101-4495, United States
50 Hampshire Street, 5th Floor, Cambridge, MA 02139, United States
The Boulevard, Langford Lane, Kidlington, Oxford OX5 1GB, United Kingdom

Library of Congress Cataloging-in-Publication Data
A catalog record for this book is available from the Library of Congress

British Library Cataloguing-in-Publication Data
A catalogue record for this book is available from the British Library

ISBN: 978-0-12-811762-0

For information on all Academic Press publications visit our website at https://www.elsevier.com/books-and-journals

Publisher: John Fedor
Acquisition Editor: Linda Versteeg-buschman
Editorial Project Manager: Pat Gonzalez
Production Project Manager: Kiruthika Govindaraju
Cover Designer: Christian Bilbow

Typeset by TNQ Books and Journals

Contents

Contributors xiii
Preface xv

1. **Immunoassays: An Overview**
Sandeep K. Vashist and John H.T. Luong

1. **Overview of Immunoassays** 1
2. **Antibody Structure** 3
3. **Need for Immunoassays** 5
   3.1 Clinical 5
   3.2 Industrial 5
   3.3 Environment and Security 7
   3.4 Food 7
   3.5 Personalized Healthcare 8
4. **Immunoassay Formats** 8
5. **Conclusions and Future Trends** 12
   References 14

2. **Antibody Immobilization and Surface Functionalization Chemistries for Immunodiagnostics**
Sandeep K. Vashist and John H.T. Luong

1. **Introduction** 19
2. **Surface Functionalization Chemistries** 20
   2.1 Hydroxyl Groups 20
   2.2 Amino Groups 21
   2.3 Carboxyl Groups 22
   2.4 Sulfhydryl Groups 22
   2.5 Epoxy Groups 23
3. **Antibody Immobilization Chemistries** 23
   3.1 Covalent 23
   3.2 Oriented 32
   3.3 Noncovalent 34
   3.4 Site-Specific 36
   3.5 Peptide Nucleic Acid and Deoxyribonucleic Acid–Directed 37
   3.6 Recombinant Antibody 37

4. **Surface Characterization** 37
5. **Conclusions, Challenges, and Future Trends** 38
References 41

3. **Monoclonal Antibody Generation by Phage Display: History, State-of-the-Art, and Future**

Christian Hentrich, Francisco Ylera, Christian Frisch, Andre Ten Haaf and Achim Knappik

1. **Introduction** 47
   1.1 History of the Development of Antibody Phage Display 47
   1.2 Antibody Formats Used for Phage Display 48
   1.3 Further Recombinant Antibody Formats 50
2. **Phage Display Selection** 51
   2.1 Advantages of Recombinant Antibody Selection 53
   2.2 Guided Selection 56
   2.3 Affinity Improvement 57
   2.4 Other Selection Technologies 58
3. **Antibody Libraries** 58
   3.1 Immune Libraries 59
   3.2 Naïve Natural Libraries 60
   3.3 Naïve Semisynthetic Libraries 61
   3.4 Naïve Synthetic Libraries 61
   3.5 Special Library Designs 62
   3.6 Synthetic Libraries From Nonhuman Species 63
4. ***In Vitro* Selection of Antibodies for Specific Applications** 63
   4.1 Tissue Panning for Immunohistochemistry Antibodies 64
   4.2 Sandwich Pair Selection, Complex-Specific Antibodies, and Drug Monitoring 64
   4.3 Fully Human Controls in Diagnostic Immunoassays 65
   4.4 Site-Specific Conjugation 67
5. **Conclusion and Outlook** 69
   5.1 Future 70
References 72

4. **Bioanalytical Requirements and Regulatory Guidelines for Immunoassays**

Sandeep K. Vashist and John H.T. Luong

1. **Introduction** 81
2. **Bioanalytical Requirements for an Immunoassay** 82
   2.1 Accuracy 82
   2.2 Precision 82
   2.3 Selectivity 82
   2.4 Sensitivity 82
   2.5 Reproducibility 83
   2.6 Stability 83

2.7 Recovery 84
2.8 Calibration Curve 84
2.9 Bioanalytical Performance Parameters 85
3. **Critiques and Outlook** 89
4. **Conclusions** 92
References 93

5. Enzyme-Linked Immunoassays

Sandeep K. Vashist and John H.T. Luong

1. **Introduction** 97
2. **Conventional Enzyme-Linked Immunoassays** 98
2.1 Enzyme-Linked Immunosorbent Assay 98
2.2 Competitive Enzyme-Linked Immunoassay 101
3. **Emerging Enzyme-Linked Immunoassays** 102
3.1 High-Sensitivity Sandwich Enzyme-Linked Immunoassay 102
3.2 Highly Simplified Rapid Sandwich Enzyme-Linked Immunoassay 104
3.3 Wash-Free Immunoassays 105
3.4 Multiplex Immunoassays 108
3.5 Nano/Micromaterial-Based Enzyme-Linked Immunoassay 111
3.6 Paper-Based Enzyme-Linked Immunoassay 113
3.7 Microfluidics-Based Enzyme-Linked Immunoassay 114
3.8 Smartphone-Based Enzyme-Linked Immunoassay 118
4. **Portable Analyzer–Based Immunoassays** 119
5. **Critiques and Outlook** 121
6. **Conclusions** 123
References 123

6. Surface Plasmon Resonance–Based Immunoassays: Approaches, Performance, and Applications

K.L.M. Moran, D. Lemass and R. O'Kennedy

1. **Introduction** 129
1.1 Surface Plasmon Resonance 129
1.2 Surface Plasmon Resonance Principles 130
1.3 Surface Plasmon Resonance–Based Biosensor Platforms 132
2. **Surface Plasmon Resonance–Based Immunoassays** 135
2.1 Antibody Introduction 135
2.2 Antibody Screening Using Surface Plasmon Resonance 138
2.3 Surface Plasmon Resonance Immunoassay Introduction 140
3. **Future Trends in Surface Plasmon Resonance–Based Immunoassays** 144
3.1 Surface Plasmon Resonance–Mass Spectrometry 144
3.2 Surface Plasmon Resonance–Point-of-Care Applications 148
4. **Conclusions** 150
Acknowledgments 152
References 152

7. Lateral Flow Immunoassays

Aart van Amerongen, Jeroen Veen, Hugo A. Arends and Marjo Koets

1. **Introduction** 157
   - 1.1 Lateral Flow Immunoassays 157
2. **Advances in Lateral Flow Immunoassays** 162
   - 2.1 Coupling to a Range of Detection Principles 162
   - 2.2 Multianalyte and Quantitative Lateral Flow Immunoassays 163
   - 2.3 Reading MultiSpot Lateral Flow Assays 166
3. **Challenges and Future Directions** 171
   - 3.1 Updated SWOT Analysis 171
   - 3.2 Combination With Amplification Procedures 172
   - 3.3 Integration of Lateral Flow Immunoassays With Paper Diagnostics 173
4. **Bibliographic and Commercial Data** 174
5. **Conclusions** 175

References 176

8. Paper-Based Immunoassays

Zhi Liu, Jie Hu, Zhiguo Qu and Feng Xu

1. **Paper-Based Immunoassays: Strategies and Detection Principles** 184
   - 1.1 Colorimetric Method 185
   - 1.2 Thermal Method 188
   - 1.3 Electrochemical Method 188
   - 1.4 Magnetic Method 189
2. **Development of the Paper-Based Immunoassays Devices** 190
   - 2.1 Sensitivity Improvement 191
   - 2.2 Automatic Detections 192
   - 2.3 Semiquantification Detection and Quantification Detection 194
3. **Conclusions** 196

References 196

9. Acoustic Wave–Based Immunoassays

Aristea Grammoustianou and Electra Gizeli

1. **Introduction** 203
2. **Clinical Diagnostics** 205
   - 2.1 Quartz Crystal Microbalance Immunosensors 205
   - 2.2 Surface Acoustic Wave Immunosensors 213
3. **Detection of Microbial Pathogens and Toxins** 213
   - 3.1 Quartz Crystal Microbalance Immunosensors 213
   - 3.2 Surface Acoustic Wave Immunosensors 216

4.	**Detection of Parasites**	216
	4.1 Quartz Crystal Microbalance Immunosensors	216
5.	**Detection of Viruses**	217
	5.1 Quartz Crystal Microbalance Immunosensors	217
	5.2 Surface Acoustic Wave Immunosensors	222
6.	**Quartz Crystal Microbalance and Surface Acoustic Wave-Based Electronic Noses**	223
7.	**Quartz Crystal Microbalance and Surface Acoustic Wave Immunoassays in Environmental Monitoring and Allergens Detection**	224
8.	**Integrated Acoustic Wave Immunosensors for Point of Care**	225
9.	**Commercial Acoustic Wave Immunosensors**	227
10.	**Market Potential and Conclusions**	228
	Acknowledgments	229
	References	229

10. Interferometry-Based Immunoassays

Panagiota Petrou, Eleni Makarona, Sotirios Kakabakos, Georgios Koukouvinos, Konstantinos Misiakos and Ioannis Raptis

1.	**Introduction: General Context**	241
2.	**Principles of Operation**	243
	2.1 Label-Free Optical Sensing	243
	2.2 Interferometric Sensors	247
3.	**Sensor Surface Functionalization**	253
	3.1 Chemical Activation of Transducers	255
	3.2 Immobilization of Recognition Molecules	256
	3.3 Elimination of Nonspecific Binding	260
4.	**Application of Interferometric Immunosensors**	260
	4.1 Mach–Zehnder Interferometers	261
	4.2 Young Interferometers	263
	4.3 Bimodal Interferometers	264
5.	**Conclusions and Future Perspectives**	265
	References	266

11. Nanomaterial- and Micromaterial-Based Immunoassays

Minli You, Zedong Li, Penghui Zhang, Dan Bai, Min Lin and Feng Xu

1.	**Introduction**	273
2.	**Micromaterial-Based Immunoassay**	274
	2.1 Fluorescent Polystyrene Microsphere	274
	2.2 Magnetic Microbeads	275
	2.3 Nanomaterial-Based Immunoassay	277
3.	**Colorimetric Immunoassay**	279
	3.1 Lateral Flow Assay	279
	3.2 Plate-Based Colorimetric Immunoassay	282

4. Electrochemical Immunoassay 285
5. Fluorescent Immunoassay 290
5.1 Heterogeneous Immunoassay 290
5.2 Fluorescence Resonance Energy Transfer Assay 293
6. Conclusion 299
References 300

12. Microcantilever-Based Sensors
Sandeep K. Vashist and John H.T. Luong

1. Introduction 305
2. Microcantilevers and Their Modes of Operation 306
2.1 Operating Modes for Cantilever Mass Sensors 307
3. Detection Methods 308
3.1 Optical 308
3.2 Piezoresistive 309
3.3 Capacitive 310
3.4 Piezoelectric 311
3.5 Interferometry 311
3.6 Optical Diffraction Grating 312
3.7 Charge-Coupled Device 312
4. Bending Behavior of Microcantilevers 312
5. Fabrication of Microcantilevers 314
6. Microcantilever-Based Sensors 315
6.1 Detection of Biomolecules 317
6.2 Detection of Gaseous Analytes 319
6.3 Detection of Chemicals and Metal Ions 320
6.4 Detection of Humidity and pH 321
6.5 Detection of Explosives and Monitoring of Ammunition 322
7. Electronic Nose 323
8. Nanocantilevers 323
9. Commercial Availability 325
10. Conclusions and Future Trends 326
References 327

13. Quartz Crystal Microbalance–Based Sensors
Sandeep K. Vashist and John H.T. Luong

1. Introduction 333
2. Detection of Biomolecules 335
3. Detection of Bacteria 343
4. Detection of Volatile Organic Compounds 345
5. Detection of Chemical Analytes 347
6. Detection of Gaseous Analytes 348
7. Special Analytical Applications 349
8. Other Analytical Applications 351
9. Conclusions and Future Trends 352
References 353

14. Electrochemical Immunosensors: Fundamentals and Applications in Clinical Diagnostics

Kuldeep Mahato, Suveen Kumar, Ananya Srivastava, Pawan K. Maurya, Renu Singh and Pranjal Chandra

1.	**Introduction**	359
2.	**Electrochemical Techniques in Immunosensing**	360
	2.1 Voltammetry	361
	2.2 Amperometry	362
	2.3 Potentiometry	363
	2.4 Electrochemiluminescence	363
	2.5 Electrochemical Impedance Spectroscopy	364
	2.6 Photoelectrochemical Techniques	365
3.	**Advanced Approaches in Electrochemical Immunosensors**	366
	3.1 Electrochemical Lateral Flow Assay	366
	3.2 Approaches Based on Receptor–Ligand–Receptor Binding	367
	3.3 Colorimetric Immunosensing Approaches	368
	3.4 Redox Cycling–Based Approaches	370
4.	**Biorecognition Elements in Immunosensors**	372
	4.1 Antibodies: The Primary Recognition Elements of Immunoassays	373
	4.2 Truncated (Fragmented) Antibodies	374
	4.3 Aptamer–Antibodies Hybrid	375
5.	**Electrochemical Immunosensors**	376
	5.1 Label-Free Protein-Based Immunosensors	377
	5.2 Bacterial/Cancer Cells–Based Immunosensors	381
	5.3 Enzyme-Labeled Immunosensors	388
	5.4 Redox-Labeled Immunosensors	397
	5.5 Aptamer-Based Immunosensors	399
6.	**Conclusions and Outlook**	401
	Acknowledgments	403
	References	403

15. Lab-on-a-Chip (LOC) Immunoassays

Sandeep K. Vashist and John H.T. Luong

1.	**Introduction**	415
2.	**Lab-on-a-Chip Immunoassays**	416
	2.1 Microfluidic	416
	2.2 Paper	420
	2.3 Cellphone	421
3.	**Critiques and Outlook**	425
4.	**Conclusions**	427
	References	428

16. Smartphone-Based Immunoassays

Sandeep K. Vashist and John H.T. Luong

1.	**Introduction**	433
2.	**Signal Detection in Smartphone-Based Immunoassays**	434
	2.1 Colorimetric, Fluorescent, and Luminescent Detection	434
	2.2 Electrochemical Detection	435
	2.3 Surface Plasmon Resonance Detection	439
	2.4 Lateral Flow Assays	439
	2.5 Microscopy	442
	2.6 Cytometry	443
3.	**Applications of Smartphone-Based Immunoassays**	444
	3.1 Detection of Biomolecules and Microorganisms	444
	3.2 Detection of Other Analytes	448
	3.3 Detection of Blood Glucose	449
4.	**Conclusions, Challenges, and Future Trends**	449
	References	450

17. Immunoassays: Future Prospects and Possibilities

Sandeep K. Vashist and John H.T. Luong

1.	**Immunoassays: Trends and Prospects**	455
2.	**Evolving Healthcare**	457
3.	**Challenges**	459
4.	**Future Directions**	461
	References	463

Index 467

Contributors

Hugo A. Arends, HAN University of Applied Sciences, Arnhem, The Netherlands

Dan Bai, Xi'an Jiaotong University, Xi'an, P.R. China

Pranjal Chandra, Indian Institute of Technology Guwahati, Guwahati, India

Christian Frisch, Bio-Rad, Puchheim, Germany

Electra Gizeli, University of Crete, Heraklion, Greece; Institute of Molecular Biology and Biotechnology, Foundation for Research and Technology-Hellas, Heraklion, Greece

Aristea Grammoustianou, University of Crete, Heraklion, Greece; Institute of Molecular Biology and Biotechnology, Foundation for Research and Technology-Hellas, Heraklion, Greece

Christian Hentrich, Bio-Rad, Puchheim, Germany

Jie Hu, Xi'an Jiaotong University, Xi'an, P.R. China

Sotirios Kakabakos, Institute of Nuclear & Radiological Sciences & Technology, Energy & Safety, Athens, Greece

Achim Knappik, Bio-Rad, Puchheim, Germany

Marjo Koets, Wageningen University & Research, Wageningen, The Netherlands

Georgios Koukouvinos, Institute of Nuclear & Radiological Sciences & Technology, Energy & Safety, Athens, Greece

Suveen Kumar, Ulsan National Institute of Science and Technology (UNIST), Ulsan, Republic of Korea

D. Lemass, Dublin City University, Dublin, Ireland

Zedong Li, Xi'an Jiaotong University, Xi'an, P.R. China

Min Lin, Xi'an Jiaotong University, Xi'an, P.R. China

Zhi Liu, Xi'an Jiaotong University, Xi'an, P.R. China

John H.T. Luong, University College Cork, Cork, Ireland

Kuldeep Mahato, Indian Institute of Technology Guwahati, Guwahati, India

Eleni Makarona, Institute of Nanoscience and Nanotechnology, Athens, Greece

Pawan K. Maurya, Universidade Federal de Sao Paulo-UNIFESP, Sao Paulo, Brazil

Konstantinos Misiakos, Institute of Nanoscience and Nanotechnology, Athens, Greece

K.L.M. Moran, Dublin City University, Dublin, Ireland

R. O'Kennedy, Dublin City University, Dublin, Ireland

Panagiota Petrou, Institute of Nuclear & Radiological Sciences & Technology, Energy & Safety, Athens, Greece

Zhiguo Qu, Xi'an Jiaotong University, Xi'an, P.R. China

Ioannis Raptis, Institute of Nanoscience and Nanotechnology, Athens, Greece

Renu Singh, University of Minnesota, Saint Paul, MN, United States

Ananya Srivastava, Indian Institute of Technology Delhi, New Delhi, India

Andre Ten Haaf, Bio-Rad, Puchheim, Germany

Aart van Amerongen, Wageningen University & Research, Wageningen, The Netherlands

Sandeep K. Vashist, IDS Immunodiagnostic Systems Deutschland GmbH, Frankfurt am Main, Germany

Jeroen Veen, HAN University of Applied Sciences, Arnhem, The Netherlands

Feng Xu, Xi'an Jiaotong University, Xi'an, P.R. China

Francisco Ylera, Bio-Rad, Puchheim, Germany

Minli You, Xi'an Jiaotong University, Xi'an, P.R. China

Penghui Zhang, Xi'an Jiaotong University, Xi'an, P.R. China

Preface

As the smart eyes of healthcare professionals, immunoassays play a prominent role in clinical diagnostics to confirm suspected diseases, enabling clinical decisions and subsequent treatments. Ideally, immunoassays must have high detection selectivity and specificity together with a high degree of automation, low cost, and fast analysis time. In clinical settings, pertinent treatments based on fast and reliable diagnosis favor the patient's prognosis and prevent the disease spreading. Other important applications of immunoassays include veterinary diagnostics, food analysis, environmental monitoring, defense and security, and other bioanalytical settings. To date, immunoassays are available for a plethora of analytes such as disease biomarkers, pathogens, drug impurities, environmental contaminants, allergens, food adulterants, drugs of abuse, and various biomolecules.

This book unravels the role and technology evolution of immunoassays in bioanalytical sciences. During the last four decades, immunoassays have been transformed from conventional enzyme-linked immunosorbent assays to more advanced formats including smartphone-based point-of-care devices. Such advanced formats stem from rapid bioanalytical procedures, novel biosensing schemes, fully integrated lab-on-a-chip platforms, prolonged biomolecular storage strategies, device miniaturization and interfacing, and the emerging smart system technologies equipped with personalized mobile healthcare tools.

OBJECTIVES OF THE BOOK

This book aims to provide the comprehensive details of various types of immunoassays, which have been performed in healthcare, industrial, environmental, and other bioanalytical settings. The multifarious aspects of immunoassays are discussed to cover the principle of operation, characteristics, pros and cons, and potential bioanalytical applications.

SCOPE OF THE BOOK

This book outlines various aspects of immunoassays to encompass antibody/protein immobilization, surface functionalization, point-of-care devices, and regulatory requirements. Various types of immunoassays together with their bioanalytical performance and characteristics are discussed in depth.

TARGET AUDIENCE

This book aims to provide a thorough understanding and comprehensive view of immunoassays and in vitro diagnostics (IVD) to healthcare professionals, biomedical engineers/scientists, researchers, healthcare economists, policy makers, investors, professionals in preventive healthcare, and persons interested in personalized healthcare. This book serves as a very useful resource and teaching aid for professionals and researchers in immunoassays, IVD, biosensors, biomedical diagnostics, medicine, bioengineering, and multidisciplinary scientific disciplines.

BOOK ORGANIZATION

This book targets the various immunoassay formats developed to date and the multifarious aspects of immunoassays. The first two chapters provide an overview of immunoassays, antibody immobilization, and surface functionalization chemistries. The third chapter provides an in-depth expert insight into monoclonal antibody generation. The bioanalytical requirements and regulatory requirements for immunoassays are discussed in the fourth chapter. The subsequent chapters 5–16 discuss the various types of immunoassays developed to date in comprehensive detail. They provide the principles of operation, advantages, potentials, challenges, limitations, bioanalytical applications, and future trends of each immunoassay. The detection platforms are based on enzyme-linked immunoassay, surface plasmon resonance, lateral flow, paper, acoustic wave, interferometry, nano- and micromaterials, microcantilevers, quartz crystal microbalance, electrochemical, lab-on-a-chip technologies, and smartphone readout. The concluding chapter discusses the future prospects of immunoassays.

We highly appreciate the efforts devoted by the authors in drafting the outstanding book chapters. The knowledge and insight provided by this book on immunoassays will lend considerable support to more effective healthcare management as it provides the most comprehensive and updated information about various types of immunoassays and its varied aspects including the immunoanalytical platforms/systems/devices. Immunoassays play an instrumental role in clinical decision-making by enabling rapid and highly sensitive detection of disease biomarkers and analytes. Moreover, they are critically important in the biopharmaceutical industry for the manufacture of improved drugs with lesser impurities. Additionally, they have a prominent role in environmental monitoring. Therefore, they form an integral part of bioanalytical sciences in healthcare, industry, environmental, and other bioanalytical settings. The information provided in this book will empower the targeted end-users with the desired knowledge, competent skills, and comprehensive resource pertaining to the immunoassays. It will empower the targeted users to effectively manage the bioanalytical applications and lead to critically improved outcomes in industrial, healthcare, research, and other settings.

The guided insight and knowledge of immunoassays and varied aspects of immunoassays provided in this book will act as a potential resource for professionals, researchers, and students, who are working or would like to work in the field of bioanalytical sciences.

Sandeep Kumar Vashist
Aachen, Germany
John H.T. Luong
Cork, Ireland

Chapter 1

Immunoassays: An Overview

Sandeep K. Vashist[1], John H.T. Luong[2]
[1]*IDS Immunodiagnostic Systems Deutschland GmbH, Frankfurt am Main, Germany;* [2]*University College Cork, Cork, Ireland*

1. OVERVIEW OF IMMUNOASSAYS

Immunoassays (IAs) play a critical role in various bioanalytical settings, such as clinical diagnostics, biopharmaceutical analysis, environmental monitoring, security, and food testing. During 1995–2017, a wide range of IAs have been developed to provide the quantitative, semiquantitative, or qualitative detection of analytes. The precise early-stage detection of analytes is an essential requirement for all bioanalytical settings to effectively monitor and manage the quality of the biopharmaceutical drugs, foods, and environment. It is even more critical to effectively diagnose, monitor, and manage the patients' health. Considering the prominent role that IAs play in the clinical decision-making, they are indispensable for healthcare settings.

Tremendous advances in IA formats, bioanalytical platforms, immunoanalytical systems, and complementary technologies have led to various emerging IA formats. Initially, most of the IAs, developed several decades ago, were based on radioimmunoassay (RIA) and enzyme-linked immunosorbent assay (ELISA). Consequently, RIA- and ELISA-based kits for a wide range of analytes became commercially available as the antibodies were generated against such analytes. Owing to extremely high sensitivity, specificity, precision, and throughput of ELISA, this format has served as the gold standard for a plethora of analytes. The last two decades has seen tremendous innovation in ELISA technology. Automated assays are equipped with robotic workstations while the microtiter plate (MTP) readers have also improved drastically in terms of technology, features, and cost-effectiveness. The conventional 96-well MTP formats are transformed into higher-throughput 384- and 1536-well formats. The bioanalytical performance of ELISA has also improved significantly by better antibody immobilization chemistries [23,61,78–80], development of ultrasensitive enzyme substrates, use of micromaterial- or nanomaterial-based signal enhancement strategies [20,71,74], and novel IA concepts [38]. The current trend in ELISA deviates toward wash-free ELISA

Handbook of Immunoassay Technologies. https://doi.org/10.1016/B978-0-12-811762-0.00001-3

such as AlphaLISA by Perkin Elmer [8], which has critically reduced the IA duration and complexity.

Advanced bioengineering, microfluidic, and biosensor technologies [24,31,35,55,62] have further led to the development of prominent immunoanalytical systems including surface plasmon resonance (SPR)-based BiaCore instruments for label-free IA from GE Healthcare (previously Pharmacia Biosensor AB, Biacore AB Corporation). These systems have become the global standard for developing rapid IA for the detection of analytes and screening of immunological components based on the determination of biomolecular interactions [27,54]. Although GE Healthcare is still the leader in this domain, other competitors have developed SPR-based immunoanalytical systems.

The advent of microfluidics has started the quest for novel IA platforms and formats, which enable rapid IA using the minimal volume of reagents [77]. An example is the Optimiser ELISA by Siloam Biosciences, USA, which involves the conversion of the conventional 96-well MTP-based ELISA into microfluidic ELISA [29]. The analysis time is only a few minutes as this hybrid system employs significantly reduced number of steps and a microfluidic IA protocol. Similarly, a large number of prospective lab-on-a-chip technologies and formats have been developed toward the development of various microfluidic IA [41]. The most widely used microfluidic IA format is the lateral flow IA (LFIA). Although most of the LFIAs are qualitative, many prospective quantitative LFIAs have been demonstrated in recent years such as those based on smartphone (SP) detection [42]. Another prospective development route has been the development of electrochemical IA [55].

There is also an emerging trend toward the development of multiplexed IA that holds a tremendous potential in "in vitro" diagnostics and bioanalytical sciences [32] as exemplified by the two most widely used formats: multiplex bead-based Luminex assays by Thermo Fisher Scientific and multiplex IA by Meso Scale Discovery.

Other new biosensing concepts are based on microgravimetry (cantilevers and quartz crystal microbalance) [21,57,60,70], interferometry [39], and acoustic waves [33]. Although such formats have been demonstrated in various laboratories, they could not gather much commercial attention. Of interest is the deployment of mass spectroscopy–based IAs, which have been extensively used as the clinical and industrial IA standard for a large number of analytes. Albeit it requires highly sophisticated and costly instrumentation along with highly skilled analysts, this technique offers unmatched high sensitivity, specificity, and analytical performance. Accordingly, it is being used as the established clinically validated IAs for various analytes, which makes it essential to use these IAs to determine the precision of newly developed IAs. The development of low-cost IA formats for the developing nations has received considerable attention [40] as attested by paper-based IA formats for a number of IAs [20,26]. Despite being low-cost and easy to produce in large numbers, they still lack high analytical sensitivity comparable with that of the established IAs [63].

The most recent trend is the development of SP-based IA (SPIA) formats and devices, which outperform the central laboratory IA and instruments [62,63,66]. SPIA and devices have been developed for LFIA, electrochemical IA, colorimetric IA, chemiluminescent IA, fluorescent IA, SPR-based IA, lab-on-a-chip–based IA, and other IA formats. The use of SPs enhances the capabilities of diagnostics and enables the analysis at the point-of-care settings, such as decentralized and remote locations, and personalized settings. The IA readings together with the time and date stamp and spatiotemporal information can be stored securely on the SP and the centralized server by Cloud computing via wireless or Internet connectivity. The readings can be immediately processed on the SP or the centralized server and the results are transmitted back to the user. SPs have emerged as the ideal point-of-care device for IA as they are extensive outreach and capabilities considering over 7 billion SP users, i.e., ~96% of the world's population [3].

Other parallel improvements are antibody immobilization chemistries, rapid IA strategies, portable readout devices, microfluidic and lab-on-a-chip technologies, innovative diagnostic platforms, and assisting complementary technologies. These ongoing advances have considerably improved the bioanalytical performance of IA and would further lead to continuous improvements in IA during the coming years.

2. ANTIBODY STRUCTURE

The antibody, also known as immunoglobulin, is the most critical component of an IA as it provides the desired high specificity and sensitivity for the analyte of interrogation. The screening and generation of the specific antibody are the preliminary tasks before the development of IA for an analyte. Among five classes of antibodies that are present in humans, the most predominantly used Ab in immunodiagnostics is immunoglobulin G (IgG). IgG accounts for ~75% of total serum immunoglobulin and protects the newborn during the initial months as it can cross the placenta in humans. An IgG is a Y-shaped glycoprotein consisting of four polypeptide chains with two identical heavy chains (H) inside and two identical light chains (L) outside (Fig. 1.1). Each L chain contains a variable domain V_L and a constant domain C_L while each H chain has a variable domain, V_H, and three constant domains C_H1, C_H2, and C_H3. The amino terminal ends of the polypeptide chains, located at the top of Y-shaped Ab, are referred to as the variable (V) regions as they have significant variation in amino acid composition. In contrast, the remaining Ab structure is composed of relatively constant (C) regions. The V regions of H and L chains form the

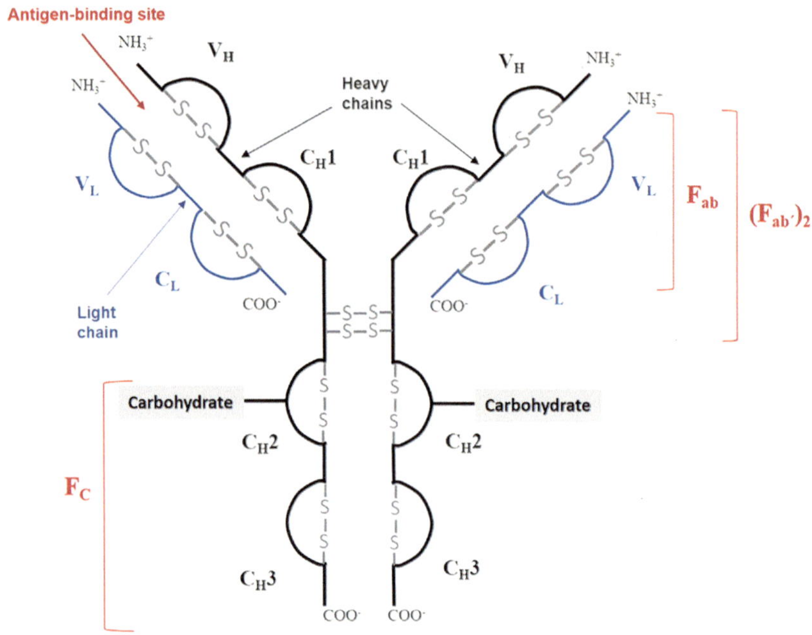

FIGURE 1.1 The structure of an antibody.

antigen-binding sites of IgG on both arms of Y-shaped Ab. Therefore, each IgG is bivalent as it contains two antigen-binding sites that are held together by covalent disulfide bonds between the H and L chains, and noncovalent interactions. The H chains at the hinge region are also held together by disulfide bonds. The total MW of an IgG is ~150 kDa, which is composed of ~440 amino acids. The heavy chain has an MW of ~50 kDa and double the number of amino acids than the light chain having MW of ~25 kDa. There are four subclasses of IgG in humans, i.e., IgG1, IgG2, IgG3, and IgG4

The other classes of Ab have different heavy chains, i.e., μ-chains for IgM; α-chains for IgA; ε-chains for IgE; and δ-chains for IgD. They function in different types of immune responses and at particular stages of the immune response. The polypeptide protein sequences responsible for these differences are located primarily in the Fc fragment. However, there are only two main types of light chains, i.e., kappa (κ) and lambda (λ). These antibodies have a varying number of Y-like monomers. IgM is a pentamer containing 10 light chains, 10 heavy chains, and 10 antigen-binding sites. The IgM monomers are bound together by disulfide bonds and a joining (J) chain. It is the most efficient complement-fixing Ab that constitutes ~10% of normal human serum Ab content and is majorly involved in the primary immune response to most antigens. IgA constituting ~15% of the total serum Ab exists in both monomeric and dimeric forms. It is predominant in saliva and tears as a dimer, where it protects against local infections and prevents foreign substances from entering into the circulatory system. IgD and IgE are prevalent in serum at a very low concentration where IgD acts as a receptor for antigens on mature B lymphocytes while IgE prevents the invasion of parasites.

3. NEED FOR IMMUNOASSAYS

3.1 Clinical

A large number of diseases are diagnosed exclusively based on the determination of disease biomarkers' concentrations by IA. The monitoring and management of the disease and the effectiveness of the therapeutic regimen are also done via IA. Therefore, IAs play a critical role in the clinical decision-making. Most of the IA in developed countries are fully automated and performed using commercial clinical analyzers, such as those from Roche, Abbott, and Siemens. However, manual ELISA is still mainly used in developing nations due to the lack of resources and infrastructure. Most automated IAs have the sample-to-analysis time of about half an hour, a critically required feature for clinical diagnostics. In brief, commercial automated IAs are based on disease biomarkers, as specified in Table 1.1. Nevertheless, there is still a need to develop such automated IAs for less prevalent disease biomarkers. To date, a large number of manual ELISAs have been developed for numerous disease biomarkers. The leading companies in the manual ELISA are R&D Systems, BD Biosciences, Bio-Rad Laboratories, Life Technologies Corporation, Millipore Corporation, Thermo Fisher Scientific Inc., BioMerieux SA, EMD Biosciences Inc., Abbott Diagnostics, BioLegend, and Ortho-Clinical Diagnostics Inc. It is essential for all IA to be clinically validated before they can be used in healthcare settings. Therefore, the development of a clinically validated IA must follow stringent bioanalytical testing guidelines.

3.2 Industrial

IAs are being used routinely in biopharmaceuticals for the analysis of biopharmaceutical analytes [18]. The most widely used IAs are those for Chinese hamster ovary protein and monocyte chemotactic protein, which need to be monitored very frequently. Similarly, several biological compounds and biomarkers are determined in industries by IAs. These analytes include insulin, cytokines, prostate-specific antigen, troponin I, insulin-like growth factor 1, matrix metallopeptidase 9, C-reactive protein, N-terminal prohormone of brain natriuretic peptide, glycated hemoglobin, and other biomolecules/disease

TABLE 1.1 Biomarkers for Various Clinical Areas That Are Determined by Automated Clinical Analyzer—Based Immunoassays

Clinical Areas	Biomarkers
Inflammation	IL-1beta, IL-10, IL-2R, IL-6, IL-8, LBP, TNF-alpha, Anti-CCP, ASLO, C3c, C4, CRP, ceruloplasmin, haptoglobulin, IgA, IgE, IgG, IgM, IgA CSF, IgM CSF, kappa light chains, kappa light chains free, lambda light chains, lambda light chains free, prealbumin, procalcitonin
Oncology	AFP, free PSA, PSA, CA 125, CA 15-3, CA 19-9, CA 72-4, calcitonin, CEA, Cyfra 21-1, beta-hCG, HE4, kappa light chains free, lambda light chains free, NSE, proGRP, TPS, TPA
Hepatology	Anti-HBs, Anti-HBs quantitative, Anti-HBc IgM, Anti-HBc total, HBsAg, HBsAg confirmatory, HBeAg
Endocrinology	ACTH, cortisol, homocysteine, insulin, calcitonin, C-peptide, hGH, IGF-1, PTH, T3, T4, TSH, thyreoglobulin
Fertility	Anti-mullerian hormone, androstenedione, DHEA, DHEA-S, estradiol, FSH, hCG, free beta-HCG, LH, PAPP-A, progesterone, 17-OH progesterone, prolactin, SHBG, testosterone, free testosterone, unconjugated estriol, estrone, DHT
Diabetes	C-peptide, insulin, microalbumin, IAA, ICA, anti-GAD
Growth	HCG, IGF-1, IGF BP-3
Thyroid	TG, anti-TG, TPO, anti-TPO, free T3, free T4, intact PTH, T-uptake, TBG, thyroglobulin, total T3, total T4, TSH, TSI
Therapeutic drug monitoring	Carbamazepine, digitoxin, digoxin, phenobarbital, phenytoin, theophylline, valproic acid
Renal	Cortisol, DHEA-S, ACTH, heptacidin 25 (bioactive)
Infectious diseases	Anti-HAV, anti-HAV IgM, anti-HBc, anti-HBc IgM, anti-HBe, HBeAg, anti-HBs, HBsAg, HCV, anti-HCV, CMV IgG, CMV IgM, herpes I and II IgG, rubella IgG, rubella IgM, Toxoplasma IgG, Toxoplasma IgM, Chagas, HIV-Ag, HSV-1 IgG, HSV-2 IgG, HSV-1 IgM, HSV-2 IgM, HTLV 1 and 2, syphilis, EBV-EBNA IgG, EBV-VCA IgG, EBV-VCA IgM, H. pylori IgG, calprotectin, Chlamydia IgG, Chlamydia IgM, Fasciola IgG, Treponema Screen
Bone	Osteocalcin, pyrilinks-D, 25-OH vitamin D
Anemia	Vitamin B_{12}, EPO, hepcidin 25 (bioactive)
Allergy	Allergen-specific IgE, allergen-specific IgG, allergen-specific IgG4, ECP, IgE total
Autoimmune	Anti-CCP, transglutaminase IgG, transglutaminase IgM
Cardiac	CK, CK-MB, D-dimer, high-sensitivity CRP, myoglobin, NT-proBNP, troponin I, troponin T high sensitivity, cystatin C

TABLE 1.1 Biomarkers for Various Clinical Areas That Are Determined by Automated Clinical Analyzer–Based Immunoassays—cont'd

Clinical Areas	Biomarkers
Coagulation	AT III, D-dimer
Neurotransmitters	Adrenalin, dopamine, serotonin, normetanephrine, metanephrine, histamine, melatonin
Hypertension	Renin (active), aldosterone, angiotensin

ACTH, adrenocorticotropic hormone; *AFP*, alpha-fetoprotein; *ASLO*, antistreptolysin O; *AT III*, antithrombin III; *CA*, carcinoma antigen; *CCP*, cyclic citrullinated peptide; *CEA*, carcinoembryonic antigen; *CK*, creatine kinase; *CK-MB*, creatine kinase MB isoenzyme; *CMV*, cytomegalovirus; *CRP*, C-reactive protein; *DHEA*, dehydroepiandrosterone; *DHEA-S*, dehydroepiandrosterone sulfate; *DHT*, dihydrotestosterone; *EBV-EBNA*, Epstein–Barr virus nuclear antigen; *EBV-VCA*, Epstein–Barr virus viral capsid antigen; *ECP*, eosinophil cationic protein; *EPO*, erythropoietin; *FSH*, follicle-stimulating hormone; *LH*, luteinizing hormone; *GAD*, glutamic acid decarboxylase; *HAV*, hepatitis A virus; *HBc*, hepatitis B core; *HBeAg*, hepatitis B e antigen; *HBs*, hepatitis B surface; *HBsAg*, hepatitis B surface antigen; *hCG*, human chorionic gonadotropin; *HCV*, hepatitis C virus; *HE4*, human epididymis protein 4; *hGH*, human growth hormone; *HIV*, human immunodeficiency virus; *HSV*, herpes simplex virus; *HTLV*, human T-lymphotropic virus; *IAA*, indoleacetic acid; *ICA*, islet cell antibodies; *IGF BP-3*, insulin-like growth factor binding protein 3; *IGF-1*, insulin-like growth factor 1; *IL*, interleukin; *LBP*, lipopolysaccharide-binding protein; *NSE*, neuron specific enolase; *NT-proBNP*, N-terminal prohormone of brain natriuretic peptide; *PAPP-A*, pregnancy-associated plasma protein A; *proGRP*, Progastrin-releasing peptide; *PSA*, prostate-specific antigen; *PTH*, parathyroid hormone; *SHBG*, sex hormone–binding globulin; *T3, 3,5′,3′*-triiodo-L-thyronine; *T4, 3,3′,5,5′*-tetraiodo-L-thyronine; *TBG*, thyroxine-binding globulin; *TG*, thyroglobulin; *TNF*, tumor necrosis factor; *TPA*, tissue polypeptide antigen; *TPO*, thrombopoietin; *TPS*, tissue polypeptide–specific antigen; *TSH*, thyroid-stimulating hormone; *TSI*, thyroid-stimulating immunoglobulins.

biomarkers. Further, the current trend in biopharmaceuticals is toward the manufacturing of biologics, i.e., biomolecules such as antibodies that are used as drugs. Therefore, there is a growing need for IAs in industry. Although most of the IAs used in industry are based on automated ELISA, there has been an evolving trend toward the continuously increasing use of real-time SPR-based IAs such as those using BiaCore instruments from GE Healthcare or similar instruments from other companies [43,54].

3.3 Environment and Security

Several IAs need to be developed for the monitoring of the polluting substances such as harmful chemicals (pesticides, mercury, arsenic, etc.), toxins, or biowarfare agents. The outbreak of these substances could lead to epidemics. An example was the use of Anthrax spores as a biowarfare agent, which led to serious security concerns.

3.4 Food

The testing of the potential adulterants, toxins, or harmful substances in foods is done by IAs [1,47]. The most common are the IA for testing the food allergens [53], which are one of the main concerns in the Western society. There have been many outbreaks of *Escherichia coli* O157: H7 contamination of food products [14,44]. Similarly, there have been potential outbreaks of botulinum toxin [5,76] and other pathogens such as *Salmonella* [15,22] and *Campylobacter* [82] in food products that have led to numerous critical investigations.

3.5 Personalized Healthcare

The current trend deviates strongly toward the development of point-of-care IAs, such as those based on a disposable lab-on-a-chip and SP readout, which can be used at any place and time for the personalized monitoring of important health parameters [62,63,67]. Many prospective IAs have also been recently developed, which enable the monitoring of analytes in just a few minutes using an easy-to-use procedure. The bioanalytical performance of such IA is comparable or even better than the IAs performed in the central laboratories using advanced instruments.

4. IMMUNOASSAY FORMATS

ELISA has been the most widely used IA format, which has been used for the development of immunodiagnostic kits for numerous analytes due to its high sensitivity, high specificity, and high throughput. The ELISA procedure has been automated by employing robotic workstations equipped with washing and dispensing modules, and automated mechanical handling and analysis. There has been continuous innovation in antibody immobilization strategies [23,61,78–80], enzyme substrates [19], signal enhancement approaches [20,71,74], and novel IA concepts [38], which have considerably improved the bioanalytical performance of ELISA. The outstanding advance is the development of wash-free ELISA, i.e., AlphaLISA by Perkin Elmer [8] (Fig. 1.2A), which has considerably reduced the IA duration and complexity of the conventional ELISA. AlphaLISA employs streptavidin-coated alpha donor beads and anti-analyte-conjugated acceptor beads. The formation of the immune complex in the presence of analyte brings the acceptor and donor beads in close proximity leading to the generation of the chemiluminescent signal. It offers high sensitivity, broad dynamic range, and analytical performance superior to that of the conventional ELISA. Another prospective development is the Optimiser ELISA by Siloam Biosciences, USA, which employs a microfluidic ELISA protocol in a novel MTP [29] (Fig. 1.2B). The IA format has a sample-to-analysis time of only a few minutes due to the microfluidic IA procedure with reduced number of steps and lesser volume requirement of IA reagents. An interesting advance is the naked-eye ELISA using gold nanoparticles with attogram per mL range detection of analyte [9,16].

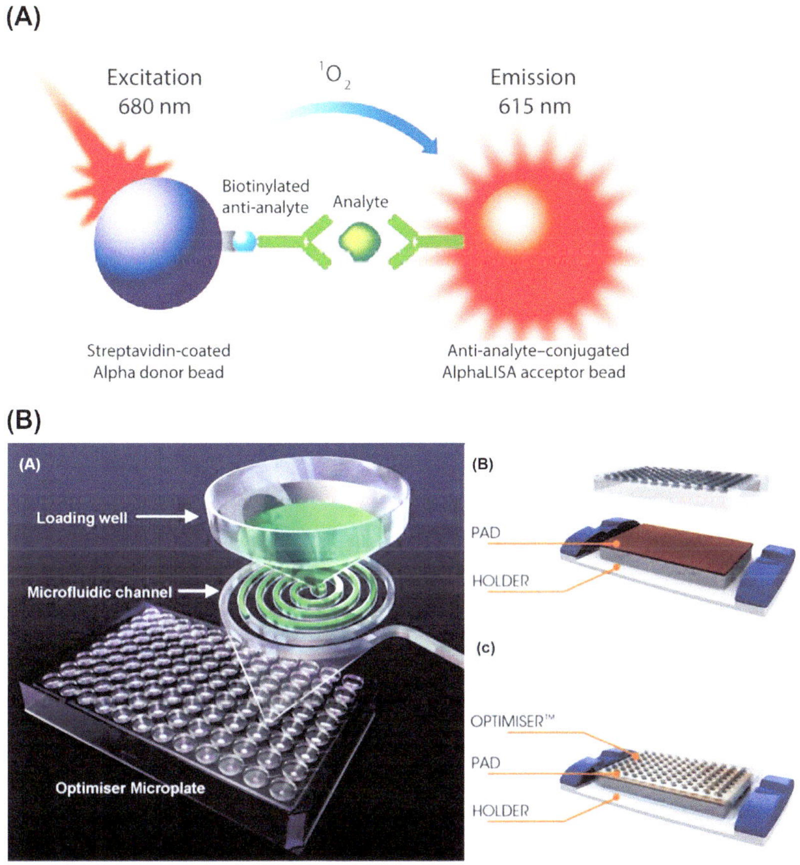

FIGURE 1.2 (A) AlphaLISA, a wash-free ELISA by Perkin Elmer [11]. (B) Microfluidic ELISA-based Optimiser microplate by Siloam Biosciences [15] for rapid ELISA. *(A) Reproduced with permission from Nature publishing group. (B) Reproduced with permission from the Royal Society of Chemistry.*

Most ELISA procedures employ a multistep IA procedure such as sandwich ELISA and competitive ELISA. However, many prospective one-step kinetics-based rapid IA procedures have been demonstrated [58,64]. The most prominent developments are the SP-based readout devices, which enables the readout of colorimetric or chemiluminescent ELISA with the same or increased sensitivity as microplate readers [65,69]. These novel readout devices have further extended the outreach of ELISA to remote and decentralized locations as they do not require any power supply and can store, transmit, and analyze data at any place and time. Of interest is the SP-based EasyELISA platform technology, which enables the one-step kinetics-based rapid IA to be performed within 30 min [72].

Microfluidics and lab-on-a-chip technologies have led to several automated rapid IA formats that employ minimal volumes of reagents [77]. LFIA is the most widely used format for the remote semiquantitative measurement of analytes in just a few minutes. Although it has been used predominantly for pregnancy testing of human chorionic gonadotropin, many prospective IAs have been developed for other potential biomarkers. The most recent developments are the LFIA formats for the quantitative analyte measurement and the SP-based integrated test reader by Cellmic, USA [42] (Fig. 1.3A). Another

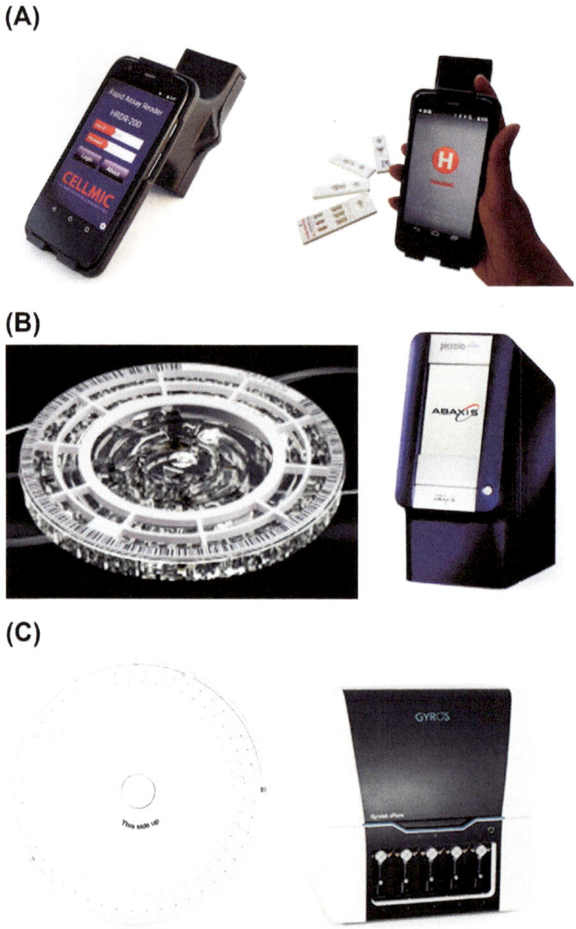

FIGURE 1.3 (A) Holomic rapid diagnostic reader (HRDR) for the readout of lateral flow immunoassays (IAs). Left: HRDR-200, Right: HRDR-300. Centrifugal microfluidics-based LabDisk platform and analyzer for fully automated IAs by (B) Abaxis Inc., USA, and (C) Gyros AB, Sweden. *(A) Reproduced with permission from Cellmic, USA. (B, C) Reproduced with permission from Abaxis Inc. and Gyros AB.*

prospective IA format is the paper-based IA [20,26], an ideal immunodiagnostic solution for the developing nations because of its low-cost and mass production. However, these formats could not match the high analytical sensitivity and performance of established clinically accredited IA that are being used in the central laboratories [63]. The most prominent microfluidic IA developed for clinical diagnostics have been developed by Abaxis Inc. (Fig. 1.3B) and Gyros AB (Fig. 1.3C). They employ centrifugal microfluidics-based LabDisk platforms and analyzers, which can automatically determine many analytes via fully automated microfluidic operations. Abaxis provides a large variety of Clinical Laboratory Improvement Amendments–waived multianalyte tests. Similarly, the IAs developed by Gyros AB are being widely used by the pharmaceutical industry as they have a broad dynamic range and short sample-to-answer time using only minimal reagent volumes.

SPR-based real-time IA using BiaCore instruments from GE Healthcare have become a global biopharmaceutical and clinical research standard [27,43,54] (Fig. 1.4). They require the gold coated SPR chips, which could also be prefunctionalized with specific biomolecules or chemical entities as required to facilitate the binding of capture antibodies. The fully automated microfluidic IA procedure enables the determination of analytes in just a few minutes. Several prospective surface functionalization and antibody immobilization chemistries [59,68], and strategies for signal-enhanced IA have been developed during the last two decades, which have significantly improved the performance of SPR-based IA. Several prospective SPR-based instruments are commercially available by IBIS Technologies, Bio-Rad Laboratories, NanoSPR Devices, Reichert Technologies, SensiQ, etc. As the eyes of analysts, in diverse bioanalytical settings, they provide a thorough understanding of the complex biomolecular interactions and enable rapid screening of antibodies and components for an IA. They also enable effective and rapid troubleshooting of an IA, which can otherwise take a lot of time and costs.

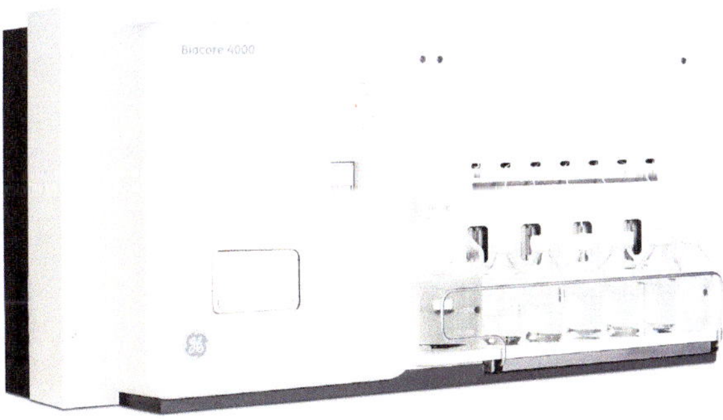

FIGURE 1.4 Surface plasmon resonance–based BiaCore systems by GE Healthcare for fully automated label-free and real-time rapid immunoassays. *Reproduced with permission from GE Healthcare.*

Several biosensing concepts, based on microcantilevers [21,57,60], quartz crystal microbalance [70], interferometry [39], electrochemistry [55], and acoustic waves [33] have also been used for the development of a wide range of IAs. Despite a good bioanalytical performance shown by such biosensors, they are still extensive research efforts required to develop commercially and clinically viable biosensor-based IAs. In particular, mass spectroscopy−based assays have been demonstrated for a large number of analytes in healthcare and industrial settings [25,36]. Considering their high analytical precision, high sensitivity and specificity, and short analysis time, they are the established standard for clinical diagnostics. In fact, they are being used for the method comparison and determination of precision of new IAs.

There is considerable interest in the multiplex detection of biomarkers [32] and the clinical scoring−based diagnosis of a particular disease [30,45,56], which can be used in predictive diagnostics. If the biomarkers responsible for a particular disease are known and weightage can be assigned to them based on their relevance in the diagnosis of a disease, the clinical score can be calculated based on the multiplex detection of biomarkers. It will provide an easy-to-use clinical indicator for the monitoring and management of a disease, and analyze the effectiveness of therapeutic treatment and/or personalized/medical intervention. Various algorithms for the clinical scoring have been demonstrated and used clinically for the diagnosis and differentiation of various diseases. Several IA formats have been demonstrated for multiplexed detection of biomarkers, such as the multiplex bead-based Luminex IA by Thermo Fisher Scientific [4,11,17,49] and multispot electrochemiluminescence-based IA by Meso Scale Discovery [11,13].

SPs have become ubiquitous and personalized and have advanced biosensors, processing speed, better connectivity, and other desired features, which make them the ideal device for point-of-care IAs [62,63,66,67]. In brief, a large number of IAs with readout by SP have been developed. Several prospective SP-based devices are based on the detection of colorimetric [65,69,73], absorbance [37,75], chemiluminescent [6,81], fluorescent [7,12], electrochemical [28,50−52], SPR [34,46], and other optical signals [10,48]. They show superior bioanalytical performances in comparison with central laboratory-based IAs.

The near future would witness continuous innovation in IA formats, diagnostic platforms, biosensing concepts, IA chemistries, and complementary technologies, which would provide an impetus to the development of next-generation IA technologies [63].

5. CONCLUSIONS AND FUTURE TRENDS

A wide range of IA formats, platforms, and devices have been developed during the last two decades for various bioanalytical applications. The most widely used IA format being used in clinical diagnostics is the automated

ELISA using commercial clinical analyzers. However, most of the clinical IAs in the developing nations with limited resources and infrastructure are still performed using manual ELISA. The last decade has witnessed an evolving trend toward the use of fully automated SPR-based IA in biopharmaceuticals and research establishments. The extensive research efforts have led to the emergence of many prospective IA formats, based on microcantilevers, QCM, micro/nanomaterials, lab-on-a-chip formats, and microfluidics. The mass spectroscopy−based clinically validated assays have been developed for a large number of analytes and serve as the established assays for determining the precision of newly developed IA. Moreover, several new biosensing concepts have emerged during the last two decades, which have been demonstrated for a wide range of IAs.

The field of point-of-care testing has also evolved drastically and is undergoing a constant change as there is an extensive need for immunodiagnostics for decentralized and remote settings. Similarly, there is a requirement for low-cost point-of-care diagnostics in the developing nations. The most prospective IA formats for these are the paper, lab-on-a-chip, and SPIA. Despite intensive efforts, the paper-based IAs have not been successful for the clinical diagnostics due to the lack of precision and sensitivity. Some lab-on-a-chip-based IAs have already become commercially successful and are being used for clinical diagnostics. As the latest trend, the SP-based devices are being seen as the ideal device for personalized healthcare monitoring and management. Many of such SP-based mobile healthcare devices are already commercial and are being used by millions of users worldwide. However, the development of a clinically accredited SPIA would still require extensive efforts before they can be used by analysts and the general population. It would require more rigorous clinical validation and testing to validate that the SPIAs are as precise as the established clinically accredited IAs being used in the central laboratory. Similarly, there are concerns about the enormous data such as data storage, data security, data transmission, and data ownership, which need to be counteracted effectively.

There is a need for multiplex IA formats so that all the biomarkers involved in the diagnosis of a disease can be quantitatively detected at the same time. Further, the clinical score for a disease can be determined by various clinical scoring algorithms. This would enable the rapid screening of a disease in a population and would facilitate the continuous monitoring and management of the disease via effective medical and personal intervention.

The continuous improvements in IA formats, bioanalytical platforms, novel biosensing concepts, portable detectors, and complementary technologies would lead to the development of several prospective IA and devices in the near future, which would change the landscape of immunodiagnostics. Such advances would pave the way to next generation of IA that would have rapid analysis time, less cost, high-throughput, and superior bioanalytical performance. This would lead to much better healthcare as diseases could be detected and treated at a very early stage, which could significantly cut down the healthcare costs.

REFERENCES

[1] E.C. Alocilja, S.M. Radke, Market analysis of biosensors for food safety, Biosens. Bioelectron. 18 (2003) 841–846.

[2] Anonymous, Antibody Production and Purification Technical Handbook, Thermo Fisher Scientific Inc., 2010 (version 2).

[3] Anonymous, Measuring the Information Society Report, 2016. http://www.itu.int/en/ITU-D/Statistics/Pages/publications/mis2016.aspx.

[4] M.E. Arellano-Garcia, S. Hu, J. Wang, et al., Multiplexed immunobead-based assay for detection of oral cancer protein biomarkers in saliva, Oral Dis. 14 (2008) 705–712.

[5] S.S. Arnon, R. Schechter, T.V. Inglesby, et al., Botulinum toxin as a biological weapon: medical and public health management, J. Am. Med. Assoc. 285 (2001) 1059–1070.

[6] R. Arts, I. Den Hartog, S.E. Zijlema, et al., Detection of antibodies in blood plasma using bioluminescent sensor proteins and a smartphone, Anal. Chem. 88 (2016) 4525–4532.

[7] A.I. Barbosa, P. Gehlot, K. Sidapra, et al., Portable smartphone quantitation of prostate specific antigen (PSA) in a fluoropolymer microfluidic device, Biosens. Bioelectron. 70 (2015) 5–14.

[8] L. Beaudet, R. Rodriguez-Suarez, M.-H. Venne, et al., AlphaLISA immunoassays: the no-wash alternative to ELISAs for research and drug discovery, Nat. Methods 5 (2008) an8an9.

[9] D. Cecchin, R. De La Rica, R. Bain, et al., Plasmonic ELISA for the detection of gp120 at ultralow concentrations with the naked eye, Nanoscale 6 (2014) 9559–9562.

[10] S. Cho, T.S. Park, T.G. Nahapetian, et al., Smartphone-based, sensitive microPAD detection of urinary tract infection and gonorrhea, Biosens. Bioelectron. 74 (2015) 601–611.

[11] F. Chowdhury, A. Williams, P. Johnson, Validation and comparison of two multiplex technologies, Luminex® and Mesoscale discovery, for human cytokine profiling, J. Immunol. Methods 340 (2009) 55–64.

[12] A.F. Coskun, R. Nagi, K. Sadeghi, et al., Albumin testing in urine using a smart-phone, Lab Chip 13 (2013) 4231–4238.

[13] D. Dabitao, J.B. Margolick, J. Lopez, et al., Multiplex measurement of proinflammatory cytokines in human serum: comparison of the meso scale discovery electrochemiluminescence assay and the cytometric bead array, J. Immunol. Methods 372 (2011) 71–77.

[14] B. Davis, R. Brogan, A widespread community outbreak of *E. coli* 0157 infection in Scotland, Public Health 109 (1995) 381–388.

[15] M.L. De Buyser, B. Dufour, M. Maire, et al., Implication of milk and milk products in food-borne diseases in France and in different industrialised countries, Int. J. Food Microbiol. 67 (2001) 1–17.

[16] R. De La Rica, M.M. Stevens, Plasmonic ELISA for the ultrasensitive detection of disease biomarkers with the naked eye, Nat. Nanotechnol. 7 (2012) 821–824.

[17] S.A. Dunbar, M.R. Hoffmeyer, Microsphere-based multiplex immunoassays: development and applications using Luminex® xMAP® technology, in: The Immunoassay Handbook: Theory and Applications of Ligand Binding, ELISA and Related Techniques vol. 157, 2013.

[18] J.W. Findlay, W.C. Smith, J.W. Lee, et al., Validation of immunoassays for bioanalysis: a pharmaceutical industry perspective, J. Pharm. Biomed. Anal. 21 (2000) 1249–1273.

[19] A. Frey, B. Meckelein, D. Externest, et al., A stable and highly sensitive 3,3',5,5'-tetramethylbenzidine-based substrate reagent for enzyme-linked immunosorbent assays, J. Immunol. Methods 233 (2000) 47–56.

[20] X. Ge, A.M. Asiri, D. Du, et al., Nanomaterial-enhanced paper-based biosensors, Trends Anal. Chem. 58 (2014) 31–39.

[21] K.M. Goeders, J.S. Colton, L.A. Bottomley, Microcantilevers: sensing chemical interactions via mechanical motion, Chem. Rev. 108 (2008) 522–542.

[22] T.W. Hennessy, C.W. Hedberg, L. Slutsker, et al., A national outbreak of *Salmonella enteritidis* infections from ice cream. The Investigation Team, N. Engl. J. Med. 334 (1996) 1281–1286.

[23] G.T. Hermanson, Bioconjugate Techniques, Academic press, 2013.

[24] A.E. Herr, A.V. Hatch, D.J. Throckmorton, et al., Microfluidic immunoassays as rapid saliva-based clinical diagnostics, Proc. Natl. Acad. Sci. U.S.A. 104 (2007) 5268–5273.

[25] A.N. Hoofnagle, M.H. Wener, The fundamental flaws of immunoassays and potential solutions using tandem mass spectrometry, J. Immunol. Methods 347 (2009) 3–11.

[26] J. Hu, S. Wang, L. Wang, et al., Advances in paper-based point-of-care diagnostics, Biosens. Bioelectron. 54 (2014) 585–597.

[27] L. Jason-Moller, M. Murphy, J. Bruno, Overview of Biacore systems and their applications, Curr. Protoc. Protein Sci. 45 (2006), 19.13:19.13.1–19.13.14.

[28] H. Jiang, A. Sun, V. Govindaraj, et al., An audio jack based electrochemical impedance spectroscopy sensor for point-of-care diagnostics, IEEE Sens. J. 17 (2016) 589–597.

[29] J. Kai, A. Puntambekar, N. Santiago, et al., A novel microfluidic microplate as the next generation assay platform for enzyme linked immunoassays (ELISA), Lab Chip 12 (2012) 4257–4262.

[30] K. Kalantar-Zadeh, G. Block, C.J. Mcallister, et al., Appetite and inflammation, nutrition, anemia, and clinical outcome in hemodialysis patients, Am. J. Clin. Nutr. 80 (2004) 299–307.

[31] M. Karle, S.K. Vashist, R. Zengerle, et al., Microfluidic solutions enabling continuous processing and monitoring of biological samples: a review, Anal. Chim. Acta 929 (2016) 1–22.

[32] S.F. Kingsmore, Multiplexed protein measurement: technologies and applications of protein and antibody arrays, Nat. Rev. Drug Discov. 5 (2006) 310–320.

[33] K. Lange, B.E. Rapp, M. Rapp, Surface acoustic wave biosensors: a review, Anal. Bioanal. Chem. 391 (2008) 1509–1519.

[34] K.L. Lee, M.L. You, C.H. Tsai, et al., Nanoplasmonic biochips for rapid label-free detection of imidacloprid pesticides with a smartphone, Biosens. Bioelectron. 75 (2016) 88–95.

[35] C.-C. Lin, J.-H. Wang, H.-W. Wu, et al., Microfluidic immunoassays, J. Lab. Autom. 15 (2010) 253–274.

[36] R. Liu, P. Wu, L. Yang, et al., Inductively coupled plasma mass spectrometry based immunoassay: a review, Mass Spectrom. Rev. 33 (2014) 373–393.

[37] K.D. Long, H. Yu, B.T. Cunningham, Smartphone instrument for portable enzyme-linked immunosorbent assays, Biomed. Opt. Express 5 (2014) 3792–3806.

[38] P.B. Luppa, L.J. Sokoll, D.W. Chan, Immunosensors-principles and applications to clinical chemistry, Clin. Chim. Acta 314 (2001) 1–26.

[39] E. Makarona, P. Petrou, S. Kakabakos, et al., Point-of-need bioanalytics based on planar optical interferometry, Biotechnol. Adv. 34 (2016) 209–233.

[40] A.W. Martinez, S.T. Phillips, G.M. Whitesides, et al., Diagnostics for the developing world: microfluidic paper-based analytical devices, Anal. Chem. 82 (2009) 3–10.

[41] M.I. Mohammed, M.P. Desmulliez, Lab-on-a-chip based immunosensor principles and technologies for the detection of cardiac biomarkers: a review, Lab Chip 11 (2011) 569–595.

[42] O. Mudanyali, S. Dimitrov, U. Sikora, et al., Integrated rapid-diagnostic-test reader platform on a cellphone, Lab Chip 12 (2012) 2678–2686.

[43] D.G. Myszka, R.L. Rich, Implementing surface plasmon resonance biosensors in drug discovery, Pharm. Sci. Technol. Today 3 (2000) 310–317.

[44] S.M. Ostroff, P.M. Griffin, R.V. Tauxe, et al., A statewide outbreak of *Escherichia coli* 0157: H7 infections in Washington State, Am. J. Epidemiol. 132 (1990) 239–247.

[45] G.I. Papachristou, D.J. Papachristou, H. Avula, et al., Obesity increases the severity of acute pancreatitis: performance of APACHE-O score and correlation with the inflammatory response, Pancreatology 6 (2006) 279–285.

[46] P. Preechaburana, M.C. Gonzalez, A. Suska, et al., Surface plasmon resonance chemical sensing on cell phones, Angew. Chem. Int. Ed. Engl. 51 (2012) 11585–11588.

[47] U. Samarajeewa, C.I. Wei, T.S. Huang, et al., Application of immunoassay in the food industry, Crit. Rev. Food Sci. Nutr. 29 (1991) 403–434.

[48] T. San Park, W. Li, K.E. Mccracken, et al., Smartphone quantifies *Salmonella* from paper microfluidics, Lab Chip 13 (2013) 4832–4840.

[49] J.F.D. Siawaya, T. Roberts, C. Babb, et al., An evaluation of commercial fluorescent bead-based luminex cytokine assays, PLoS One 3 (2008) e2535.

[50] A. Sun, A. Venkatesh, D.A. Hall, A multi-technique reconfigurable electrochemical biosensor: enabling personal health monitoring in mobile devices, IEEE Trans. Biomed. Circuits Syst. 10 (2016) 945–954.

[51] A. Sun, T. Wambach, A. Venkatesh, et al., A low-cost smartphone-based electrochemical biosensor for point-of-care diagnostics, in: Biomedical Circuits and Systems Conference (BioCAS), 2014 IEEE, 2014, pp. 312–315.

[52] A.C. Sun, C. Yao, A.G. Venkatesh, et al., An efficient power harvesting mobile phone-based electrochemical biosensor for point-of-care health monitoring, Sens. Actuators B Chem. 235 (2016) 126–135.

[53] S.L. Taylor, J.A. Nordlee, L.M. Niemann, et al., Allergen immunoassays-considerations for use of naturally incurred standards, Anal. Bioanal. Chem. 395 (2009) 83–92.

[54] P. Thillaivinayagalingam, J. Gommeaux, M. Mcloughlin, et al., Biopharmaceutical production: applications of surface plasmon resonance biosensors, J. Chromatogr. B Analyt. Technol. Biomed. Life Sci. 878 (2010) 149–153.

[55] A.P. Turner, Biosensors: sense and sensibility, Chem. Soc. Rev. 42 (2013) 3184–3196.

[56] K.J. Van Zee, T. Kohno, E. Fischer, et al., Tumor necrosis factor soluble receptors circulate during experimental and clinical inflammation and can protect against excessive tumor necrosis factor alpha in vitro and in vivo, Proc. Natl. Acad. Sci. U.S.A. 89 (1992) 4845–4849.

[57] S.K. Vashist, A review of microcantilevers for sensing applications, AzoNano J. Nanotech. 3 (2007) 1–18.

[58] S.K. Vashist, G. Czilwik, T. Van Oordt, et al., One-step kinetics-based immunoassay for the highly sensitive detection of C-reactive protein in less than 30 min, Anal. Biochem. 456 (2014) 32–37.

[59] S.K. Vashist, C.K. Dixit, B.D. Maccraith, et al., Effect of antibody immobilization strategies on the analytical performance of a surface plasmon resonance-based immunoassay, Analyst 136 (2011) 4431–4436.

[60] S.K. Vashist, H. Holthöfer, Microcantilevers for sensing applications, Meas. Control 43 (2010) 84–88.

[61] S.K. Vashist, E. Lam, S. Hrapovic, et al., Immobilization of antibodies and enzymes on 3-aminopropyltriethoxysilane-functionalized bioanalytical platforms for biosensors and diagnostics, Chem. Rev. 114 (2014) 11083–11130.

[62] S.K. Vashist, J.H.T. Luong, Trends in in vitro diagnostics and mobile healthcare, Biotechnol. Adv. 34 (2016) 137–138.

[63] S.K. Vashist, P.B. Luppa, L.Y. Yeo, et al., Emerging technologies for next-generation point-of-care testing, Trends Biotechnol. 33 (2015) 692–705.

[64] S.K. Vashist, E. Marion Schneider, J.H.T. Luong, Rapid sandwich ELISA-based in vitro diagnostic procedure for the highly-sensitive detection of Human fetuin A, Biosens. Bioelectron. 67 (2015) 73–78.

[65] S.K. Vashist, E. Marion Schneider, R. Zengerle, et al., Graphene-based rapid and highly-sensitive immunoassay for C-reactive protein using a smartphone-based colorimetric reader, Biosens. Bioelectron. 66 (2015) 169–176.

[66] S.K. Vashist, O. Mudanyali, E.M. Schneider, et al., Cellphone-based devices for bioanalytical sciences, Anal. Bioanal. Chem. 406 (2014) 3263–3277.

[67] S.K. Vashist, E.M. Schneider, J.H.T. Luong, Commercial smartphone-based devices and smart applications for personalized healthcare monitoring and management, Diagnostics 4 (2014) 104–128.

[68] S.K. Vashist, E.M. Schneider, J.H.T. Luong, Surface plasmon resonance-based immunoassay for human fetuin A, Analyst 139 (2014) 2237–2242.

[69] S.K. Vashist, T. Van Oordt, E.M. Schneider, et al., A smartphone-based colorimetric reader for bioanalytical applications using the screen-based bottom illumination provided by gadgets, Biosens. Bioelectron. 67 (2015) 248–255.

[70] S.K. Vashist, P. Vashist, Recent advances in quartz crystal microbalance-based sensors, J. Sens. 2011 (2011) 1–13.

[71] S.K. Vashist, A.G. Venkatesh, K. Mitsakakis, et al., Nanotechnology-based biosensors and diagnostics: technology push versus industrial/healthcare requirements, BioNanoScience 2 (2012) 115–126.

[72] A. Venkatesh, F.V. Stetten, F.V. Stetten, et al., Smartphone EasyELISA-A point-of-care platform for cost-effective and easy-to-use in vitro diagnostics, MikroSystemTechnik 2015 (2015) 26–28.

[73] A.G. Venkatesh, T. Van Oordt, E.M. Schneider, et al., A smartphone-based colorimetric reader for human C-reactive protein immunoassay, in: A. Rasooly, B. Prickril (Eds.), Biosensors and Biodetection: Methods and Protocols Volume 1: Optical-Based Detectors, Springer New York, New York, NY, 2017, pp. 343–356.

[74] J. Wang, Nanomaterial-based amplified transduction of biomolecular interactions, Small 1 (2005) 1036–1043.

[75] L.J. Wang, Y.C. Chang, R. Sun, et al., A multichannel smartphone optical biosensor for high-throughput point-of-care diagnostics, Biosens. Bioelectron. 87 (2017) 686–692.

[76] L.M. Wein, Y. Liu, Analyzing a bioterror attack on the food supply: the case of botulinum toxin in milk, Proc. Natl. Acad. Sci. U.S.A. 102 (2005) 9984–9989.

[77] G.M. Whitesides, The origins and the future of microfluidics, Nature 442 (2006) 368–373.

[78] D. Wild, The Immunoassay Handbook: Theory and Applications of Ligand Binding, ELISA and Related Techniques, Newnes, 2013.

[79] L.S. Wong, F. Khan, J. Micklefield, Selective covalent protein immobilization: strategies and applications, Chem. Rev. 109 (2009) 4025–4053.

[80] J. Youngájeong, B. Hyunáchung, Recent advances in immobilization methods of antibodies on solid supports, Analyst 133 (2008) 697–701.

[81] M. Zangheri, L. Cevenini, L. Anfossi, et al., A simple and compact smartphone accessory for quantitative chemiluminescence-based lateral flow immunoassay for salivary cortisol detection, Biosens. Bioelectron. 64 (2015) 63–68.

[82] C. Zhao, B. Ge, J. De Villena, et al., Prevalence of *Campylobacter* spp., *Escherichia coli*, and *Salmonella serovars* in retail chicken, turkey, pork, and beef from the Greater Washington, DC, Area. Appl. Environ. Microbiol. 67 (2001) 5431–5436.

Chapter 2

Antibody Immobilization and Surface Functionalization Chemistries for Immunodiagnostics

Sandeep K. Vashist[1], John H.T. Luong[2]
[1]*IDS Immunodiagnostic Systems Deutschland GmbH, Frankfurt am Main, Germany;* [2]*University College Cork, Cork, Ireland*

1. INTRODUCTION

In immunoassays (IAs), immobilized antibodies are preferred over their free counterpart to curtail the number of detection steps and time, as well as to improve high detection sensitivity and reproducibility. The performance of an IA is critically dependent on the characteristics of the immobilized antibody (Ab) as this immobilization step will dictate detection sensitivity, reproducibility, robustness, and other bioanalytical parameters. The support material or substrate should have highly reactive functional groups, mechanical stability, rigidity, and nontoxicity. Ideally, the Ab must also be strongly retained to the surface in a functionally oriented fashion such that its antigen-binding sites are free to bind the antigen. Ideal support properties must be accessible at a low cost, hydrophilic, inert toward biomolecules, and biocompatible. Albeit there is no universal support, suited for all biomolecules, a large number of bioanalytical platforms or substrates have been used in IAs. The most widely used are polystyrene microtiter plates, gold (Au) substrates, magnetic beads, nanoparticles, and transparent substrates, such as cyclic polyolefins. Consequently, the surface of such materials must be functionalized to form covalent bonding with the detection Ab.

A wide range of Ab immobilization chemistries have been developed during the last three decades. Initially, the detection Ab was simply adsorbed onto the substrate after a prolonged incubation by passive adsorption. Subsequently, various functionalization and cross-linking strategies were developed to circumvent critical shortcomings of the passive adsorption procedure, leading to the emergence of a wide range of Ab immobilization chemistries. These include

Handbook of Immunoassay Technologies. https://doi.org/10.1016/B978-0-12-811762-0.00002-5

covalent, oriented, noncovalent, site-specific, affinity or entrapment-based, peptide nucleic acid (PNA) and deoxyribonucleic acid (DNA)-directed, and highly simplified Ab immobilization chemistries. Characterization tools at the nanoscale level provide valuable, qualitative, and quantitative information, including morphological visualization of the immobilized Ab. This important step enables the assessment of the efficacy of an immobilization technique and the development of future immobilization strategies.

This chapter provides an overview of the various surface functionalization, Ab immobilization chemistries, and surface characterization together with the guided insights pertaining to their suitability for various immunodiagnostic applications.

2. SURFACE FUNCTIONALIZATION CHEMISTRIES

Immobilization of biomolecules has been around for over 100 years, but recent trends foster the procedures that can be performed under mild conditions without the use of toxic materials or the production of toxic by-products. The characteristics of immobilized biomolecules have been debated concerning their conformational change, denaturation, or changes in kinetic properties. The immobilization of biomolecules on diversified substrates has been described extensively in the literature. Thus, only most common surface functionalization chemistries that are used for the modification of substrates in immunodiagnostics are discussed here.

2.1 Hydroxyl Groups

Hydroxyl groups can be induced on the surface by treatment with KOH, NaOH, or piranha solution (H_2SO_4: 30% $H_2O_2 = 3:1$ [v/v]). The treatment with KOH is very popular as it has been demonstrated on a wide range of substrates, such as cyclic polyolefins, polystyrene, polyethylene, PMMA, PDMS, and glass [20,21,71,73,75–77]. NaOH has also been used to induce hydroxyl groups on the substrate. In particular, piranha solution has been mainly used for the induction of hydroxyl groups on gold (Au) surface [70]. In the past, the substrate with −OH groups, e.g., Sepharose, can be treated with cyanogen bromide to form reactive cyclic amidocarbamate, which forms a covalent bond with the amino group of the antibody. Similarly, cellulose is subjected to ethyl chloroformate to form a cyclic reactive, which forms a covalent bond with the amino group of the antibody. Other activation methods involve *p*-toluenesulfonyl chloride or derivatives of sulfonic esters [50]. Silanes readily react with surfaces through surface hydroxyl groups to form a strongly bound coating that includes both covalent bonds and multiple van der Waals interactions. In brief, the silanes hydrolyze in solution to form silanols, which readily attach to the surface by their reactions to the surface OH to form Si−O−Si (siloxane) linkages. Hydrogen-bonded vicinal silanols react more readily with silane coupling agents compared with isolated or free hydroxyls. Effectiveness of silane on silica, quartz, glass, and aluminum is excellent.

Good silanization can be achieved for inorganic oxides (Fe_2O_3, TiO_2, and Cr_2O_3); however, the effectiveness of silanization is poor for graphite and carbon black. Silanes with different functional groups such as amino (monoamine, diamine, triamine, secondary amine, etc.), carboxyl, anhydride and azide are commercially available and useful for the covalent attachment of antibodies on substrates with −OH groups.

2.2 Amino Groups

The induction of amino groups on the surface has been one of the most widely used surface functionalization chemistry. The amino groups can be induced on thiolated substrates by treatment with ethyleneimine or 2-bromoethylamine. Further, the carboxylated substrate can be induced with amino groups by treatment with 1-ethyl-3-(3-dimethylaminopropyl) carbodiimide hydrochloride (EDC) and ethylenediamine.

The most extensively used surface functionalization chemistry for the induction of amino groups on a wide range of substrate is the treatment with APTES via wet chemical procedures or chemical vapor deposition (CVD) [73]. APTES can be deposited on solid materials, electrode materials, nanomaterials, and nanocomposites under variable conditions of concentration, solvent, temperature, and time. In addition, curing conditions such as air/heat drying might be necessary depending on the intended application [73]. Silanization occurs on any substrate, normally with chemically active hydroxyl groups for silane grafting [33]. Some substrates (e.g., silica, agarose) already have hydroxyl groups while others require a pretreatment step using KOH/NaOH, acid, piranha solution, or plasma treatment to introduce hydroxyl groups as mentioned previously. APTES reacts with the free hydroxyls of an oxidized substrate by S_N2 exchange with the loss of ethanol. APTES bonds to a substrate in three different ways. Surface attachment and condensation lead to horizontal polymerization when a surface-bound APTES molecule forms siloxane with its neighboring surface-bound APTES [33,82]. In vertical polymerization, the surface-bound APTES reacts with a nearby APTES in solution [5]. In the third scenario, where the substrate possesses surface hydroxyl groups, the amine of APTES forms a hydrogen bond with the metal surface or becomes protonated by abstracting protons from the surface [26]. Therefore, a mixture of neutral amine and protonated and hydrogen-bonded amines coexists on the solid surface; the former orients away from the surface for biomolecule immobilization, whereas the latter is close to the surface. The amine group is transformed with various curing conditions and substrate-dependent. For the substrate without oxide groups, amines react quickly with carbon dioxide of the atmosphere to form bicarbonates and then imines in the presence of heating.

The silanization mechanism is complex with adsorption and chemical sorption contributing to the rate-determining steps. Initially, silanization is fast and governed by the Langmuir model leading to a SAM followed by a slow kinetic saturation process [24,57]. Silanization can be done by wet chemistry or CVD.

The former is performed in a liquid medium and classified as organic-, aqueous-, or evaporation-based, depending on the choice of solvents. Silanization is initiated via the hydrolysis of ethoxy terminal groups of APTES by moisture (organic phase and evaporation-based approaches) or by water (aqueous phase). The CVD approach is based on the deposition of silane vapor on a substrate surface.

Different polymeric substrates can be pretreated with 1% (w/v) KOH and oxygen plasma (e.g., using a reactive ion etcher) to generate the desired hydroxyl groups for silanization. Subsequently, these resulting polymers are incubated with 2% (v/v) APTES in deionized water for 1 h (Fig. 2.1) [20]. This simple treatment obviates the use of complicated oxygen plasma for silanization [71]. Recently, this procedure has been proven in surface plasmon resonance (SPR)-based IAs using Au-coated SPR chips [19,72]. The Au SPR chip is simply treated with a piranha solution for 2 min to generate the desired surface hydroxyl groups and incubated in 2% (v/v) APTES in water for 1 h. A typical mixture of the piranha solution is 3:1 (v/v) concentrated sulfuric acid to 30% H_2O_2 solution.

CVD is applicable for mass production of silanized substrates. It obviates the limitations of other approaches such as the presence of trace water in solution and the purity of silane. An activated substrate is simply placed in an environment saturated with silane vapor, leading to substrate silanization. As the silane polymers have a negligible vapor pressure at the reaction temperatures, they do not get deposited on the target, thereby resulting in smooth monolayers. However, the attachment mechanism of silane molecules to the substrate is uncertain. The APTES layer thickness on silica is 5 Å and 6 Å at the deposition temperature of 70 and 90 °C, respectively, corresponding to the length of an APTES molecule, illustrating the formation of a monolayer on the substrate [86]. Plasma-enhanced CVD (PECVD) has been adopted as a potential CVD technique for surface functionalization of point-of-care diagnostic platforms [25].

2.3 Carboxyl Groups

Carboxyl groups can be created on the amine-functionalized substrate by treatment with succinic anhydride. Similarly, the thiolated substrates can be induced with carboxyl groups by S-carboxymethylation of the thiol group with iodoacetate. Further, the aminated substrates can be converted to carboxylated ones by treatment with succinic anhydride, glutaric anhydride, or citraconic anhydride. The hydroxyl groups on the substrate can be converted into carboxyl groups by treatment with chloroacetic acid.

2.4 Sulfhydryl Groups

The thiol groups can be induced on amine-functionalized substrates by treatment with N-acetylhomocysteine thiolactone, S-acetylthioglycolic acid, S-acetylmer captosuccinic anhydride, thiobutyrolactone, 2-iminothiolane, thiol-containing

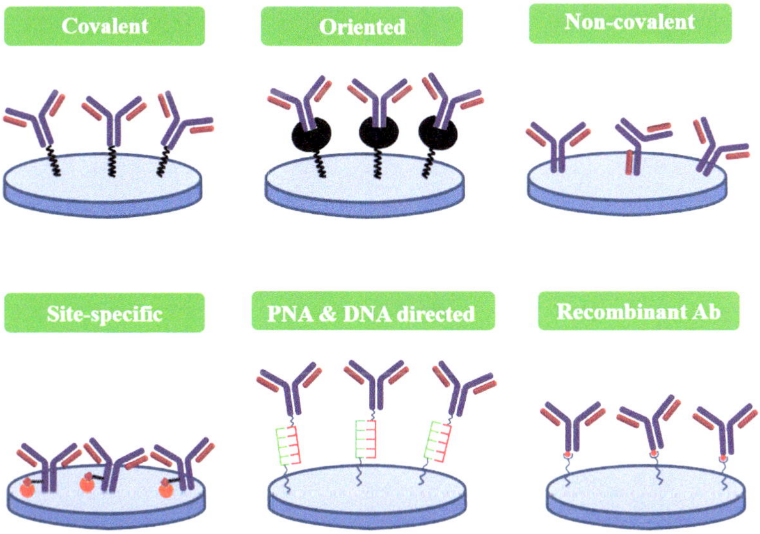

FIGURE 2.1 Various antibody (Ab) immobilization chemistries being employed for immunoassays.

succinimidyl derivatives or thiol-containing imidoesters. The hydroxylated substrates can be induced with sulfhydryl groups by sequential treatments with tosyl chloride and thioacetate.

2.5 Epoxy Groups

Epoxy groups can be induced on the substrate by treatment with (3-glycid oxypropyl)-trimethoxysilane (GOPS). They are very stable at neutral pH and, under aqueous reaction conditions, are reactive toward nucleophiles such as amines, thiols, or acids. The aminated substrate can also be induced with epoxy groups by treatment with oligo(poly)ethyleneglycol diglycidyl ether.

3. ANTIBODY IMMOBILIZATION CHEMISTRIES

A wide range of Ab immobilization chemistries have been developed (Fig. 2.1). The main characteristics of these chemistries along with their pros and cons are discussed in Table 2.1.

3.1 Covalent

Various chemistries have been developed for the covalent immobilization of Ab to the substrate. The most commonly used chemistries are based on the use of cross-linkers that cross-links the functional groups present on the Ab to the functional groups present or induced on the substrate. Tables 2.2 and 2.3 provide an overview of various cross-linking chemistries.

TABLE 2.1 Main Characteristics of Various Antibody (Ab) Immobilization Chemistries

Immobilization Chemistry	Advantages	Disadvantages
Covalent	• Prolonged stability • Leach-proof binding • Ab modification is not required • Commercial availability	• Random orientation • Might mask antigen binding by binding to antigen-binding site • Require cross-linkers and additional process steps • Some cross-linking agents are susceptible to hydrolysis
Oriented	• Oriented Ab immobilization • Bound Ab has maximum functionality as there is no masking of antigen binding sites • No requirement of Ab modification • FBPs bind strongly to Au-coated substrates • FBPs can be bound to almost any substrate by their amino or carboxyl groups	• Requires FBPs • Unblocked FBPs after Ab binding can lead to nonspecific binding • Not suitable for sandwich IA • Additional process steps • Stability is less compared with covalently bound Ab
Noncovalent	• Simple one-step Ab immobilization by passive adsorption due to physicochemical phenomenon, such as hydrophobic and electrostatic interactions • No Ab modification • No cross-linking agents required • High Ab immobilization density	• Ab binding is not strong as the bound Ab can leach off • Random Ab orientation • Ab immobilization density cannot be controlled • Ab can change their conformation and denature • Reduced Ab functional activity with time • High nonspecific binding

Site-specific	• Site-specific • Highly selective Ab immobilization	• Requires Ab modification • Ab modification can decrease its functional activity • Ab immobilization is not leach-proof
PNA and DNA-directed	• Oriented Ab immobilization • Site-specific	• Requires bioconjugation of DNA/PNA
Recombinant antibody	• Oriented Ab as His or cysteine tag is directly fused with recombinant Ab • Genetically engineered Ab	• Limited commercial availability • Requires custom manufacture

DNA, deoxyribonucleic acid; *FBP*, fragment crystallizable (Fc)-binding protein; *IA*, immunoassay; *PNA*, peptide nucleic acid.

TABLE 2.2 Most Used Cross-linker Reactive Groups for Antibody (Ab) Immobilization

Reactivity	Target Functional Group	Reactive Chemical Groups
Amine-reactive	$-NH_2$	NHS ester Imidoester Epoxide Isothiocyanate Aldehyde Pentafluorophenyl ester Hydroxymethyl phosphine
Carboxyl-to-amine −reactive	−COOH	Carbodiimide (e.g., EDC)
Sulfhydryl-reactive	−SH	Maleimide Haloacetyl (bromo- or iodo-) Pyridyldisulfide Thiosulfonate Vinylsulfone
Aldehyde-reactive, i.e., oxidized sugars (carbonyls)	−CHO	Hydrazide Alkoxyamine
Photoreactive, i.e., nonselective, random insertion	Random	Diazirine Aryl azide
Hydroxyl (nonaqueous)-reactive	−OH	Isocyanate
Azide-reactive	$-N_3$	Phosphine

Most covalent chemistries are based on the generation of amino, carboxyl, and epoxy groups on the substrate, which is followed by the hetero- or homobifunctional cross-linking of Ab via its amino or carboxyl groups on its surface. The most widely used chemistry is the heterobifunctional cross-linking using EDC and *N*-hydroxysuccinimide (NHS)/*N*-hydroxysulfosuccinimide (SNHS), where EDC of the EDC-NHS/SNHS complex first binds the carboxyl-terminal of the Ab (Fig. 2.2A) [20]. EDC-NHS/SNHS is a zero-length cross-linker, which leaves the activated surface after the formation of an amide bond between the amine-functionalized substrate and the carboxyl group on the Ab. Although most immobilization chemistries recommend using EDC-SNHS/EDC-NHS, EDC alone has also been used [69]. The NHS ester reacts with the amine group of lysine and forms a stable amide bond. It is unstable in aqueous conditions as the ester hydrolysis competes with the

TABLE 2.3 Commercially Available Cross-linkers for Antibody (Ab) Immobilization

Reactivity (Cross-linker Type)	Cross-linkers	Cross-linking Chemistry
Amine-reactive (homobifunctional)	• Glutaraldehyde • Disuccinimidyl suberate • 1,4-Butanediol diglycidyl ether • 1,4-Phenylene diisothiocyanate • Terephthaldialdehyde • Dimethyl suberimidate • Divinyl sulfone • Disuccinimidyl carbonate • Disuccinimidyl tartrate • Sulfodisuccinimidyl tartrate • Bis(sulfosuccinimidyl)suberate • Dimethyl adipimidate • Dimethyl pimelimidate • 4,4′-Diisothiocyanatostilbene-2,2′-disulfonic acid • Dithiobis(succinimidyl propionate) • 3,3′-Dithiobis(succinimidyl propionate) • Ethylene glycol bis(succinimidyl succinate) • Bis(2-[succinimidooxycarbonyloxy]ethyl) sulfone • Glutaric anhydride	Cross-links amino group of Ab to amino groups on substrate
Amine-reactive (heterobifunctional)	• N-succinimidyl S-acetylthioacetate	Reacts with the primary amines of Ab and adds a protected sulfhydryl group, which can be used for binding to substrates

Continued

TABLE 2.3 Commercially Available Cross-linkers for Antibody (Ab) Immobilization—cont'd

Reactivity (Cross-linker Type)	Cross-linkers	Cross-linking Chemistry
Thiol-reactive (homobifunctional)	• 2,2′-Dipyridyl disulfide • N,N′-carbonyldiimidazole • N,N′-bis(3-maleimidopropyl)-1,3-propanediamine • 1,4-Di-(3′-[2′pyridyldithio]-propionamido) butane	Binds thiol groups of Ab to the thiolated substrates
Sulfhydryl- and hydroxy-reactive (heterobifunctional)	• p-Maleimidophenyl isocyanate)	The maleimide group in the cross-linker binds to the thiolated Ab while the isocyanate group binds to hydroxyl groups on the substrate
Amine- and thiol-reactive (heterobifunctional)	• Sulfosuccinimidyl 4-(*N*-maleimidomethyl) cyclohexane-1-carboxylate • 2-Maleimidoacetic acid NHS • 3-Maleimidopropionic acid NHS • 4-Maleimidobutyric acid NHS • 6-Maleimidocaproic acid NHS • Bromoacetic acid NHS • Iodoacetic acid NHS • 6-Iodoacetamidocaproic acid NHS • m-Maleimidobenzoyl-N-hydroxysuccinimide ester • m-Maleimidobenzoyl-N-hydroxysulfosuccinimide ester • N-γ-maleimidobutyryloxysuccinimide ester	Binds first to amino groups of Ab and then to the sulfhydryl groups on the substrate or vice versa i.e., binding first to the aminated substrate and then to thiolated Ab

	• N-γ-maleimidobutyryloxysulfosuccinimide ester • N-(ε-maleimidocaproic acid) hydrazide • N-(ε-maleimidocaproyloxy) succinimide ester • N-(ε-maleimidocaproyloxy) sulfo succinimide ester • N-(ρ-maleimidophenyl) isocyanate • Succinimidyl 4-(ρ-maleimidophenyl) butyrate • Sulfosuccinimidyl 4-(ρ-maleimidophenyl) butyrate • N-succinimidyl(4-iodoacetyl) aminobenzoate • N-sulfosuccinimidyl(4-iodoacetyl) aminobenzoate • Maleimide PEG N-hydroxysuccinimide	
Amine- and carboxyl-reactive	• EDC • EDC-NHS • EDC-sulfoNHS	Binds first to the carboxyl groups of Ab and then cross-links to the amino groups on the substrate, and vice versa

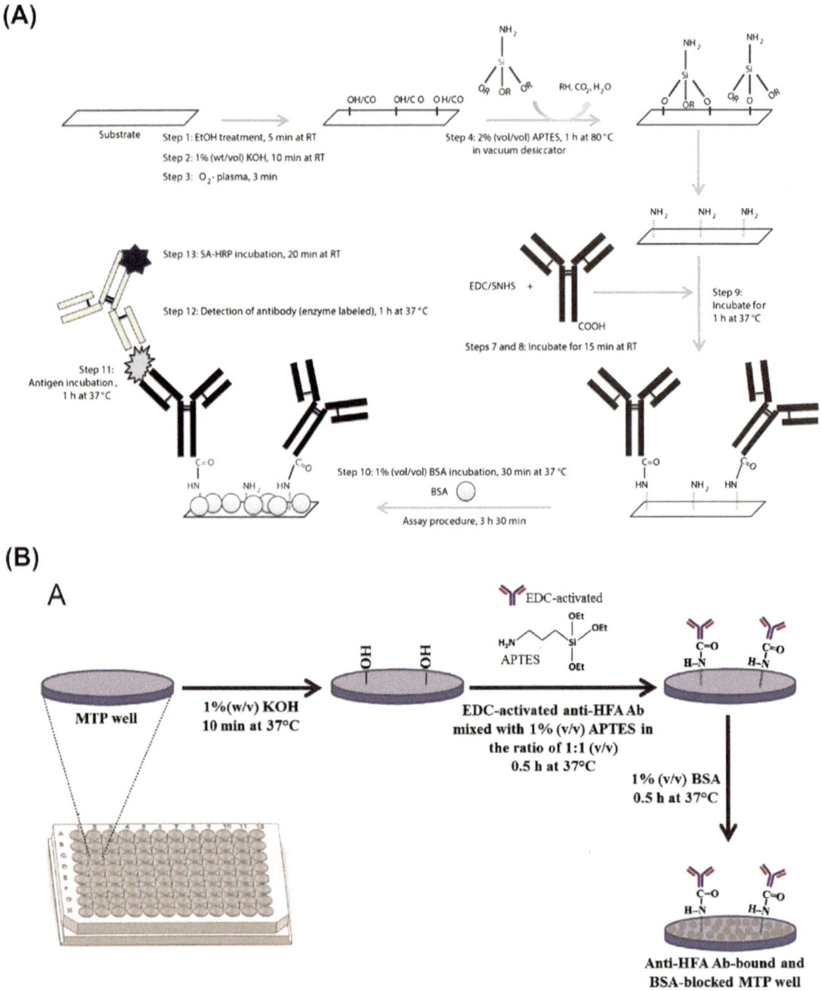

FIGURE 2.2 Covalent antibody (Ab) immobilization chemistries. (A) 1-Ethyl-3-(3-dimethyl aminopropyl) carbodiimide hydrochloride (EDC) and sulfoNHS activated Ab is bound to APTES-functionalized substrate by a multistep procedure. (B) A simplified procedure based on the hetero-bifunctional cross-linking of EDC-activated Ab diluted in APTES to the hydroxylated substrate. *(A) Reproduced with permission from Nature Publishing Group, C.K. Dixit, S.K. Vashist, B.D. Maccraith, et al. Multisubstrate-compatible ELISA procedures for rapid and high-sensitivity immunoassays. Nat. Protoc. 6 (2011) 439–445. (B) Reproduced with permission from Elsevier B.V, S.K. Vashist, E.M. Schneider, J.H.T. Luong. Rapid sandwich ELISA-based in vitro diagnostic procedure for the highly-sensitive detection of human fetuin A. Biosens. Bioelectron. 67 (2015) 73–78.*

covalent binding of the biomolecule. The NHS derivatives in solution at pH 8.0 have a half-life of only 1 h [80]. Of interest is the highly simplified covalent Ab immobilization chemistry, which involves the dispensing of EDC-activated Ab diluted in APTES on the hydroxylated substrate [75] (Fig. 2.2B).

It leads to the rapid leach-proof binding of Ab to the substrate just prior to IAs in less than 30 min, which is preferred for immunodiagnostics.

One of the most widely used approaches is the binding of thiolated Ab to the gold-coated substrates, which leads to the leach-proof covalent binding of the Ab. This strategy has an advantage that it is rapid and employs minimal steps. However, it requires the modification of the Ab, which needs to be thiolated by treatment with Traut's reagent before it can be immobilized.

In another approach, glutaraldehyde (GLD), a homobifunctional cross-linking agent, cross-links the amino groups available on the amine-functionalized surface to the amino groups on the Ab. The immobilization procedure involves the formation of imine bonds by the aldehyde groups induced on a GLD-functionalized substrate with the amine groups of lysine in Ab, which leads to the formation of reversible Schiff bases. In a low ionic medium, Ab immobilization involves an initial ionic exchange with the amino groups on the substrate, which is followed by the covalent binding. But in a high ionic medium, Ab is directly bound covalently with slower immobilization kinetics. However, the binding of Ab by their amino groups might lead to their random orientation as it can mask the antigen-binding sites of Ab that are present near the amino terminal. But this limitation has been tackled effectively using heterobifunctional cross-linkers and fragment crystallizable (F_c)-binding proteins (FBPs).

A prospective chemistry is the covalent binding of Ab by their amino groups on amine-functionalized platforms using 1,4-phenylene diisothiocyanate (PDITC) [54], a homobifunctional cross-linker having two isothiocyanate groups on a phenyl ring that can bind to amines. The immobilization procedure involves the binding of one isothiocyanate group of PDITC to the free amino groups on the amine-functionalized surface, which results in the formation of a thiourea linkage. This is followed by the binding of the other isothiocyanate group of PDITC to the amino groups of Ab.

Ab could be bound covalently to the epoxy group-functionalized substrates via a two-step procedure, i.e., adsorption followed by covalent reaction between nucleophilic groups (amino or thiol) on Ab and epoxy groups on the surface. However, high ionic strength is necessary to promote Ab binding via this chemistry, which could denature proteins/fragile enzymes [44], thereby limiting its widespread use.

Another prospective chemistry is the covalent binding of Ab to the amino-functionalized substrate via a hydrazone bond [15]. The procedure involves the binding of Ab to the aldehyde moiety of succinimidyl 4-formylbenzoate (S-4FB), which is followed by its covalent binding to the amine-functionalized substrate via a hydrazone bond that is formed by the activation of succinimidyl 6-hydra zinonicotinamide acetone hydrazine.

Further, the "Click" chemistry has also been widely used for the covalent binding of Ab to the substrate [41]. It involves the formation of Cu(I)-catalyzed triazole formation between azide- and alkyne-tagged partners. But

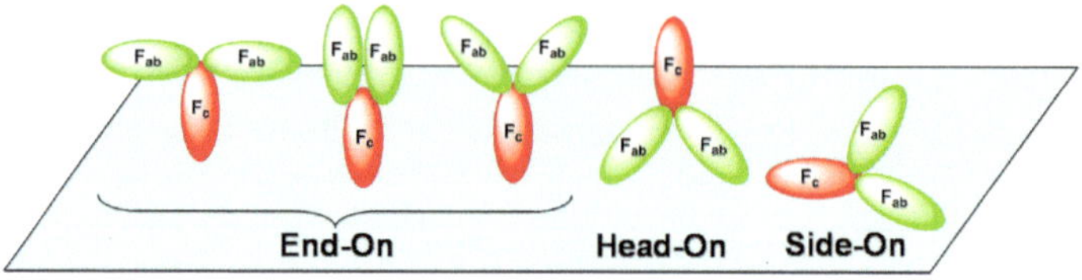

FIGURE 2.3 Various possible orientations of immobilized antibodies (Ab) on a substrate. *Reproduced with permission from the American Chemical Society, S.K. Vashist, E. Lam, S. Hrapovic, et al. Immobilization of antibodies and enzymes on 3-aminopropyltriethoxysilane-functionalized bioanalytical platforms for biosensors and diagnostics. Chem. Rev. 114 (2014) 11083–11130.*

Ab needs to be labeled with azide or alkyne before it can be immobilized using this chemistry. Moreover, Cu(I) is cytotoxic and require careful handling.

3.2 Oriented

The orientated immobilization of capture Ab on the substrate is very important as the bound Ab can have several possible orientations, i.e., end-on, head-on, and side-on, as shown in Fig. 2.3 [73]. The end-on orientation ensures the functionally oriented Ab immobilization while the side-on is only partially functional as one of the antigen-binding sites is masked. However, the head-on orientation is completely nonfunctional as both the antigen-binding sites are masked. Most of the oriented immobilization chemistries employ FBP, such as Protein A [11,64], Protein G [1,2], and Protein A/G [62], for the oriented binding of Ab to the substrate. They bind to the F_c region of Ab, which leaves the antigen-binding sites at the fragment antigen-binding (F_{ab}) region of Ab free for binding to antigen molecules. The native Protein A is a cell wall component that is produced by several strains of *Streptococcus aureus* while the recombinant Protein A is produced in *Escherichia coli*. It has a molecular weight of ~46 kDa and an isoelectric point of 5.16. It contains five high affinity ($K_a = 10^8\ mol^{-1}$) binding sites that can bind to the F_c region of immunoglobulin G (IgG) from several species such as rabbit and human. However, it has been demonstrated that a Protein A molecule can bind to a maximum of two Abs due to spatial constraints [83]. But the binding of Protein A to IgG is not equivalent for all species and even for the subclasses of a particular IgG. Although Protein A adsorbs very strongly to the Au substrate, it can be bound covalently to the amino-functionalized substrates by hetero-bifunctional cross-linking. As an example, Protein A was bound covalently to the amine-functionalized Au-coated SPR chip via EDC-SNHS [72], which was found to have superior bioanalytical performance than other covalent and noncovalent Ab immobilization chemistries.

Protein G, a bacterial cell wall protein from group G *Streptococci*, is another FBP. The native Protein G has two IgG binding domains and sites for nonspecific binding to albumin and cell surface. But the nonspecific binding

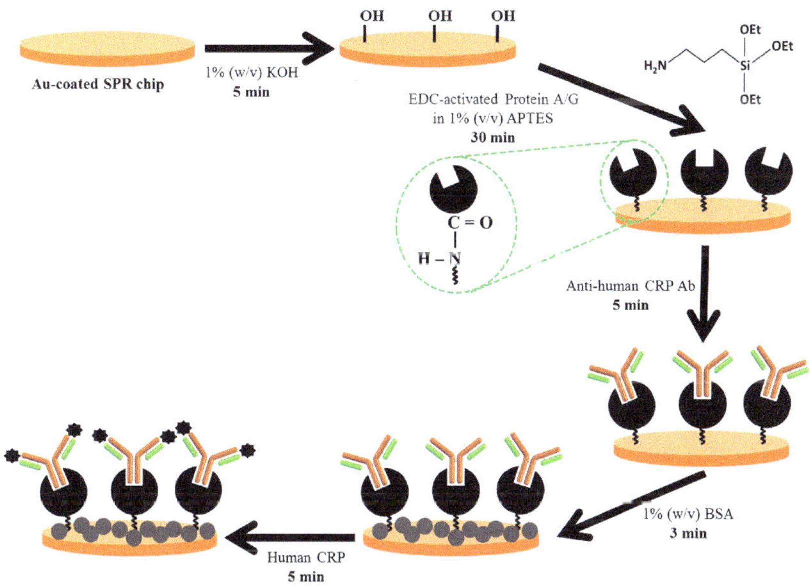

FIGURE 2.4 Oriented antibody (Ab) immobilization on the Au-coated surface plasmon resonance (SPR) chip using covalently bound Protein A/G. *Reproduced with permission from Royal Society of Chemistry, S.K. Vashist, E.M. Schneider, J.H.T. Luong. Surface plasmon resonance-based immunoassay for human C-reactive protein. Analyst 140 (2015) 4445–4452.*

sites are eliminated from recombinant Protein G (MW = ~22 kDa and pI = 4.5), which is produced commercially from *E. coli*. Protein G has higher affinity than Protein A for most mammalian IgGs, but it does not show any binding to human IgM, IgA, and IgD. Although the tertiary structures of Protein A and G are very similar, they have different amino acid compositions. Proteins A and G bind to the Ab optimally at pH of 8.2 and 5, respectively, although the binding is still sufficient in the pH range of 7.0–7.6. However, as there are inherent differences in the binding properties of these FBPs, it has to be prechecked whether they can bind to the specific capture Ab of an IA. This limitation has been solved by Protein A/G, which is prepared by the gene fusion of the F_c-binding domains of Proteins A and G. Protein A/G is superior to other FBPs as it binds with a high affinity to all human IgG subclasses with the exception of IgA, IgE, IgM, and to a lesser extent IgD. Moreover, the binding of Protein A/G to Ab is less pH-dependent as the optimum Ab binding occurs at pH 5–8. An interesting Ab immobilization chemistry was the covalent binding of EDC-activated Protein A/G diluted in APTES to the hydroxylated Au-coated SPR chip, as shown in Fig. 2.4 [76].

Apart from the FBPs, protein L (MW = ~36 kDa and pI = 4.5) isolated from *Peptostreptococcus magnus* has been employed as an intermediate protein [36,49]. Another prominent FBP is Protein L, which binds to the kappa light chains of a wide range of IgG classes and subclasses and does not interfere with the antigen-binding sites of Ab. It is an ideal FBP that binds the F_{ab} fragments and single chain variable fragments.

FBPs have been extensively used for direct binding assays based on SPR, microcantilevers, and quartz crystal microbalance. However, they have not been widely used for sandwich IA as the secondary Ab can bind the vacant FBP, thereby leading to a higher background signal due to increased nonspecific binding. Additionally, the blocking of these vacant FBPs would increase the cost of analysis.

3.3 Noncovalent

3.3.1 Adsorption

Ab molecules can be attached to various substrates by hydrophobic, electrostatic, or van der Waals forces [53,67,79]. Ab adsorbs to hydrophobic surfaces by the release of water dipoles from the hydrophobic surface to the bulk solution. Ab concentration plays an important role in the steady-state adsorption of Ab but only on hydrophilic surfaces [13] not on hydrophobic surfaces [68]. But the adsorption of Ab induces a change in conformation in Ab, which can decrease their functional activity. Antibodies are soft proteins that adsorb on various substrates by spreading. The adsorption of Ab is higher on hydrophobic substrates in comparison with the hydrophilic substrates. However, the immobilization of Ab by passive adsorption has several limitations. It binds the Ab in a randomly oriented fashion, where the bound Ab can leach out of the surface by desorption or can get denatured due to the change in their conformation by spreading on the surface.

The Ab immobilization density has been observed to be higher for adsorption than for covalent binding. Ab has also been bound to GOPS and APTES functionalized substrates by passive adsorption that involves strong ionic interactions. IgG has a molecular weight of 156 kDa and dimensions of $14.5 \times 8.5 \times 4$ nm while the area covered by an IgG molecule with side-on orientation is minimal and has Ab immobilization density of $\sim$2.6 mg/m^2 [39]. The structure of Ab is unaffected by incubation of IgG with the silanized substrate. The most simplified chemistry involved the dispensing of capture Ab diluted in a particular concentration of APTES onto the polystyrene microtiter plate (MTP) and incubating it for half an hour [74] (Fig. 2.5A). It leads to the strong leach-proof binding of the capture Ab on the MTP's surface in just one-step, which demonstrated superior bioanalytical performance in IA compared with Ab immobilized by covalent means. The same chemistry was applied to the immobilization of Ab on the Au SPR chip, where it also showed superior bioanalytical performance in IA [77] (Fig. 2.5B).

3.3.2 Affinity

The most widely used affinity-based, noncovalent Ab immobilization strategy is based on the highly specific and strong biomolecular interactions between

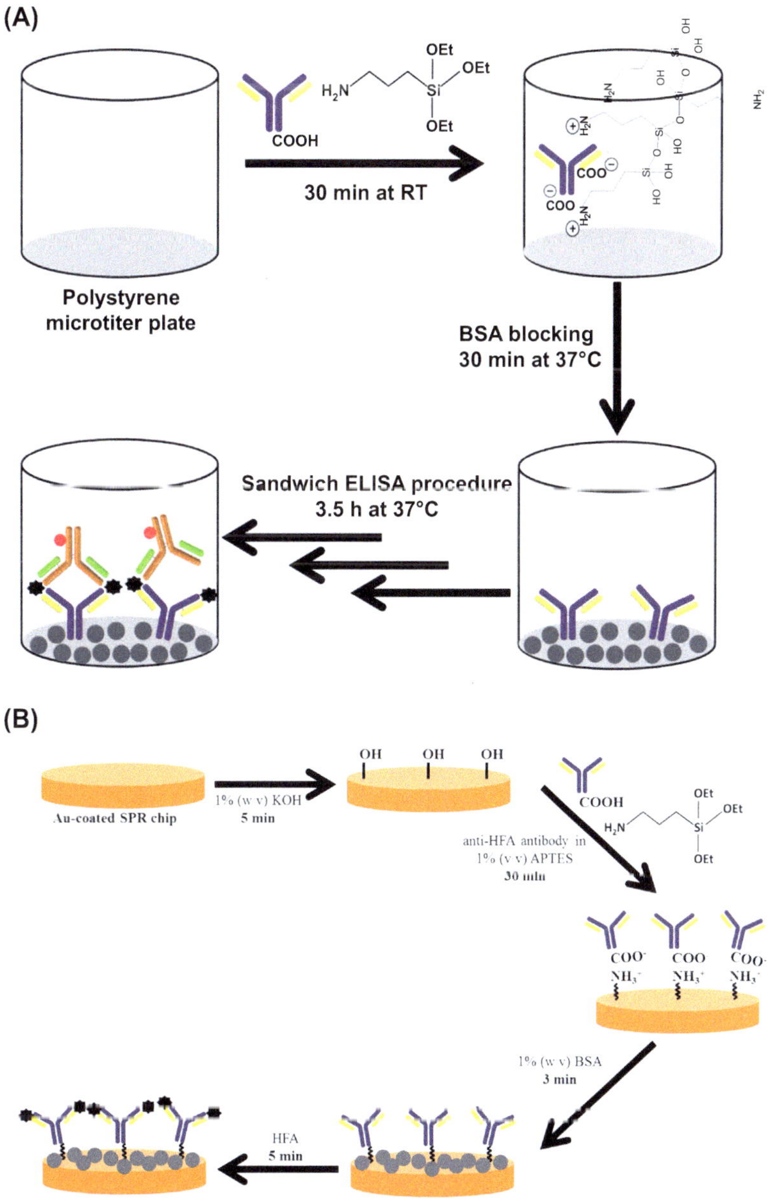

FIGURE 2.5 Noncovalent antibody (Ab) immobilization chemistries based on adsorption. (A) One-step Ab immobilization chemistry–based immunoassay (IA) that involves the dispensing of Ab diluted in APTES onto the polystyrene microtiter plate (MTP) followed by incubation for 30 min. (B) Surface plasmon resonance (SPR)-based IA based on the immobilization of Ab on an Au-coated SPR chip by the same approach. *(A) Reproduced with permission from Nature Publishing group, S.K. Vashist, E. Marion Schneider, E. Lam, et al. One-step antibody immobilization-based rapid and highly-sensitive sandwich ELISA procedure for potential in vitro diagnostics. Sci. Rep. 4 (2014) 4407. (B) Reproduced with permission from the Royal Society of Chemistry, S.K. Vashist, E.M. Schneider, J.H.T. Luong. Surface plasmon resonance-based immunoassay for human fetuin A. Analyst 139 (2014) 2237–2242.*

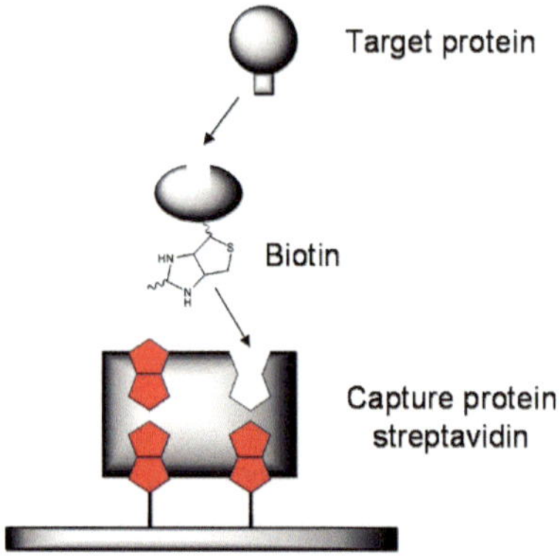

FIGURE 2.6 The immobilization of a protein by affinity-based noncovalent antibody (Ab) immobilization via streptavidin–biotin interaction. *Reproduced with permission from the American Chemical Society, S.K. Vashist, E. Lam, S. Hrapovic, et al. Immobilization of antibodies and enzymes on 3-aminopropyltriethoxysilane-functionalized bioanalytical platforms for biosensors and diagnostics. Chem. Rev. 114 (2014) 11083–11130.*

avidin and biotin ($K_d \sim 10^{-15}$ M), where an avidin molecule has binding sites for four biotin molecules. Avidin could be covalently bound to amine-functionalized substrates by cross-linking its amino or carboxyl group using GLD- or EDC-mediated cross-linking, respectively. Thereafter, biotinylated Ab could be bound to the avidin-functionalized substrate (Fig. 2.6). Several protocols have been developed by researchers for the biotinylation of Ab [8,23,27–30], which does not affect the functional activity of Ab as biotin is a small molecule with a molecular weight of 244 Da. Another prospective Ab immobilization strategy could be the use of hexahistidine tag for the immobilization of histidine-tagged Ab onto the substrate [40,59,60]. But the preparation of histidine-tagged Ab involves many process steps and reagents, thereby making it laborious, time-consuming, and costly [45]. Further, the specificity and stability of histidine tag are lower in comparison with other affinity methods.

3.4 Site-Specific

Site-specific Ab immobilization chemistries involve the insertion or chemical binding of an amino acid, an oligopeptide or an affinity tag at the specific site of an Ab or its derivatives, which is followed by its binding to the complementary affinity molecule bound to a surface [4,51]. The commercial affinity tags that are being used widely are polyhistidine [18], avidin–biotin [35], and FLAG tags [32,47,84]. The polyhistidine-tagged Ab can be immobilized only if the substrate is modified with the metal–NTA complex. Similarly, the surface needs to be modified with avidin before the biotinylated Ab can be bound. Another prospective chemistry is the immobilization of cysteine-modified Ab on Au surfaces via self-assembly [7,10]. Cysteine is a free amino acid containing a sulfhydryl group, which can be induced or incorporated on the Ab away from its antigen recognition site so that it can be used by sulfhydryl cross-linkers for immobilizing Ab on various substrates [14,55]. The cysteine tag has been used for the leach-proof binding of Ab on cyanogen bromide (CNBr)-activated substrate [38,46]. Despite a wide range of affinity tags [37] that have been developed and used for Ab immobilization, the site-specific immobilization chemistries are not so popular. Further, the tagging of Ab could decrease its functional activity that could result in reduced binding efficiency [58].

3.5 Peptide Nucleic Acid and Deoxyribonucleic Acid–Directed

PNAs are chemically synthesized, single-stranded nucleic acid that contains an amino acid derivative of alkylamide instead of the usual phosphodiester backbone [22,48]. They form nonionic and achiral PNA-PNA, PNA-RNA, and PNA-DNA, which are resistant to enzymatic hydrolysis and temperature. PNA-conjugated Ab was immobilized on the surfaces functionalized with antisense oligonucleotide or PNA by hybridization [61,63]. The distance of the Ab from the substrate could be controlled by PNA monomers, which provide strong site-specific and orientated Ab immobilization [31,43,56]. Similarly, the oligonucleotide tags could be employed for the binding of Ab as they have a distinct advantage that they can be separated from their complementary strands simply by changing the pH or divalent ion strength [85]. A novel approach for DNA-directed Ab immobilization involved the binding of DNA-conjugated Ab to the single-stranded DNA bound to the substrate by DNA hybridization [12] (Fig. 2.7).

3.6 Recombinant Antibody

Several prospective approaches have been employed for the development of recombinant antibody where an appropriate affinity tag, such as histidine or cysteine tag, is genetically engineered into the Ab [65,66]. Such recombinant Ab can be immobilized on the substrates by the same Ab immobilization chemistries as mentioned in Section 3.4 for affinity tags.

4. SURFACE CHARACTERIZATION

The key issues whether immobilization preserves the protein function or induces losses of biological activity together with its structural changes arise. Another question is related to the density of functional groups on the support and the distribution pattern of the biomolecules. Therefore, techniques capable of probing the activity and the structure of biomolecules immobilized on surfaces are valuable to refine the best immobilization strategy for different "biomolecule-support" pairs. Various analytical techniques emerge and are valuable for probing the formation of individual biomolecules, the homogeneity of the surface coverage, the chemical compositions, the orientation on surfaces, and functionality of immobilized biomolecules. Such powerful tools can be used in combination to address many challenging issues in biomolecule immobilization (Table 2.4).

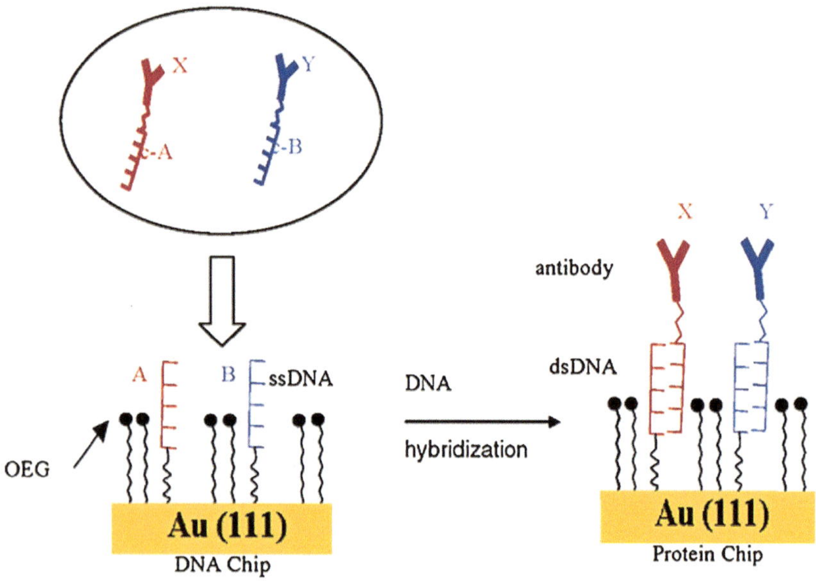

FIGURE 2.7 DNA-directed antibody (Ab) immobilization where the DNA-conjugated Ab binds to the single-stranded DNA that is bound to the substrate. *Reproduced with permission from the American Chemical Society, C. Boozer, J. Ladd, S. Chen, et al. DNA-directed protein immobilization for simultaneous detection of multiple analytes by surface plasmon resonance biosensor. Anal. Chem. 78 (2006) 1515–1519.*

5. CONCLUSIONS, CHALLENGES, AND FUTURE TRENDS

A large number of Ab immobilization chemistries are available for immunodiagnostics. But the screening of an optimal Ab immobilization chemistry is critical because it significantly impacts the bioanalytical performance of an IA. The Ab immobilization chemistry should be simple, rapid, cost-effective, easy-to-use, and reproducible. It should lead to leach-proof Ab immobilization in a functionally oriented manner. The Ab immobilization chemistry should be selected taking into consideration the particular bioanalytical application and the substrate. Although passive adsorption–based Ab immobilization chemistries have been the most widely used strategies, the last two decades have witnessed tremendous evolution that has led to the emergence of multifarious Ab immobilization chemistries. The covalent Ab immobilization chemistries based on the use of various surface-functionalized substrates and cross-linkers have taken a central place and are the most widely used chemistries. Further, the FBPs have been widely used for biosensors and immunodiagnostic applications, especially those employing gold-coated substrates. Ab immobilization chemistry based on the biotin–avidin interactions has also been used for several bioanalytical applications. Of interest is the use of a highly simplified one-step Ab immobilization chemistry where Ab is

TABLE 2.4 Analytical and Imaging Techniques for Probing Biomolecules Immobilized on Supports

Technique	Key Features	References
Thermal gravimetric analysis (TGA)	• To study thermal stability of biomolecules and their supports before and after immobilization. • To decipher new biomolecule structures and their supports after immobilization. • To estimate the quantity of functional groups on the support.	[34,78]
Scanning electron microscopy (SEM)	• SEM samples are commonly used to observe the morphology of biomolecules immobilized on the support. However, it only provides limited information about the internal structure of immobilized biomolecules.	[81]
Field emission SEM (FESEM)	• FESEM is capable of probing very small topographic or morphological details on the surface.	[17]
Transmission electron microscopy (TEM)	• Unlike FESEM, TEM provides only two-dimensional images of the solid structure. It also provides the particle size of the nanoscale supports, such as nanoparticles, carbon nanotubes, and graphene. It is also capable of probing the distribution of biomolecules onto the support material, which is related to the accessibility of the immobilized biomolecules. TEM resolution is improving and approaching the nanometer length.	[52]
X-ray photoelectron spectroscopy (XPS)	• XPS provides a total elemental analysis (carbon, oxygen, nitrogen but not hydrogen and helium) at 10 to 200 Å.	[3,16]

Continued

TABLE 2.4 Analytical and Imaging Techniques for Probing Biomolecules Immobilized on Supports—cont'd

Technique	Key Features	References
	The binding energy and intensity of various photoelectron peaks are used to identify the elements contained in the sample surface, i.e., the surface structure and composition of graft-modified and biomolecule-functionalized layers. • It is capable of providing pertinent information to assess the depth distributions and immobilized layer thicknesses.	
Atomic force microscopy (AFM)	• AFM is used to visualize the surface topography of biomolecules on the carrier surface. AFM also provides other important features including adhesion and elasticity of the immobilized biomolecules. • Protein–protein and protein–surface interactions and surface-induced conformational changes of biomolecules can be measured by force spectroscopy measurements.	[42]
Time-of-flight secondary ion mass spectroscopy (TOF-SIMS)	• A powerful tool to provide the composition, structure, orientation, and spatial distribution of the biomolecules and chemical structures on the surface. • It is useful for the submicron-scale mapping of protein-adsorbed materials, chemical mapping information, i.e., the imaging of the distribution pattern of the biomolecules.	[9]

bound to the substrate simply by diluting and incubating the Ab with APTES solution.

The immobilization chemistries such as those based on the silanization should take into account the formation of multilayers. The deposition of APTES prepared in water is recommended as it results in thin APTES layers that bind to the negatively charged substrate via electrostatic interactions. However prolonged incubation for many hours could result in a thick multilayer.

The amine cross-linking based Ab immobilization chemistries can lead to the formation of multiple bonds as numerous lysine residues are present on Ab in addition to the free amino terminals located near the antigen binding sites. Therefore, they could lead to conformational changes in the Ab and could also block its antigen-binding sites. But this could be obviated by employing intermediate proteins such as FBPs, and avidin-biotin interactions.

The oriented immobilization of Ab is recommended as the covalent and electrostatic Ab immobilization chemistries do not effectively control the orientation of the Ab. Therefore, FBPs such as Protein A, G or A/G could be employed for the orientated Ab immobilization as they bind specifically to the F_c region of the Ab [6].

Nanomaterial-functionalized or nanostructured substrates are the upcoming trend in immunodiagnostics as they provide substantially increased surface areas, which leads to ultrahigh Ab immobilization density. The increased Ab immobilization provides higher sensitivity for analyte detection. The recent developments in surface functionalization chemistries have made it possible to induce the desired functional groups on any nanomaterial-functionalized substrate. The surface of the substrate can be effectively modified with the desired functional group by chemical treatment or CVD, which facilitates the immobilization of Ab. Nowadays, there is a large variety of hetero- or homobifunctional cross-linkers, which can be used to cross-link the functional groups induced or present on the substrates to the functional groups, such as carboxyl, amino, sulfhydryl, hydroxyl or carbonyl, present on the Ab. Further, it might be possible in near future to incorporate the desired functional group within the smart materials used for making the substrates. The continuous development of complementary technologies in the coming years will further lead to critically improved, rapid, and simplified Ab immobilization chemistries that will result in the superior bioanalytical performance of IAs.

REFERENCES

[1] B. Akerstrom, L. Bjorck, A physicochemical study of protein G, a molecule with unique immunoglobulin G-binding properties, J. Biol. Chem. 261 (1986) 10240–10247.

[2] B. Akerstrom, T. Brodin, K. Reis, et al., Protein G: a powerful tool for binding and detection of monoclonal and polyclonal antibodies, J. Immunol. 135 (1985) 2589–2592.

[3] S.A. Al-Bataineh, L.G. Britcher, H.J. Griesser, Rapid radiation degradation in the XPS analysis of antibacterial coatings of brominated furanones, Surf. Interf. Anal. 38 (2006) 1512–1518.

[4] J. Arnau, C. Lauritzen, G.E. Petersen, et al., Current strategies for the use of affinity tags and tag removal for the purification of recombinant proteins, Protein Expr. Purif. 48 (2006) 1–13.

[5] E. Asenath Smith, W. Chen, How to prevent the loss of surface functionality derived from aminosilanes, Langmuir 24 (2008) 12405–12409.

[6] S. Babacan, P. Pivarnik, S. Letcher, et al., Evaluation of antibody immobilization methods for piezoelectric biosensor application, Biosens. Bioelectron. 15 (2000) 615–621.

[7] C.D. Bain, E.B. Troughton, Y.T. Tao, et al., Formation of monolayer films by the spontaneous assembly of organic thiols from solution onto gold, J. Am. Chem. Soc. 111 (1989) 321–335.

[8] B. Barat, A.M. Wu, Metabolic biotinylation of recombinant antibody by biotin ligase retained in the endoplasmic reticulum, Biomol. Eng. 24 (2007) 283–291.

[9] A.M. Belu, D.J. Graham, D.G. Castner, Time-of-flight secondary ion mass spectrometry: techniques and applications for the characterization of biomaterial surfaces, Biomaterials 24 (2003) 3635–3653.

[10] S.K. Bhatia, L.C. Shriver-Lake, K.J. Prior, et al., Use of thiol-terminal silanes and heterobifunctional crosslinkers for immobilization of antibodies on silica surfaces, Anal. Biochem. 178 (1989) 408–413.

[11] I. Björk, B.Å. Petersson, J. Sjöquist, Some physicochemical properties of protein A from *Staphylococcus aureus*, Eur. J. Biochem. 29 (1972) 579–584.

[12] C. Boozer, J. Ladd, S. Chen, et al., DNA-directed protein immobilization for simultaneous detection of multiple analytes by surface plasmon resonance biosensor, Anal. Chem. 78 (2006) 1515–1519.

[13] M.G. Bremer, J. Duval, W. Norde, et al., Electrostatic interactions between immunoglobulin (IgG) molecules and a charged sorbent, Colloids Surf. A Physicochem. Eng. Asp. 250 (2004) 29–42.

[14] K.L. Brogan, K.N. Wolfe, P.A. Jones, et al., Direct oriented immobilization of F (ab′) antibody fragments on gold, Anal. Chim. Acta 496 (2003) 73–80.

[15] J.Y. Byeon, F.T. Limpoco, R.C. Bailey, Efficient bioconjugation of protein capture agents to biosensor surfaces using aniline-catalyzed hydrazone ligation, Langmuir 26 (2010) 15430–15435.

[16] J. Chastain, R.C. King, J. Moulder, Handbook of X-ray Photoelectron Spectroscopy: A Reference Book of Standard Spectra for Identification and Interpretation of XPS Data, Physical Electronics, Eden Prairie, MN, 1995.

[17] S. Chatterjee, D. Yadav, L. Barbora, et al., Silk-Cocoon matrix immobilized lipase catalyzed transesterification of sunflower oil for production of biodiesel, J. Catal. (2014).

[18] F. Darain, C. Ban, Y.B. Shim, Development of a new and simple method for the detection of histidine-tagged proteins, Biosens. Bioelectron. 20 (2004) 857–863.

[19] C.K. Dixit, S.K. Vashist, B.D. Maccraith, et al., Evaluation of apparent non-specific protein loss due to adsorption on sample tube surfaces and/or altered immunogenicity, Analyst 136 (2011) 1406–1411.

[20] C.K. Dixit, S.K. Vashist, B.D. Maccraith, et al., Multisubstrate-compatible ELISA procedures for rapid and high-sensitivity immunoassays, Nat. Protoc. 6 (2011) 439–445.

[21] C.K. Dixit, S.K. Vashist, F.T. O'neill, et al., Development of a high sensitivity rapid sandwich ELISA procedure and its comparison with the conventional approach, Anal. Chem. 82 (2010) 7049–7052.

[22] M. Egholm, O. Buchardt, L. Christensen, et al., PNA hybridizes to complementary oligonucleotides obeying the Watson-Crick hydrogen-bonding rules, Nature 365 (1993) 566–568.

[23] A.B. Emerman, Z.R. Zhang, O. Chakrabarti, et al., Compartment-restricted biotinylation reveals novel features of prion protein metabolism in vivo, Mol. Biol. Cell 21 (2010) 4325–4337.

[24] A.Y. Fadeev, R. Helmy, S. Marcinko, Self-assembled monolayers of organosilicon hydrides supported on titanium, zirconium, and hafnium dioxides, Langmuir 18 (2002) 7521–7529.

[25] R.P. Gandhiraman, V. Gubala, C.C. O'mahony, et al., PECVD coatings for functionalization of point-of-care biosensor surfaces, Vacuum 86 (2012) 547–555.

[26] R.P. Gandhiraman, C. Volcke, V. Gubala, et al., High efficiency amine functionalization of cycloolefin polymer surfaces for biodiagnostics, J. Mater. Chem. 20 (2010) 4116–4127.

[27] G.T. Hermanson, Bioconjugate techniques, Academic press, 1996.

[28] J.D. Hirsch, L. Eslamizar, B.J. Filanoski, et al., Easily reversible desthiobiotin binding to streptavidin, avidin, and other biotin-binding proteins: uses for protein labeling, detection, and isolation, Anal. Biochem. 308 (2002) 343–357.

[29] K. Hofmann, G. Titus, J.A. Montibeller, et al., Avidin binding of carboxyl-substituted biotin and analogs, Biochem 21 (1982) 978–984.

[30] K. Hofmann, S.W. Wood, C.C. Brinton, et al., Iminobiotin affinity columns and their application to retrieval of streptavidin, Proc. Natl. Acad. Sci. USA 77 (1980) 4666–4668.

[31] F. Höök, A. Ray, B. Nordén, et al., Characterization of PNA and DNA immobilization and subsequent hybridization with DNA using acoustic-shear-wave attenuation measurements, Langmuir 17 (2001) 8305–8312.

[32] T.P. Hopp, K.S. Prickett, V.L. Price, et al., A short polypeptide marker sequence useful for recombinant protein identification and purification, Biotech 6 (1988) 1204–1210.

[33] J.A. Howarter, J.P. Youngblood, Optimization of silica silanization by 3-aminopropyl triethoxysilane, Langmuir 22 (2006) 11142–11147.

[34] L. Huang, Z.-M. Cheng, Immobilization of lipase on chemically modified bimodal ceramic foams for olive oil hydrolysis, Chem. Eng. J. 144 (2008) 103–109.

[35] C.J. Jones, R.W. Stoddart, A post-embedding avidin-biotin peroxidase system to demonstrate the light and electron microscopic localization of lectin binding sites in rat kidney tubules, Histochem. J. 18 (1986) 371–379.

[36] W. Kastern, U. Sjobring, L. Bjorck, Structure of peptostreptococcal protein L and identification of a repeated immunoglobulin light chain-binding domain, J. Biol. Chem. 267 (1992) 12820–12825.

[37] M.E. Kimple, A.L. Brill, R.L. Pasker, Overview of affinity tags for protein purification, Curr. Protoc. Protein Sci. 73 (9) (2013) 9.

[38] S.M. Kipriyanov, M. Little, Affinity purification of tagged recombinant proteins using immobilized single chain Fv fragments, Anal. Biochem. 244 (1997) 189–191.

[39] K.B. Lee, S.J. Park, C.A. Mirkin, et al., Protein nanoarrays generated by dip pen nanolithography, Science 295 (2002) 1702–1705.

[40] C. Ley, D. Holtmann, K.M. Mangold, et al., Immobilization of histidine-tagged proteins on electrodes, Colloids Surf. B Biointerfaces 88 (2011) 539–551.

[41] J.-F. Lutz, Z. Zarafshani, Efficient construction of therapeutics, bioconjugates, biomaterials and bioactive surfaces using azide–alkyne "click" chemistry, Adv. Drug Del. Rev. 60 (2008) 958–970.

[42] C. Marcuello, R. De Miguel, C. Gomez-Moreno, et al., An efficient method for enzyme immobilization evidenced by atomic force microscopy, Protein Eng. Des. Sel. 25 (2012) 715–723.

[43] M. Masuko, Hybridization of an immobilized PNA probe with its complementary oligodeoxyribonucleotide on the surface of silica glass, in: Nucleic Acids Symposium Series, Oxford University Press, p, 2003, pp. 145–146.

[44] C. Mateo, G. Fernández-Lorente, O. Abian, et al., Multifunctional epoxy supports: a new tool to improve the covalent immobilization of proteins. The promotion of physical adsorptions of proteins on the supports before their covalent linkage, Biomacromolecules 1 (2000) 739–745.

[45] P. Millner, H. Hays, A. Vakurov, et al., Nanostructured transducer surfaces for electrochemical biosensor construction—Interfacing the sensing component with the electrode, in: Seminars in Cell & Developmental Biology, Elsevier, 2009, pp. 34–40.

[46] J. Morfill, K. Blank, C. Zahnd, et al., Affinity-matured recombinant antibody fragments analyzed by single-molecule force spectroscopy, Biophys. J. 93 (2007) 3583–3590.

[47] H. Nakajima, P.K. Brindle, M. Handa, et al., Functional interaction of STAT5 and nuclear receptor co-repressor SMRT: implications in negative regulation of STAT5-dependent transcription, EMBO J. 20 (2001) 6836–6844.

[48] P.E. Nielsen, M. Egholm, R.H. Berg, et al., Sequence-selective recognition of DNA by strand displacement with a thymine-substituted polyamide, Science 254 (1991) 1497–1500.

[49] B.H. Nilson, L. Lögdberg, W. Kastern, et al., Purification of antibodies using protein L-binding framework structures in the light chain variable domain, J. Immunol. Methods 164 (1993) 33–40.

[50] K. Nilsson, K. Mosbach, Immobilization of enzymes and affinity ligands to various hydroxyl group carrying supports using highly reactive sulfonyl chlorides, Biochem. Biophys. Res. Commun. 102 (1981) 449–457.

[51] T. Nogi, T. Sangawa, S. Tabata, et al., Novel affinity tag system using structurally defined antibody-tag interaction: application to single-step protein purification, Protein Sci. 17 (2008) 2120–2126.

[52] H. Noureddini, X. Gao, Characterization of sol-gel immobilized lipases, J. Solgel Sci. Technol. 41 (2007) 31–41.

[53] M. Rabe, D. Verdes, S. Seeger, Understanding protein adsorption phenomena at solid surfaces, Adv. Colloid Interf. Sci. 162 (2011) 87–106.

[54] J. Raj, G. Herzog, M. Manning, et al., Surface immobilisation of antibody on cyclic olefin copolymer for sandwich immunoassay, Biosens. Bioelectron. 24 (2009) 2654–2658.

[55] S.V. Rao, K.W. Anderson, L.G. Bachas, Oriented immobilization of proteins, Mikrochim. Acta 128 (1998) 127–143.

[56] A. Ray, B. Norden, Peptide nucleic acid (PNA): its medical and biotechnical applications and promise for the future, FASEB J. 14 (2000) 1041–1060.

[57] N. Rozlosnik, M.C. Gerstenberg, N.B. Larsen, Effect of solvents and concentration on the formation of a self-assembled monolayer of octadecylsiloxane on silicon (001), Langmuir 19 (2003) 1182–1188.

[58] D. Saerens, L. Huang, K. Bonroy, et al., Antibody fragments as probe in biosensor development, Sensors 8 (2008) 4669–4686.

[59] F. Schroper, A. Baumann, A. Offenhausser, et al., Bidirectional immobilization of affinity-tagged cytochrome c on electrode surfaces, Chem. Commun. 46 (2010) 5295–5297.

[60] F. Schröper, A. Baumann, A. Offenhäusser, et al., Direct electrochemistry of novel affinity-tag immobilized recombinant horse heart cytochrome c, Biosens. Bioelectron. 34 (2012) 171–177.

[61] H. Sheng, B.C. Ye, Different strategies of covalent attachment of oligonucleotide probe onto glass beads and the hybridization properties, Appl. Biochem. Biotechnol. 152 (2009) 54–65.

[62] W.D. Sikkema, An Fc-binding protein, Am. Biotechnol. Lab 7 (1989) 42.

[63] T.A. Silva, L.F. Ferreira, L.M. Souza, et al., New approach to immobilization and specific-sequence detection of nucleic acids based on poly (4-hydroxyphenylacetic acid), Mater. Sci. Eng. C 29 (2009) 539–545.

[64] J. Sjoquist, B. Meloun, H. Hjelm, Protein A isolated from *Staphylococcus aureus* after digestion with lysostaphin, Eur. J. Biochem. 29 (1972) 572–578.

[65] C. Steinhauer, C. Wingren, F. Khan, et al., Improved affinity coupling for antibody microarrays: engineering of double-(His)6-tagged single framework recombinant antibody fragments, Proteomics 6 (2006) 4227–4234.

[66] K. Terpe, Overview of tag protein fusions: from molecular and biochemical fundamentals to commercial systems, Appl. Microbiol. Biotechnol. 60 (2003) 523–533.

[67] T.S. Tsapikouni, Y.F. Missirlis, Protein–material interactions: from micro-to-nano scale, Mater. Sci. Eng. B 152 (2008) 2–7.

[68] M. Van Der Veen, M.C. Stuart, W. Norde, Spreading of proteins and its effect on adsorption and desorption kinetics, Colloids Surf. B Biointerfaces 54 (2007) 136–142.

[69] S.K. Vashist, Comparison of 1-Ethyl-3-(3-Dimethylaminopropyl) carbodiimide based strategies to crosslink antibodies on amine-functionalized platforms for immunodiagnostic applications, Diagnostics 2 (2012) 23–33.

[70] S.K. Vashist, Effect of antibody modifications on its biomolecular binding as determined by surface plasmon resonance, Anal. Biochem. 421 (2012) 336–338.

[71] S.K. Vashist, A sub-picogram sensitive rapid chemiluminescent immunoassay for the detection of human fetuin A, Biosens. Bioelectron. 40 (2013) 297–302.

[72] S.K. Vashist, C.K. Dixit, B.D. Maccraith, et al., Effect of antibody immobilization strategies on the analytical performance of a surface plasmon resonance-based immunoassay, Analyst 136 (2011) 4431–4436.

[73] S.K. Vashist, E. Lam, S. Hrapovic, et al., Immobilization of antibodies and enzymes on 3-aminopropyltriethoxysilane-functionalized bioanalytical platforms for biosensors and diagnostics, Chem. Rev. 114 (2014) 11083–11130.

[74] S.K. Vashist, E. Marion Schneider, E. Lam, et al., One-step antibody immobilization-based rapid and highly-sensitive sandwich ELISA procedure for potential in vitro diagnostics, Sci. Rep. 4 (2014) 4407.

[75] S.K. Vashist, E.M. Schneider, J.H.T. Luong, Rapid sandwich ELISA-based in vitro diagnostic procedure for the highly-sensitive detection of human fetuin A, Biosens. Bioelectron. 67 (2015) 73–78.

[76] S.K. Vashist, E.M. Schneider, J.H.T. Luong, Surface plasmon resonance-based immunoassay for human C-reactive protein, Analyst 140 (2015) 4445–4452.

[77] S.K. Vashist, E.M. Schneider, J.H.T. Luong, Surface plasmon resonance-based immunoassay for human fetuin A, Analyst 139 (2014) 2237–2242.

[78] M.L. Verma, M. Naebe, C.J. Barrow, et al., Enzyme immobilisation on amino-functionalised multi-walled carbon nanotubes: structural and biocatalytic characterisation, PLoS One 8 (2013) e73642.

[79] E.A. Vogler, Protein adsorption in three dimensions, Biomaterials 33 (2012) 1201–1237.

[80] P. Wagner, M. Hegner, P. Kernen, et al., Covalent immobilization of native biomolecules onto Au(111) via N-hydroxysuccinimide ester functionalized self-assembled monolayers for scanning probe microscopy, Biophys. J. 70 (1996) 2052–2066.

[81] J. Wang, J.A. Carlisle, Covalent immobilization of glucose oxidase on conducting ultrananocrystalline diamond thin films, Diam. Relat. Mater. 15 (2006) 279–284.

[82] M. Yamaura, R. Camilo, L. Sampaio, et al., Preparation and characterization of (3-aminopropyl) triethoxysilane-coated magnetite nanoparticles, J. Magn. Magn. Mater. 279 (2004) 210–217.

[83] L. Yang, M. Biswas, P. Chen, Study of binding between protein A and immunoglobulin G using a surface tension probe, Biophys. J. 84 (2003) 509–522.
[84] L. Zhang, Y. Liu, T. Chen, Label-free amperometric immunosensor based on antibody immobilized on a positively charged gold nanoparticle/L-cysteine-modified gold electrode, Microchim. Acta 164 (2009) 161–166.
[85] M. Zhong, J. Fang, Y. Wei, Site specific and reversible protein immobilization facilitated by a DNA binding fusion tag, Bioconjug. Chem. 21 (2010) 1177–1182.
[86] M. Zhu, M.Z. Lerum, W. Chen, How to prepare reproducible, homogeneous, and hydrolytically stable aminosilane-derived layers on silica, Langmuir 28 (2011) 416–423.

Chapter 3

Monoclonal Antibody Generation by Phage Display: History, State-of-the-Art, and Future

Christian Hentrich, Francisco Ylera, Christian Frisch, Andre Ten Haaf, Achim Knappik
Bio-Rad, Puchheim, Germany

1. INTRODUCTION

Antibodies are indispensable detection reagents in research and diagnostics and represent the fastest-growing class of biological therapeutics on the market [1,2]. Of the approximately 50 antibodies approved by the U.S. Food and Drug Administration by 2016, 6 were generated or improved by phage display. Many more phage display–derived antibodies are currently in late clinical trials [3].

Research antibodies are used in standard methods such as enzyme-linked immunosorbent assays (ELISAs), Western blotting, immunohistochemistry (IHC), flow cytometry, or affinity chromatography. Furthermore, antibodies are important tools to decode the human proteome [4] or to measure clinical parameters such as pharmacokinetics during drug development [5,6].

For most of these applications, the use of recombinant antibodies is highly advantageous or even necessary. Several *in vitro* technologies have been developed to generate such recombinant antibodies. Of these, phage display is the most widely used method. It is based on the presentation of antibody fragments on the surface of phages that contain the corresponding DNA. The phage thus directly links phenotype and genotype of the antibody, enabling selection *in vitro*.

1.1 History of the Development of Antibody Phage Display

Phage display was invented in 1985 when George Smith demonstrated that an exogenous protein could be expressed on the surface of the filamentous phage

Handbook of Immunoassay Technologies. https://doi.org/10.1016/B978-0-12-811762-0.00003-7

M13 [7]. Smith inserted the DNA coding for a fragment of the EcoRI endonuclease into the phage gene III, which codes for a protein at the tip of the phage, the minor capsid protein III. He was able to detect the EcoRI-pIII fusion protein on the surface of phages with polyclonal antibodies targeting the EcoRI endonuclease. Additionally, he showed that phages bearing the fusion protein on their surface could be selected from a mixture containing a large excess of wild-type phages. In 1989, Richard Lerner and colleagues were able to isolate antibody fragments from a library of randomly combined murine variable heavy and variable light chains through phage display [8], thereby presenting an alternative to hybridoma technology for the generation of monoclonal antibodies. One year later, McCafferty *et al.* expressed a single-chain variable fragment against lysozyme on the surface of bacteriophages and showed specific binding of this phage antibody in an ELISA [9]. Furthermore, the phage antibody was purified from a mixture of wild-type phages using a column of lysozyme-sepharose. The first human phage display antibody libraries were created in the early 1990s [10,11]. Currently, phage display has proven to be the most widely used *in vitro* antibody selection technology [1].

1.2 Antibody Formats Used for Phage Display

Antibodies are at the core of the immune system and have evolved to bind a large variety of substances with high affinity and specificity. Additionally, they have useful properties for drug development such as long *in vivo* half-life and Fc-mediated functions such as complement-dependent cytotoxicity. Antibodies are encoded by two genes, heavy chain and light chain, which complicates the development of antibody libraries and the application of selection technologies. Also, due to their complexity, full-length antibodies cannot be produced in bacteria in sufficient amounts in a functional form, which is one reason why antibody fragments are used for *Escherichia coli*–based phage display systems (Fig. 3.1). The most commonly used format in phage display is the single-chain variable fragment (scFv), which consists of the variable heavy chain and the variable light chain connected by a flexible linker. The primary reason for its wide use is the simplicity of the cloning process, which enables fast and easy library generation. Due to the small protein size ($\sim$25 kDa), scFvs are well tolerated by bacteria and less likely to be degraded, leading to a high display rate [12]. However, scFvs are less stable in comparison with Fab fragments and tend to form dimers, a tendency that is increased by shorter linkers. Although dimerization can be reduced by increasing the length of the linker to more than 20 amino acids, it is difficult to eliminate dimer formation completely and obtain pure monomers [12]. Furthermore, some scFvs exhibit a decreased affinity in comparison with the corresponding Fab fragment [13].

Fab fragments consist of the light chain (V_L–C_L) and the Fd-domain (V_H–C_{H1}) of the heavy chain of an antibody. During bacterial expression,

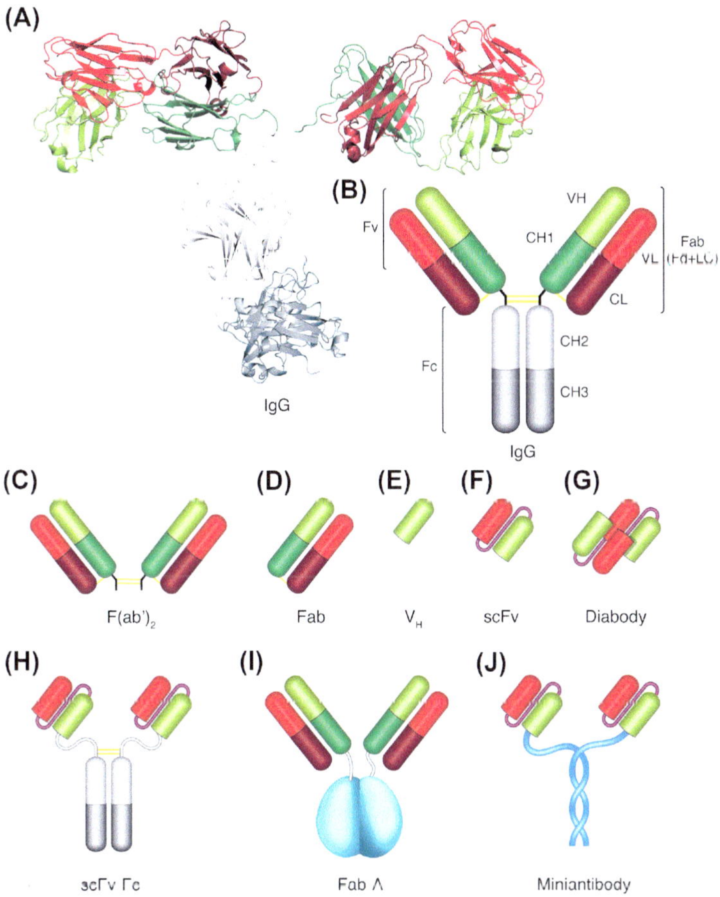

FIGURE 3.1 Recombinant antibody formats. (A) Crystal structure of a full-length IgG antibody (rendered from PDB 1IGT). (B)–(J): Schematic representation of different antibody formats. Domains are colored as in (A). (B) Full-length IgG antibody. Disulfide bonds are dark yellow. (C) Bivalent F(ab′)$_2$ (D) Monovalent Fab. The disulfide bond is optional and not required for the formation of the V_L–C_L/V_H–C_{H1} heterodimer. (E) Single-domain antibody, such as VHH. (F) Single-chain variable Fragment (scFv). (G) Diabodies are noncovalent dimers of scFvs, which spontaneously form depending on the linker length between V_H and V_L. Another form of diabodies is two scFvs connected with a short linker (not shown). (H) scFv-Fc are scFvs dimerized by the Fc domain. (I) Fab-A is created by genetic fusion of the Fab Fd gene with the alkaline phosphatase (PhoA) gene and coexpressing the light chain gene. As bacterial alkaline phosphatase is a homodimeric enzyme, the construct shows bivalent binding and can be directly detected using a colorimetric substrate. (J) Miniantibodies are scFvs or Fabs connected via a flexible linker to self-associating structures such as helix bundles or leucine zippers. Tetravalency and bispecificity can be generated by this way as well.

these two chains are synthesized separately and secreted into the periplasm where they fold to form heterodimers. Besides higher stability, Fabs also possess better pharmacokinetic and pharmacodynamic qualities than scFvs [14]. In addition, they are easier to convert into full-length antibodies [12].

Many groups have developed alternative protein scaffolds, derived from natural binding proteins or other protein domains, as the basis of library design for the selection of binding molecules. Such scaffolds are usually much smaller than antibodies, highly expressible in bacteria, very stable, encoded in one gene sequence, and well suited for *in vitro* library generation and selection methodologies. Such scaffolds are structurally well-defined proteins that are able to present surface loops of varying sequence and length without significant changes in framework structure [15].

The use of single immunoglobulin domains as a scaffold started with the discovery of a subclass of antibodies in camelids that are devoid of light chains [16]. It was found that the V_H domain of these antibodies (termed VHH) has an increased surface solubility in the area that faces the V_L domain in ordinary antibodies and is stable without V_L. The first libraries of soluble human V_H domains were built on a "camelized" human V_H domain [17]. This principle has been further developed to the so-called nanobodies, of which several are in clinical trials for drug development [18]. Another example of single-domain antibodies (sdAbs) is derived from the IgNAR (new antigen receptor) antibody found in cartilaginous fish, e.g., sharks. These antibodies also consist of heavy chains only, with a stable single domain responsible for antigen binding, referred to as vNAR. Both natural VHH and vNAR domains possess several loop structures that differ from conventional antibodies and can, therefore, form unique complementarity-determining regions (CDRs).

A multitude of nonantibody-based scaffolds has been exploited over the last years. Only a few examples of prominent nonantibody scaffolds will be mentioned here, as they have been reviewed in detail elsewhere [19]: "Affibodies" are based on the Z domain of the IgG-binding staphylococcal protein A [20]. "Anticalins" are derived from the diverse set of lipocalin proteins [21]. "DARPins," or designed ankyrin repeat proteins, use a scaffold where a varying number of structural motifs (repeats) are stacked to form the repeat protein domain [22]. "Avimers" consist of two or more peptide sequences connected by linker peptides [23]. "Affimers" are derived from the cysteine protease inhibitor family of cystatins [24], and "Monobodies" use the 10th fibronectin type III domain of human fibronectin as a scaffold [25].

1.3 Further Recombinant Antibody Formats

Antibody formats used for phage display are typically monovalent and thus can be used for determination of individual binding interaction (the intrinsic affinity) between an antibody and its antigen without interfering avidity effects [26]. Monovalent antibodies are also well suited for crystallization and protein structure studies [27]. For certain applications such as Western blotting, IHC, or flow cytometry, multivalency can, however, be beneficial, as multivalent binding increases the functional affinity (avidity), thereby leading to improved performance in various assays. The selected monovalent fragments from phage display can serve as building blocks for multivalent fragments, and a variety of different formats have been investigated, such as miniantibodies [28], diabodies [29], or Fab-A [30] (Fig. 3.1). An advantage common to all recombinant antibody formats is that affinity or epitope tags such as the hexa-histidine or the FLAG tag can easily be added at will. However, many of the well-established secondary reagents for ordinary IgG detection (mostly polyclonal antisera binding to full-length IgG from the various species) do not work with such antibody fragments due to the lack of the Fc domain and need to be replaced by either anti-tag antibodies, depending on the peptide tags attached to the antibody fragments, or antibodies specific for the constant part of the Fab fragment or for the multimerization domain, if present.

2. PHAGE DISPLAY SELECTION

Bacteriophages are viruses that infect bacteria. From the many different classes of bacteriophages, the nonlytic filamentous phage is the most often used for phage display, primarily the M13 or Fd strains (Fig. 3.2). For phage display, proteins to be selected are covalently linked to the surface of the phage, and fusions to all five coat proteins have been used for this purpose. However, by far the most frequently used protein in phage display is pIII. The pIII protein is essential for infection of bacteria and consists of three domains: two N-terminal domains (N1, N2) and a C-terminal domain, which is anchored in the phage coat. To initiate infection, the N2-domain interacts with the F pilus of bacteria, which leads to retraction of the pilus and allows the N1-domain to bind the inner membrane protein TolA [31]. The antibody sequence was initially cloned directly into the phage DNA [9], leading to

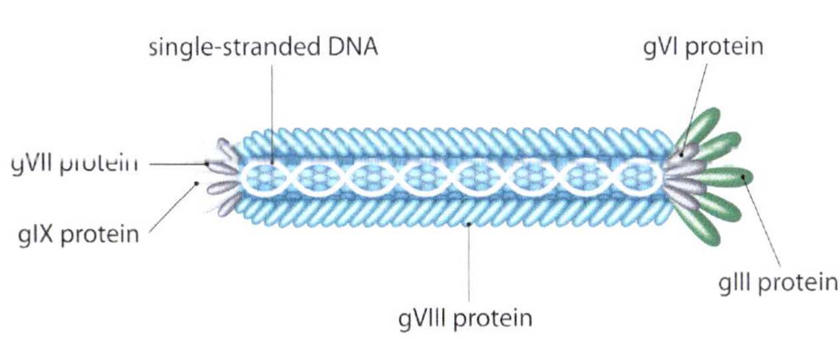

FIGURE 3.2 Scheme of filamentous Fd/M13 phage. Filamentous phages contain small, single-stranded, 6–7 kilobase genomes, which encode 11 genes. The elongated body of the phage consists of thousands of copies of the major coat protein pVIII (blue). One end of the phage consists of five copies of each of the two minor coat proteins pVII and pIX, and the other end consists of five copies of each of the two proteins pIII and pVI.

phages with antibodies displayed on every pIII protein. As an alternative to full-length phage genomes, small plasmids termed phagemids were developed. Phagemids contain the pIII-antibody fusion, an antibiotic resistance marker, origins of replication for phage and bacterial DNA polymerases, and a phage morphogenetic signal for the packaging of the phagemid into the phage [32]. The missing genes for phage production are supplied by infection with a helper phage, a full-length phage with a mutated packaging signal. Due to this mutation, predominantly the phagemid, and not the helper phage DNA, is packaged into the phage coat. On average, these phages contain only one antibody on the surface since the wild-type pIII from the helper phage is incorporated faster than the fusion protein from the phagemid. The resulting monovalent antibody display allows the selection for antibody binding strength as it is not influenced by any avidity effects. Multivalent display can be achieved using a special helper phage that does not contain a wild-type pIII; therefore, all five copies of pIII have to originate from the phagemid [33]. A further positive effect of using phagemids, where the displayed protein is genetically isolated from the remaining phage genes, is a higher genetic stability of the library and thereby a lower tendency to lose antibody genes [34]. Instead of a genetic fusion between an antibody and pIII, it is also possible to connect the antibody to the phage via a disulfide bond. In this system termed CysDisplay, the pIII on the phagemid carries a cysteine at the N-terminus, and the antibody has an additional cysteine at the C-terminus of the heavy chain [35]. On export to the periplasm, they form a disulfide bond and are incorporated into the phage coat. The advantage of CysDisplay is the simple and quantitative elution of the specific phage antibodies by cleavage of the disulfide bond with a reducing agent such as dithiothreitol, a method that is independent of antibody affinity. This avoids bias toward low-affinity antibodies, which are more readily eluted by conventional pH shift elution. The same advantage can be achieved by a genetic fusion containing a protease cleavable linker [36].

During the selection process (Fig. 3.3), the phages presenting the diverse antibodies from the library are incubated with an antigen, which is either immobilized on magnetic beads [37], on polystyrene surfaces [10], or on columns [38], or is used in solution as biotinylated antigen and later captured by immobilized streptavidin [39]. Nonspecific phages are removed by extensive washing, and specific phages are eluted and used for infection of bacteria. Infected bacteria are amplified and produce new phages displaying antibodies after a secondary infection with helper phages. Depending on the size of the library and stringency of the selection, two to four selection rounds are usually required to sufficiently enrich the specific antibodies. Plating the infected bacteria either directly or after subcloning of the antibody DNA into an expression vector allows isolation of individual clones, each expressing one monoclonal antibody. The subsequent screening step, often done by ELISA, flow cytometry on cells, or on antigen-coated beads, allows the identification

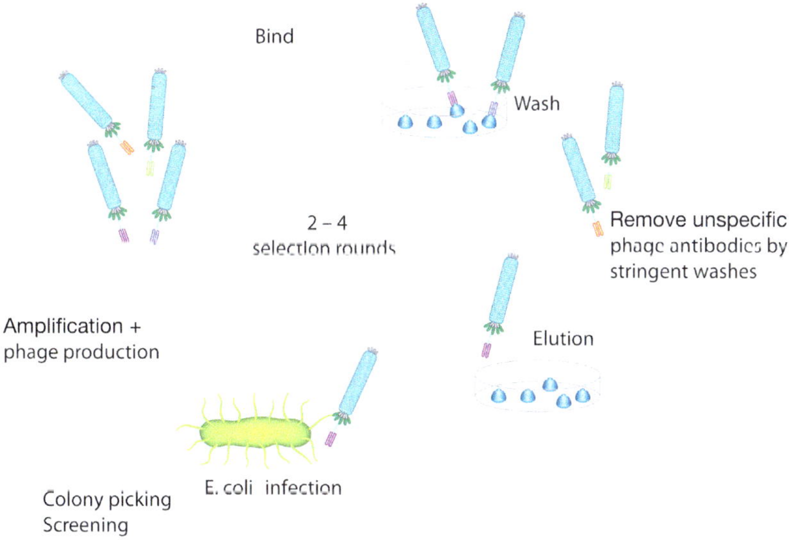

FIGURE 3.3 Principle of phage display selection. The phage antibody library is incubated with an immobilized antigen. Unspecific phage antibodies are washed off and specific antibodies are eluted. Amplification of eluted phage antibodies is achieved by infection of *Escherichia coli*, which depends on the interaction of phage protein III with a bacterial F-pilus. From the infected bacteria, new phage antibodies are produced for the next selection round. Usually, two to four selection rounds are required to enrich the specific antibodies sufficiently. Plating the infected bacteria separates the individual clones for identification of hits in a subsequent screening step.

of those clones with the desired cross-reactivity profile and binding strength. Finally, the selected clones are sequenced to obtain the unique antibody sequences.

2.1 Advantages of Recombinant Antibody Selection

The development of animal-free *in vitro* selection processes was motivated by the need for developing human antibody therapeutics. Thus many economically successful libraries were generated by biotech and pharma companies (Table 3.1). However, the *in vitro* process has more advantages than being able to generate human antibodies for therapy. The use of highly diverse antibody libraries can deliver antibodies for almost any desired antigen. *In vitro* selection enables generation of antibodies against highly conserved or self-antigens [40], conformational variants [41], low immunogenic antigens [42], and also toxic components [43], which is not possible by *in vivo* immunization of animals. In addition to proteins, peptides, or haptens, cell lines [44], tissue slides [45], or virus particles [46] can be used as a starting material for the antibody generation process. While the antibody generation through animal immunization and hybridoma technology is laborious and slow (it can take up

TABLE 3.1 Examples of Phage Display Antibody Libraries

Library Name or *First Author*	Institute/ Company	Species	Library Type	Format	# Germ Line Genes	Diversity	Best Affinity	References
Burton	Scripps	Human	Immune	Fab	natural	1.0×10^7	10 nM	[130]
Rader	Scripps	Rabbit	Immune	Fab	rabbit-human hybrid	2.0×10^7	0.4 nM	[85]
Lorimer	NIH	Mouse	Immune	scFv	natural	8.0×10^6	11 nM	[131]
Schier	UCSF	Human	Naïve natural	scFv	natural	3.0×10^7	16 nM	[132]
CAT1	CAT/ MedImmune	Human	Naïve natural	scFv	natural	1.4×10^{10}	0.3 nM	[133]
CAT2	CAT/ MedImmune	Human	Naïve natural	scFv	natural	1.3×10^{11}	n/a	[122]
Glanville	Pfizer	Human	Naïve natural	scFv	natural	3.1×10^{10}	0.1 nM	[134]
HAL9/10	TU Braunschweig/ Yumab	Human	Naïve natural	scFv	natural	1.5×10^{10}	n/a	[135]
Hoogenboom	MRC	Human	Naïve semisynthetic	scFv	$49\ V_H + 1\ V_L$	1.0×10^7	700 nM	[11]
Barbas III	Scripps	Human	Naïve semisynthetic	Fab	$1\ V_H + 1\ V_L$	5.0×10^7	100 nM	[136]
Nissim	MRC	Human	Naïve semisynthetic	scFv	$50\ V_H + 1\ V_L$	1.0×10^8	n/a	[66]

Pini	Universita' di Siena	Human	Naïve semisynthetic	scFv	1 V_H + 1 V_L	3.0×10^8	19 nM	[65]
Lib-1-Fab/Lib-2-Fab	Genentech/ Roche	Human	Naïve semisynthetic	Fab	1 V_H + 1 V_L	1.0×10^{10}	0.5 nM	[137]
FAB-300	Dyax	Human	Naïve semisynthetic	Fab	1 V_H + natural V_L	1.0×10^{10}	0.2 nM	[138]
PHILOtop	ETH Zürich	Mouse	Naïve semisynthetic	scFv	1 V_H + 1 V_L	1.5×10^9	100 nM	[83]
HuCAL1	MorphoSys	Human	Naïve synthetic	scFv	7 V_H + 7 V_L	2.0×10^9	1.0 nM	[70]
HuCAL GOLD	MorphoSys	Human	Naïve synthetic	Fab	7 V_H + 7 V_L	1.6×10^{10}	0.04 nM	[35]
HuCAL PLATINUM	MorphoSys	Human	Naïve synthetic	Fab	7 V_H + 6 V_L	4.5×10^{10}	0.002 nM	[71]
Ylanthia	MorphoSys	Human	Naïve synthetic	Fab	36 V_H/V_L pairs	1.3×10^{11}	0.7 nM	[72]
Batonick	AxioMx/Abcam	Human	Naïve synthetic	scFv	1 V_H + 1 V_L	1.0×10^8	n/a	[139]
NaLi–H1	Institut Curie	Llama	Naïve synthetic	VHH	1 VHH	3.0×10^9	0.05	[82]

scFv, single-chain variable fragment.Table of notable phage display libraries. The first column lists the library name or the first author of the publication in italics (if the library was not named). Best affinity in column 8 refers to the best reported monovalent affinity in the original paper, without the use of any further affinity maturation.

to 12 months until monoclonal antibodies are produced and tested [47]) the complete development of new monoclonal antibodies by phage display can be achieved in about 2 months including the expression, purification, and testing of antibodies such as scFv or Fab fragments in *E. coli*. A principal advantage of *in vitro* selection technologies is the direct access to the genetic information of the antibody, which allows downstream applications such as further maturation and enables a faster conversion into different antibody formats such as full-length antibodies or scFv-Fc fusion proteins. In recent years, concerted efforts toward antibody validation and quality of antibody generation have made the power of recombinant antibody technologies evident. Since the DNA sequence is known, gene synthesis technology guarantees infinite access to the antibody, and reproducibility with consistent quality is assured. Furthermore, phage display is more readily automatable and thus can be used for high-throughput antibody generation endeavors as well as routine use [42].

2.2 Guided Selection

The term guided selection summarizes several *in vitro* strategies that allow the selection of antibodies with rare properties or specificities that are extremely difficult to find with animal-based methods. *In vitro* selection allows for the modification of selection conditions to match those of the later application with regard to buffer, additives, pH, or temperature [48]. It is possible to prevent the selection of undesirable antibodies through a strategy termed blocking: by adding closely related antigens in excess to the library, cross-reactive antibodies bind to the more abundant protein in solution and will be washed away (Fig. 3.4A). This blocking process is regularly used for the generation of antibodies that can differentiate between closely related antigens. Examples generated by guided selection include antibodies against the Parkinson-related protein DJ-1 that are specific for the oxidized cysteine-106 form [49], antibodies that distinguish proteins differing by a few amino acids in length [50], or antibodies directed against the active site of a protease [51]. Another example is the straightforward generation of anti-idiotypic antibodies, which specifically recognize the binding site of a therapeutic antibody but do not cross-react with other human antibodies even at high concentrations present in serum samples [5,6,52]. Such antibodies are difficult to generate with immunization strategies and usually require screening of thousands of candidates. If desired, antibodies binding to more than one antigen can be selected by alternating antigens in different selection rounds. This can be used to drive the antibody selection to certain epitopes that are shared by the different antigens, for instance, to develop cross-species-specific antibodies. Generation of antibodies specific to protein–protein or protein–hapten complexes is also extremely difficult to achieve *in vivo*. With phage display, however, it is possible to use the protein complex as the antigen in combination with a matching blocking strategy. The resulting antibodies

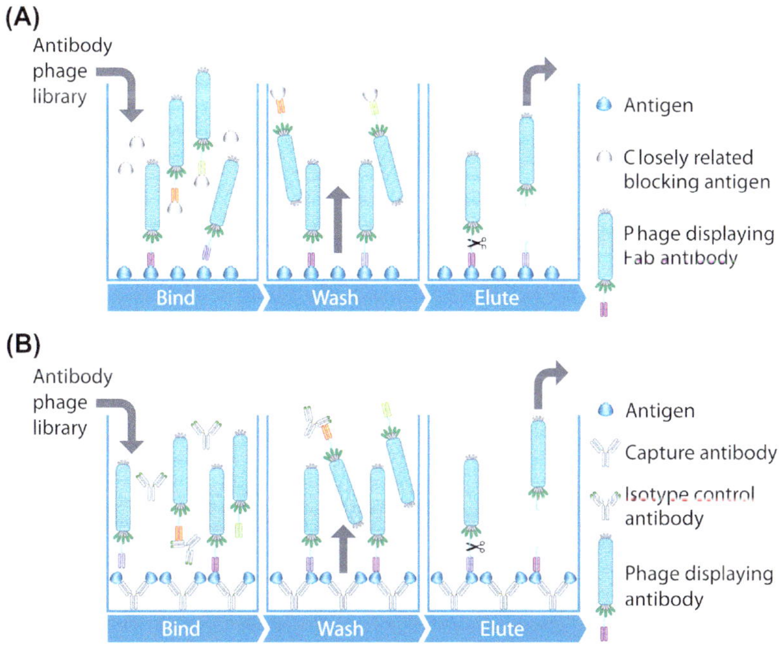

FIGURE 3.4 Guided selection strategies. (A) Blocking strategy to prevent selection of cross-reactive antibodies. Closely related antigens, which should not be detected by the antibody, are added in excess to the antibody phage solution; all cross-reactive antibodies bind to the more abundant blocking antigen and are washed away leaving only the specific antibodies bound to the immobilized antigen. (B) Guided selection strategy for identification of detection antibodies. Phage display panning is performed on capture antibody–antigen complex; an isotype-matched antibody is used for blocking capture antibody–specific phages. For a selection of complex-specific antibodies, blocking or preincubation of the antibody phage library needs to be performed separately with antigen and capture reagent.

bind the protein complex only but not the individual proteins alone. This strategy in particular has been applied to select antibodies against therapeutic antibody–target complexes [52].

2.3 Affinity Improvement

While the immune system generally does not create antibodies with monovalent affinities better than ~100 pM [53], affinities of recombinant antibodies can often be improved beyond this limit using *in vitro* evolution strategies referred to as *in vitro* affinity maturation. For this process, a known antibody sequence is diversified either completely, by random mutagenesis using a technique such as error-prone PCR, or the diversity is limited to certain CDRs, which are exchanged with CDR libraries or are randomly mutagenized. These derivative libraries are then used for additional rounds of selection. Affinity improvement of 5000-fold has been achieved with this strategy, leading to antibodies with affinities in the low or even subpicomolar range [39,54].

2.4 Other Selection Technologies

In the last 20 years, other selection technologies that circumvent some of the limitations of phage display have been developed. For instance, bacterial expression systems are not ideal for eukaryotic proteins and consequently some antibodies suffer from poor expression. The selection and amplification process is supposed to be based on antibody affinity alone but is strongly biased by uneven bacterial growth and phage production [55]. Furthermore, the large phage surface can lead to unspecific binding and enrichment of irrelevant clones. An important parameter for the generation of high-affinity antibodies is the library size, which is limited for phage display to about 10^{11} candidates. Alternative selection systems overcome some of these shortcomings but often have their own limitations.

Besides the display on phages, antibodies have also been displayed on bacteria, mammalian cells, yeast, and other viruses. Yeast display is becoming increasingly popular, as it is a eukaryotic system with improved antibody folding capabilities [56]—even full-length IgG can be displayed on the surface of yeast cells [57]. Another advantage is the selection of the complete library by fluorescence-activated cell sorting (FACS) after incubation with fluorescently labeled antigens. Antibodies with low femtomolar affinities have been generated using this approach [58]. Yeast display has a limited library size of 10^8 to 10^{10} due to lower transformation efficiency.

Conversely, very large library sizes of up to 10^{15} can be achieved with *in vitro* selection technologies such as ribosome display or mRNA display. These technologies avoid the use of any cells, and the genotype–phenotype linkage is achieved by stabilizing the RNA–ribosome–protein complex at the end of the RNA translation for ribosome display [59] or by covalently coupling the RNA to the translated protein via the incorporation of a puromycin linker [60]. Amplification of the selection output is achieved by RT-PCR and can easily be combined with error-prone PCR to increase the diversity further. Over the last 25 years, enormous efforts have been put into developing and improving these selection technologies, leading to robust procedures for the generation of recombinant antibodies.

Together with phage display, these technologies lead to an increasing acceptance of recombinant antibodies in the market, making recombinant antibodies a viable alternative to animal-derived antibodies.

3. ANTIBODY LIBRARIES

In vivo, the diversity of the antibody repertoire is generated through genetic rearrangement of antibody germ line genes and consequent somatic hypermutation. The aim in constructing antibody libraries for phage display is to

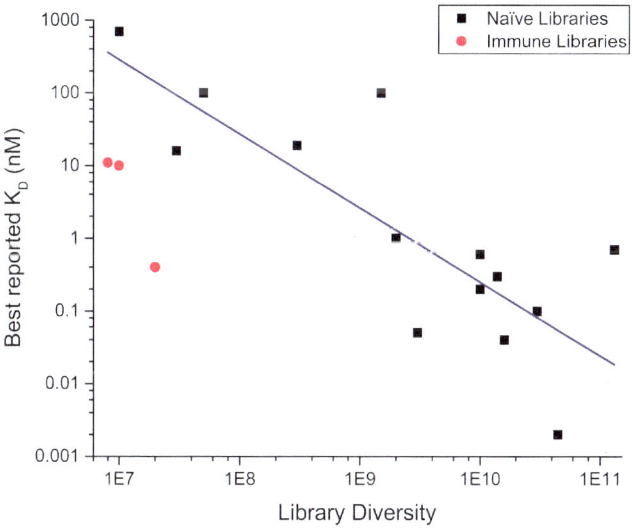

FIGURE 3.5 Library size versus best selected antibody affinity. Scatter plot of the antibody library diversities against the best reported affinities (from the original publication, without additional maturation) for the libraries listed in Table 3.1. Naïve libraries are represented by *black dots*; immune libraries are in red. The *blue line* is a linear fit of the naïve libraries in double-logarithmic space.

mimic this diversity in the form of antibody fragments displayed on phages. Antibody libraries have now been constructed for more than 25 years and have been extensively reviewed [2,34,61].

The probability of finding high-affinity antibodies increases as a function of library size (Table 3.1; Fig. 3.5). In phage display, the limiting factors for library size are the transformation efficiency of the library DNA into *E. coli* and liquid volume handling capacity for bacterial cultures. The resulting maximal achievable library size using full-length phages is $\sim 10^7$. As transformation efficiency is dependent on the size of the transformed DNA [62], larger libraries (up to 10^{11}; see Table 3.1) can be achieved with the smaller phagemid DNA, which is therefore used by all modern antibody libraries. In the following sections, different classes of antibody libraries commonly used in phage display will be described.

3.1 Immune Libraries

Phage display with immune libraries shares similarities with the conventional generation of antibodies. The initial steps in library generation consist of immunizing an animal with an antigen and isolating the mRNA from B lymphocytes. The mRNA is then reverse transcribed into cDNA, and the variable regions of expressed antibodies are amplified via PCR and cloned into a phage display vector [2].

Using phage display with immune libraries offers several distinct advantages over hybridoma-based methods of monoclonal antibody generation. Compared with hybridoma screening, the antibody selection is, at least in part, affinity based [55]. This is especially relevant with antigens that are only weakly immunogenic, where phage display can still find antibodies [63]. Immune libraries can be generated from any animal and even humans, whereas conventional monoclonal generation is limited to species where hybridoma technology works well. Antibodies selected by phage display can also be easily converted into other formats and allow the possibility of further affinity maturation. The generation of antibodies against self-antigens is normally not possible with immune libraries, although occasionally knockout animals have been used successfully for that purpose [64].

Immune libraries also have some limitations. The codon usage is not optimized for bacterial tRNAs, which may lead to poor expression and poor display of otherwise strongly binding antibodies. As phage display selects on a combination of affinity and expressibility, such strongly binding yet poorly expressed antibodies are often lost during phage display [55]. The second principal disadvantage of immune libraries is the reshuffling of light and heavy chains during library preparation. Depending on the number of different heavy and light chain sequences in the library and the overall combinatorial library diversity, this may lead to the absence of strong biologically selected antibodies from the library. Other practical disadvantages of immune libraries are the initial need to immunize animals or the complicated sourcing of immune human lymphocytes.

Table 3.1 contains several examples of immune phage display libraries. Due to the *in vivo* affinity maturation, the achieved affinities are one to two orders of magnitude stronger than those obtained from similarly sized naïve libraries.

3.2 Naïve Natural Libraries

Naïve natural libraries are universal antibody libraries generated from B-cells of nonimmunized donors and eliminate the need to construct new libraries for each antigen. However, since the host did not form a specific immune response against the antigen of interest, the antibodies generally have lower affinities than those generated during *in vivo* affinity maturation. Therefore, to find good antibodies against diverse antigens, these libraries need to be very large. This makes the generation of naïve natural libraries very laborious, which is why these libraries have been constructed almost exclusively by commercial companies and from human B-cells, with the aim of generating biotherapeutics.

Besides having the same advantages as immune libraries, naïve natural libraries offer the advantage of absolute freedom in antigen choice. Several antibodies selected by phage display from human naïve libraries have already been approved as drugs, such as raxibacumab, ramucirumab, necitumumab, or belimumab [3]. See Table 3.1 for notable examples of naïve natural libraries.

3.3 Naïve Semisynthetic Libraries

Naïve semisynthetic libraries are usually libraries that have been isolated from nonimmune hosts and where one or several CDRs were exchanged with synthetic peptides or were randomly mutated. For example, Pini *et al.* randomly mutated 8 positions in the HCDR3 and LCDR3 of one human framework to create a diverse library [65], whereas Nissim *et al.* used randomized oligonucleotides to generate diversity in the HCDR3 of a single V_H gene while keeping the complete diversity of light chain germ lines [66]. The semisynthetic approach is a way to achieve high diversity without requiring a large number of donors and can generate specificities not normally included in natural repertoires. These libraries can cover the entire repertoire of germ lines and, since only a few of the CDRs are artificial, have overall low immunogenicity in hosts. Several biotherapeutics developed through semisynthetic libraries are currently in late clinical trials [3].

3.4 Naïve Synthetic Libraries

The principal advantage of naïve synthetic libraries over semisynthetic libraries is that the biophysical parameters and codon usage of the framework region can be optimized for expressibility and stability. Synthetic libraries usually closely follow one or several germ line frameworks but optimize individual amino acids for beneficial properties. The CDRs are usually integrated as exchangeable modules and can be made in several ways. Random nucleotides or limited diversity degenerate codons can be used with conventional DNA synthesis but have the disadvantage of either randomly including stop codons or being limited to a small subset of amino acids. In both approaches, the composition of amino acids at each position cannot be freely adjusted and is furthermore not codon optimized. Advanced DNA synthesis methods such as trinucleotide phosphoramidites (TRIM) [67], "Slonomics" [68], or chip-based DNA photolithography [69] offer the ability to precisely define the frequency of each amino acid at each position with optimized codons. A further advantage of synthetic libraries is that the CDRs can be of higher diversity as well as different in composition than biologically occurring CDRs, hence offering a potentially larger paratope space.

Naïve synthetic libraries have been produced by academics and commercial companies and are widely in use. We will discuss in detail different phage display libraries developed by biopharmaceutical company MorphoSys to illustrate fundamental design principles of synthetic libraries.

For the HuCAL 1 library [70], seven V_H and seven V_L consensus framework sequences were chosen to represent the entire diversity of human antibody germ lines. The framework genes were designed as modules to enable incorporation of diversity in all six CDR regions. The sequences were codon optimized for expressibility and assembled into the scFv format. Both CDR3s were synthesized by TRIM trinucleotide synthesis to match the natural amino acid distribution closely at each position. The final library size was 2×10^9.

In its next major version, the HuCAL GOLD library [35], all six CDRs were diversified by trinucleotide synthesis, and the Fab format was used. The framework genes from the original HuCAL library were kept, but the occurrence of the synthetic human framework VH4 in the library was increased. The final library size was 1.6×10^{10}.

For the subsequent HuCAL PLATINUM library [71], the framework genes were further optimized for eukaryotic expression without compromising prokaryotic expression. Potential N-glycosylation sites were removed from the library at both framework and CDR positions. The inclusion of V_H and V_L frameworks was adjusted based on results from the previous libraries. Finally, the HCDR3 was further diversified in length with differently defined compositions depending on CDR length.

In Morphosys' most recent library, Ylanthia [72], heavy and light chain pairings were set based on biophysical parameters from a wide range of germ line combinations that had been analyzed. The CDR1s and CDR2s were kept constant within each V_H/V_L pairing, whereas the LCDR3s were diversified by TRIM synthesis and the HCDR3s by Slonomics gene synthesis. The aim of this library was the improvement of the developability of biotherapeutics.

Several antibodies generated from synthetic phage display libraries are in late clinical trials, and synthetic libraries are also used commercially for the generation of research and diagnostic antibodies (Note added in proof: Guselkumab, an antibody selected from the HuCAL GOLD library, was approved by the FDA in July 2017).

3.5 Special Library Designs

Specialized library designs have been used for several purposes. For example, reduced complexity libraries have been used to investigate the nature of paratope formation. Sidhu and coworkers constructed libraries with CDRs consisting of only four or even only two different amino acids in nonconserved CDR positions and could show that these libraries still yielded fully functional antibodies, emphasizing the importance of tyrosine, in particular, in epitope recognition [73,74]. In other cases, hybrid human−animal libraries have been used for the stepwise humanization of animal antibodies, leading to the generation of one approved drug and several in clinical trials [75].

Knowledge about specific ligand-binding motifs can also be useful in the design of specialized libraries. For example, it was known that integrin receptors bind arginine-glycine-aspartate (RGD) tripeptides and thus a library containing RGD in the HCDR3 was constructed, yielding strongly binding inhibitory antibodies against a difficult target [76]. One related example is the generation of a library with increased histidine content in the CDRs [77]. The library was used to select antibodies with pH-sensitive binding, which often relies on the presence of ionizable histidines in the binding pocket. Such antibodies may have applications such as the half-life extension of therapeutic proteins or use as pH-switchable affinity chromatography material.

Another reason for specialized library design is to increase the likelihood of finding high-affinity antibodies against certain classes of antigens, such as haptens, peptides, or phosphopeptides. An approach to design such libraries is to analyze known antibodies to said antigen classes and to incorporate motifs, amino acid distributions, or CDR lengths specific to such antibodies into the library [78]. For example, critical residues interacting with the phosphate of phosphopeptides in several antibodies were identified by crystallography. These residues were kept constant in the generation of new antibody libraries that could indeed be used to isolate phospho-specific antibodies [79]. In another example, the binding mode of an anti-fluorescein antibody was used as a template for a hapten-specific library that yielded antibodies specific to several soluble haptens such as testosterone [80]. Overall, the resulting affinities from these approaches were modest so far, and these designs were probably stifled by the relatively low number of available antibody–antigen pairs used for analysis in library construction.

3.6 Synthetic Libraries From Nonhuman Species

As previously mentioned, almost all noteworthy naïve libraries were generated for the purpose of biotherapeutics development and are thus human. There are, however, some exceptions. For example, a semisynthetic chicken scFv library was generated for diagnostic purposes [81], and there exist naïve libraries for the selection of humanized llama single-domain VHH antibodies [82]. Other purposes for naïve animal libraries are target validation and preclinical tests in animal models without raising a cross-species immune reaction, which has led to the development of naïve mouse libraries [83]. All these libraries have in common that they are at least an order of magnitude smaller than large human naïve libraries. A distinct setup has been used with rabbit antibodies, which are thought to be of especially high affinity [84]. In a first step, a rabbit immune library was screened against a human antigen. In a second step, a new library was generated by grafting the selected CDRs onto diversified human frameworks. This new library was then used in selection for hybrid rabbit–human antibodies with therapeutic potential [85].

4. *IN VITRO* SELECTION OF ANTIBODIES FOR SPECIFIC APPLICATIONS

There are several applications for which *in vitro* selection offers decisive advantages or is even the only choice. Some of these methods and applications will be shown as examples, highlighting the differences between immunization and phage display for antibody generation.

4.1 Tissue Panning for Immunohistochemistry Antibodies

The standard procedure for preservation and preparation of tissue sections for IHC is a fixation with formalin and subsequent paraffin embedding. The vast majority of archival tissue is stored in this form. However, treatment with formalin leads to cross-linking and (partial) denaturation of proteins. Most likely for this reason, only a small fraction of monoclonal antibodies binding the native antigen will also bind it in formalin-fixed paraffin embedded (FFPE) tissue. It is therefore desirable to use FFPE tissue directly for selection of antibodies, which is possible using *in vitro* technologies. There are several examples where phage display antibody selection has been successfully performed for generation of IHC-positive antibodies [45,86,87]. For example, a differential guided selection against substructures of tissues using pre-adsorption of phages on fresh murine thymic tissue has been performed [87], and binding phages were isolated from panning on microdissected frozen tissue fragments. In another study, FFPE-reactive antibodies to disease-specific target structures have been selected from the HuCAL phage display library by subtractive panning on FFPE tissue [86] using normal lymphatic tissue for preadsorption and malignant lymphatic tissue for subsequent selection of binding antibodies. In a similar study by ten Haaf *et al.* [45], disease-specific antibodies in the scFv format were selected from panning on small cell lung cancer FFPE tissue using healthy lung tissue for subtraction.

4.2 Sandwich Pair Selection, Complex-Specific Antibodies, and Drug Monitoring

An antibody generation strategy that can only be performed using *in vitro* technology in a controlled way is the selection on complexes, e.g., protein–protein, protein–peptide, or protein–hapten complexes. There are two main goals for performing such a strategy: the selection of the second antibody of a matched pair for building a sandwich ELISA, or the generation of complex-specific antibodies.

Generation of a pair of antibodies against a given target can be achieved by a simple selection on target antigen and testing of all resulting antibodies against each other. Sometimes no pairs can be identified using this approach, e.g., due to a dominant epitope and/or small size of the antigen. In this situation it is very difficult to generate a matching antibody with traditional animal-based technologies. Using *in vitro* technologies, guided selection strategies can be performed with the aim of selecting a matching detection antibody or an antibody–antigen complex-specific antibody. In a first step, the original pool of antibodies is analyzed for the best capture reagents, and these are used for a panning on the antibody–antigen complex. In phage display selection, the capture antibody is immobilized and then incubated with the target protein. Selection is performed on the capture antibody–antigen complex using a subclass-matched control antibody for blocking to avoid selection of clones binding to the capture antibody (Fig. 3.4B). Selection can be performed on the complex for all panning rounds, or complexed and free antigen can be alternated if antibodies binding to the complex only are not desired.

Another recent application of this selection strategy is the generation of antibodies that bind to an antibody—antigen complex and do not bind the antibody or the antigen alone [52]. Anti-idiotypic antibodies, i.e., antibodies binding to the idiotope of another antibody, are commonly used to study the pharmacokinetics of drug antibodies in clinical trials. Almost all anti-idiotypic antibodies are directed to the CDRs of the therapeutic antibody, detect free drug antibody, not bound to its target, and are inhibitory [6,26]. These antibodies were recently termed Type 1 anti-idiotypic antibodies [88]. The second type of anti-idiotypic antibody can detect the drug antibody whether or not it is bound to its target and is noninhibitory. Such antibodies are called Type 2 antibodies and can be generated using guided selection phage display panning on the antibody—target complex. The third type of antibody that binds the drug—target complex exclusively, and not to free drug or target alone, enables the direct detection and quantification of the bound drug. This antibody specificity termed Type 3 allows the design of more robust assays with higher sensitivity (Fig. 3.6).

Complex-specific antibodies are also useful for building simple, sensitive, and robust noncompetitive assays for quantification of small antigens such as haptens, which do not provide sufficient epitopes for the independent binding of two antibodies. A noncompetitive assay built using an antibody specific for an antibody—hapten complex avoids the problem of false positive results that arise from the use of a competitive assay format [89]. Antibodies specific for a morphine—antibody complex have been generated by phage display using guided selection, and the specificity was confirmed in a homogeneous noncompetitive immunoassay [90].

4.3 Fully Human Controls in Diagnostic Immunoassays

Recombinant antibodies are highly advantageous for the development of standardization material for a range of diagnostic immunoassays where the patient's antibody response is measured. Reliability and reproducibility are critical factors in the performance of such assays, which can vary between laboratories according to factors such as assay reagents used, variation in the level of automation, operator experience, or sample throughput. The quality and stability of assay components are critical to success. An important choice during assay development is the source of control and/or calibrator material and its standardization. Lack of consistency of this assay component makes it difficult to compare results from clinical studies performed by different personnel using different kits at different times. The classical choice for

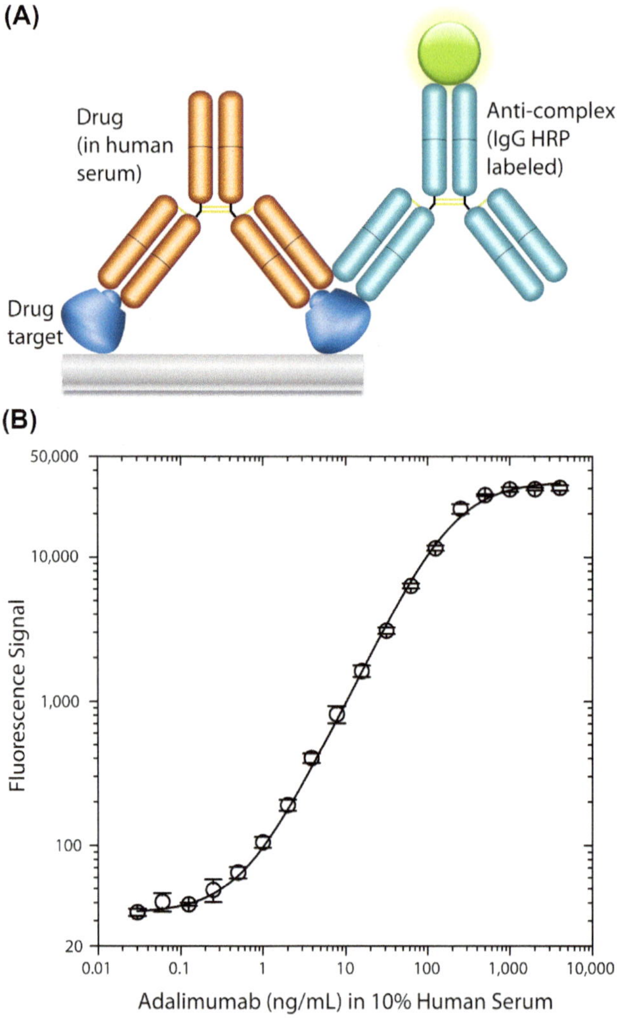

FIGURE 3.6 Pharmacokinetic assay using complex-specific recombinant antibody. (A) Assay setup: drug target is immobilized and captures drug from human serum. Detection is performed with HRP labeled anticomplex antibody. (B) Example data for adalimumab/TNFα complex–specific antibody: TNFα was coated at 5 μg/mL on a microtiter plate overnight. After washing and blocking with 5% BSA in PBST, adalimumab in the given concentrations was added in PBST plus 10% human serum. Detection was performed using HRP conjugated antiadalimumab/TNFα complex–specific antibody (in hIgG1 format, catalog #HCA207, Bio-Rad Laboratories, Oxford, UK), 2 μg/mL in HISPEC Assay Diluent (cat. #BUF049A, Bio-Rad Laboratories, Oxford, UK), plus fluorogenic peroxidase substrate.

controls and calibrators in autoimmune assays are human disease state sera. However, for standardization, there are known drawbacks to using sera, including obtaining them in sufficient quantity and ensuring consistency of long-term supply [91,92], as a new batch of control serum requires new calibration of the assay. This challenge of standardization for an autoimmune antibody assay can be addressed by using an alternative to human disease state sera as the calibrator material. Sera can be replaced by recombinant phage display antibodies with the desired specificity and reformatted to the human isotype of choice. This alternative is being increasingly accepted and adopted, especially when human disease state sera of sufficiently high titer are unavailable or difficult to source in sufficient quantity, as in the case of rare diseases. For anticardiolipin antibody assays, several highly specific antibodies have been selected from the HuCAL library by phage display, converted into human IgA, and assessed for performance in existing antiphospholipid syndrome ELISAs [93]. The binding curves generated using the recombinant IgAs compared favorably with commercial, serum-based assays. In another study, Golden *et al.* [94] established a human recombinant antibody in the IgG4 format as a positive control for serological diagnostic tests for detection of the neglected tropical disease onchocerciasis. A positive control antibody generated by phage display was also used by Whelan *et al.* [95] for assessing the antibody responses against factor VIII. The recombinant antibody was converted into human IgG1, IgG2, IgG3, IgG4, IgA, and IgM to serve as a control for the different immunoglobulin isotypes and IgG subclasses.

4.4 Site-Specific Conjugation

Availability of the antibody genes makes it possible to engineer functional groups into the antibody molecule [96]. This allows attachment of dyes, enzymes, or other labels such as biotin with defined stoichiometry at specific positions, leading to more consistent production batches than random labeling, which often generates a heterogeneous mixture of products or even affects the antibody binding site. When the functional group is an enzyme or a fluorescent protein, it can be added by genetic fusion. Both bacterial alkaline phosphatase [30] and green fluorescent protein [97] have been fused to antibody fragments expressed in *E. coli* and used for direct detection. Other labels have been site-specifically attached using specific chemistry, self-labeling antibody-enzyme fusions, or by enzymatic posttranslational antibody modification (see Table 3.2). Site-specific chemical labeling is possible by the incorporation of reactive surface–exposed cysteines [98] or unnatural amino acids [99]. Self-labeling enzymatic reactions are based on the genetic fusion of antibodies to certain enzymes that catalyze the covalent linkage of their substrate to the enzyme. One example is the SNAP-tag [100], which is a derivative of the O^6-alkylguanine transferase (AGT) that catalyzes the irreversible transfer of the alkyl group of O^6-benzylguanine derivatives to an active site cysteine. Another example is the SpyTag/SpyCatcher system, which consists of the CnaB2 domain from the fibronectin-binding protein FbaB from *Streptococcus pyogenes*. The domain was split into a 13 residue peptide (SpyTag), which can be fused to a recombinant protein, and the complementary remaining domain (SpyCatcher). These two parts spontaneously reconstitute to form an isopeptide bond [101].

TABLE 3.2 Examples of Technologies for Site-Specific Conjugations

Principle	Technology	Description	References
Nonenzymatic reactions	Cys-tag	Genetic addition of one or more cysteine residues at the C-terminus of the heavy chain. Free thiol group of purified antibody can be used to site-specifically cross-link enzymes or dyes using maleimide cross-linker chemistry	[98]
	Unnatural amino acids	Incorporation of amino acids with specific functional groups for coupling by amber codon suppression or selenocysteine incorporation	[99]
	SpyTag	A peptide with 13 residues (SpyTag) that rapidly forms a covalent bond with its protein partner (138 amino acids, 15 kDa, SpyCatcher). SpyTag can be fused to either terminus or internally	[101]
Enzymatic reactions	Sortase	Genetic fusion of a short peptide (the sorting motif LPXTG) to the C-terminus of recombinant antibodies, which can be specifically modified by adding a labeled ligation partner in the presence of the sortase enzyme	[104]
	BCCP-tag	Genetic fusion of a 23 amino acid acceptor peptide to the C-terminus of recombinant antibodies, which can be specifically biotinylated by the BirA enzyme	[102]
	SNAP-tag	Genetic fusion of the 182 residue polypeptide SNAP-tag (19.4 kDa), a derivative of the DNA repair enzyme O(6)-alkylguanine DNA alkyltransferase that allows the covalent coupling of benzylguanine (BG)-modified substrates such as fluorescent dyes	[100]
High-affinity noncovalent attachments	Strep-tag XT	Strep-Tactin XT is a new Strep-Tactin variant with higher affinity for Strep-tag II fusion proteins (nM affinity) and especially Twin-Strep-tag fusion proteins (pM affinity)	[105]
	Meditope technology	A peptide (meditope) was found to bind specifically to antibodies that contain certain key residues in their H and L chains outside the antibody paratope (meditope enablement). The peptide can be functionalized for attachment of small molecules such as dyes	[106]

Examples of methods that have been used for site-specific conjugations of recombinant antibodies.

The site-specific posttranslational modification of antibodies by enzymes is typically achieved by fusing a short peptide tag (derived from the consensus sequence of the enzyme substrate) to a recombinant antibody. The bond-forming enzyme then covalently attaches a small-molecule label to the tag. Examples are the BCCP-tag [102] where biotin is attached to a 23 amino acid "acceptor peptide" by biotin ligase (BirA), or the use of tags for trans-glutaminase [103] or sortase [104] enzymes that catalyze protein cross-linking.

Covalent labeling is not always necessary, as there are known high-affinity noncovalent interactions between peptide tags and labeled binding partners. Two such interactions are the Strep-Tactin XT system [105], which consists of the Strep-tag and a derivative of streptavidin that binds the tag with high-affinity, and the meditope technology, where a peptide binds specifically with high affinity to antibodies that contain certain key residues in their H and L chains outside the antibody paratope [106].

Several of these methods are of increasing interest for the generation of antibody–drug conjugates [107]. In addition, such technologies can also be used for directed immobilization of antibodies on surfaces for the development of immunosensors and immunoassays [108].

5. CONCLUSION AND OUTLOOK

In this chapter, phage display has been discussed as a powerful *in vitro* selection technology to isolate specific antibodies from antibody libraries. Several advantages of this technology compared with animal immunizations have become apparent, such as the availability of antibody sequences. Electronically stored gene sequences ensure the information is not lost, as can happen with hybridoma cell lines. As gene synthesis becomes a low-priced commodity, it will be more efficient in the future to email an antibody sequence to another laboratory instead of shipping a cell line or a purified antibody product.

The recombinant nature of such antibodies also allows for a higher batch-to-batch consistency, especially when compared with traditional polyclonal antibody sera [109]. Advanced technologies for site-specific labeling allow the same for batch-to-batch consistent antibody conjugates (see Table 3.2).

The technologies described in this chapter lead to antibodies without the need to sacrifice animals [110]. In Europe, two laws regulate the use of animals in research. These laws require that alternatives to animals should be used when "reasonably and practicably available." While antibody generation by immunization is still widely used and leads to the majority of new research antibodies appearing on the market, certain antibody production techniques that use animals, such as the ascites method, are either restricted or banned in many European countries and others, such as Australia and Canada [111]. Production of recombinant antibodies (as antibody fragments) is mostly performed using the bacterial host *E. coli* with typical yields between 1 and 10 mg per liter from overnight culture in shaking flasks [112]. High-yield bacterial production has been reported by using high–cell density fermentation [113] or cell-wall mutated *E. coli* strains [28]. Full-length antibodies are typically produced in mammalian cell lines either by transient transfection of an antibody-containing plasmid or by generating a stable cell line that has integrated the antibody gene in its genome and was selected for high expression titers. The production of antibody drugs by such methods has stimulated the research for high expression methods [114], and titers of several grams per liter can routinely be achieved. Directed gene integration techniques such as CRISPR-Cas9 will undoubtedly further improve these methods and will also lead to faster cell line development. An increasing number of traditionally generated monoclonal antibodies are being cloned and sequenced. This guards against the loss of an antibody due to an unstable hybridoma and also enables such antibody expression methods to be exploited.

Multiplexing of immunoassays such as Western blotting or flow cytometry can be achieved with primary antibodies from different species or of different isotype/subtype in combination with corresponding differently fluorescence-labeled secondary antibodies [115]. However, the number of different isotypes, subtypes, and host species for immunization is limited, and the required number of different primary antibodies is not always available. With recombinant antibodies, antibody variable domains selected from phage display libraries can be attached to Fc domains from a variety of different species or isotypes by genetic engineering without any loss in binding affinity, enabling the use of well-established isotype-specific secondary reagents from animals such as mouse, rat, rabbit, or goat.

Antibodies for flow cytometry may bind to Fc-receptors from cells in the sample, thereby creating background staining. Recombinant antibody technology can provide a solution by either producing the antibodies without the Fc part or by including mutations that suppress Fc-receptor binding (LALA, TM, or N297A), which have been investigated on human antibody Fc regions but can also be transferred to other species [116].

5.1 Future

What are the pending challenges and future directions of antibody generation by phage display? In the following section challenges regarding the phage display process will be discussed, and in conclusion, the future and the overall potential of this technology outside the therapeutic antibody field will be highlighted.

The design and generation of antibody libraries plus the *in vitro* selection process are at first glance more laborious (and therefore more expensive) than the immunization of animals followed by hybridoma generation. However, the *in vitro* selection process can be automated [117], enabling the processing of many selections in parallel, which considerably reduces cost. Methods to further accelerate phage selections have also been developed [118]. After selection, the screening process with *in vitro* selected recombinant antibodies is usually quicker and less expensive compared with hybridoma screening as the expression takes place in a bacterial host. Nevertheless, the throughput of screening is an important aspect since phage display selection campaigns with a large library and multiple panning strategies can deliver thousands of ELISA-positive antibodies. Ideally, the screening setup should not only allow identification of hits but simultaneously measure parameters such as affinity [119], whereby the monovalent nature of antibody fragments is an intrinsic advantage. The final validation of selected antibodies in the intended applications is an elaborate manual process regardless of the source of the antibody [120].

Monovalent affinities from antibodies against protein antigens directly selected by phage display from the best naïve libraries are typically in 10^{-8} to 10^{-10} M range. It is estimated that this range is similar to that achieved through the natural affinity maturation processes of the metazoan immune systems after antigen challenge [121]. Higher affinities are obtainable by *in vitro* affinity maturation [39]. Other ways to improve average affinities have been connected to even larger functional library sizes [71,72,122], developing methods for accelerated affinity maturations [54,119], better selection [35], and screening systems to identify very high-affinity antibodies.

As the field has concentrated on advancing antibody drug development, the potential for research and diagnostic antibody generation by phage display has not been fully explored. To date, the vast majority of commercially available research antibodies are generated by animal immunization, and the initial promise that *in vitro* technologies will ultimately replace animal-produced antibodies altogether has not been realized [123].

Antibody library technology has focused on human antibodies for obvious reasons, and only recently has the antibody repertoire from other species gained attention [124], partly due to the power of next-generation sequencing technologies, and partly because antibodies from other species are believed to add diversity in terms of epitope coverage or higher affinities [125]. Also, nontherapeutic applications do not require the close similarity of the antibody sequence to human antibodies to avoid immunogenicity [126], which means the design of antibody libraries and especially the CDR length and composition in synthetic libraries do not have the restriction of being close to human germ line, opening the door for more creativity in designing new libraries.

The fact that antibody phage display is an *in vitro* process allows the selection to be performed under harsh conditions, where only antibodies that can withstand such conditions can survive the selection process, as the filamentous phage used is itself very stable [127]. Antibodies have been successfully selected under such conditions, and the limits of such methods have not yet been fully explored. Furthermore, the guided selection process allows selection for rare specificities such as conformation-specific antibodies or antibodies binding to protein complexes.

Once the antibody has been selected, the availability of its DNA sequence and the ease of production using *E. coli* facilitate antibody engineering. A few of the numerous engineering successes such as multivalency, genetic fusions with reporter enzymes, or site-specific labeling have been described in this chapter. Another accomplishment of antibody engineering is the development of bispecific antibodies that combine two different antigen binding sites in one molecule [128]; these specialized biotherapeutics are now entering the drug market. This format may also have advantages in research and diagnostic applications, as it may offer increased specificity when the two arms recognize different epitopes on the same antigen. Further engineering to develop multispecific antibodies with up to five specificities in one molecule has been described [129].

For the future, we envision that the methods and technologies described in this chapter will be adopted on a much wider scale both by researchers and commercial antibody suppliers since the advantages are so manifold. The generation, selection, and engineering of antibodies and antibody-like molecules for research and diagnostics have just begun.

REFERENCES

[1] T. Schirrmann, T. Meyer, M. Schutte, A. Frenzel, M. Hust, Phage display for the generation of antibodies for proteome research, diagnostics and therapy, Molecules 16 (1) (2011) 412–426.

[2] S.S. Sidhu, C.R. Geyer, Phage Display in Biotechnology and Drug Discovery, second ed., CRC Press/Taylor & Francis Group, Boca Raton, 2015 xv, 567.

[3] A. Frenzel, T. Schirrmann, M. Hust, Phage display-derived human antibodies in clinical development and therapy, MAbs (2016) 0.

[4] M. Mersmann, D. Meier, J. Mersmann, S. Helmsing, P. Nilsson, S. Graslund, et al., Towards proteome scale antibody selections using phage display, Nat. Biotechnol. 27 (2) (2010) 118–128.

[5] R.I. Shapiro, T. Plavina, B.R. Schlain, R.B. Pepinsky, E.A. Garber, M. Jarpe, et al., Development and validation of immunoassays to quantify the half-antibody exchange of an IgG4 antibody, natalizumab (Tysabri(R)) with endogenous IgG4, J. Pharm. Biomed. Anal. 55 (1) (2011) 168–175.

[6] M. Tornetta, D. Fisher, K. O'Neil, D. Geng, A. Schantz, M. Brigham-Burke, et al., Isolation of human anti-idiotypic antibodies by phage display for clinical immune response assays, J. Immunol. Methods 328 (1–2) (2007) 34–44.

[7] G.P. Smith, Filamentous fusion phage: novel expression vectors that display cloned antigens on the virion surface, Science 228 (4705) (1985) 1315–1317.

[8] W.D. Huse, L. Sastry, S.A. Iverson, A.S. Kang, M. Alting-Mees, D.R. Burton, et al., Generation of a large combinatorial library of the immunoglobulin repertoire in phage lambda, Science 246 (4935) (1989) 1275–1281.

[9] J. McCafferty, A.D. Griffiths, G. Winter, D.J. Chiswell, Phage antibodies: filamentous phage displaying antibody variable domains, Nature 348 (6301) (1990) 552–554.

[10] C.F. Barbas 3rd, A.S. Kang, R.A. Lerner, S.J. Benkovic, Assembly of combinatorial antibody libraries on phage surfaces: the gene III site, Proc. Natl. Acad. Sci. USA 88 (18) (1991) 7978–7982.

[11] H.R. Hoogenboom, G. Winter, By-passing immunisation. Human antibodies from synthetic repertoires of germline VH gene segments rearranged in vitro, J. Mol. Biol. 227 (2) (1992) 381–388.

[12] A.R. Bradbury, J.D. Marks, Antibodies from phage antibody libraries, J. Immunol. Methods 290 (1–2) (2004) 29–49.

[13] R.E. Bird, B.W. Walker, Single chain antibody variable regions, Trends Biotechnol. 9 (4) (1991) 132–137.

[14] N.V. Tikunova, V.V. Morozova, Phage display on the base of filamentous bacteriophages: application for recombinant antibodies selection, Acta Nat. 1 (3) (2009) 20–28.

[15] J. Ku, P.G. Schultz, Alternate protein frameworks for molecular recognition, Proc. Natl. Acad. Sci. USA 92 (14) (1995) 6552–6556.

[16] C. Hamers-Casterman, T. Atarhouch, S. Muyldermans, G. Robinson, C. Hamers, E.B. Songa, et al., Naturally occurring antibodies devoid of light chains, Nature 363 (6428) (1993) 446–448.

[17] J. Davies, L. Riechmann, Antibody VH domains as small recognition units, Biotechnology (NY) 13 (5) (1995) 475–479.

[18] S. Steeland, R.E. Vandenbroucke, C. Libert, Nanobodies as therapeutics: big opportunities for small antibodies, Drug Discov. Today 21 (7) (2016) 1076–1113.

[19] R. Vazquez-Lombardi, T.G. Phan, C. Zimmermann, D. Lowe, L. Jermutus, D. Christ, Challenges and opportunities for non-antibody scaffold drugs, Drug Discov. Today 20 (10) (2015) 1271–1283.

[20] K. Nord, E. Gunneriusson, J. Ringdahl, S. Stahl, M. Uhlen, P.A. Nygren, Binding proteins selected from combinatorial libraries of an alpha-helical bacterial receptor domain, Nat. Biotechnol. 15 (8) (1997) 772–777.

[21] G. Beste, F.S. Schmidt, T. Stibora, A. Skerra, Small antibody-like proteins with prescribed ligand specificities derived from the lipocalin fold, Proc. Natl. Acad. Sci. USA 96 (5) (1999) 1898–1903.

[22] M.T. Stumpp, H.K. Binz, P. Amstutz, DARPins: a new generation of protein therapeutics, Drug Discov. Today 13 (15–16) (2008) 695–701.

[23] J. Silverman, Q. Liu, A. Bakker, W. To, A. Duguay, B.M. Alba, et al., Multivalent avimer proteins evolved by exon shuffling of a family of human receptor domains, Nat. Biotechnol. 23 (12) (2005) 1556–1561.

[24] L.K. Stadler, T. Hoffmann, D.C. Tomlinson, Q. Song, T. Lee, M. Busby, et al., Structure-function studies of an engineered scaffold protein derived from Stefin A. II: development and applications of the SQT variant, Protein Eng. Des. Sel. 24 (9) (2011) 751–763.

[25] A. Koide, S. Koide, Monobodies: antibody mimics based on the scaffold of the fibronectin type III domain, Methods Mol. Biol. 352 (2007) 95–109.

[26] F. Ylera, S. Harth, D. Waldherr, C. Frisch, A. Knappik, Off-rate screening for selection of high-affinity anti-drug antibodies, Anal. Biochem. 441 (2) (2013) 208–213.

[27] L.C. Kovari, C. Momany, M.G. Rossmann, The use of antibody fragments for crystallization and structure determinations, Structure 3 (12) (1995) 1291–1293.

[28] P. Pack, M. Kujau, V. Schroeckh, U. Knupfer, R. Wenderoth, D. Riesenberg, et al., Improved bivalent miniantibodies, with identical avidity as whole antibodies, produced by high cell density fermentation of *Escherichia coli*, Biotechnology (NY) 11 (11) (1993) 1271–1277.

[29] P. Holliger, T. Prospero, G. Winter, "Diabodies": small bivalent and bispecific antibody fragments, Proc. Natl. Acad. Sci. USA 90 (14) (1993) 6444–6448.

[30] E. Weiss, G. Orfanoudakis, Application of an alkaline phosphatase fusion protein system suitable for efficient screening and production of Fab-enzyme conjugates in *Escherichia coli*, J. Biotechnol. 33 (1) (1994) 43–53.

[31] R. Calendar, The Bacteriophages, second ed., Oxford University Press, Oxford, New York, 2006 xiii, 746.

[32] H. Qi, H. Lu, H.J. Qiu, V. Petrenko, A. Liu, Phagemid vectors for phage display: properties, characteristics and construction, J. Mol. Biol. 417 (3) (2012) 129–143.

[33] S. Rondot, J. Koch, F. Breitling, S. Dubel, A helper phage to improve single-chain antibody presentation in phage display, Nat. Biotechnol. 19 (1) (2001) 75–78.

[34] S. Miersch, S.S. Sidhu, Synthetic antibodies: concepts, potential and practical considerations, Methods 57 (4) (2012) 486–498.

[35] C. Rothe, S. Urlinger, C. Lohning, J. Prassler, Y. Stark, U. Jager, et al., The human combinatorial antibody library HuCAL GOLD combines diversification of all six CDRs according to the natural immune system with a novel display method for efficient selection of high-affinity antibodies, J. Mol. Biol. 376 (4) (2008) 1182–1200.

[36] R.L. Ward, M.A. Clark, J. Lees, N.J. Hawkins, Retrieval of human antibodies from phage-display libraries using enzymatic cleavage, J. Immunol. Methods 189 (1) (1996) 73–82.

[37] A. Moghaddam, T. Borgen, J. Stacy, L. Kausmally, B. Simonsen, O.J. Marvik, et al., Identification of scFv antibody fragments that specifically recognise the heroin metabolite 6-monoacetylmorphine but not morphine, J. Immunol. Methods 280 (1–2) (2003) 139–155.

[38] F. Breitling, S. Dubel, T. Seehaus, I. Klewinghaus, M. Little, A surface expression vector for antibody screening, Gene 104 (2) (1991) 147–153.

[39] S. Steidl, O. Ratsch, B. Brocks, M. Durr, E. Thomassen-Wolf, In vitro affinity maturation of human GM-CSF antibodies by targeted CDR-diversification, Mol. Immunol. 46 (1) (2008) 135–144.

[40] A.D. Griffiths, M. Malmqvist, J.D. Marks, J.M. Bye, M.J. Embleton, J. Mccafferty, et al., Human anti-self antibodies with high specificity from phage display libraries, EMBO J. 12 (2) (1993) 725–734.

[41] C. Nizak, S. Monier, E. del Nery, S. Moutel, B. Goud, F. Perez, Recombinant antibodies to the small GTPase Rab6 as conformation sensors, Science 300 (5621) (2003) 984–987.

[42] Z. Konthur, M. Hust, S. Dubel, Perspectives for systematic in vitro antibody generation, Gene 364 (2005) 19–29.

[43] H. Wang, R. Yu, T. Fang, T. Yu, X. Chi, X. Zhang, et al., Tetanus neurotoxin neutralizing antibodies screened from a human immune scFv antibody phage display library, Toxins (Basel) 8 (9) (2016).

[44] J. Fitting, T. Blume, A. ten Haaf, W. Blau, S. Gattenlohner, M.K. Tur, et al., Phage display-based generation of novel internalizing antibody fragments for immunotoxin-based treatment of acute myeloid leukemia, MAbs 7 (2) (2015) 390–402.

[45] A. ten Haaf, S. Pscherer, K. Fries, S. Barth, S. Gattenlohner, M.K. Tur, Phage display-based on-slide selection of tumor-specific antibodies on formalin-fixed paraffin-embedded human tissue biopsies, Immunol. Lett. 166 (2) (2015) 65–78.

[46] A. ten Haaf, J. Kohl, S. Pscherer, H.P. Hamann, H.U. Eskens, M. Bastian, et al., Development of a monoclonal sandwich ELISA for direct detection of bluetongue virus 8 in infected animals, J. Virol. Methods 243 (2017) 172–176.

[47] R. Ohara, A. Knappik, K. Shimada, C. Frisch, F. Ylera, H. Koga, Antibodies for proteomic research: comparison of traditional immunization with recombinant antibody technology, Proteomics 6 (9) (2006) 2638–2646.

[48] S. Jung, A. Honegger, A. Pluckthun, Selection for improved protein stability by phage display, J. Mol. Biol. 294 (1) (1999) 163–180.

[49] H. Ooe, S.M. Iguchi-Ariga, H. Ariga, Establishment of specific antibodies that recognize C106-oxidized DJ-1, Neurosci. Lett. 404 (1–2) (2006) 166–169.

[50] A. Casrouge, A. Bisiaux, L. Stephen, M. Schmolz, J. Mapes, C. Pfister, et al., Discrimination of agonist and antagonist forms of CXCL10 in biological samples, Clin. Exp. Immunol. 167 (1) (2012) 137–148.

[51] S.A. Shiryaev, I.A. Radichev, B.I. Ratnikov, A.E. Aleshin, K. Gawlik, B. Stec, et al., Isolation and characterization of selective and potent human Fab inhibitors directed to the active-site region of the two-component NS2B-NS3 proteinase of West Nile virus, Biochem. J. 427 (3) (2010) 369–376.

[52] S. Härtle, C. Frisch, A. Knappik, inventors, A.G. Morphosys, assignee, Complex-Specific Antibodies and Antibody Fragments and Its Use, 2015. Patent US20150197579A1.

[53] F.D. Batista, M.S. Neuberger, Affinity dependence of the B cell response to antigen: a threshold, a ceiling, and the importance of off-rate, Immunity 8 (6) (1998) 751–759.

[54] J. Prassler, S. Steidl, S. Urlinger, In vitro affinity maturation of HuCAL antibodies: complementarity determining region exchange and RapMAT technology, Immunotherapy 1 (4) (2009) 571–583.

[55] J. McCafferty, Phage display: factors affecting panning efficiency, in: B.K. Kay, J. Winter, J. McCafferty (Eds.), Phage Display of Peptides and Proteins: A Laboratory Manual, Academic Press, San Diego, 1996, pp. 261–276.

[56] E.T. Boder, K.D. Wittrup, Yeast surface display for screening combinatorial polypeptide libraries, Nat. Biotechnol. 15 (6) (1997) 553–557.

[57] L. Rhiel, S. Krah, R. Gunther, S. Becker, H. Kolmar, B. Hock, REAL-Select: full-length antibody display and library screening by surface capture on yeast cells, PLoS One 9 (12) (2014) e114887.

[58] E.T. Boder, K.S. Midelfort, K.D. Wittrup, Directed evolution of antibody fragments with monovalent femtomolar antigen-binding affinity, Proc. Natl. Acad. Sci. USA 97 (20) (2000) 10701–10705.

[59] C. Schaffitzel, J. Hanes, L. Jermutus, A. Pluckthun, Ribosome display: an in vitro method for selection and evolution of antibodies from libraries, J. Immunol. Methods 231 (1–2) (1999) 119–135.

[60] D. Lipovsek, A. Pluckthun, In-vitro protein evolution by ribosome display and mRNA display, J. Immunol. Methods 290 (1–2) (2004) 51–67.

[61] H. Zhou, Y.L. Zhang, G. Lu, H. Ji, C.P. Rodi, Recombinant antibody libraries and selection technologies, Nat. Biotechnol. 28 (5) (2011) 448–452.

[62] D. Hanahan, Studies on transformation of *Escherichia coli* with plasmids, J. Mol. Biol. 166 (4) (1983) 557–580.

[63] Q. Dinh, N.P. Weng, M. Kiso, H. Ishida, A. Hasegawa, D.M. Marcus, High affinity antibodies against Lex and sialyl Lex from a phage display library, J. Immunol. Baltim. (MD, 1950) 157 (2) (1996) 732–738.

[64] B. Brocks, P. Garin-Chesa, E. Behrle, J.E. Park, W.J. Rettig, K. Pfizenmaier, et al., Species-crossreactive scFv against the tumor stroma marker "fibroblast activation protein" selected by phage display from an immunized FAP-/- knock-out mouse, Mol. Med. 7 (7) (2001) 461–469.

[65] A. Pini, F. Viti, A. Santucci, B. Carnemolla, L. Zardi, P. Neri, et al., Design and use of a phage display library – human antibodies with subnanomolar affinity against a marker of angiogenesis eluted from a two-dimensional gel, J. Biol. Chem. 273 (34) (1998) 21769–21776.

[66] A. Nissim, H.R. Hoogenboom, I.M. Tomlinson, G. Flynn, C. Midgley, D. Lane, et al., Antibody fragments from a 'single pot' phage display library as immunochemical reagents, EMBO J. 13 (3) (1994) 692–698.

[67] B. Virnekas, L. Ge, A. Pluckthun, K.C. Schneider, G. Wellnhofer, S.E. Moroney, Trinucleotide phosphoramidites: ideal reagents for the synthesis of mixed oligonucleotides for random mutagenesis, Nucleic Acids Res. 22 (25) (1994) 5600–5607.

[68] J. Van den Brulle, M. Fischer, T. Langmann, G. Horn, T. Waldmann, S. Arnold, et al., A novel solid phase technology for high-throughput gene synthesis, Biotechniques 45 (3) (2008) 340–343.

[69] X. Bai, J. Kim, S. Kang, W. Kim, H. Shim, A novel human scFv library with non-combinatorial synthetic CDR diversity, PLoS One 10 (10) (2015) e0141045.

[70] A. Knappik, L. Ge, A. Honegger, P. Pack, M. Fischer, G. Wellnhofer, et al., Fully synthetic human combinatorial antibody libraries (HuCAL) based on modular consensus frameworks and CDRs randomized with trinucleotides, J. Mol. Biol. 296 (1) (2000) 57–86.

[71] J. Prassler, S. Thiel, C. Pracht, A. Polzer, S. Peters, M. Bauer, et al., HuCAL PLATINUM, a synthetic Fab library optimized for sequence diversity and superior performance in mammalian expression systems, J. Mol. Biol. 413 (1) (2011) 261–278.

[72] T. Tiller, I. Schuster, D. Deppe, K. Siegers, R. Strohner, T. Herrmann, et al., A fully synthetic human Fab antibody library based on fixed VH/VL framework pairings with favorable biophysical properties, MAbs 5 (3) (2013) 445–470.

[73] F.A. Fellouse, C. Wiesmann, S.S. Sidhu, Synthetic antibodies from a four-amino-acid code: a dominant role for tyrosine in antigen recognition, Proc. Natl. Acad. Sci. USA 101 (34) (2004) 12467–12472.

[74] F.A. Fellouse, B. Li, D.M. Compaan, A.A. Peden, S.G. Hymowitz, S.S. Sidhu, Molecular recognition by a binary code, J. Mol. Biol. 348 (5) (2005) 1153–1162.

[75] C. Rader, D.A. Cheresh, C.F. Barbas, A phage display approach for rapid antibody humanization: designed combinatorial V gene libraries, Proc. Natl. Acad. Sci. USA 95 (1998) 8910–8915.

[76] C.F. Barbas 3rd, L.R. Languino, J.W. Smith, High-affinity self-reactive human antibodies by design and selection: targeting the integrin ligand binding site, Proc. Natl. Acad. Sci. USA 90 (21) (1993) 10003–10007.

[77] C. Schroter, R. Gunther, L. Rhiel, S. Becker, L. Toleikis, A. Doerner, et al., A generic approach to engineer antibody pH-switches using combinatorial histidine scanning libraries and yeast display, MAbs 7 (1) (2015) 138–151.

[78] G. Raghunathan, J. Smart, J. Williams, J.C. Almagro, Antigen-binding site anatomy and somatic mutations in antibodies that recognize different types of antigens, J. Mol. Recogn. 25 (3) (2012) 103–113.

[79] J.T. Koerber, N.D. Thomsen, B.T. Hannigan, W.F. Degrado, J.A. Wells, Nature-inspired design of motif-specific antibody scaffolds, Nat. Biotechnol. 31 (10) (2013) 916–921.

[80] H. Persson, H. Wallmark, A. Ljungars, J. Hallborn, M. Ohlin, In vitro evolution of an antibody fragment population to find high-affinity hapten binders, Protein Eng. Des. Selection 21 (8) (2008) 485–493.

[81] W. van Wyngaardt, T. Malatji, C. Mashau, J. Fehrsen, F. Jordaan, D. Miltiadou, et al., A large semi-synthetic single-chain Fv phage display library based on chicken immunoglobulin genes, BMC Biotechnol. 4 (2004) 6.

[82] S. Moutel, N. Bery, V. Bernard, L. Keller, E. Lemesre, A. de Marco, et al., NaLi-H1: a universal synthetic library of humanized nanobodies providing highly functional antibodies and intrabodies, eLife (2016) 5.

[83] R. Sommavilla, V. Lovato, A. Villa, D. Sgier, D. Neri, Design and construction of a naive mouse antibody phage display library, J. Immunol. Methods 353 (1–2) (2010) 31–43.

[84] J. Weber, H. Peng, C. Rader, From rabbit antibody repertoires to rabbit monoclonal antibodies, Exp. Mol. Med. 49 (3) (2017) e305.

[85] C. Rader, G. Ritter, S. Nathan, M. Elia, I. Gout, A.A. Jungbluth, et al., The rabbit antibody repertoire as a novel source for the generation of therapeutic human antibodies, J. Biol. Chem. 275 (18) (2000) 13668–13676.

[86] T. Jarutat, C. Nickels, C. Frisch, F. Stellmacher, K.P. Hofig, A. Knappik, et al., Selection of vimentin-specific antibodies from the HuCAL® phage display library by subtractive panning on formalin-fixed, paraffin-embedded tissue, Biol. Chem. 388 (6) (2007) 651–658.

[87] W. Van Ewijk, J. de Kruif, W.T. Germeraad, P. Berendes, C. Ropke, P.P. Platenburg, et al., Subtractive isolation of phage-displayed single-chain antibodies to thymic stromal cells by using intact thymic fragments, Proc. Natl. Acad. Sci. USA 94 (8) (1997) 3903–3908.

[88] A. Beck, P.J. Carter, H.P. Gerber, A.A. Lugovskoy, T. Wurch, J.R. Junutula, et al., 8(th) annual European antibody congress 2012: November 27–28, 2012, Geneva, Switzerland, MAbs 5 (3) (2013) 339–357.

[89] S. Kerrigan, W.H. Phillips Jr Jr., Comparison of ELISAs for opiates, methamphetamine, cocaine metabolite, benzodiazepines, phencyclidine, and cannabinoids in whole blood and urine, Clin. Chem. 47 (3) (2001) 540–547.

[90] T. Pulli, M. Hoyhtya, H. Soderlund, K. Takkinen, One-step homogeneous immunoassay for small analytes, Anal. Chem. 77 (8) (2005) 2637–2642.

[91] J.H. Simmons, Development, application, and quality control of serology assays used for diagnostic monitoring of laboratory nonhuman primates, ILAR J. 49 (2) (2008) 157–169.

[92] R.H. Jacobson, Validation of serological assays for diagnosis of infectious diseases, Rev. Sci. Tech. 17 (2) (1998) 469–526.

[93] A. Knappik, F. Capuano, C. Frisch, F. Ylera, F. Bonelli, Development of recombinant human IgA for anticardiolipin antibodies assay standardization, Ann. NY Acad. Sci. 1173 (2009) 190–198.

[94] A. Golden, E.J. Stevens, L. Yokobe, D. Faulx, M. Kalnoky, R. Peck, et al., A recombinant positive control for serology diagnostic tests supporting elimination of *Onchocerca volvulus*, PLoS Negl. Trop. Dis. 10 (1) (2016) e0004292.

[95] S.F. Whelan, C.J. Hofbauer, F.M. Horling, P. Allacher, M.J. Wolfsegger, J. Oldenburg, et al., Distinct characteristics of antibody responses against factor VIII in healthy individuals and in different cohorts of hemophilia A patients, Blood 121 (6) (2013) 1039–1048.

[96] D. Rabuka, Chemoenzymatic methods for site-specific protein modification, Curr. Opin. Chem. Biol. 14 (6) (2010) 790–796.

[97] P. Oelschlaeger, S. Srikant-Iyer, S. Lange, J. Schmitt, R. Schmid, Fluorophor-linked immunosorbent assay: a time- and cost-saving method for the characterization of antibody fragments using a fusion protein of a single-chain antibody fragment and enhanced green fluorescent protein, Anal. Biochem. 309 (1) (2002) 27.

[98] H. Albrecht, P.A. Burke, A. Natarajan, C.Y. Xiong, M. Kalicinsky, G.L. DeNardo, et al., Production of soluble scFvs with C-terminal-free thiol for site-specific conjugation or stable dimeric scFvs on demand, Bioconjug. Chem. 15 (1) (2004) 16–26.

[99] T.J. Hallam, E. Wold, A. Wahl, V.V. Smider, Antibody conjugates with unnatural amino acids, Mol. Pharm. 12 (6) (2015) 1848–1862.

[100] A.F. Hussain, M. Amoury, S. Barth, SNAP-tag technology: a powerful tool for site specific conjugation of therapeutic and imaging agents, Curr. Pharm. Des. 19 (30) (2013) 5437–5442.

[101] S.C. Reddington, M. Howarth, Secrets of a covalent interaction for biomaterials and biotechnology: SpyTag and SpyCatcher, Curr. Opin. Chem. Biol. 29 (2015) 94–99.

[102] M. Czerwinski, A. Krop-Watorek, K. Wasniowska, D. Smolarek, S.L. Spitalnik, Construction of an agglutination tool: recombinant Fab fragments biotinylated in vitro, Nat. Biotechnol. 26 (5) (2009) 215–221.

[103] V. Siegmund, S. Schmelz, S. Dickgiesser, J. Beck, A. Ebenig, H. Fittler, et al., Locked by design: a conformationally constrained transglutaminase tag enables efficient site-specific conjugation, Angew. Chemie Int. Ed. Engl. (2015).

[104] L. Schmohl, D. Schwarzer, Sortase-mediated ligations for the site-specific modification of proteins, Curr. Opin.Chem. Biol. 22 (2014) 122–128.

[105] A. Yeliseev, L. Zoubak, T.G. Schmidt, Application of Strep-Tactin XT for affinity purification of Twin-Strep-tagged CB2, a G protein-coupled cannabinoid receptor, Protein Expr. Purif. 131 (2016) 109–118.

[106] K.N. Avery, C. Zer, K.P. Bzymek, J.C. Williams, Development of a high affinity, non-covalent biologic to add functionality to Fabs, Sci. Rep. 5 (2015) 7817.

[107] P. Agarwal, C.R. Bertozzi, Site-specific antibody-drug conjugates: the nexus of bioorthogonal chemistry, protein engineering, and drug development, Bioconjug. Chem. 26 (2) (2015) 176–192.

[108] E. Steen Redeker, D.T. Ta, D. Cortens, B. Billen, W. Guedens, P. Adriaensens, Protein engineering for directed immobilization, Bioconj. Chem. 24 (11) (2013) 1761–1777.

[109] A. Bradbury, A. Pluckthun, Reproducibility: standardize antibodies used in research, Nature 518 (7537) (2015) 27–29.

[110] A.C. Gray, S.S. Sidhu, P.C. Chandrasekera, C.F.M. Hendriksen, C.A.K. Borrebaeck, Animal-friendly affinity reagents: replacing the needless in the Haystack, Trends Biotechnol. (2016).

[111] K. Groff, J. Brown, A.J. Clippinger, Modern affinity reagents: recombinant antibodies and aptamers, Biotechnol. Adv. (2015).

[112] A. Knappik, R. Brundiers, Recombinant Antibody Expression and Purification, 2010, pp. 1–16.

[113] P. Carter, R.F. Kelley, M.L. Rodrigues, B. Snedecor, M. Covarrubias, M.D. Velligan, et al., High level *Escherichia coli* expression and production of a bivalent humanized antibody fragment, Biotechnol. (NY) 10 (2) (1992) 163–167.

[114] J. Li, W. Gu, D.G. Edmondson, C. Lu, N. Vijayasankaran, B. Figueroa, et al., Generation of a cholesterol-independent, non-GS NS0 cell line through chemical treatment and application for high titer antibody production, Biotechnol. Bioeng. 109 (7) (2012) 1685–1692.

[115] J.C. Gingrich, D.R. Davis, Q. Nguyen, Multiplex detection and quantitation of proteins on western blots using fluorescent probes, Biotechniques 29 (3) (2000) 636–642.

[116] E. Arduin, S. Arora, P.R. Bamert, T. Kuiper, S. Popp, S. Geisse, et al., Highly reduced binding to high and low affinity mouse Fc gamma receptors by L234A/L235A and N297A Fc mutations engineered into mouse IgG2a, Mol. Immunol. 63 (2) (2015) 456–463.

[117] B. Krebs, R. Rauchenberger, S. Reiffert, C. Rothe, M. Tesar, E. Thomassen, et al., High-throughput generation and engineering of recombinant human antibodies, J. Immunol. Methods 254 (1–2) (2001) 67–84.

[118] S. Hogan, K. Rookey, R. Ladner, URSA: ultra rapid selection of antibodies from an antibody phage display library, Biotechniques 38 (4) (2005) 536–538.

[119] M. Tornetta, R. Reddy, J.C. Wheeler, Selection and maturation of antibodies by phage display through fusion to pIX, Methods 58 (1) (2012) 34–39.

[120] M. Uhlen, A. Bandrowski, S. Carr, A. Edwards, J. Ellenberg, E. Lundberg, et al., A proposal for validation of antibodies, Nat. Methods (2016).

[121] P.T. Beernink, S. Giuntini, I. Costa, A.H. Lucas, D.M. Granoff, Functional analysis of the human antibody response to meningococcal factor H binding protein, MBio 6 (3) (2015) e00842.

[122] C. Lloyd, D. Lowe, B. Edwards, Modelling the human immune response: performance of a 10^{11} human antibody repertoire against a broad panel of therapeutically relevant antigens, Protein Eng. Des. Sel. 22 (3) (March 1, 2009) 159–168. Published online: October 29, 2008, see https://academic.oup.com/peds/article-lookup/doi/10.1093/protein/gzn058.

[123] G. Winter, C. Milstein, Man-made antibodies, Nature 349 (6307) (1991) 293–299.

[124] J.J. Lavinder, K.H. Hoi, S.T. Reddy, Y. Wine, G. Georgiou, Systematic characterization and comparative analysis of the rabbit immunoglobulin repertoire, PLoS One 9 (6) (2014) e101322.

[125] Y.N. Abdiche, R. Harriman, X. Deng, Y.A. Yeung, A. Miles, W. Morishige, et al., Assessing kinetic and epitopic diversity across orthogonal monoclonal antibody generation platforms, MAbs 8 (2) (2016) 264–277.

[126] F.A. Harding, M.M. Stickler, J. Razo, R.B. Dubridge, The immunogenicity of humanized and fully human antibodies: residual immunogenicity resides in the CDR regions, MAbs 2 (3) (2010).

[127] W.O. Salivar, H. Tzagoloff, D. Pratt, Some physical-chemical and biological properties of the rod-shaped coliphage M13, Virology 24 (1964) 359–371.

[128] U. Brinkmann, R.E. Kontermann, The making of bispecific antibodies, MAbs 9 (2) (2017) 182–212.

[129] D.W. LaFleur, D. Abramyan, P. Kanakaraj, R.G. Smith, R.R. Shah, G. Wang, et al., Monoclonal antibody therapeutics with up to five specificities: functional enhancement through fusion of target-specific peptides, MAbs 5 (2) (2013) 208–218.

[130] D.R. Burton, C.F. Barbas 3rd, M.A. Persson, S. Koenig, R.M. Chanock, R.A. Lerner, A large array of human monoclonal antibodies to type 1 human immunodeficiency virus from combinatorial libraries of asymptomatic seropositive individuals, Proc. Natl. Acad. Sci. USA 88 (22) (1991) 10134–10137.

[131] I.A.J. Lorimer, A. Kepplerhafkemeyer, R.A. Beers, C.N. Pegram, D.D. Bigner, I. Pastan, Recombinant immunotoxins specific for a mutant epidermal growth factor receptor – targeting with a single chain antibody variable domain isolated by phage display, Proc. Natl. Acad. Sci. USA 93 (25) (1996) 14815–14820.

[132] R. Schier, J.D. Marks, E.J. Wolf, G. Apell, C. Wong, J.E. McCartney, et al., In vitro and in vivo characterization of a human anti-c-erbB-2 single-chain Fv isolated from a filamentous phage antibody library, Immunotechnology 1 (1) (1995) 73–81.

[133] T.J. Vaughan, A.J. Williams, K. Pritchard, J.K. Osbourn, A.R. Pope, J.C. Earnshaw, et al., Human antibodies with sub-nanomolar affinities isolated from a large non-immunized phage display library, Nat. Biotechnol. 14 (3) (1996) 309–314.

[134] J. Glanville, W. Zhai, J. Berka, D. Telman, G. Huerta, G.R. Mehta, et al., Precise determination of the diversity of a combinatorial antibody library gives insight into the human immunoglobulin repertoire, Proc. Natl. Acad. Sci. USA 106 (48) (2009) 20216–20221.

[135] J. Kugler, S. Wilke, D. Meier, F. Tomszak, A. Frenzel, T. Schirrmann, et al., Generation and analysis of the improved human HAL9/10 antibody phage display libraries, BMC Biotechnol. 15 (2015) 10.

[136] C.F. Barbas, J.D. Bain, D.M. Hoekstra, R.A. Lerner, Semisynthetic combinatorial antibody libraries: a chemical solution to the diversity problem, Proc. Natl. Acad. Sci. USA 89 (10) (1992) 4457–4461.

[137] C.V. Lee, W.C. Liang, M.S. Dennis, C. Eigenbrot, S.S. Sidhu, G. Fuh, High-affinity human antibodies from phage-displayed synthetic Fab libraries with a single framework scaffold, J. Mol. Biol. 340 (5) (2004) 1073–1093.

[138] R.M. Hoet, E.H. Cohen, R.B. Kent, K. Rookey, S. Schoonbroodt, S. Hogan, et al., Generation of high-affinity human antibodies by combining donor-derived and synthetic complementarity-determining-region diversity, Nat. Biotechnol. 23 (3) (2005) 344–348.

[139] M. Batonick, E.G. Holland, V. Busygina, D. Alderman, B.K. Kay, M.P. Weiner, et al., Platform for high-throughput antibody selection using synthetically-designed antibody libraries, Nat. Biotechnol. 33 (5 Pt A) (2016) 565–573.

Chapter 4

Bioanalytical Requirements and Regulatory Guidelines for Immunoassays

Sandeep K. Vashist[1], John H.T. Luong[2]
[1]*IDS Immunodiagnostic Systems Deutschland GmbH, Frankfurt am Main, Germany;* [2]*University College Cork, Cork, Ireland*

1. INTRODUCTION

The growing concerns for public health and safety and the continuously evolving immunoassay (IA) technologies have led to stringent regulatory guidelines for validating the IA performance. The most prominent requirements are accuracy, precision, selectivity, sensitivity, reproducibility, and stability. Accordingly, the IA developers must adhere to a highly streamlined design control procedure as outlined by the regulatory authorities to manufacture IAs [1−4]. However, numerous approved IAs [5−11] are still subject to deficient and conflicting bioanalytical performances. Thus, the validity of the regulatory requirements [12,13] becomes questionable, and this point must be tackled and revised rather extensively. Also, the revised regulatory guidelines must reflect a rapid increase in IA formats and technologies [14], upcoming next-generation technologies [15], changing the landscape of healthcare monitoring and management, and new concepts in healthcare delivery [16,17].

Besides the regulatory approved IAs [18−21], there are some growing concerns about the clinical utility with respect to the current trend for medical value IAs [22,23]. The development costs for medical value IAs are exorbitant, over threefold higher than that of standard IAs, which can only be afforded by the industrial in vitro diagnostic (IVD) giants. Nevertheless, there are significant improvements in IAs that are codeveloped with pertinent healthcare professionals by considering the clinical inputs and user evaluation at every development stage.

This chapter provides an overview of the IA bioanalytical requirements, including the healthcare needs and applications. The advances, the challenges, and the trends in this field have also been discussed apart from providing the guided insights into bioanalytical testing.

Handbook of Immunoassay Technologies. https://doi.org/10.1016/B978-0-12-811762-0.00004-9

2. BIOANALYTICAL REQUIREMENTS FOR AN IMMUNOASSAY

2.1 Accuracy

The accuracy of an IA shows how close the mean test results obtained by the developed IA are w.r.t. the actual target concentration, i.e., nominal value. The mean test results should be within 15% of the nominal value for the entire detection range. The only exception is the lower limit of quantification (LLOQ), where the mean value should be within 20% of the nominal value. The accuracy is determined by replicate analysis of quality controls (QCs) of an IA that contain known analyte concentrations. The procedure involves the measurement of minimal five determinations per analyte concentration and three analyte concentrations in the expected concentration range.

2.2 Precision

The precision indicates the closeness of individual test results of an analyte when the developed IA is repeatedly used for the determination of an analyte in multiple aliquots of a single homogeneous volume of the biological matrix. The precision for the various analyte concentrations in the detection range should be within 15% of the coefficient of variation (CV), whereas the LLOQ should be within 20% of the CV. The precision must be estimated from a minimum of five determinations per analyte concentration and at least three analyte concentrations in the range of expected study sample concentrations. The "within-run (intrabatch)" precision is derived from a single analytical run, whereas the "between-run (interbatch)" precision is obtained from multiple analytical runs over time that may involve different analysts, equipment, reagents, and laboratories. The samples with concentrations higher than the upper limit of the standard curve should be diluted accordingly.

2.3 Selectivity

The selectivity of an IA is its ability to differentiate and quantify the desired analyte in the intended biological sample together with potentially interfering substances, such as endogenous matrix components, metabolites, decomposition products, and pharmacological substances. For a multiplex IA with parallel sensing, the selectivity must be evaluated for each analyte. In this case, the selectivity should employ the blank samples of the biological matrix from minimum six sources with each blank sample tested for interference. The selectivity should be ensured at the LLOQ.

2.4 Sensitivity

The sensitivity, also known as the LLOQ, is the lowest concentration of an analyte that can be measured with acceptable accuracy and precision.

2.5 Reproducibility

The reproducibility is obtained from the replicate determinations of analyte concentrations in an IA employing the desired QCs and possibly incurred samples.

2.6 Stability

It refers to the ability of IA reagents to retain its original performance and properties over a period of time when stored under defined conditions. The stability of an analyte in a particular matrix and container system should be determined under specific conditions for the specified time intervals and is valid only for the specified matrix and container system. Various types of stabilities need to be determined based on the type of IA. It is real-time or accelerated stability based on whether the analyte is stored at a specific temperature (usually 4°C) or 37°C, respectively. The stability analysis should be similar to the experimental conditions that reflect the actual conditions encountered during "real" sample handling and analysis at the clinical site. The analyte stability during sample collection, handling, and transport should also be determined. The onboard stability of the IA reagents in the cartridge should be done for the automated IA used in clinical analyzers. It demonstrates the stability of the IA reagents or samples once they are opened and stored on board the instrument. Further, the in-use stability must be determined to unravel the stability of all IA reagents or components after opening the container and storing as specified in the instructions for use (IFU). The stability of the sample during multiple freezing and thawing cycles also needs to be known as it is critical for the developed IAs.

The stability determination should employ only the samples prepared from a freshly prepared stock solution of the analyte in a suitable biological matrix devoid of analyte and interferences. The stability of an IA reagent is satisfactory when the IA results fall within 15% of the nominal concentrations of these reagents. It involves triplicate analysis at each of the low and high analyte concentrations.

The onboard stability of the reagent cartridge and test calibration should be at least 28 days while the onboard stability of IA kit controls, kit calibrators, calibration verifiers, and real samples should be at least 8 h. The in-use stability of IA cartridge, kit calibrators, and kit controls should be at least 28 days. The accelerated stability should be performed for at least 28 days for all IA components. The accelerated stability data obtained at 37°C and 25°C provide an estimate of the stability of IA reagents at 4°C, as shown in Table 4.1. These values are determined based on the temperature coefficient Q_{10}, the increased reaction rate at a temperature increase of 10°C. Further, the freezing/thawing stability of the samples should be analyzed for up to three cycles of freezing/thawing.

TABLE 4.1 Determination of Stability Estimate of immunoassay (IA) Kit at 4°C Based on the Accelerated Stability Testing Performed at 37°C or 25°C

No. of Days of Testing	Estimate of Days of Stability of IA Kit at 4°C Based on Testing at	
	37°C	25°C
7	68	30
14	137	60
21	207	90
28	276	120
35	345	150
42	414	180
49	483	210
56	552	240

2.7 Recovery

The recovery of an analyte in an IA demonstrates its analyte extraction efficiency, which is determined from the detection signal obtained from the amount of the analyte added to the biological matrix in comparison with the detection signal obtained for the actual concentration of the analyte in a solvent. It is recommended to have percentage recovery in the range of 90–110, but it is still acceptable to have values within 80–100. In this determination, the detection signals for the extracted samples at low, medium, and high concentrations are compared with those obtained from unextracted standards that represent 100% recovery. The sample recovery experiments are conducted for each matrix type if the IA is intended for multiple sample matrices. Such experiments determine the presence and extent of the matrix effect for a particular sample type.

2.8 Calibration Curve

A calibration curve deciphers the reproducible relationship between the signal responses and the known analyte concentrations. For the multiplex IA, a calibration curve should be generated for each analyte in the sample. Based on the concentration range of an IA, the calibration standards are prepared by spiking known analyte concentrations in the same matrix as the biological samples of interest. If some biological matrices are difficult to obtain, e.g., cerebrospinal fluid, substitute matrices are an alternative. A blank sample (no analyte or internal standard), a zero sample (no analyte but contains internal standard), and at least six analyte samples within the range of IA including LLOQ (with the analyte and the internal standard) are used to generate a calibration curve. The calibration standards should be within 15% of nominal concentrations for all concentrations above the LLOQ, whereas the calibration standard should be within 20% of nominal concentration at LLOQ. At least 75% of nonzero calibration standards including the one at LLOQ should meet these criteria.

2.9 Bioanalytical Performance Parameters

2.9.1 Limit of Blank

The limit of blank (LOB) is the highest measurement result that is likely to be observed with a stated probability (typically at 95% certainty) for a blank sample containing no analyte (Fig. 4.1) [24]. The blank samples should be natural, such as native serum, plasma, and urine, instead of an artificial buffer matrix. LOB = Mean of blank samples + 1.645 × standard deviation (SD) of blank samples.

2.9.2 Limit of Detection

The limit of detection (LOD) is defined as the lowest concentration of an analyte in a sample that can be consistently detected with a stated probability (typically at 95% certainty) [24]. The LOB is first determined followed by the

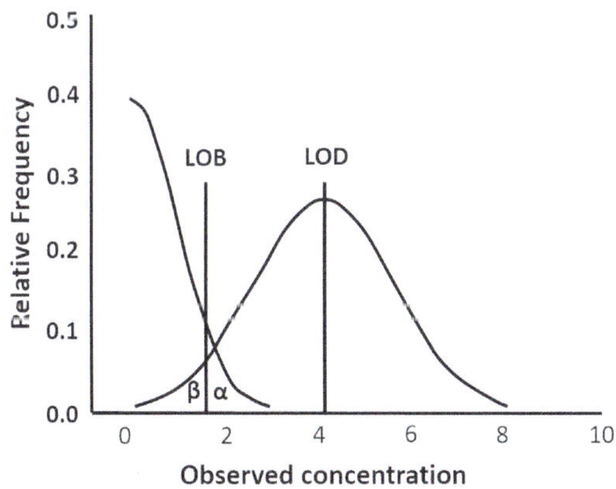

FIGURE 4.1 The distribution of results for the blank sample with no analyte (*left curve*) has 95% of the measurement results below the limit of blank (LOB) (taking $\alpha = 0.05$). On the other hand, the distribution of results for the positive sample with low analyte concentration at the limit of detection (LOD) has 95% of the results above the LOB (taking $\beta = 0.05$).

LOD using the replicate samples with very low analyte concentrations. LOD = LOB + 1.645 × SD of samples with low concentration samples. Such samples should be natural such as native serum, plasma, and urine as artificial or spiked samples may lead to varying results.

2.9.3 Limit of Quantification

The Limit of Quantification (LOQ) is the lowest analyte concentration that can be quantitatively detected with a stated accuracy and precision [24]. However, the determination of LOQ depends on the predefined acceptance criteria and performance requirements set by the IA developers. Although such criteria and performances are not internationally adopted, it is of importance to consider the clinical utility of the IA to define such performance requirements.

The LLOQ is the lowest calibration standard on the calibration curve where the detection response for the analyte should be at least five times over the blank. The detection response should be discrete, identifiable, and reproducible. The precision of the determined concentration should be within 20% of the CV while its accuracy should be within 20% of the nominal concentration. The upper limit of quantification (ULOQ) is the highest calibration standard on the calibration curve, where the analyte response is reproducible, and the precision and accuracy are within 15% of the CV and 15% of the nominal concentration, respectively.

The calibration curve should not be extrapolated below the LLOQ or above the ULOQ to determine the analyte concentration in unknown samples. The samples with concentrations higher than ULOQ must be diluted using the same biological matrix as an actual sample while those with concentrations below the LLOQ should be mentioned as the zero concentration.

2.9.4 Lower Limit of the Linear Interval

The lower limit of the linear interval is the lowest analyte concentration at which the IA detection signal shows a linear relationship with the analyte concentration.

2.9.5 Lower Limit of the Measuring Interval

The lower limit of the measuring interval is the lowest analyte concentration at which all the specified performance characteristics of an IA, such as linearity, bias, and imprecision, are met.

2.9.6 Errors

As the sum of the random and systematic errors, the accuracy error represents the deviation of the determined analyte concentration from the actual analyte concentration. The systematic error is the difference between the actual analyte concentration and the mean of an infinite number of measurements carried out under repeatable conditions. The total analytical error is a combination of bias and imprecision based on a particular error model [25]. Total analytical error = Bias + 1.65 × imprecision.

2.9.7 Carryover

The carryover is the error in the determination of an actual analyte concentration in a sample when samples with very high concentrations of the analyte are run before samples with very low analyte concentrations. It is the amount of analyte carried by the measuring system from an IA into subsequent IA reactions, which introduces error in the analyte determination in the subsequent IA [26]. The carryover is calculated as the mean of the High-Low results minus the mean of the Low-Low results. High-Low result depicts a low analyte concentration determination result that immediately follows a high analyte concentration determination result. In contrast, Low-Low result shows a low analyte concentration determination result that immediately follows a low analyte concentration determination result.

2.9.8 Interference

Interference is defined as the significant bias in the determined analyte concentration from an IA, which may lead to clinically imprecise results [27,28]. The interference in an IA can result from various factors such as the presence of nonspecific components in the sample, such as the physiological biochemical substances (red blood cells, hemoglobin, bilirubin, etc.) and drugs/metabolites. Appropriate measures should be taken to circumvent the interference by changing the assay formulation and/or assay format [29–31]. But if there is confirmed interference, it should be stated clearly in the IFU. If there is a bias of more than 10% in an IA due to interfering substance, the concentration of the interfering substances at which less than 10% bias is achieved has to be established by titration. This new concentration of interfering substances tested has to be mentioned in the IFU.

2.9.9 Quality Controls

The QC samples are responsible for accepting or rejecting an IA run. It is obligatory to take at least three concentrations of QCs in duplicate, where low QC should be within three times the LLOQ, middle QC should be in the midrange, and high QC should be at the high end of the concentration range. The IA run is acceptable if minimum 67% of QCs and 50% of QCs at each level are within 15% of their nominal concentrations. The calibration standards and QCs should be prepared from separate stock solutions. But if the stock solution is verified for stability and accuracy, they could also be prepared from same spiking stock solutions.

2.9.10 Linear Range

The linear range is the particular analyte concentration range on the calibration curve, where the detection signal is proportional to the analyte concentration in the sample. The linearity studies should be performed for each sample matrix of the analyte if the assay is intended for the determination of an analyte in multiple sample types such as serum, plasma, and urine.

2.9.11 Analytical Measurement Range

The analytical measurement range is the range of analyte concentrations that an IA can determine in the patient sample without employing any sample pretreatment such as dilution and concentration.

2.9.12 Clinically Reportable Range

The clinically reportable range is defined as the range of analyte concentrations that an IA can determine in the patient sample employing sample pretreatment steps, such as dilution and concentration.

2.9.13 Bias

The bias is the difference between the actual result (accepted reference value) and the expected test result. It is usually measured by the determination of an analyte in the sample at the beginning and the end of a long series of tests using the same assay components under the same ambient conditions.

2.9.14 Hook Effect

The Hook effect is observed in sandwich IA, where at very high concentrations of the analyte, the assay signal is saturated and leveled off [32]. This phenomenon occurs due to the elevated number of analyte molecules that bind to both the capture and detection Ab, thereby preventing them from forming sandwich immune complexes. In fact, the signal response might decrease at extremely high concentrations and fall in the calibration curve range. This effect then leads to a misleading lower analyte concentration while the actual analyte concentration is much higher. Therefore, in sandwich IAs with the observed the Hook effect, the maximum allowed concentration is the one that corresponds to the onset of the saturated signal.

2.9.15 Method Comparison

The developed IA is compared with an established IA, which results in the measurement of correlation and bias based on the testing of patient samples. The results of method comparison are required for regulatory submissions and also used by companies for preparing their marketing literature and claims. However, the two IAs must have some common analytical measurement ranges for comparison. The guidelines for the method comparison must be taken into account to perform such studies. It is recommended to take the analyte samples covering the reference range of an IA and also take samples lower and higher than the reference range. It is important to have native patient samples only, if possible, for method comparison. The samples must be pristine, i.e., not subject to treatment by heat, pH, stripping, or filtering. About half of the samples should be above the reference range while half should be within and below the reference range. At least 120 samples should be tested for both IAs for the United States Food and Drug Administration (FDA) 510K submission. However, a large sample size is needed to improve statistical confidence. Samples with a prolonged storage should be obviated as FDA recommends avoiding the storage of samples over 12 months.

2.9.16 Cross-reactivity

Cross-reactivity is defined as the percentage of the measured concentration of cross-reactant over the absolute concentration of the cross-reactant. It is an indicator of the reaction between an antibody and a different but similar antigen. It must be proven that the IA shows no cross-reactivity against the substances mentioned in the product design specifications and meet the acceptance criteria. If the cross-reactivity against a particular substance cannot be overcome, this point must be stated clearly in the IFU.

3. CRITIQUES AND OUTLOOK

The variability of IA results is IA-dependent and affected by instrument, day, IA run, reagent lot, calibration lot, calibration cycle, operator, consumables, and laboratory. The IA developers can identify and investigate the sources of variation in IAs by taking the feedback from experts in the field, literature survey, and pertinent information from competitors' products.

Most of the IA developments to date by researchers and industrial giants have emphasized the improvements achieved in an IA based on the decrease in LOD, the most exaggerated parameter for the determination of an IA analytical performance. Notice also that the improvement in LOD does not improve the analyte detection if the linear range of the IA remains the same. The linear range of the IA should cover the clinically relevant cutoff and the pathophysiological range of the biomarker. The clinical cutoff value should be above the LLOQ of the IA so that the analyte concentration could be determined reliably.

It is recommended to follow an improvised reporting scheme for IA results. If the developed IA results for the analyte concentration in the samples is <LOD, the report should mention it as "*not detected; result* < *LOD*" instead of "*not detected*." But if the determined analyte concentration is between LOD and LOQ, the report should mention it as "analyte detected; interpret IA result with caution due to high IA imprecision" instead of "analyte detected." The determined analyte concentration > LOQ should be reported as "report the determined analyte concentration."

Sandwich IAs are most prone to the Hook effect [32], resulting in falsely low analyte concentrations at very analyte high concentrations. For a sandwich IA with the Hook effect, the IA linear range must be restricted to the analyte concentration at the onset of the plateau signal. During the optimization phase of automated analyzer—based IAs, it needs to avoid "the carryover," which likely provides falsely elevated values for samples with low analyte concentrations.

It has also been seen that the end-users employ the developed IA for a sample matrix that is not specified in the manufacturer's IFU. They also often develop their customized operating, handling, and storage procedures, which might be quite different from the detailed instructions, protocols, and application notes provided by the IA developers. Such "modus operandi" should be avoided since they often result in significant errors in analysis.

The IA must be calibrated using the manufacturer's instructions at the start of the measurement. The calibration must also be done periodically as designated by the manufacturer. Any attempt to deviate from the manufacturer's protocol often ends up with significant errors in IA measurements.

The raw materials play a critical role in the reproducibility of manufacturing lots. The variability in the raw materials can deteriorate the IA performance, thereby making it imprecise [33,34]. Therefore, only high-quality raw materials with desired specifications should be used to ensure IAs with good accuracy, precision, and reproducibility.

The performance of the developed IA must be assessed during the design control phases by simulating the actual end-user scenario and conditions. Most IAs are developed in the standard laboratories with controlled ambient conditions using streamlined standard operational procedures. There is a lack of developmental data about how the assay would behave in the actual clinical scenario as the clinical testing, analysis, and sample handling guidelines might be very different from those employed by the IA developers. The developed IA must have random access capability, i.e., the running of IAs on a particular analyzer would not impact the bioanalytical performance of the developed IA [35,36]. This needs to be checked at an early development stage to ascertain the reproducibility of IA in the clinical settings. The IA developers must also keep an eye on the clinical testing and medical decision-making guidelines and procedures as there might be critical testing requirements for the specific IA and need to simultaneously determine additional analytes in the same clinical sample.

The IA developers must test all the possible interferences for an IA as specified by the regulatory authorities [27,28,37]. The list and concentrations of interfering substances vary from time to time. Thus, an FDA presubmission at the feasibility stage is preferred to consider all plausible interferences. If the interference is confirmed with a desired high concentration of an interfering substance, it must be indicated in the IFU. However, the selection of particular high concentrations of cross-reacting substances is still problematical even for the regulatory authorities. As an example, FDA recommends the concentration of cross-reacting substances be sufficiently high but does not stipulate their exact high concentrations. Consequently, varying high concentrations of cross-reacting substances are used by different IA developers. Nowadays, there is a trend toward the continuously increasing concentration of biotin [38–41], which needs to be tested for potential interference. The US FDA is recommending the use of biotin in micrograms per mL, which is hard to achieve for most IA developers. In contrast, high-dose biotin therapy can lead to false assay results [42]. Roche recalled its Elecsys B-CrossLaps/serum IA in 2013 as its claims for biotin interference and sample stability stated in the IFU could not be reproduced [43]. In this context, the interfering and cross-reacting substances should be checked by the IA developers at an early development stage before finalizing the IA format and protocol.

There has been a significant decrease in the error rates in IAs from 1995 to 2017 but the pre- and postanalytical steps still account for highest error prevalences [44,45]. These need to be counteracted effectively by IA developers [46].

Most IAs are developed based on the correlation with the predicate device specified by the regulatory authorities for a particular IA. The developed IA is not compared with other clinically accredited IA that are being used in healthcare. Therefore, if the healthcare setting is using an IA that is different from the predicate and the developed IA is not correlating with the clinically accredited IA, this would be a potential concern. The IA developers need to incorporate the most commonly used IAs in healthcare together with the predicate IA for evaluating the precision of the developed IA during the developmental stages. The last two decades have witnessed continuously growing reports from healthcare professionals concerning the conflicting results obtained from various IAs developed by different companies for a particular analyte. Therefore, the IA developers must go for the guided development toward medical value IAs and participate in external quality assessment and assurance schemes [47–49] to develop clinically viable IAs. The validation of IAs should be done stringently by taking the regulatory guidelines into consideration [50].

The IA developers must be fully aware of the clinical testing procedures, diagnostic algorithms, decision-making, and sampling guidelines to develop a robust assay. Extensive patient samples must be tested to evaluate the performance of the developed IAs. Nevertheless, the gap with respect to imprecision still exists between the developed IAs and clinically accredited assays, such as those based on LC-MS [5,18–21].

Another challenge is to keep abreast of the latest developments in the clinical testing and regulatory requirements for a particular biomarker. If the international standard for the clinical analyte is changed, e.g., the case with sex hormone–binding globulin, the IA should be developed using the new international standard only to eliminate the variability in assays [51,52]. The data integrity must be periodically checked to evince that the instruments are operating in stable conditions during development. The instruments must be properly maintained, calibrated, and under a stringent quality control during the entire course of development.

The training of the operators should be extensive and completed before the collection of data for the developed IA. Extensive training should cover the actual sample preparation, calibration, IA procedure, handling of instruments, operational and maintenance procedures, calibration, monitoring, and data analysis. A troubleshooting table for a particular IA must be provided so that the IA can be verified more systematically when needed.

The quality and reproducible operations of an IA analyzer is another important task. Any defect in the IA analyzer would adversely affect its subsequent operations and results. Many units of Beckman Coulter Inc (USA) analyzers were recalled as they suffered from a defective pulley drive that resulted in inadequate washing or mixing of IA components in the reaction vessel [53].

The IA developers must continuously monitor the performance of launched IAs. For any commercially sold manufacturing lots of an IA with unsatisfactory bioanalytical performances that might cause adverse health consequences, these particular lots should be recalled immediately by the developer. An example is the recall of Elecsys troponin I and Elecsys troponin I STAT IAs in 2012 by Roche Diagnostics [54]. The particular lots of these two IAs demonstrated inadequate analytical performances as it can lead to incorrect, falsely low results (up to 50% lower concentration of troponin I) that can lead to fatal health consequences.

Another litigation was filed in 2010 against Beckman Coulter Inc. for the discrepancies between the troponin tests on UniCel DxI and Access IA systems. The tests on DxI system showed a positive bias [55]. However, Beckman asked its customers to take certain actions or discontinue the use of troponin test on the DxI system. FDA mentioned that Beckman has made certain modifications to its troponin test on DxI and Access systems without obtaining an appropriate product clearance from the FDA. This news led to the critical decrease in the value of Beckman shares and earnings in 2010. Complaints were filed that Beckman did not maintain appropriate controls for the product quality and regulatory compliance.

4. CONCLUSIONS

The IAs have to meet the bioanalytical requirements stipulated by the regulatory authorities before their products can be launched in the market and used by industrial, healthcare, and bioanalytical settings. The constantly increasing immunoassay technologies and formats along with the growing concerns about public health and safety have stressed the need for stringent bioanalytical requirements. The IVD companies must constantly verify the bioanalytical performance characteristics of IA for each manufacturing lot to provide desired high-quality IAs with high reliability for clinical decision-making.

Despite matching the stringent bioanalytical requirements, the developed IAs do not always demonstrate the bioanalytical performance characteristics desired by the healthcare. Similarly, the IA could be seriously impacted by the variability in raw materials and manufacturing lots. Most of the previous IAs followed the standard IVD development procedure. However, the current trend followed by the industrial IVD giants is the medical value IVD development. It is imperative for IVD manufacturers to work with healthcare professionals by taking their clinical inputs and requirements into consideration at all development steps. This joint endeavor would lead to the development of bioanalytical superior and clinically viable IAs with a tremendous medical value and an ease of acceptance by healthcare settings. Lastly, new bioanalytical requirements must take into account the increasing trend toward smartphone-based IA, mobile healthcare, and personalized medicine.

REFERENCES

[1] E. Trullols, I. Ruisanchez, F.X. Rius, Validation of qualitative analytical methods, Trends Anal. Chem. 23 (2) (2004) 137–145.

[2] S.L.R. Ellison, T. Fearn, Characterising the performance of qualitative analytical methods: statistics and terminology, Trends Anal. Chem. 24 (6) (2005) 468–476.

[3] U.H. Stenman, Immunoassay standardization: is it possible, who is responsible, who is capable? Clin. Chem. 47 (5) (2001) 815–820.

[4] M.A. Valentin, et al., Validation of immunoassay for protein biomarkers: bioanalytical study plan implementation to support pre-clinical and clinical studies, J. Pharm. Biomed. Anal. 55 (5) (2011) 869–877.

[5] J. Taieb, et al., Testosterone measured by 10 immunoassays and by isotope-dilution gas chromatography-mass spectrometry in sera from 116 men, women, and children, Clin. Chem. 49 (8) (2003) 1381–1395.

[6] A.C. Celniker, et al., Variability in the quantitation of circulating growth hormone using commercial immunoassays, J. Clin. Endocrinol. Metab. 68 (2) (1989) 469–476.

[7] P.A. Marchildon, et al., Evaluation of three commercial enzyme immunoassays compared with the 13C urea breath test for detection of *Helicobacter pylori* infection, J. Clin. Microbiol. 34 (5) (1996) 1147–1152.

[8] A. Ledur, et al., Variable estimates of cytokine levels produced by commercial ELISA kits: results using international cytokine standards, J. Immunol. Methods 186 (2) (1995) 171–179.

[9] V. Moal, et al., Low serum testosterone assayed by liquid chromatography-tandem mass spectrometry. Comparison with five immunoassay techniques, Clin. Chim. Acta. 386 (1–2) (2007) 12–19.

[10] C.M. Sturgeon, et al., Differences in recognition of the 1st WHO international reference reagents for hCG-related isoforms by diagnostic immunoassays for human chorionic gonadotropin, Clin. Chem. 55 (8) (2009) 1484–1491.

[11] E.M. Davelaar, et al., Comparison of seven immunoassays for the quantification of CA 125 antigen in serum, Clin. Chem. 44 (7) (1998) 1417–1422.

[12] Guidance for industry–bioanalytical method validation, 2013. Available from: https://www.fda.gov/downloads/Drugs/Guidances/ucm368107.pdf.

[13] Guideline on bioanalytical method validation, 2011. Available from: http://www.ema.europa.eu/docs/en_GB/document_library/Scientific_guideline/2011/08/WC500109686.pdf.
[14] C.A. Marquette, L.J. Blum, State of the art and recent advances in immunoanalytical systems, Biosens. Bioelectron. 21 (8) (2006) 1424–1433.
[15] S.K. Vashist, et al., Emerging technologies for next-generation point-of-care testing, Trends Biotechnol. 33 (11) (2015) 692–705.
[16] M. Strandberg-Larsen, A. Krasnik, Measurement of integrated healthcare delivery: a systematic review of methods and future research directions, Int. J. Integr. Care 9 (1) (2009) e01.
[17] P. Varkey, M.K. Reller, R.K. Resar, Basics of quality improvement in health care, Mayo Clin. Proc. 82 (2007). Elsevier.
[18] D. Weismann, et al., Measurements of plasma metanephrines by immunoassay versus LC-MS/MS for diagnosis of pheochromocytoma, Eur. J. Endocrinol. (2015) 251–260.
[19] S.K.G. Grebe, R.J. Singh, LC-MS/MS in the clinical laboratory–where to from here? Clin. Biochem. Rev. 32 (1) (2011) 5.
[20] R. West, et al., Comparison of clonazepam compliance by measurement of urinary concentration by immunoassay and LC-MS/MS in pain management population, Pain Physician 13 (1) (2010) 71–78.
[21] H.N. Bui, et al., Dynamics of serum testosterone during the menstrual cycle evaluated by daily measurements with an ID-LC-MS/MS method and a 2nd generation automated immunoassay, Steroids 78 (1) (2013) 96–101.
[22] U.P. Rohr, et al., The value of in vitro diagnostic testing in medical practice: a status report, PLoS One 11 (3) (2016) e0149856.
[23] I. Miller, et al., Market access challenges in the EU for high medical value diagnostic tests, Pers Med. 8 (2) (2011) 137–148.
[24] D.A. Armbruster, T. Pry, Limit of blank, limit of detection and limit of quantitation, Clin. Biochem. Rev. 29 (Suppl. 1) (2008) S49–S52.
[25] J.S. Krouwer, Setting performance goals and evaluating total analytical error for diagnostic assays, Clin. Chem. 48 (6 Pt 1) (2002) 919–927.
[26] D.A. Armbruster, D.B. Alexander, Sample to sample carryover: a source of analytical laboratory error and its relevance to integrated clinical chemistry/immunoassay systems, Clin. Chim. Acta 373 (1–2) (2006) 37–43.
[27] L.J. Kricka, Interferences in immunoassay—still a threat, Clin. Chem. 46 (2000).
[28] J. Tate, G. Ward, Interferences in immunoassay, Clin. Biochem. Rev. 25 (2) (2004) 105–120.
[29] A.A.A. Ismail, A radical approach is needed to eliminate interference from endogenous antibodies in immunoassays, Clin. Chem. 51 (1) (2005) 25–26.
[30] J. Bjerner, et al., Immunometric assay interference: incidence and prevention, Clin. Chem. 48 (4) (2002) 613–621.
[31] H. Niu, et al., A biotin-drug extraction and acid dissociation (BEAD) procedure to eliminate matrix and drug interference in a protein complex anti-drug antibody (ADA) isotype specific assay, J. Immunol. Methods 446 (2017) 30–36.
[32] S.A. Fernando, G.S. Wilson, Studies of the 'Hook'effect in the one-step sandwich immunoassay, J. Immunol. Methods 151 (1–2) (1992) 47–66.
[33] Class 2 Device Recall CoatACount Direct Androstenedione, 2014. Available from: https://www.accessdata.fda.gov/scripts/cdrh/cfdocs/cfRES/res.cfm?id=128287.
[34] Class 2 Device Recall IMMULITE/IMMULITE 1000 Systems Androstenedione, 2014. Available from: https://www.accessdata.fda.gov/scripts/cdrh/cfdocs/cfRes/res.cfm?id=124991.

[35] G. Carey, S.C. Lewis, Method of Handling Reagents in a Random Access Protocol, 2002 (Google Patents).

[36] P.L. Gebrian, T.P. Evers, Random Access Reagent Delivery System for Use in an Automatic Clinical Analyzer, 2007 (Google Patents).

[37] A.A.A. Ismail, et al., Interference in immunoassay is an underestimated problem, Ann. Clin. Biochem. 39 (4) (2002) 366–373.

[38] S. Kummer, D. Hermsen, F. Distelmaier, Biotin treatment mimicking Graves' disease, N. Engl. J. Med. 375 (7) (2016) 704–706.

[39] A. Clerico, M. Plebani, Biotin interference on immunoassay methods: sporadic cases or hidden epidemic? Clin. Chem. Lab. Med. 55 (6) (2017) 777–779.

[40] E.W. Holmes, J. Garbincius, K. McKenna, Biotin supplements: a hair-raising challenge to the analytical accuracy of clinical immunoassays, FASEB J. 31 (1 Suppl.) (2017) 795.5.

[41] K.Y. Chun, Biotin interference in diagnostic tests, Clin. Chem. 63 (2) (2017) 619–620.

[42] M.L. Piketty, et al., High-dose biotin therapy leading to false biochemical endocrine profiles: validation of a simple method to overcome biotin interference, Clin. Chem. Lab. Med. 55 (6) (2017) 817–825.

[43] Class 2 Device Recall ELECSYS CrossLaps/Serum Immunoassay 2013, Available from: https://www.accessdata.fda.gov/scripts/cdrh/cfdocs/cfres/res.cfm?id=118059.

[44] P. Carraro, M. Plebani, Errors in a stat laboratory: types and frequencies 10 years later, Clin. Chem. 53 (7) (2007) 1338–1342.

[45] Y. Ismail, A.A. Ismail, A.A.A. Ismail, Erroneous laboratory results: what clinicians need to know, Clin. Med. 7 (4) (2007) 357–361.

[46] M. Plebani, The detection and prevention of errors in laboratory medicine, Ann. Clin. Biochem. 47 (Pt 2) (2010) 101–110.

[47] F. Ceriotti, The role of External Quality Assessment Schemes in monitoring and improving the standardization process, Clin. Chim. Acta 432 (2014) 77–81.

[48] C. Perich, et al., External quality assurance programs as a tool for verifying standardization of measurement procedures: pilot collaboration in Europe, Clin. Chim. Acta 432 (2014) 82–89.

[49] I. Taverniers, M. De Loose, E. Van Bockstaele, Trends in quality in the analytical laboratory. II. Analytical method validation and quality assurance, Trends Anal. Chem. 23 (8) (2004) 535–552.

[50] J.W. Findlay, et al., Validation of immunoassays for bioanalysis: a pharmaceutical industry perspective, J. Pharm. Biomed. Anal. 21 (6) (2000) 1249–1273.

[51] M. Thaler, et al., Steroid binding properties of the 2nd WHO International Standard for sex hormone-binding globulin, Clin. Chem. Lab. Med. 49 (5) (2011) 869–872.

[52] M. Jin, M.H. Wener, D.D. Bankson, Evaluation of automated sex hormone binding globulin immunoassays, Clin. Biochem. 39 (1) (2006) 91–94.

[53] Class 2 Device Recall Access Immunoassay System 2012, Available from: https://www.accessdata.fda.gov/scripts/cdrh/cfdocs/cfRES/res.cfm?id=114387.

[54] Roche Recalls Two Troponin I Assays 2012, Available from: http://www.medscape.com/viewarticle/762613.

[55] 2010. Available from: http://www.bernardmgross.com/sites/default/files/cases/2014/05/filed-complt-122.pdf.

Chapter 5

Enzyme-Linked Immunoassays

Sandeep K. Vashist[1], John H.T. Luong[2]
[1]*IDS Immunodiagnostic Systems Deutschland GmbH, Frankfurt am Main, Germany;* [2]*University College Cork, Cork, Ireland*

1. INTRODUCTION

As an integral part of healthcare, in vitro diagnostics (IVD) enables the healthcare professionals to screen, diagnose, treat, monitor, and manage the diseases. A wide range of enzyme-linked immunoassays (EIAs), the most widely used IVD assays in clinical diagnostics, have been developed. Their deployment has contributed to considerable advances in healthcare in terms of sustainable and better health outcomes, detection of diseases at early stages, lower healthcare costs, and advanced analytical capabilities. The EIA turnaround time (TAT) has significantly decreased from ~24 h to a few hours. Indeed, most of the current EIAs being used in clinical diagnostics have a sample-to-answer time of less than an hour. Of notice is the microfluidic lab-on-a-chip (LOC) formats [1,2] with an immunoassay (IA) duration of less than 30 min. Furthermore, the costs of analysis and the need for highly skilled analysts have decreased significantly during the last three decades. The current EIA formats employ a minimal number of IA steps, an easy-to-use procedure, reduced reagent volumes, and an automated operational procedure. The readout devices and instruments required for performing EIA have witnessed an evolving trend over the years. Most initial EIA formats were bulky, expensive, and semiautomated instruments, including microplate readers. However, a wide range of fully automated and portable clinical analyzers have been developed during the last two decades. Most of the fully automated point-of care (POC) analyzers in clinical diagnostics, developed by Roche, Siemens, and Abbott, are integrated with advanced technologies, features, and information technology tools, to provide a complete system for healthcare needs. The last decade (2007–17) has also witnessed the emergence of portable immunoanalytical devices that enable a fully automated EIA on LOC platforms and integrated readers. The most recent development of smartphone (SP) readers [3–11] for EIA can read out detection signals generated in EIAs such as colorimetric, spectroscopic, chemiluminescence, fluorescence, surface plasmon resonance, Mie scattering, and electrochemical.

Handbook of Immunoassay Technologies. https://doi.org/10.1016/B978-0-12-811762-0.00005-0

The highly sensitive and specific detection of biomarkers at an early stage of a disease is an essential requirement in healthcare as it enables the healthcare professionals to precisely diagnose the patient's condition and start an effective treatment regimen at the earliest. As EIAs determine most of the biomarkers, they form an indispensable part of clinical decision-making and monitoring the efficacy of the treatment regimen. To date, numerous EIA formats have been demonstrated with considerable improvements in sensitivity, IA duration, IA procedure, cost-effectiveness, simplicity, and automation. Emerging nano- and micromaterials further increase the sensitivity of EIA [12] while new detection concepts result in an analytically superior EIA for multiplex detection of biomarkers [13,14]. The advent of microfluidic protocols [2,15–17] and LOC platforms [18] has contributed to rapid and fully automated POC EIAs for the remote and decentralized settings. Similarly, several prospective low-cost formats, mainly paper-based EIAs [19,20], have also been advocated in the developing nations with limited resources. The recent trend toward SP-based EIAs using LOC platforms and SP readers are paving the way to prospective next-generation EIA formats [21] equipped with mobile healthcare, telemedicine, and personalized healthcare monitoring and management tools.

This chapter provides an overview of current EIA formats and the trend for the next-generation development. The chapter also provides some guided insights pertaining to the suitability of EIA formats into various bioanalytical and diagnostic settings and applications.

2. CONVENTIONAL ENZYME-LINKED IMMUNOASSAYS

2.1 Enzyme-Linked Immunosorbent Assay

2.1.1 Direct Enzyme-Linked Immunosorbent Assay

The simplest direct enzyme-linked immunosorbent assay (ELISA) has been employed for the IA of highly defined antigens. The first step involves the passive antigen adsorption onto the microtiter plate (MTP) wells by incubation for a specific duration ranging from a few hours to overnight (Fig. 5.1). After intensive washing to remove the unbound excess antigen, nonspecific protein binding sites are blocked by a blocking buffer, e.g., bovine serum albumin (BSA). The passively adsorbed antigen molecules are then detected by incubating with an enzyme-labeled antibody (Ab) that is specific to the antigen, to form an immune complex. After the unbound Ab is washed away by repeated washing, a specific substrate solution is provided for the enzyme labeled to the Ab. The enzyme catalyzes the conversion of substrate molecules into the colorimetric product, and the enzyme–substrate reaction is stopped by the addition of a stop solution after a certain time. The absorbance of the resulting colorimetric product solution is measured at a specific wavelength. The buffer used for diluting the antigen should not contain proteins that might interfere

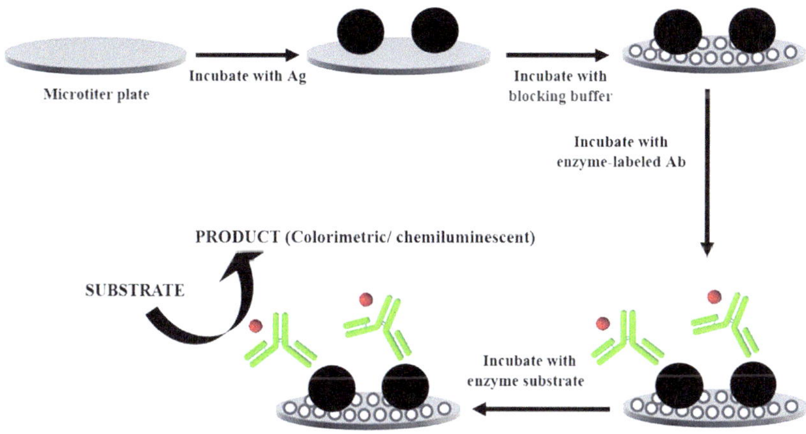

FIGURE 5.1 Schematic of direct enzyme-linked immunosorbent assay.

with the specific binding of the target antigen to the MTP. All the incubation steps are performed in incubators at 37°C, which is the most widely used temperature in bioanalytical settings.

2.1.2 Indirect Enzyme-Linked Immunosorbent Assay

The initial steps in indirect ELISA are very similar to direct ELISA as the antigens are bound to the MTP wells, i.e., the solid phase and the nonspecific protein binding sites on the MTP are covered by treatment with a blocking buffer. However, in indirect ELISA, the unlabeled detection Ab is used for the specific binding to the coated antigen molecules (Fig. 5.2). The detection Abs are then detected by an extra step that involves incubation with enzyme-labeled antispecies Ab against the immunoglobulins of the species in which the detection Abs are produced. As an example, if the detection Abs are produced in a goat, the enzyme-labeled antispecies Ab must be antigoat immunoglobulins. Therefore, antispecies Ab could be selected from many commercially available immunoglobulins. The remaining steps of enzyme–substrate reaction, stopping the reaction, and readout of the colorimetric signal are the same as indirect ELISA. Indirect ELISA has been widely used by the IVD industry to examine a large number of antisera for binding to the target antigen using a specific antispecies Ab. However, the varying nonspecific binding in individual sera might lead to variability in IA results.

2.1.3 Sandwich Enzyme-Linked Immunosorbent Assay

Sandwich ELISA is the most commonly used EIA format for the development of a large number of commercial IVD kits for the quantification of clinical biomarkers. However, this format is less applicable for very small biomarkers with limited epitopes on their surface for binding to the detection Ab. The

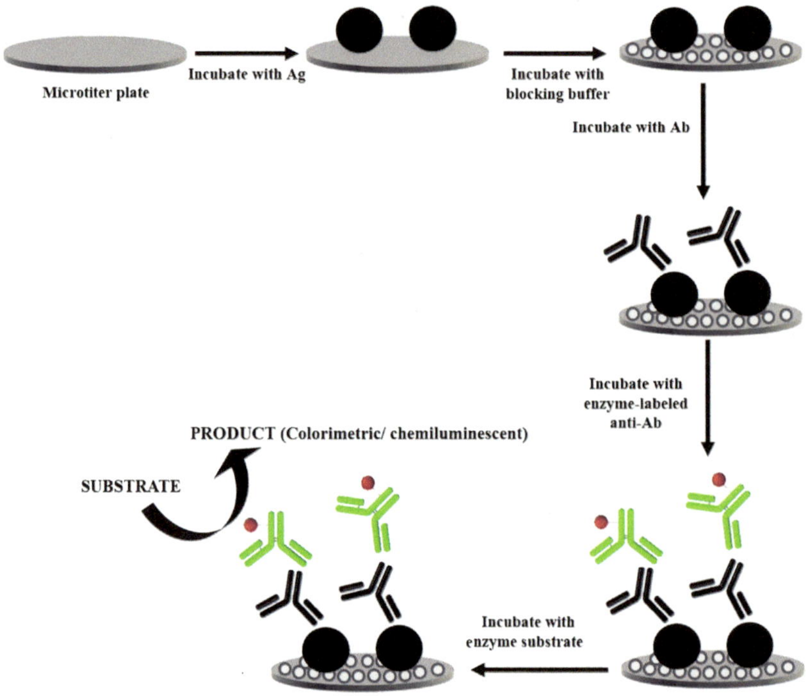

FIGURE 5.2 Schematic of indirect enzyme-linked immunosorbent assay.

commercial sandwich ELISA kit contains a capture Ab–bound preblocked MTP or Ab-bound 96-well plate strips, analyte standards, biotinylated detection Ab, horseradish peroxidase (HRP)-labeled streptavidin (SA-HRP), an enzyme substrate kit, and a stop solution. The substrate kit often consists of 3,3′,5,5′-tetramethylbenzidine (TMB) and 0.02% H_2O_2 as the enzyme substrates for HRP and a stop acid solution (usually sulfuric acid). Commercial sandwich ELISA kits provide highly sensitive and selective detection of analytes in "real-world" sample matrices such as human serum, plasma, urine, cerebrospinal fluid, or whole blood.

The sandwich ELISA procedure involves the binding of the target analyte to the capture Ab-bound MTP or plate strips to form an immune complex (Fig. 5.3). Nonspecifically bound and excess analytes are removed from the MTP wells by multiple washings. The binding of biotinylated detection Ab to the analyte forms a sandwich immune complex (SIMC). The nonspecifically bound and excess detection Ab was then removed by multiple washings. Subsequently, SA-HRP is introduced to the MTP, which binds specifically to the biotinylated detection Ab via highly specific and strong SA-biotin interactions. After multiple washings, HRP conjugated to SA in SA-HRP is allowed to react with the enzyme substrate, e.g., TMB is mixed with H_2O_2 to

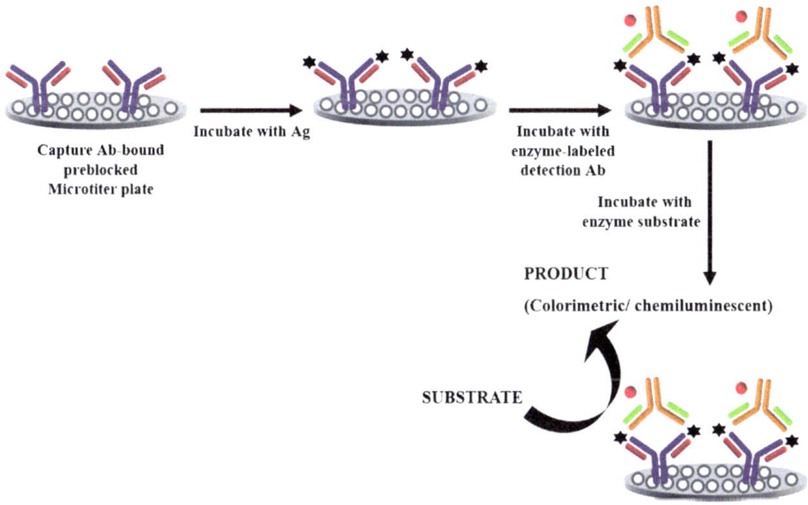

FIGURE 5.3 Schematic of sandwich enzyme-linked immunosorbent assay.

generate a blue product. The resulting absorbance is measured specifically at 450 nm with 540 nm or 650 nm as a reference wavelength. The sandwich ELISA procedure has been automated using commercial robotic stations or analyzers for high-throughput analysis of clinical patient samples.

2.2 Competitive Enzyme-Linked Immunoassay

The competitive EIA involves the quantification of a substance by its ability to compete with the analyte to be determined in the ELISA. The EIA formats could be developed based on direct, indirect, or sandwich ELISAs. The most common is the direct competitive ELISA on Ag-coated MTP, where the presence of Ag in the sample would prevent the binding of enzyme-labeled Ab to the Ag-coated MTP well, as the Ag—Ab immune complex is removed during the subsequent washing steps (Fig. 5.4). The presence of Ag in the sample would lead to competition, thereby preventing any colorimetric product formation, whereas the absence of Ag in the sample would have no competition, which might lead to the usual colorimetric assay signal. A similar competitive ELISA procedure is used for the detection of Ab in the sample using Ag-prebound MTP, where the presence of Ab in the sample would prevent the binding of enzyme-labeled Ab to the Ag-coated MTP, thereby leading to no signal response. Similar competitive ELISA formats are used for indirect and sandwich ELISA formats. In principle, competitive ELISA formats can be developed for any analyte. However, they are the most preferred EIA format for small molecules such as steroids with insufficient epitopes on their surface.

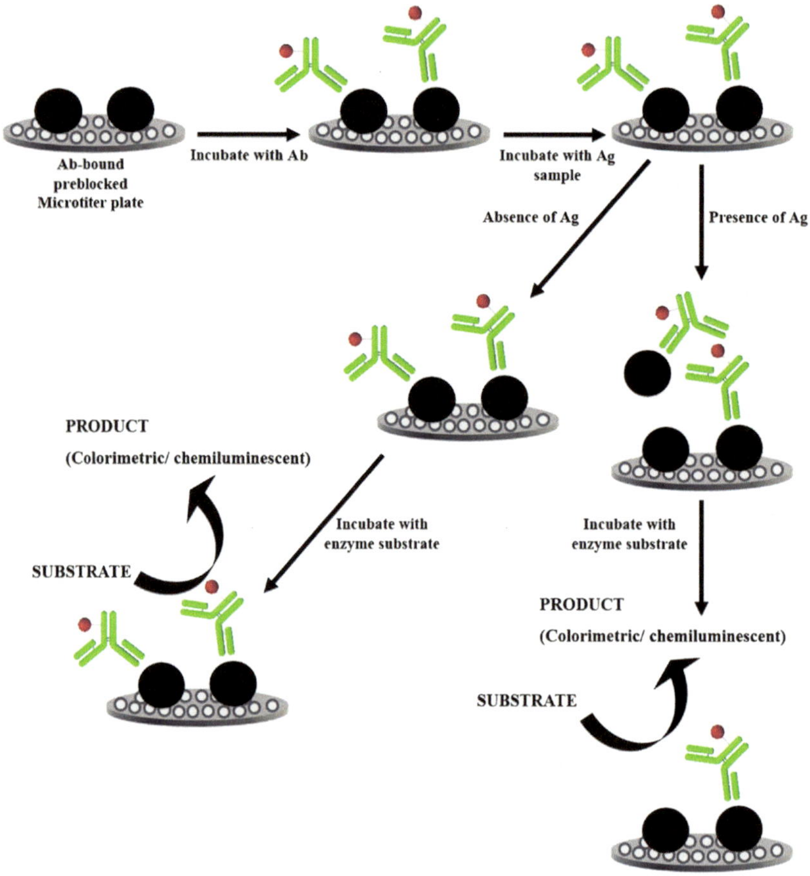

FIGURE 5.4 Schematic of competitive enzyme-linked immunoassay.

3. EMERGING ENZYME-LINKED IMMUNOASSAYS

3.1 High-Sensitivity Sandwich Enzyme-Linked Immunoassay

Among several high-sensitivity sandwich EIA formats, one high-sensitivity EIA format is based on a novel immobilization strategy to provide the leach-proof covalent binding of capture Ab to a 3-aminopropyltriethoxysilane (APTES)-functionalized MTP [22] (Fig. 5.5). The procedure involves the generation of hydroxyl groups on the MTP surface by treatment with KOH and oxygen plasma. The generated hydroxyl groups bind the ethoxy groups of APTES, resulting in an APTES-functionalized MTP with free amino groups for binding of Abs covalently. The free amino groups on the APTES-functionalized MTP are often conjugated to the carboxyl groups present on the fragment crystallizable (F_c) region of the Ab via heterobifunctional

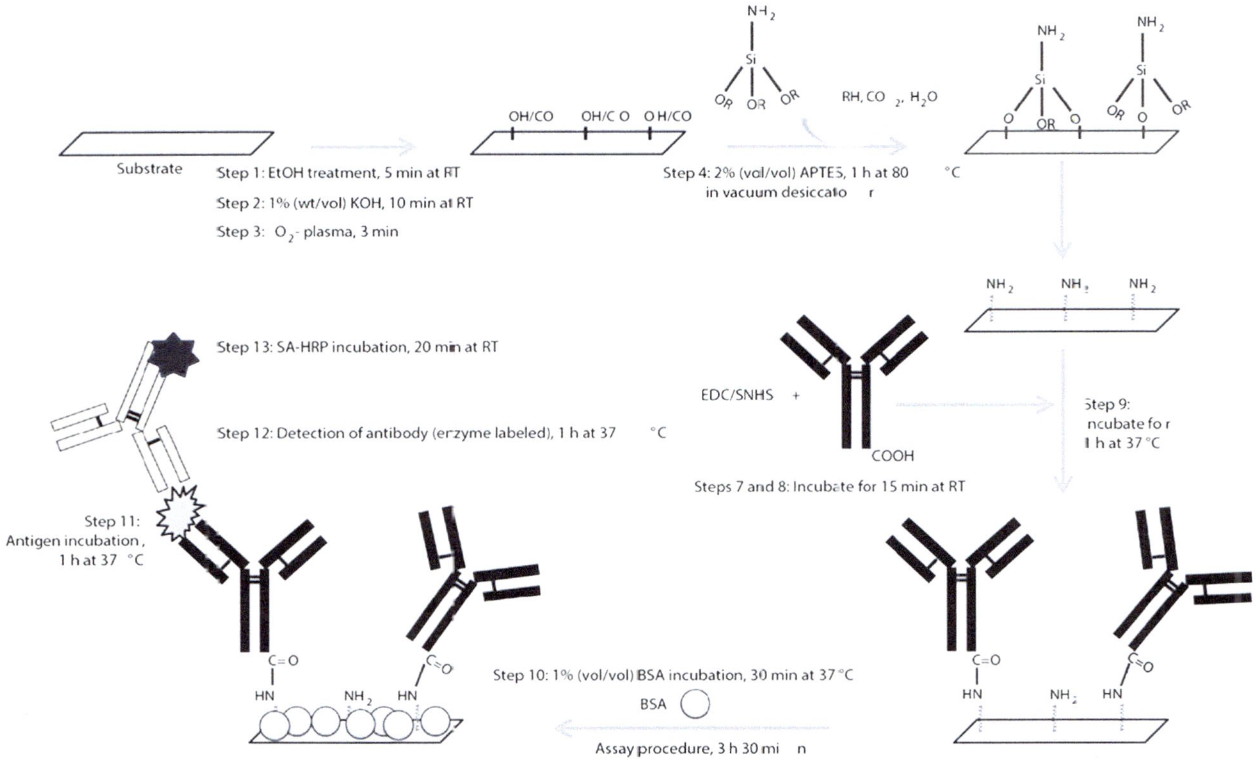

FIGURE 5.5 High-sensitivity sandwich enzyme-linked immunoassay developed using a leach-proof Ab immobilization strategy, which involves sequentially the treatment of the surface with KOH and oxygen plasma, 3-aminopropyltriethoxysilane functionalization, immobilization of capture Ab, blocking, Ag detection, binding of biotinylated detection Ab, binding of SA-HRP, enzyme–substrate reaction, and readout of absorbance of the colorimetric signal. *Reproduced with permission from Nature America Inc. C.K. Dixit, et al., Multisubstrate-compatible ELISA procedures for rapid and high-sensitivity immunoassays, Nat. Protoc. 6 (4) (2011) 439–445.*

cross-linking based on 1-ethyl-3-(3-dimethylaminopropyl)carbodiimide (EDC) and *N*-hydroxysulfosuccinimide. This immobilization strategy is analytically superior for human fetuin A (HFA) as the resulting sandwich EIA had ~16-fold higher analytical sensitivity (i.e., 39 pg/mL vs. 625 pg/mL for conventional ELISA) and ~3-fold shorter immunoassay duration (i.e., ~6 h vs. ~20 h for conventional ELISA) than the commercial sandwich ELISA kit based on the conventional ELISA procedure. Being multisubstrate-compatible, the Ab immobilization strategy can perform sandwich EIA on a wide range of substrates: inert, hydrophilic, and hydrophobic [23]. The detection sensitivity is further improved by chemiluminescent detection using the SuperSignal ELISA femto maximum sensitivity substrate for HRP consisting of luminol and stable peroxide [24]. Indeed, the chemiluminescent sandwich EIA offers a 125-fold higher sensitivity (i.e., 5 pg/mL) than the commercial sandwich ELISA kit.

A more recent approach employs a novel one-step Ab immobilization chemistry—based sandwich EIA. The procedure involves the leach-proof binding of capture Ab simply by dispensing the capture Ab admixed with a particular concentration of APTES onto a polystyrene MTP followed by 30 min of incubation [25] (Fig. 5.6). The leach-proof binding of capture Ab to the MTP is attained via strong ionic and hydrophobic interactions. The developed sandwich EIA offers a 51-fold higher sensitivity than the commercial sandwich ELISA.

3.2 Highly Simplified Rapid Sandwich Enzyme-Linked Immunoassay

Among prospective highly simplified rapid sandwich EIA formats developed during the last decade, one potential EIA format involves the one-step kinetics-based simplified, rapid sandwich EIA with minimal process steps to determine analytes with a TAT of 30 min [26] (Fig. 5.7). The procedure is based on the preparation of capture anti-HFA Ab—bound MTP by dispensing EDC-activated anti-HFA Ab admixed with APTES into the KOH-pretreated MTP. It leads to the covalent binding of Ab to the MTP in 30 min, followed by the blocking of nonspecific protein binding sites by treatment with BSA. The resulting MTP is then incubated with SA-HRP-bound biotinylated detection Ab and patient samples to form an SIMC in just 15 min if the target analyte is present in the sample. After the removal of excess nonspecific and unbound components, the analyte concentration in the sample is determined by the HRP-based TMB substrate reaction. In another approach, a one-step chemiluminescent IA (CIA) is developed based on the formation of an SIMC among capture monoclonal anti-CRP Ab—bound polystyrene beads, C-reactive protein (CRP), and acridinium-labeled detection monoclonal anti-CRP Ab [27]. The results obtained by the developed IA are correlated well with those obtained by high-sensitivity latex-enhanced nephelometry.

Another approach employs agarose microparticles for signal-enhanced sandwich EIA, where agarose—APTES-modified MTPs are used for one-step

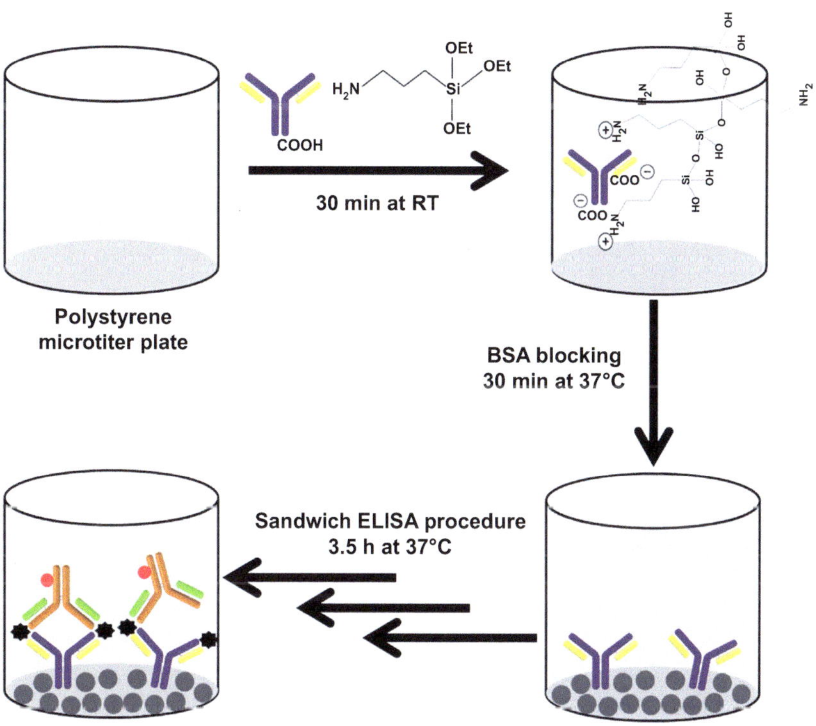

FIGURE 5.6 High-sensitivity sandwich enzyme-linked immunoassay employing a novel one-step antibody (Ab) immobilization chemistry, involving the leach-proof binding of capture Ab onto a polystyrene microtiter plate by simply incubating with capture Ab admixed with 3-aminopropyltriethoxysilane. *Reproduced with permission from Nature Publishing Group. S.K. Vashist, et al., One-step antibody immobilization-based rapid and highly-sensitive sandwich ELISA procedure for potential in vitro diagnostics, Sci. Rep. 4 (2014) 4407.*

kinetics-based sandwich EIA [28] (Fig. 5.8). The agarose−APTES complex binds covalently to the hydroxyl groups generated on the KOH-treated MTP. The EDC-activated Ab is then attached covalently to the free amino groups of APTES present on the MTP's surface via carbodiimide coupling. The signal-enhanced sandwich EIA demonstrates 10-fold higher analytical sensitivity (0.02 ng/mL) and broader linear range (1−243 ng/mL) for the detection of HFA in comparison to one-step kinetics-based sandwich EIA. The superior analytical performance is attributed to the signal enhancement provided by agarose microparticles, enabling a higher Ab loading due to the significantly increased effective surface area.

3.3 Wash-Free Immunoassays

AlphaLISA [29] is a highly prospective wash-free sandwich IA from Perkin Elmer. It involves the interaction of streptavidin-coated Alpha donor beads

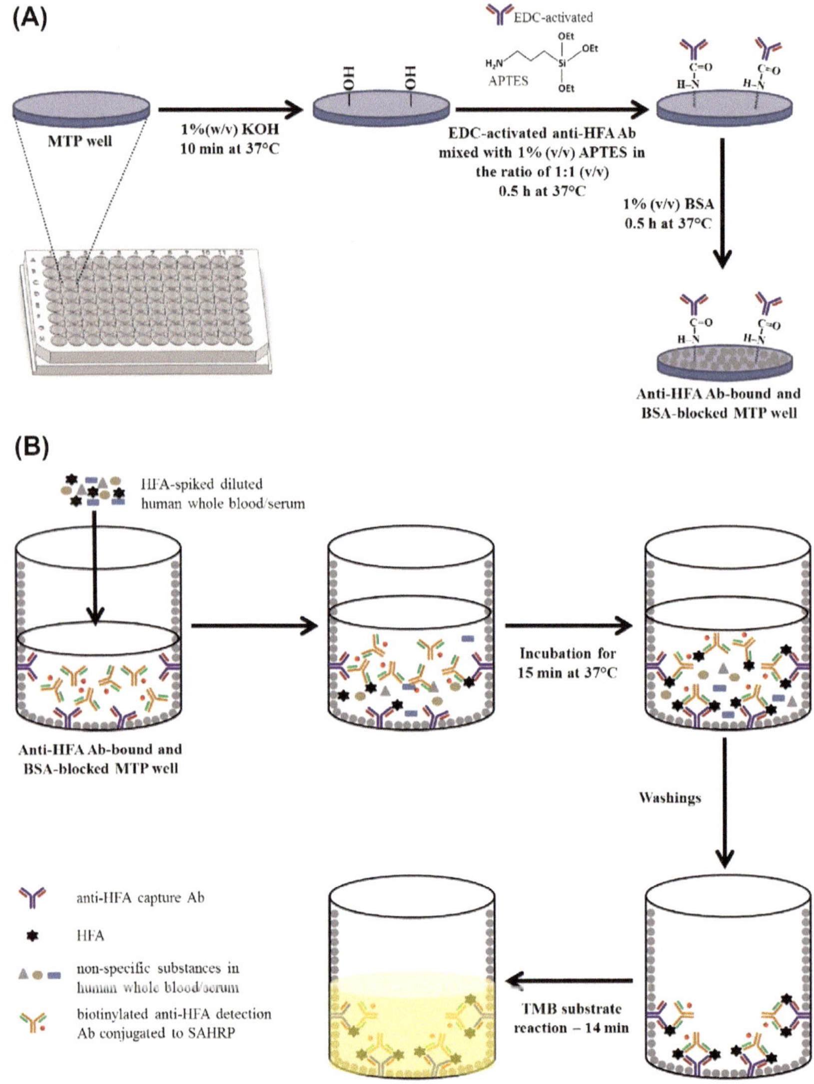

FIGURE 5.7 Highly simplified rapid sandwich enzyme-linked immunoassay (EIA). (A) Capture antibody (Ab) is bound covalently to KOH-functionalized polystyrene microtiter plate (MTP) by simply incubating with 1-ethyl-3-(3-dimethylaminopropyl)carbodiimide-activated Ab admixed with 3-aminopropyltriethoxysilane. The nonspecific protein binding sites are then blocked by treatment with bovine serum albumin (BSA). (B) Thereafter, a one-step kinetics-based rapid sandwich EIA is performed that enables analyte detection in just 30 min. *Reproduced with permission from Elsevier B.V. S.K. Vashist, E.M. Schneider, J.H.T. Luong, Rapid sandwich ELISA-based in vitro diagnostic procedure for the highly-sensitive detection of human fetuin A, Biosens. Bioelectron. 67 (2015) 73–78.*

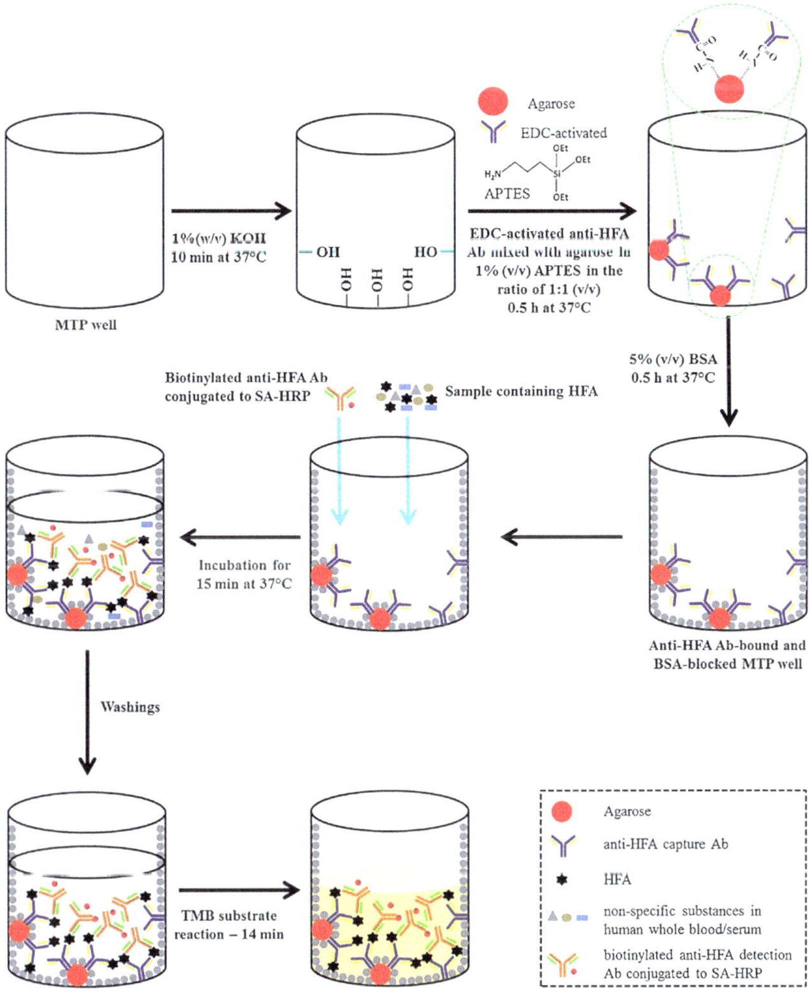

FIGURE 5.8 Highly simplified rapid sandwich enzyme-linked immunoassay (EIA) based on modifying the surface of a KOH-treated polystyrene microtiter plate with agarose and 3-aminopropyltriethoxysilane, and one-step kinetics-based rapid sandwich EIA. *Reproduced with permission from Elsevier B.V. S.K. Vashist, E.M. Schneider, J.H.T. Luong, A rapid sandwich immunoassay for human fetuin A using agarose-3-aminopropyltriethoxysilane modified microtiter plate, Anal. Chim. Acta 883 (2015) 74–80.*

bound to biotinylated antianalyte Ab with AlphaLISA acceptor beads bound to another antianalyte Ab in the presence of an analyte to form an SIMC and generate a chemiluminescent signal as the donor and acceptor beads are in proximity (Fig. 5.9). This striking development in immunodiagnostics has

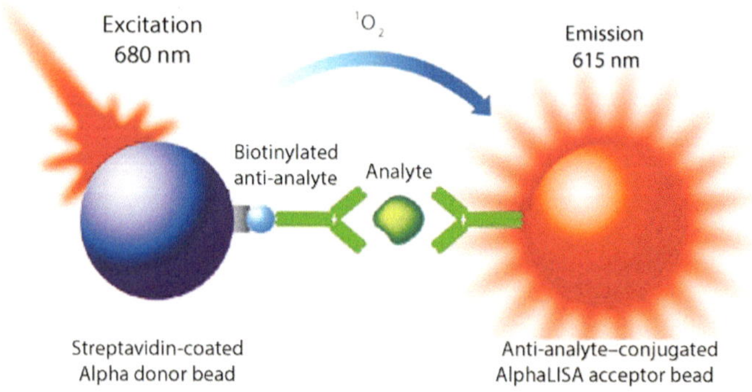

FIGURE 5.9 AlphaLISA, a wash-free IA, developed by Perkin Elmer, which involves the binding of acceptor beads conjugated to capture antibody against the analyte to biotinylated detection antibody-bound streptavidin-coated donor beads in the presence of an analyte to form a sandwich immune complex. Excitation of donor beads at 680 nm in the presence of two bound beads in proximity generates singlet oxygen molecules that trigger a series of chemical reactions in the acceptor beads, resulting in a sharp emission peak at 615 nm. *Reproduced with permission from the Nature Publishing Group. L. Beaudet, et al., AlphaLISA immunoassays: the no-wash alternative to ELISAs for research and drug discovery, Nat. Methods 5 (12) (2008) an8–an9.*

critically reduced the TAT and complexity, including its distinct wash-free feature. Moreover, it exhibits high sensitivity, a broad dynamic range, and superior analytical performance in comparison with the conventional ELISA.

Another prospective wash-free multiplex IA is the electrochemiluminescent ELISA developed by Meso Scale Diagnostics LLC [30]. This highly sensitive wash-free sandwich IA enables the detection of analytes in complex sample matrices on carbon electrode surface-based microwell plates (Fig. 5.10). The IA employs SULFO-TAG-labeled detection Ab, which emits light upon electrochemical stimulation. It enables high-throughput detection of multiple analytes in a microwell.

Of note is an exciting wash-free IA using chemiluminescence resonance energy transfer (CRET) with graphene as an energy acceptor and luminol as a donor [31] (Fig. 5.11). The detection of CRP in human serum using an IA procedure involves the formation of an SIMC of CRP with anti-CRP Ab–bound covalently to graphene nanosheets and HRP-labeled anti-CRP detection Ab, followed by the reaction with luminol and H_2O_2. The CRET-based IA has the same sensitivity and high precision as the conventional sandwich ELISA.

3.4 Multiplex Immunoassays

The xMAP technology by Luminex Corp., USA [14], widely used for the development of multiplex immunoassays [32–34], detects many analytes in an

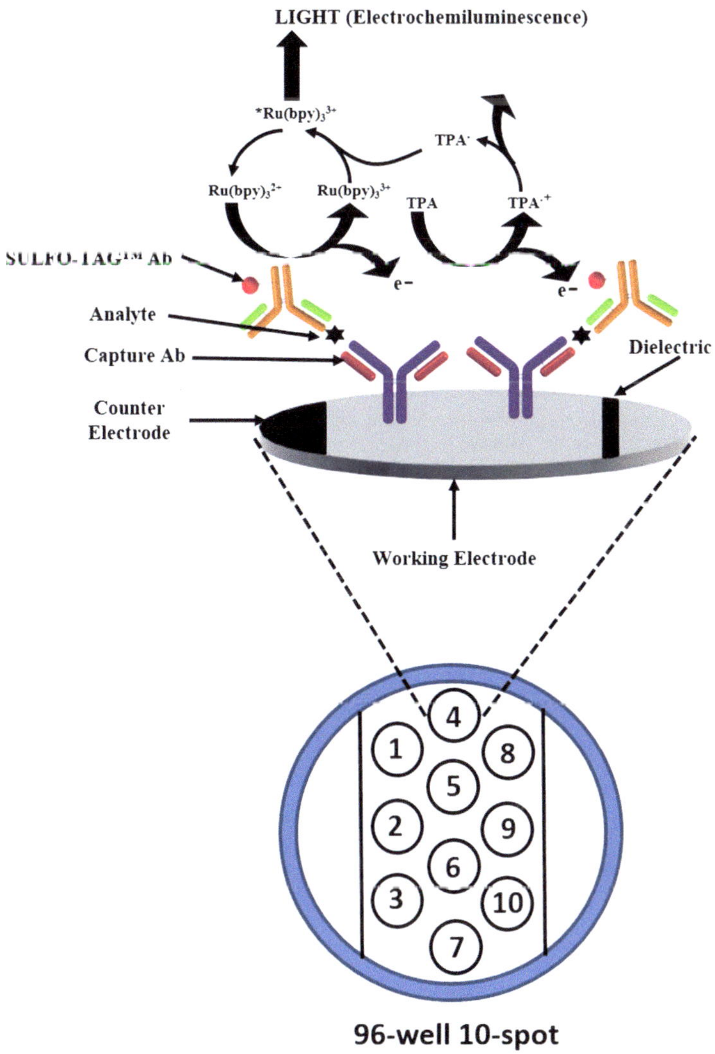

FIGURE 5.10 Electrochemiluminescent enzyme-linked immunosorbent assay–based wash-free sandwich IA for the highly sensitive detection of analytes in complex sample matrices. Upon electrochemical stimulation, light is emitted from the carbon electrode surface based microwell plates with specific SULFO-TAG labels on the detection antibody.

MTP well in one go using the 500 distinctly colored bead sets developed by the company. The immunoassay involves the capture of an analyte in the sample by capture Ab-bound, color-coded, micro-sized polystyrene beads known as microspheres, followed by the binding of biotinylated detection Ab to SA-labeled fluorescent dye (Fig. 5.12). The analyzer, containing multiple

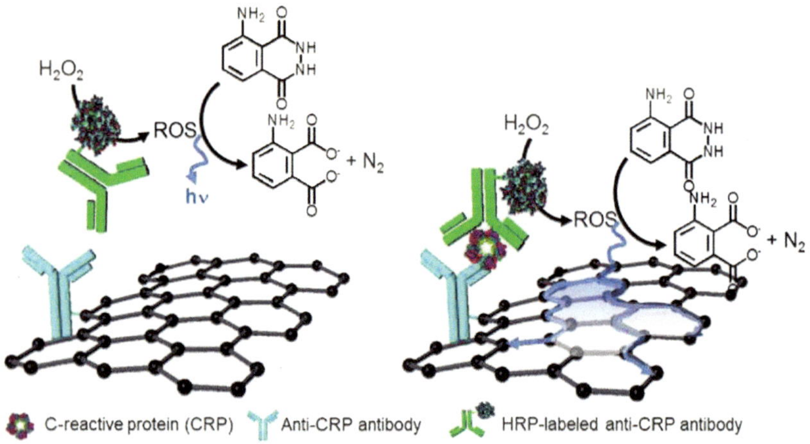

FIGURE 5.11 Graphene-based chemiluminescence resonance energy transfer platform—based wash-free IA for the detection of C-reactive protein (CRP). Anti-CRP Ab-conjugated graphene and luminol that are excited by HRP-catalyzed oxidation are used as an acceptor and a donor, respectively. *Reproduced with permission from the American Chemical Society. J.S. Lee, et al., Graphene-based chemiluminescence resonance energy transfer for homogeneous immunoassay, ACS Nano 6 (4) (2012) 2978–2983.*

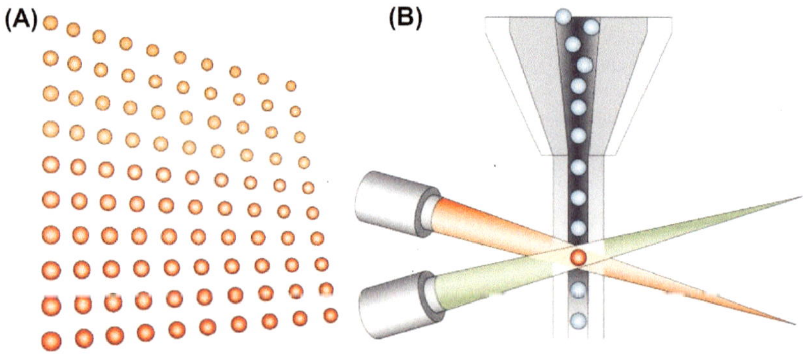

FIGURE 5.12 Luminex *x*MAP multiplex immunoassay technology. (A) Polystyrene beads are internally colored with two different fluorescent dyes: red and infrared with up to 100 distinct bead regions generated by using different concentrations of red and infrared dyes. Each bead region, bound to a different capture antibody (Ab), detects its particular analyte, followed by the subsequent binding of a biotinylated detection Ab and streptavidin-conjugated phycoerythrin (reporter dye). (B) The beads are identified individually in a rapidly flowing fluid stream that passes by two laser beams: red classification laser (635 nm) or light-emitting diode (LED) reveals the color code of the bead region, and green reporter laser (532 nm) or LED determines the analyte concentration by measuring the reporter fluorescence intensity. *Reproduced with permission from the Nature Publishing Group. K. Braeckmans, et al., Encoding microcarriers: present and future technologies, Nat. Rev. Drug Discov. 1 (6) (2002) 447–456.*

lasers or light-emitting diodes (LEDs) and high-speed digital-signal processors, reads the MTP and provides the results of the multiplex IA in the wells based on the analysis of the fluorescent signals from each individual microsphere particle. The readout involves the excitation of the microsphere's internal dyes by the laser or LED that identifies the specific microsphere set. A second laser or LED is then used for exciting the fluorescent dye on the detection Ab. High-speed digital signal processors are used for the identification of each individual microsphere and providing quantified results for detected analytes in the IA based on the readout of the fluorescent signals for the respective analytes. The company also offers an option of employing magnetic beads to facilitate the rapid separation using a magnetic separator. Thus, the technology is useful for the development of automated analyzer-based multiplex IA. Luminex Corporation has developed two flow cytometry-based analyzers, i.e., Luminex 100/200, with integrated fluidics, optics, lasers, and high-speed digital signal processors. The analyzers enable multiplexing of up to 100 analytes in a single MTP well. The xMAP-based multiplex IA offers reduced TAT, costs, and labor along with high throughput, high accuracy, and reproducibility.

Another prospective multiplex IA is the electrochemiluminescent ELISA developed by Meso Scale Diagnostics LLC [30], as described in the previous section with the analytical performance comparable to that developed by Luminex [13,35,36].

3.5 Nano/Micromaterial-Based Enzyme-Linked Immunoassay

The nano/micromaterial based EIA is the most preferred IA format for fully automated clinical analyzer—based EIA [12,37]. To date, popular magnetic beads [38,39] play an important role toward the commercialization of analyzer-based EIA by Roche, Siemens, Abbott, Immunodiagnostics Systems, DiaSorin, etc. for immunoassays [40,41]. In brief, magnetic beads enable the realization of various IA formats as they could be readily functionalized and conjugated to biomolecules. Moreover, they facilitate magnetically assisted washing of beads and various required microfluidic operations in an automated IA [42].

Of notice is a graphene nanoplatelet (GNP)-based IA using signal enhancement provided by GNP functionalization due to their significantly increased surface area to achieve high Ab immobilization density [43]. The procedure involves the surface modification of KOH-treated polystyrene MTP by incubating with a mixture of GNP and APTES. The leach-proof covalent binding of EDC-activated capture antihuman CRP Ab to APTES-functionalized GNP leads to an Ab-bound GNPs-functionalized surface (Fig. 5.13). The one-step kinetics-based IA is performed where the colorimetric signal is detected using an SP-based colorimetric reader (SBCR). The developed sandwich EIA specifically detects CRP in biological samples

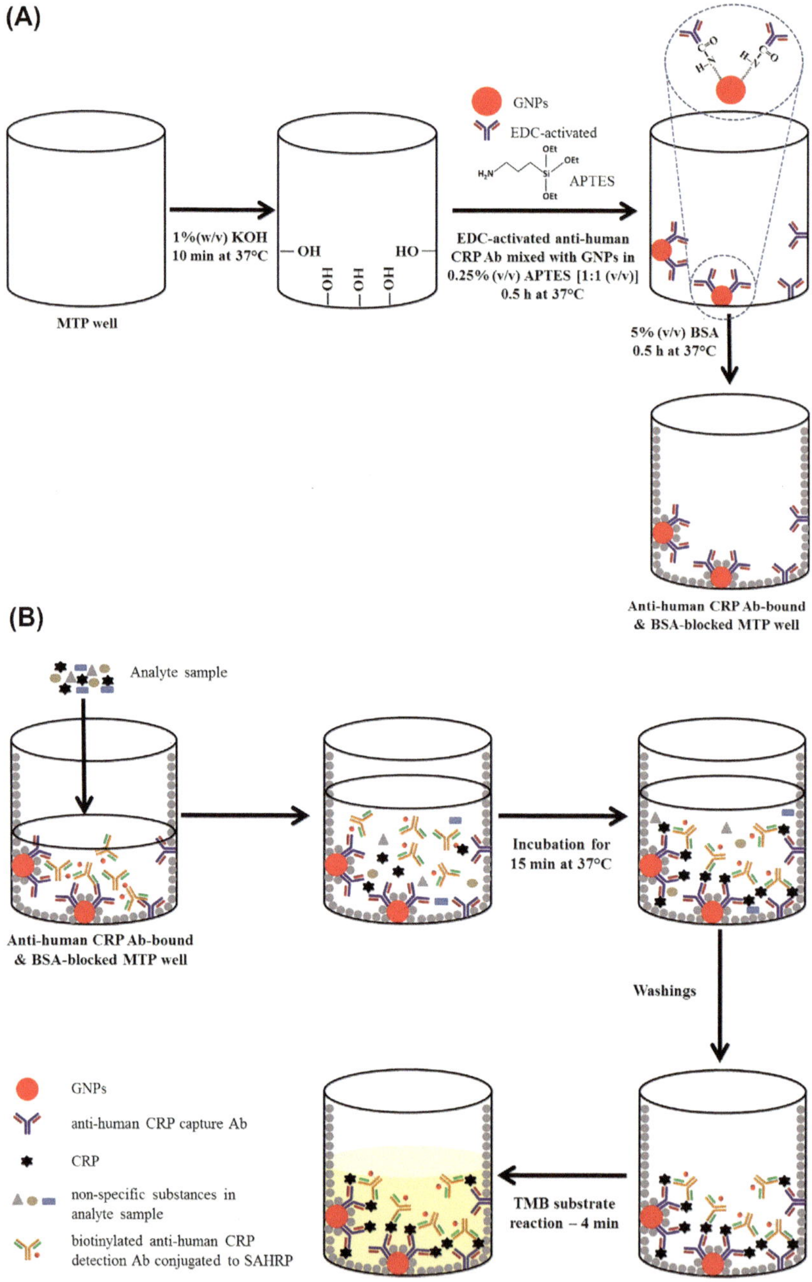

FIGURE 5.13 Graphene-based rapid and highly simplified sandwich enzyme-linked immunoassay (EIA) based on (A) modifying the surface of a KOH-treated polystyrene microtiter plate with graphene and 3-aminopropyltriethoxysilane, and (B) employing a one-step kinetics-based rapid sandwich EIA. *Reproduced with permission from Elsevier B.V. S.K. Vashist, et al., Graphene-based rapid and highly-sensitive immunoassay for C-reactive protein using a smartphone-based colorimetric reader, Biosens. Bioelectron. 66 (2015) 169–176.*

without incurring any interference with other endogenous biomarkers present in the clinical samples. The Ab-bound MTPs stored at 4°C in phosphate buffered saline (PBS) retains their functional activity for up to 6 weeks.

A fluorescence-based sandwich IA called ArcDia TPX is based on monoclonal capture Ab-bound polystyrene microspheres (3 μm in diameter) to form an SIMC with the analyte and secondary antianalyte $F(ab')_2$ fragment bound to fluorescent nanoparticles (NPs) [44]. The fluorescent NPs are excited by two-photon excitation, resulting in the generation of fluorescence only in a diffraction-limited focal volume of laser illumination. The confinement of fluorescent NP reporters on the surface of microspheres leads to an enhanced fluorescence signal.

3.6 Paper-Based Enzyme-Linked Immunoassay

Paper-based EIAs have been envisioned as an ideal low-cost diagnostic platform for the developing countries with resource-limited healthcare and diagnostic settings [19,20,45]. Most paper-based assays are based on LFA, dipstick, and microfluidic paper—based analytical devices (μPADs) [46,47]. Among various paper-based EIAs developed during the last two decades, filter paper, chromatography paper, nitrocellulose membrane, and paper/polymer or paper/nanomaterial composites are the most widely used paper substrates [19]. In brief, porosity, surface chemistry, and optical properties are critical for paper-based EIA [48]. Paper can be patterned to fabricate two-dimensional (2D) or three-dimensional (3D) μPADs [20,49]. Three-dimensional μPADs, formed by stacking layers of the 2D patterned paper, are prospective for multiplexing [50]. The sensitivity of conventional paper-based assays can be improved by signal enhancement strategies based on the use of enzymes [51] and nanomaterials [52]. The readout of colorimetric paper-based assays can be done via bare eyes by comparing the color on the paper against a predetermined score chart [53]. However, cameras, scanners, commercial test strip readers, and handheld colorimeters have also been used for the precise measurement of the color intensity of paper-based assay. The most recent development is the highly precise colorimetric readout via SP-based image analysis [7] as shown by a compact and lightweight integrated rapid diagnostic test reader [6]. The current efforts are inclined toward multiplex analyte detection [54,55], fully integrated POC paper-based diagnostic platforms with sample pretreatment and separation [56], and paper-based biosensing [57].

Of importance is the paper-based ELISA using 96-microzone paper plate, fabricated by patterning hydrophobic polymer in the hydrophilic paper [58] for the detection of rabbit immunoglobulin G (IgG) and the quantification of Ab against HIV-1 envelope protein gp41 in human serum. Rabbit IgG is bound to the paper, blocked with blocking buffer, and provided with goat antirabbit IgG labeled with alkaline phosphatase. The enzyme substrate specific for alkaline phosphatase (i.e., 5-bromo-4-chloro-3-indolyl phosphate and nitro blue tetrazolium) is then provided. The resulting solution has a dark purple color in the presence of rabbit IgG. The ELISA for the quantification of Ab against gp41 involves sequentially the binding of HIV-1 antigen (gp41) to the paper test zone, blocking buffer treatment, binding of human anti-HIV-1 to the HIV-1 antigen, binding of goat antihuman IgG to human anti-HIV-1, and the enzyme–substrate reaction (as mentioned for the detection of rabbit IgG). The developed paper-based ELISAs are cost-effective and faster but less sensitive than the conventional ELISA. They require only 3 μL of sample, <20 μL of reagents, and can be performed in less than an hour. The colorimetric signal can be measured by a low-cost desktop scanner. The test zones on the paper plate are 5 mm in diameter, which can be washed with the washing buffer from the top by pressing the bottom of the paper against a blotting paper.

Another prospective paper-based chemiluminescent ELISA for tumor biomarkers is based on chitosan-functionalized μPADs, fabricated by wax-screen printing [59]. It detects alpha-fetoprotein, cancer antigen 125, and carcinoembryonic antigen with high sensitivity in the clinically relevant range using a sandwich ELISA procedure. Similarly, a paper-based sandwich ELISA detects *Escherichia coli* down to 10^5 cells/mL in 5 h [60] with colorimetric readout performed by an SP camera. A similar interesting approach is the paper-based sandwich ELISA for the rapid and cost-effective quantitative detection of T7 bacteriophages in the range of $100-10^9$ pfu/mL [61]. The procedure involves sequentially the physisorption of capture Ab on paper, blocking with BSA, rinsing with PBS buffer, T7 sample detection, treatment with HRP-labeled secondary Ab, and colorimetric enzyme-substrate reaction. The brown colored complex formed on the paper is washed and dried before the measurement of its color intensity via readout by scanner and image analysis.

3.7 Microfluidics-Based Enzyme-Linked Immunoassay

3.7.1 Optimiser Enzyme-Linked Immunosorbent Assay

Optimiser ELISA [62,63] is a novel microfluidic MTP-based EIA (Fig. 5.14) developed by Siloam Biosciences Inc. (USA) that requires minimal reagent volumes and critically reduced number of steps. It enables rapid and highly sensitive EIA with a sample-to-answer time of just a few minutes. The Optimiser MTP is fabricated from polystyrene using injection molding technology. It employs a standard 96-well MTP layout where each MTP well is connected to a spiral microfluidic channel via a through-hole at the bottom. The microchannels provide a similar surface area but an increased surface-area-to-volume ratio (about 50-fold higher) in comparison with conventional MTP. It uses only 4.5 μL of sample and reagents compared with the 100 μL used in conventional MTP-based EIA. The IA is performed using various flow steps, where the liquid dispensed into the MTP is drawn into the microchannel by capillary forces exerted by the absorbent pad at the other end of the

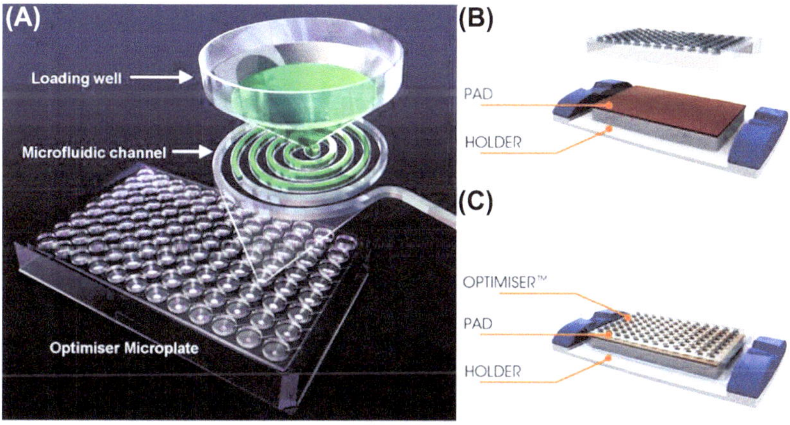

FIGURE 5.14 (A) Microfluidic enzyme-linked immunosorbent assay on Optimiser microplates developed by Siloam Biosciences Inc., USA. (B) and (C) Illustrate the various components of Optimiser microplate. *Reproduced with permission from the Royal Society of Chemistry. J. Kai, et al., A novel microfluidic microplate as the next generation assay platform for enzyme linked immunoassays (ELISA), Lab Chip 12 (21) (2012) 4257–4262.*

microchannel. However, after the liquid is drained out via the through-hole, the capillary forces on the ends of capillary channel keeps the liquid trapped in the microchannel. Subsequently, the addition of another liquid breaks the capillary barrier and restarts the capillary action of the absorbent pad. These sequential microfluidic steps enable the completion of the microfluidic Optimiser ELISA. The Optimiser ELISA procedure involves the sequential dispensing of capture Ab, blocking buffer, target analyte, fluorescently labeled detection Ab, and wash buffer. Finally, the determination of an analyte in the sample is done by fluorescence readout of Optimiser MTP in fluorescence plate reader. The Optimiser MTP–based IAs display the same or better sensitivity and the same high precision as the conventional ELISA. However, it employs critically reduced volumes of sample and reagents (i.e., 5 μL) and has a reduced TAT of about 2 h.

3.7.2 Centrifugal Microfluidics-Based Immunoassay

A prospective sandwich IA format has been developed by Abaxis Inc., based on a centrifugal microfluidics-based LabDisk (CML) platform and a POC Piccolo Xpress whole blood chemistry analyzer [64] (Fig. 5.15A). The analyzer can perform up to 14 different tests on a single reagents-integrated and barcoded LabDisk with a diameter of 8 cm, where the analyzer can simultaneously monitor multiple reactions. Abaxis provides several Clinical Laboratory Improvement Amendments–waived multianalyte tests and other tests for biomarkers, toxins, nucleic acid, pathogens, and other analytes [65]. Another widely used CML platform developed by Gyros Protein Technologies, Sweden, has been extensively utilized by the biopharmaceutical industry

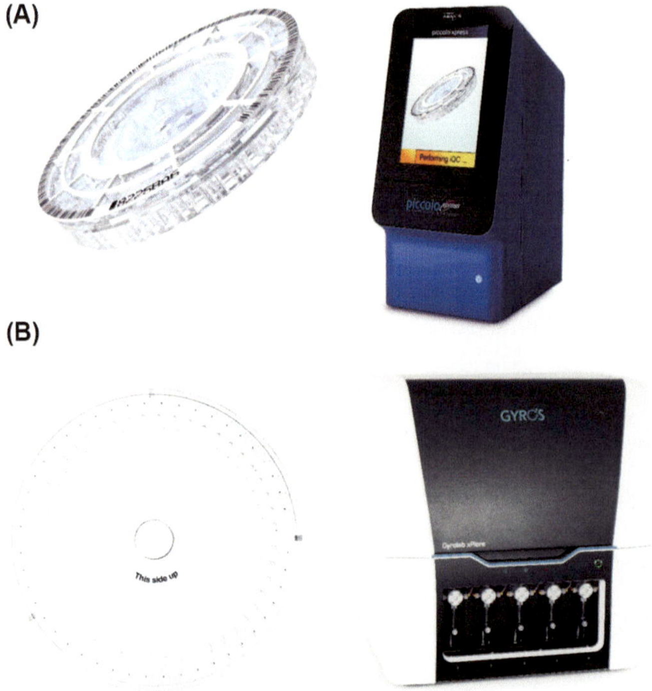

FIGURE 5.15 (A) (left) LabDisk, a centrifugal microfluidics platform, for the simultaneous detection of multiple analytes; (right) Piccolo Xpress chemistry analyzer for fully automated LabDisk-based IA. (B) (left) LabDisk centrifugal microfluidics platform for fully automated IA; (right) Gyrolab Xplore system enabling fully automated IA in a LabDisk. *(A) Reproduced with permission from Elsevier B.V. S.K. Vashist, et al., Emerging technologies for next-generation point-of-care testing, Trends Biotechnol. 33 (11) (2015) 692–705. (B) Reproduced with permission from Gyros Protein Technologies, Sweden.*

and clinical researchers. It employs a Gyrolab instrument (Gyrolab xPlore or Gyrolab xP workstation) [66] and centrifugal microfluidics-based Gyrolab Bioaffy CDs [67] (Fig. 5.15B). Gyrolab xPlore runs a single CD-based IA while Gyrolab xP workstation can run up to five disks simultaneously. Several IAs have been developed using a generic IA procedure that employs the binding of the SA-coated bead to a biotinylated capture Ab, which is followed by the binding of analyte and its detection using fluorescent-labeled detection Ab. The Gyrolab IA has a broader dynamic range and rapid TAT and requires low sample volume in comparison with conventional EIA.

Among several CML-based sandwich EIA formats, of notice is a rapid CML-based sandwich EIA with a sample-to-answer time of just 25 min [68] (Fig. 5.16). This setup employs capture antihuman CRP Ab-bound

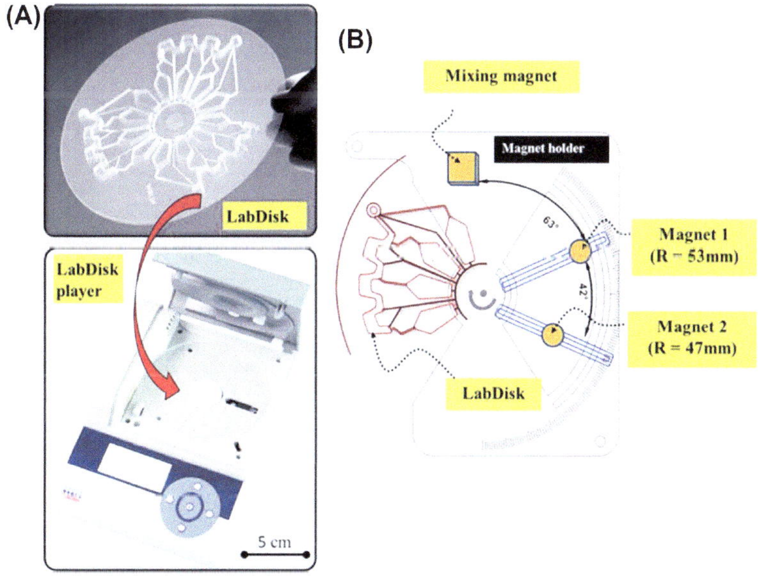

FIGURE 5.16 Centrifugal microfluidics-based enzyme-linked immunoassay (EIA). (A) The LabDisk including three separate EIA structures is mounted on the rotational axis of the LabDisk player. (B) A magnet holder is then placed above the LabDisk. The first magnetic stack (magnet 1) is used to transport the Dynabeads into the air gap toward the receiving chamber, while the second magnetic stack (magnet 2) is used for pulling the microparticles from the air gap into the receiving chamber liquid. A third magnet (mixing magnet) is attached to the holder that is used for homogenization of beads in the detection chamber before the detection of the chemiluminescent signal. *Reproduced with permission from The Royal Society of Chemistry. G. Czilwik, et al., Magnetic chemiluminescent immunoassay for human C-reactive protein on the centrifugal microfluidics platform, RSC Adv. 5 (76) (2015) 61906–61912.*

superparamagnetic Dynabeads (M-280 tosyl activated) to detect CRP in a sandwich EIA. BSA-bound dummy Dynabeads are employed to prevent the loss of capture Ab-bound Dynabeads when they are transferred between different reaction chambers on the CML. The SIMC in the immunoreaction chamber is taken to the subsequent wash chambers for removing the nonspecifically bound substances. The SIMC formed on Dynabeads is then transported to the detection chamber, where the enzyme–substrate reaction leads to the generation of a chemiluminescent signal that is read by the portable miniaturized detector integrated into the LabDisk reader. Considering its comparable sensitivity and automated operational procedure in comparison to the manual one-step kinetics-based magnetic CIA and sandwich ELISA [69], this EIA format could be used for prospective POC applications in decentralized and remote settings.

3.8 Smartphone-Based Enzyme-Linked Immunoassay

A wide range of SP-based EIAs have been developed during the recent years based on the SP readout of colorimetric, chemiluminescent, or fluorescence signals [7,8,21]. An interesting approach is the use of an SP-based colorimetric reader [9], where a 96-well MTP is simply placed on top of a gadget's screen saver and imaged by SP's back camera employing a setup comprising a dark hood and a base holder (Fig. 5.17). The gadget's screen saver illuminates the 96-well MTP from the bottom by white light, and an image processing algorithm determines the pixel intensity of the SP-captured images. The one-step kinetics-based EIA for CRP performed using the developed SBCR exhibits comparable sensitivity, precision, and analytical performance similar to that of central laboratory's microplate reader–based EIA [9]. The same group developed the SBCR-based rapid HFA IA [9] with high sensitivity, precision, and superior analytical performance [70]. Apparently, SBCR is a highly cost-effective device for the readout of EIA, which is bulky and expensive and requires a continuous power supply.

An SP-based IA has been developed for the detection of *E. coli* based on the use of a quantum dot (QD)-based sandwich IA and anti-*E. coli* O157:H7 Ab-functionalized capillary array [11]. The SP attachment is comprised of battery-powered ultraviolet LEDs butt-coupled to the capillary array that enables uniform excitation of QD-labeled *E. coli* particles and an additional lens with a focal length of 15 mm for SP-based fluorescent imaging.

An SP-based fluorimeter [71] measures the full emission spectrum of a light emitter using a cradle. The transmission diffraction grating integrated with optics collimates the light emitted from the sample onto the grating,

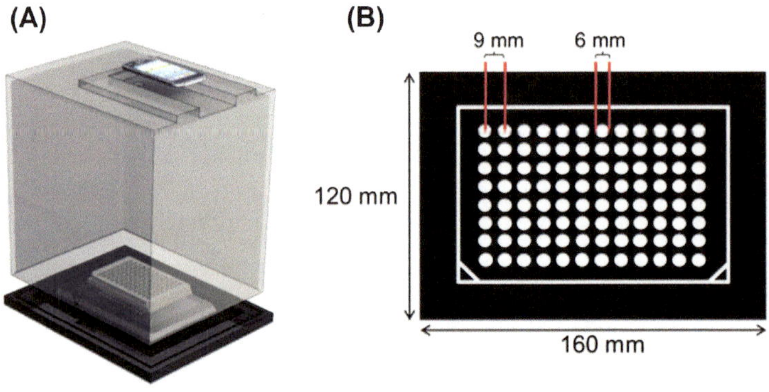

FIGURE 5.17 (A) Smartphone-based colorimetric reader developed using a Samsung Galaxy SIII mini, a gadget (iPad mini, iPad4 or iPhone 5s), and a custom-made dark hood and base holder assembly. (B) Screen saver used for the screen-based bottom illumination of the 96-well microtiter plate in iPad Mini. *Reproduced with permission from Elsevier B.V. S.K. Vashist, et al., A smartphone-based colorimetric reader for bioanalytical applications using the screen-based bottom illumination provided by gadgets, Biosens. Bioelectron. 67 (2015) 248–255.*

where the emission spectrum of the light is measured by the SP's rear camera. The same group developed the spectrometer using the SP's camera that is employed for the measurement of ELISA [72]. The procedure involves the alignment of a diffraction grating and a collimating lens between a light source and the CP camera using a custom cradle.

In another approach, a low-cost SP dongle has been developed and used for a triplex IA that detects various infectious diseases (i.e., HIV, syphilis, and active syphilis infection) simultaneously in just 15 min at POC settings [73]. The procedure involves the interfacing of the device to the SP via an audio jack connector and the use of frequency shift keying data transmission. All the reagents are prestored in a reagent cassette while the test cassette contains five detection zones.

4. PORTABLE ANALYZER–BASED IMMUNOASSAYS

Various portable analyzers, enabling EIA at POC settings, have been devel oped during the last decade. There is a continuously increasing trend to detect the critical analytes at the POC settings so that the medical intervention, treatment, and management of the disease can be started at a very early stage. This could lead to a significant decrease in healthcare costs apart from preventing the spread of infectious diseases. The most prominent portable analyzers are developed by Eurolyser Diagnostica, Samsung, Orion Diagnostica, Alere, Diasorin, and other companies.

CUBE (Fig. 5.18A) and SMART analyzers from Eurolyser Diagnostica [74] detect several clinically relevant analytes in a few minutes using only 5–20 μL of serum, plasma, or whole blood. The procedure involves the collection of a finger-pricked blood sample by the capillary-integrated dipping attachment that is bound to a sample cartridge's cap. The cartridge is then inserted into the sample chamber at the front side, and the measurement is done after closing the door. These analyzers are equipped with mobile healthcare tools for point-of-care testing (POCT) to perform various IAs such as for CRP, hsCRP, cystatin C, and troponin I.

Samsung LABGEO IB10 [75] is the handheld POC analyzer from Samsung, which enables fully automated IA for the rapid detection of analytes in just 20 min. There is no sample pretreatment for the separation of plasma and up to three analytes can be detected in a single run. The device employs a CML platform with all the reagents prestored in the LabDisk chambers, which are stable at room temperature for a month. The developed IA demonstrates high precision and correlation with clinically accredited central laboratory analyzer–based IAs. Other applications of IAs include troponin I, myoglobin, CK-MB, thyroid-stimulating hormone, procalcitonin, etc. The analyzer has advanced features and is equipped with smart mobile healthcare information technology tools.

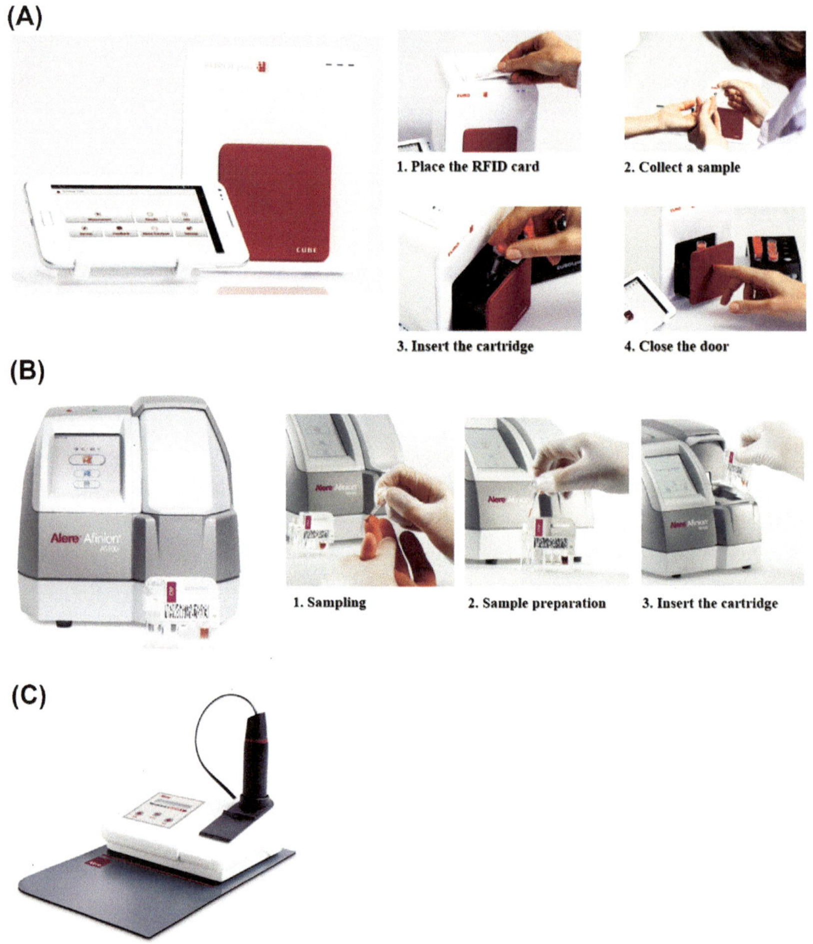

FIGURE 5.18 Portable analyzer-based immunoassay. (A) The Eurolyser CUBE POC system. (B) The Alere Afinion AS100 analyzer. (C) NycoCard. *Reproduced with permission from Elsevier B.V. S.K. Vashist, et al., Bioanalytical advances in assays for C-reactive protein, Biotechnol. Adv. 34 (3) (2016) 272–290.*

Alere Afinion AS100 analyzer [76] is another portable analyzer with an integrated sampling device, a cartridge having all the required reagents, and an integrated camera serving as a barcode reader (Fig. 5.18B). The reaction area is illuminated by LEDs and the reflected or transmitted light is detected by the integrated camera. QuikRead go [77] is another portable instrument from Orion Diagnostica, Finland, used for the rapid POC testing of various analytes in just a few minutes using ~20 μL of sample and a turbidimetric IA using NP-coated Ab. The instrument provides automatic mixing and hematocrit correction while the RBCs in the sample are hemolyzed before the measurement. Similarly, NycoCard reader II is a portable battery-charged reader with an instrument box and a reader pen (540 g) [78,79] (Fig. 5.18C). It is widely used for the determination of CRP concentration in just 2 min using only 5 μL of sample. The readout procedure involves the illumination of NycoCard with LEDs and measuring the light reflected by the sample using a photodiode.

5. CRITIQUES AND OUTLOOK

The manual ELISAs, especially the sandwich and competitive, are the most commonly used EIAs but require highly skilled analysts, expensive instruments, lengthy TAT, and a complex and laborious procedure. Therefore, they are inadequate for high-throughput settings such as central laboratories in hospitals that require fully automated EIAs for clinical diagnosis of numerous patient samples. This has led to the emergence of a broad range of clinical analyzers by the IVD companies for varying throughput needs. Consequently, various analyzer-based EIA formats have been developed for the routine clinical diagnosis of numerous diseases. Moreover, the continuous developments in the field are rapidly increasing the list of analytes and disease biomarkers, which can be measured by analyzer-based EIAs. The last two decades have also witnessed a growing clinical need for POCT in healthcare, which has further led to the emergence of highly simplified POC analyzers and rapid EIA formats with a TAT of less than an hour. These developments have resulted in critically improved and sustainable healthcare. However, such expensive instruments are more affordable to the healthcare settings in developed countries, compared with their counterparts in developing nations and remote settings.

The continuous developments in EIA formats have led to the emergence of novel IA formats, which have potential bioanalytical applications for varying needs of end-users. Various high-sensitivity EIAs have been developed for applications requiring very high sensitivity while rapid and highly simplified EIAs are appropriate for applications that require rapid healthcare intervention. In particular, the micro- and nanoparticles-based EIAs have considerably increased the sensitivity of IA, enabling the realization of various automated EIAs. The most prominent are the magnetic bead–based EIA, which are being used by most IVD companies for the development of analyzer-based IAs. Furthermore, there is an immense need for multiplex detection of biomarkers for a particular disease, which would facilitate the clinical decision-making and treatment of diseases at an early stage. The Luminex xMAP technology is widely used and appropriate for the multiplex detection of biomarkers for the diagnosis of various diseases.

Of importance are wash-free IAs, which have considerably reduced the number of steps in an IA, an impetus for the realization of highly simplified IA. The AlphaLISA by Perkin Elmer is the most prospective advancement, which might be used for the development of automated analyzer-based IA. Similarly, there have been significant advances during the last three decades in the field of microfluidics [81,82] and its use for various immunodiagnostic applications [83]. The Optimizer ELISA is an innovative EIA that enables rapid and highly sensitive detection of analytes. Similarly, CMF-based IAs and analyzers have been developed by companies such as Abaxis Inc., Gyros Protein Technologies, and Samsung. Although the improved EIA formats have the possibility of being used in the conventional clinical analyzers from leading IVD giants, the novel microfluidic EIA formats would follow a different route. However, it might be a limitation for the existing healthcare facilities to employ such novel technologies.

Many prospective portable analyzer−based EIAs have been demonstrated during the recent years by several companies such as Samsung, Alere, Eurolyser Diagnostica, Diasorin, and Orion Diagnostica. They fill the existing market gap by providing the desired clinical diagnostic technology for the physician office laboratories (POLs). They are based on innovative immunodiagnostic platforms, novel sample preparation and reagent storage strategies, lesser sample and reagents volumes, and miniaturized readout devices. Most portable clinical analyzer-based EIAs employ a plug-and-play mechanism where the sample and reagents cartridge is plugged into the analyzer and the EIA is completed in a few minutes. There would be significant developments in portable analyzers in the coming years as they facilitate the rapid clinical diagnosis at POLs, thereby enabling the physicians to make rapid decisions.

The advent of SP-based mobile healthcare has completely changed the landscape of healthcare monitoring and management and led to new concepts for healthcare delivery. The last few years have witnessed an exponentially growing trend toward SP-based EIAs with proven superior analytical performance, high precision, and cost-effectiveness in comparison with conventional EIAs being used in the central laboratories. Such SP-based devices have been developed for the detection of colorimetric, fluorescence, chemiluminescence, surface plasmon resonance, and electrochemical signals [21]. Similarly, several prospective LOC IA formats have been demonstrated. The ability of the SP to be used at any time and place without any expensive instrumentation or skilled analysts is providing an impetus to the continuous development of SP-based EIAs for POCT. The SP also provides spatiotemporal-tagged data and advanced communication features, which are a boon to healthcare as it enables rapid response to emergencies and increases healthcare outreach.

There would be significant advances and continuous evolution in EIA formats, platforms, and devices in the coming years based on the changing landscape of healthcare monitoring and management, and the upcoming of complementary technologies and novel EIA strategies. Such advanced EIAs would offer cost-effective, precise, highly sensitive, rapid, less labor-intensive, and robust EIAs, which would be useful for clinical, bioanalytical, and POC settings.

6. CONCLUSIONS

A wide range of EIA formats have been developed for applications in healthcare, industrial, POC, resource-limited, and other bioanalytical settings. Although ELISA has been the most used EIA format for the development of IVD kits, the current trend strongly deviates toward automated analyzer-based EIA and fully integrated LOC-based and portable readout–based POC EIA. There have been tremendous advances in LOC formats, microfluidic technologies, rapid EIA strategies, prolonged storage concepts, and portable readers. Furthermore, the complementary technologies, such as mobile healthcare, LOC platforms, multiplex detection concepts, novel wash-free detection strategies, microfabrication techniques, cloud computing, and clinical scoring mechanisms, have witnessed significant improvements. These striking advances and the trend toward POC EIAs equipped with personalized mobile healthcare tools will eventually pave the way toward the next-generation of low-cost and analytically superior EIAs.

REFERENCES

[1] A.E. Herr, et al., Microfluidic immunoassays as rapid saliva-based clinical diagnostics, Proc. Natl. Acad. Sci. 104 (13) (2007) 5268–5273.

[2] C.-C. Lin, et al., Microfluidic immunoassays, JALA Charlottesv Va. 15 (3) (2010) 253–274.

[3] A.F. Coskun, et al., A personalized food allergen testing platform on a cellphone, Lab Chip 13 (4) (2013) 636–640.

[4] P.B. Lillehoj, et al., Rapid electrochemical detection on a mobile phone, Lab Chip 13 (15) (2013) 2950–2955.

[5] S.K. Ludwig, et al., Cellphone-based detection platform for rbST biomarker analysis in milk extracts using a microsphere fluorescence immunoassay, Anal. Bioanal. Chem. (2014) 1–10.

[6] O. Mudanyali, et al., Integrated rapid-diagnostic-test reader platform on a cellphone, Lab Chip 12 (15) (2012) 2678–2686.

[7] S.K. Vashist, et al., Cellphone-based devices for bioanalytical sciences, Anal. Bioanal. Chem. 406 (14) (2014) 3263–3277.

[8] S.K. Vashist, E.M. Schneider, J.H.T. Luong, Commercial smartphone-based devices and smart applications for personalized healthcare monitoring and management, Diagnostics 4 (3) (2014) 104–128.

[9] S.K. Vashist, et al., A smartphone-based colorimetric reader for bioanalytical applications using the screen-based bottom illumination provided by gadgets, Biosens. Bioelectron. 67 (2015) 248–255.

[10] D.J. You, T.S. Park, J.-Y. Yoon, Cell-phone-based measurement of TSH using Mie scatter optimized lateral flow assays, Biosens. Bioelectron. 40 (1) (2013) 180–185.

[11] H. Zhu, U. Sikora, A. Ozcan, Quantum dot enabled detection of *Escherichia coli* using a cell-phone, Analyst 137 (11) (2012) 2541–2544.

[12] W. Hu, C.M. Li, Nanomaterial-based advanced immunoassays, Wiley Interdiscip. Rev. Nanomed. Nanobiotechnol. 3 (2) (2011) 119–133.

[13] F. Chowdhury, A. Williams, P. Johnson, Validation and comparison of two multiplex technologies, Luminex® and Mesoscale Discovery, for human cytokine profiling, J. Immunol. Methods 340 (1) (2009) 55–64.

[14] S.A. Dunbar, Applications of Luminex® xMAP™ technology for rapid, high-throughput multiplexed nucleic acid detection, Clin. Chim. Acta 363 (1) (2006) 71–82.

[15] A. Bange, H.B. Halsall, W.R. Heineman, Microfluidic immunosensor systems, Biosens. Bioelectron. 20 (12) (2005) 2488–2503.

[16] S. Haeberle, R. Zengerle, Microfluidic platforms for lab-on-a-chip applications, Lab Chip 7 (9) (2007) 1094–1110.

[17] E. Livak-Dahl, I. Sinn, M. Burns, Microfluidic chemical analysis systems, Annu. Rev. Chem. Biomol. Eng. 2 (2011) 325–353.

[18] D. Mark, et al., Microfluidic lab-on-a-chip platforms: requirements, characteristics and applications, Chem. Soc. Rev. 39 (3) (2010) 1153–1182.

[19] J. Hu, et al., Advances in paper-based point-of-care diagnostics, Biosens. Bioelectron. 54 (2014) 585–597.

[20] A.W. Martinez, et al., Diagnostics for the developing world: microfluidic paper-based analytical devices, Anal. Chem. 82 (1) (2009) 3–10.

[21] S.K. Vashist, et al., Emerging technologies for next-generation point-of-care testing, Trends Biotechnol. 33 (11) (2015) 692–705.

[22] C.K. Dixit, et al., Development of a high sensitivity rapid sandwich ELISA procedure and its comparison with the conventional approach, Anal. Chem. 82 (16) (2010) 7049–7052.

[23] C.K. Dixit, et al., Multisubstrate-compatible ELISA procedures for rapid and high-sensitivity immunoassays, Nat. Protoc. 6 (4) (2011) 439–445.

[24] S.K. Vashist, A sub-picogram sensitive rapid chemiluminescent immunoassay for the detection of human fetuin A, Biosens. Bioelectron. 40 (1) (2013) 297–302.

[25] S.K. Vashist, et al., One-step antibody immobilization-based rapid and highly-sensitive sandwich ELISA procedure for potential in vitro diagnostics, Sci. Rep. 4 (2014) 4407.

[26] S.K. Vashist, E.M. Schneider, J.H.T. Luong, Rapid sandwich ELISA-based in vitro diagnostic procedure for the highly-sensitive detection of human fetuin A, Biosens. Bioelectron. 67 (2015) 73–78.

[27] S.C. Shiesh, et al., Determination of C-reactive protein with an ultra-sensitivity immunochemiluminometric assay, J. Immunol. Methods 311 (1–2) (2006) 87–95.

[28] S.K. Vashist, E.M. Schneider, J.H.T. Luong, A rapid sandwich immunoassay for human fetuin A using agarose-3-aminopropyltriethoxysilane modified microtiter plate, Anal. Chim. Acta 883 (2015) 74–80.

[29] L. Beaudet, et al., AlphaLISA immunoassays: the no-wash alternative to ELISAs for research and drug discovery, Nat. Methods 5 (12) (2008) an8–an9.

[30] MSD Technology Platform, Available from: https://www.mesoscale.com/~/media/files/brochures/techbrochure.pdf, 2017.

[31] J.S. Lee, et al., Graphene-based chemiluminescence resonance energy transfer for homogeneous immunoassay, ACS Nano 6 (4) (2012) 2978–2983.

[32] K. Skogstrand, et al., Simultaneous measurement of 25 inflammatory markers and neurotrophins in neonatal dried blood spots by immunoassay with xMAP technology, Clin. Chem. 51 (10) (2005) 1854–1866.

[33] K. Kofoed, et al., Development and validation of a multiplex add-on assay for sepsis biomarkers using xMAP technology, Clin. Chem. 52 (7) (2006) 1284–1293.

[34] K. Braeckmans, et al., Encoding microcarriers: present and future technologies, Nat. Rev. Drug Discov. 1 (6) (2002) 447–456.

[35] Q. Fu, J. Zhu, J.E. Van Eyk, Comparison of multiplex immunoassay platforms, Clin. Chem. 56 (2) (2010) 314–318.
[36] E.C. Breen, et al., Multisite comparison of high-sensitivity multiplex cytokine assays, Clin. Vaccine Immunol. 18 (8) (2011) 1229–1242.
[37] X. Pei, et al., Sandwich-type immunosensors and immunoassays exploiting nanostructure labels: a review, Anal. Chim. Acta 758 (2013) 1–18.
[38] E. Paleček, M. Fojta, Magnetic beads as versatile tools for electrochemical DNA and protein biosensing, Talanta 74 (3) (2007) 276–290.
[39] H. Kuramitz, Magnetic microbead-based electrochemical immunoassays, Anal. Bioanal. Chem. 394 (1) (2009) 61–69.
[40] C.T. Lim, Y. Zhang, Bead-based microfluidic immunoassays: the next generation, Biosens. Bioelectron. 22 (7) (2007) 1197–1204.
[41] B.I. Haukanes, C. Kvam, Application of magnetic beads in bioassays, Nat. Biotechnol. 11 (1) (1993) 60–63.
[42] M.A. Gijs, Magnetic bead handling on-chip: new opportunities for analytical applications, Microfluid. Nanofluid. 1 (1) (2004) 22–40.
[43] S.K. Vashist, et al., Graphene-based rapid and highly-sensitive immunoassay for C-reactive protein using a smartphone-based colorimetric reader, Biosens. Bioelectron. 66 (2015) 169–176.
[44] J.O. Koskinen, et al., Fluorescent nanoparticles as labels for immunometric assay of C-reactive protein using two-photon excitation assay technology, Anal. Biochem. 328 (2) (2004) 210–218.
[45] M. Santhiago, et al., Microfluidic paper-based devices for bioanalytical applications, Bioanalysis 6 (1) (2014) 89–106.
[46] X. Mao, T.J. Huang, Microfluidic diagnostics for the developing world, Lab Chip 12 (8) (2012) 1412–1416.
[47] X. Li, D.R. Ballerini, W. Shen, A perspective on paper-based microfluidics: current status and future trends, Biomicrofluidics 6 (1) (2012) 11301–1130113.
[48] R. Pelton, Bioactive paper provides a low-cost platform for diagnostics, Trends Anal. Chem. 28 (8) (2009) 925–942.
[49] A. Nilghaz, et al., Flexible microfluidic cloth-based analytical devices using a low-cost wax patterning technique, Lab Chip 12 (1) (2012) 209–218.
[50] A.W. Martinez, S.T. Phillips, G.M. Whitesides, Three-dimensional microfluidic devices fabricated in layered paper and tape, Proc. Natl. Acad. Sci. 105 (50) (2008) 19606–19611.
[51] C. Parolo, A. de la Escosura-Muniz, A. Merkoci, Enhanced lateral flow immunoassay using gold nanoparticles loaded with enzymes, Biosens. Bioelectron. 40 (1) (2013) 412–416.
[52] J. Hu, et al., Oligonucleotide-linked gold nanoparticle aggregates for enhanced sensitivity in lateral flow assays, Lab Chip 13 (22) (2013) 4352–4357.
[53] M.A. Dineva, et al., Simultaneous visual detection of multiple viral amplicons by dipstick assay, J. Clin. Microbiol. 43 (8) (2005) 4015–4021.
[54] S.J. Vella, et al., Measuring markers of liver function using a micropatterned paper device designed for blood from a fingerstick, Anal. Chem. 84 (6) (2012) 2883–2891.
[55] N.R. Pollock, et al., A paper-based multiplexed transaminase test for low-cost, point-of-care liver function testing, Sci. Transl. Med. 4 (152) (2012), 152ra129.
[56] X. Yang, et al., Integrated separation of blood plasma from whole blood for microfluidic paper-based analytical devices, Lab Chip 12 (2) (2012) 274–280.
[57] A.P. Turner, Biosensors: sense and sensibility, Chem. Soc. Rev. 42 (8) (2013) 3184–3196.

[58] C.M. Cheng, et al., Paper-based ELISA, Angew. Chem. Int. Ed. Engl. 49 (28) (2010) 4771–4774.

[59] S. Wang, et al., Paper-based chemiluminescence ELISA: lab-on-paper based on chitosan modified paper device and wax-screen-printing, Biosens. Bioelectron. 31 (1) (2012) 212–218.

[60] C.M. Shih, et al., Paper-based ELISA to rapidly detect *Escherichia coli*, Talanta 145 (2015) 2–5.

[61] M.S. Khan, T. Pande, T.G. van de Ven, Qualitative and quantitative detection of T7 bacteriophages using paper based sandwich ELISA, Colloids Surf. B Biointerfaces 132 (2015) 264–270.

[62] J. Kai, et al., A novel microfluidic microplate as the next generation assay platform for enzyme linked immunoassays (ELISA), Lab Chip 12 (21) (2012) 4257–4262.

[63] C.H. Ahn, et al., A new game changer for immunoassays and IVD: microfluidics and Polymer Lab-on-a-Chips, Procedia Eng. 25 (2011) 651–656.

[64] Piccolo Xpress, Available from: http://www.abaxis.com/medical/piccolo–xpress, 2017.

[65] R. Gorkin, et al., Centrifugal microfluidics for biomedical applications, Lab Chip 10 (14) (2010) 1758–1773.

[66] Gyrolab xPlore, Available from: http://www.gyros.com/products/systems/gyrolab–xplore/, 2017.

[67] Gyrolab CDs, Available from: http://www.gyros.com/products/gyrolab–cds/, 2017.

[68] G. Czilwik, et al., Magnetic chemiluminescent immunoassay for human C-reactive protein on the centrifugal microfluidics platform, RSC Adv. 5 (76) (2015) 61906–61912.

[69] S.K. Vashist, et al., One-step kinetics-based immunoassay for the highly sensitive detection of C-reactive protein in less than 30 min, Anal. Biochem. 456 (2014) 32–37.

[70] A.G. Venkatesh, et al., Smartphone-based rapid and highly sensitive immunoassay for human fetuin A, in: Proceedings of MicroSystemTechnik Kongress, VDE Verlag, 2015, pp. 47–49. ISBN:978-3-8007-4100-7. https://www.vde-verlag.de/proceedings-en/454100006.html.

[71] H. Yu, Y. Tan, B.T. Cunningham, Smartphone fluorescence spectroscopy, Anal. Chem. 86 (17) (2014) 8805–8813.

[72] K.D. Long, H. Yu, B.T. Cunningham, Smartphone instrument for portable enzyme-linked immunosorbent assays, Biomed. Opt. Express 5 (11) (2014) 3792–3806.

[73] T. Laksanasopin, et al., A smartphone dongle for diagnosis of infectious diseases at the point of care, Sci. Transl. Med. 7 (273) (2015) 273re1.

[74] Eurolyser, Eurolyser Smart CRP Test With Integrated Capillary, 2015. Available from: http://www.eurolyser.com/de/medizinische-diagnostik/parameter/crp-test/.

[75] Samsung LABGEO IB10, Available from: http://www.samsung.com/global/business/healthcare/healthcare/in–vitro–diagnostics/BCA–IB10/DE, 2017.

[76] Alere, Alere Afinion™ AS100 Analyzer, 2015. Available from: www.alere.com/us/en/product–details/afinion–as100–analyzer.html.

[77] Orion, QuikRead go CRP, 2015. Available from: http://www.oriondiagnostica.com/Products/QuikRead–go/QuikRead–go–CRP/#.VeKks_TIDCs.

[78] Axis-Shield, NycoCard CRP Test, 2015. Available from: http://www.us.axis–shield.com/esite/esite.nsf/pub/JTHD5WRHZ2?Open.

[79] M. Soejima, Y. Koda, Evaluation of point-of-care testing of C-reactive protein in forensic autopsy cases, Forensic Sci. Int. 237 (2014) 27–29.

[80] S.K. Vashist, et al., Bioanalytical advances in assays for C-reactive protein, Biotechnol. Adv. 34 (3) (2016) 272–290.

[81] G.M. Whitesides, The origins and the future of microfluidics, Nature 442 (7101) (2006) 368–373.

[82] M. Karle, et al., Microfluidic solutions enabling continuous processing and monitoring of biological samples: a review, Anal. Chim. Acta 929 (2016) 1–22.

[83] L. Gervais, N. de Rooij, E. Delamarche, Microfluidic chips for point-of-care immunodiagnostics, Adv. Mater. 23 (24) (2011) H151–H176.

Chapter 6

Surface Plasmon Resonance–Based Immunoassays: Approaches, Performance, and Applications

K.L.M. Moran, D. Lemass, R. O'Kennedy
Dublin City University, Dublin, Ireland

1. INTRODUCTION

1.1 Surface Plasmon Resonance

Surface plasmon resonance (SPR) is a "label-free" detection method allowing analysis of biomolecular interactions. In 1990, the fundamental technologies of surface chemistry, flow processes, and optical detection were integrated into one of the world's first SPR-based analytical system, BIAcore. This system made it possible to measure biomolecular interactions in "real-time" with high sensitivity and without the need for labels [1]. Major challenges associated with the measurement of biomolecular interactions involve the use of labeling strategies (i.e., biotin, radioisotopes, fluorophores, or enzymes), which can result in steric hindrance, or structural configuration alterations that affect the labeled molecule's ability to interact with its target and reduce the complexes' affinity characteristics. Eliminating the need for specialized labeling in SPR technology applications permitted sensitive kinetic affinity measurements, and the use of native biomolecules enhanced its applicability for biologically relevant studies [2]. Since the introduction of SPR, the technology has been used successfully to determine interactions between proteins and DNA, lipid membranes, polysaccharides, and cells, as well as providing insights on receptor and drug interactions and enzyme-substrate/inhibitor complex interactions. In particular, characterization of protein–protein interactions has allowed for the generation of important kinetic information to determine the strength of binding interactions between antibodies and their target-molecules [3]. Furthermore, the development of immunoassays using specific antibodies and SPR systems has delivered extremely effective and low detection levels for

Handbook of Immunoassay Technologies. https://doi.org/10.1016/B978-0-12-811762-0.00006-2

numerous targets. SPR immunoassays are now an acceptable method for disease diagnosis, water analysis, foodborne pathogen detection, food safety monitoring, and environmental monitoring, and this chapter will focus on these uses.

1.2 Surface Plasmon Resonance Principles

The principle of SPR is based on the Kretschmann configuration where surface plasmons develop from free electrons arising at the interface of two different materials. Metal materials are useful for exploiting this phenomenon due to high electron conductivity. Glass prisms with a thinly coated metal surface are commonly used. Under conditions of total internal reflection (TIR), polarized light passing through a glass prism resonates with the oscillating electrons producing plasmons, at a precise angle called the "SPR angle" and the light reflected is at a minimum. The defined SPR angle at which resonance occurs is dependent on the refractive index (RI) of the material near the metal surface. Changes at which the SPR angle occurs or in the intensity of reflected light at the initial SPR angle can then be easily monitored. Plasmons propagate parallel to the metal surface generating an evanescent field ~300 nm from the metal surface, the depth of which is dependent on the wavelength of incident light and the refractive indices of the materials used. Consequently, small RI changes due to biomolecular interactions within this evanescent plasmon field elicit a shift in the SPR angle meaning that electrons are then no longer excited, plasmons cannot be formed, and reflected light is increased. Detection is then accomplished by monitoring binding interactions, tracking SPR angle shifts, and measuring changes in the reflected light on a detector (Fig. 6.1). The shift in SPR angle is proportional to the concentration of target-molecule binding to the immobilized biological recognition element on the metal surface [4,5].

Biosensors are analytical tools consisting of a biological recognition element, transduction system, and readout device [6]. SPR-based instruments primarily utilize two different materials with differing RIs to generate a biosensor with a metal and liquid (sample medium) interface; however, the prism and metal interface is also present. Surface plasmons can occur at both interfaces of the SPR biosensor, but the metal and liquid interface is enhanced due to the lower RI of the aqueous medium. The metal surface often forms the floor of a minute flow-cell on which the biological recognition element can be immobilized. Sample medium containing the target-molecule is then injected over the metal surface under continuous flow, forming the metal and liquid interface. If an interaction occurs between the immobilized biological recognition element and the target-molecule, the RI at the surface of the gold film changes, causing a shift in SPR angle and an increase in the intensity of reflected light observed by the detector [7]. In SPR experiments, resonance or response units (RUs) are used to describe this change, where one RU is

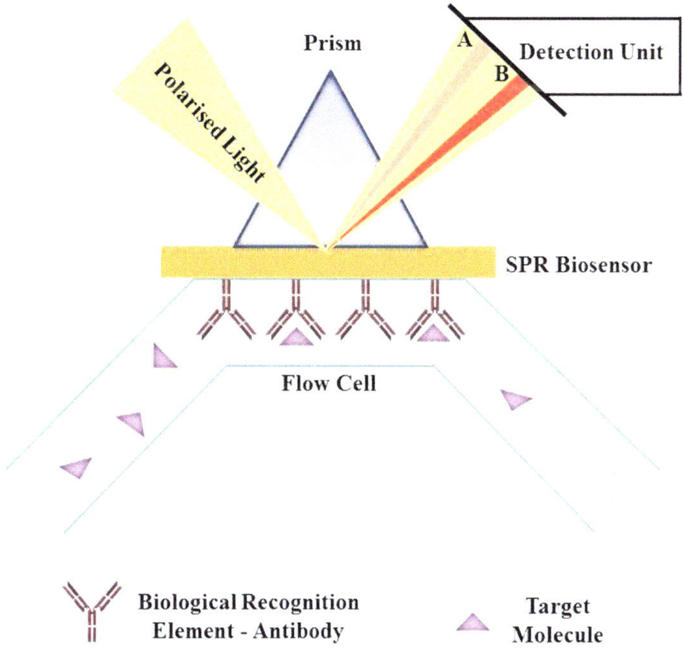

FIGURE 6.1 Surface plasmon resonance (SPR). The biological recognition element (antibody) is immobilized onto the gold surface of a biosensor. Polarized light is passed through a glass prism on the other side of the biosensor causing the oscillation of plasmons in the metal surface. The "SPR angle" is the precise angle at which this occurs and the reflected light intensity, measurable by the detection unit, is at a minimum (A). Sample medium containing the target-molecule flows over the antibody-coated biosensor surface, and as binding interactions occur, a change in refractive index leads to a shift in SPR angle and a change in the intensity of reflected light (B) measurable by the detection unit. The change in SPR angle is proportional to the mass binding at the biosensor surface.

equivalent to a critical angle shift of 10^{-4} degrees and RUs are plotted versus time in a "sensorgram." One RU can also be defined as ~1 pg/mm^2 of binding; however, more than 50 pg/mm^2 of binding may be necessary due to the difficulties associated with immobilization of high-density biological recognition elements on the sensor's surfaces. Background response is recorded and subtracted from the binding data by also injecting the sample medium containing the target-molecule over a blank metal reference surface (devoid of immobilized biological recognition element on the biosensor) and/or over a metal reference surface immobilized with a nonspecific biological recognition element. Typically an SPR-based instrument has a limit of detection of ~10 pg/mL; however, this value is dependent on numerous factors including surface concentration of the biological recognition element, the molecular weight of interacting biomolecules, and their binding affinity [2,8].

1.3 Surface Plasmon Resonance–Based Biosensor Platforms

The common attribute shared by all SPR-based biosensor platforms is the monitoring of the change in reflected light intensity caused by binding events on the surface of the sensor chip. The means by which they do this, however, vary. There are three typical approaches to generating and monitoring surface plasmons—prism, wave-guide, and grating-coupled configurations. This section describes the currently established configurations of commercial SPR-based biosensor platforms.

The Kretschmann-exploiting biosensors can be split into subgroups: fixed angle, angle scanning, and wedge-shaped beam [9]. In fixed angle, binding events are detected by the change in reflected light intensity and are the simplest form of SPR-based biosensors (Fig. 6.2A). Angle scanning devices monitor the change in the angle of resonance in "real-time" by repeatedly changing the angle at which incident light interacts with the surface. This can be achieved by using an angle-adjustable mirror to modify the incident light from its fixed source (Fig. 6.2B). Wedge-shaped beam instruments follow the change in the angle of resonance in "real-time" by shining a light beam at multiple angles simultaneously at the interface and monitoring the reflected light at all angles using an array of photodiodes (Fig. 6.2C). This approach removes the need for moving parts. However, it requires more complex detectors. A recent development in the technology is SPR imaging (SPRi). This builds on the standard SPR technique by incorporating a charge-coupled device (CCD) camera into the detection side of the configuration (Fig. 6.3A). This camera allows for a broader incident light beam as it can monitor the reflectivity of multiple points on a chip surface simultaneously [10]. Whereas earlier commercial systems probed for the resonance angle, SPRi systems monitor the reflectivity in protein spots over a broader surface, significantly increasing the throughput of simultaneous kinetic analyses [2]. Reference spots, including positive and negative controls to determine the nonspecific binding and/or cross-reactivity interactions, in addition to replicates of identical biomolecular interactions for checking the variability of the SPR biosensor surface, are also used to obtain reliable, accurate, and valid data. Mode effects caused by changes in temperature, bulk RI, or flow direction can be accounted for using these reference spots in SPRi [11].

In addition to the Kretschmann-based SPR systems above, there are two other methods, known as "grating-coupled" SPR and "waveguide" SPR. Grating-coupled SPR is similar to the traditional Kretschmann-based SPR; however, resonance is achieved without a prism. Instead, a gold (or other suitable material) film is coated onto a diffractive grating. Light is diffracted off the grating at multiple angles (Fig. 6.3B). At a specific angle, incident light will generate surface plasmons—detectable by a CCD camera (see Ref. [12]). Like other methods, binding events on the film will alter the angle of resonance. Unlike other methods, however, the configuration sends incident light through a window of the flow cell and the sample being analyzed. Hence, the

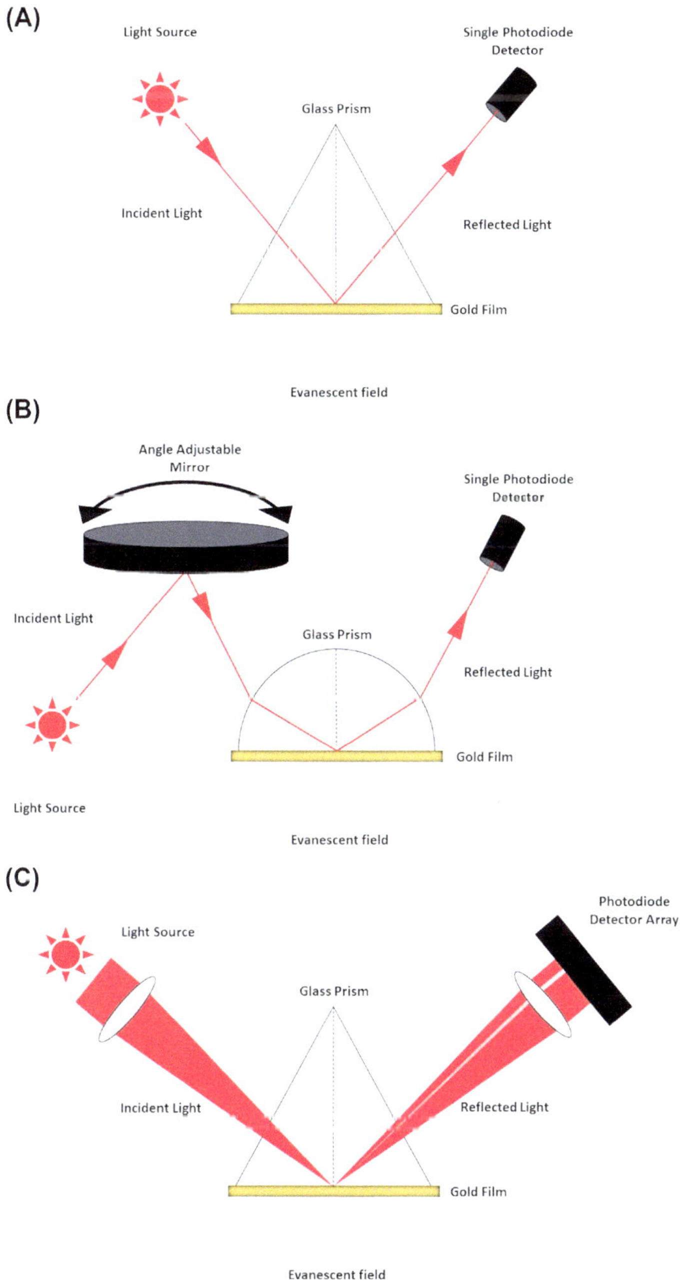

FIGURE 6.2 Fixed angle surface plasmon resonance (SPR) (A) detects changes in reflected light intensity. Angle scanning SPR (B) monitors changes in the angle of resonance by repeatedly changing the angle at which incident light interacts with the surface using adjustable mirrors to modify the incident light from its fixed source. Wedge-shaped beam SPR (C) detects the change in the resonance angle by shining a light beam at multiple angles simultaneously at the biosensor surface and monitoring the reflected light at all angles using an array of photodiodes.

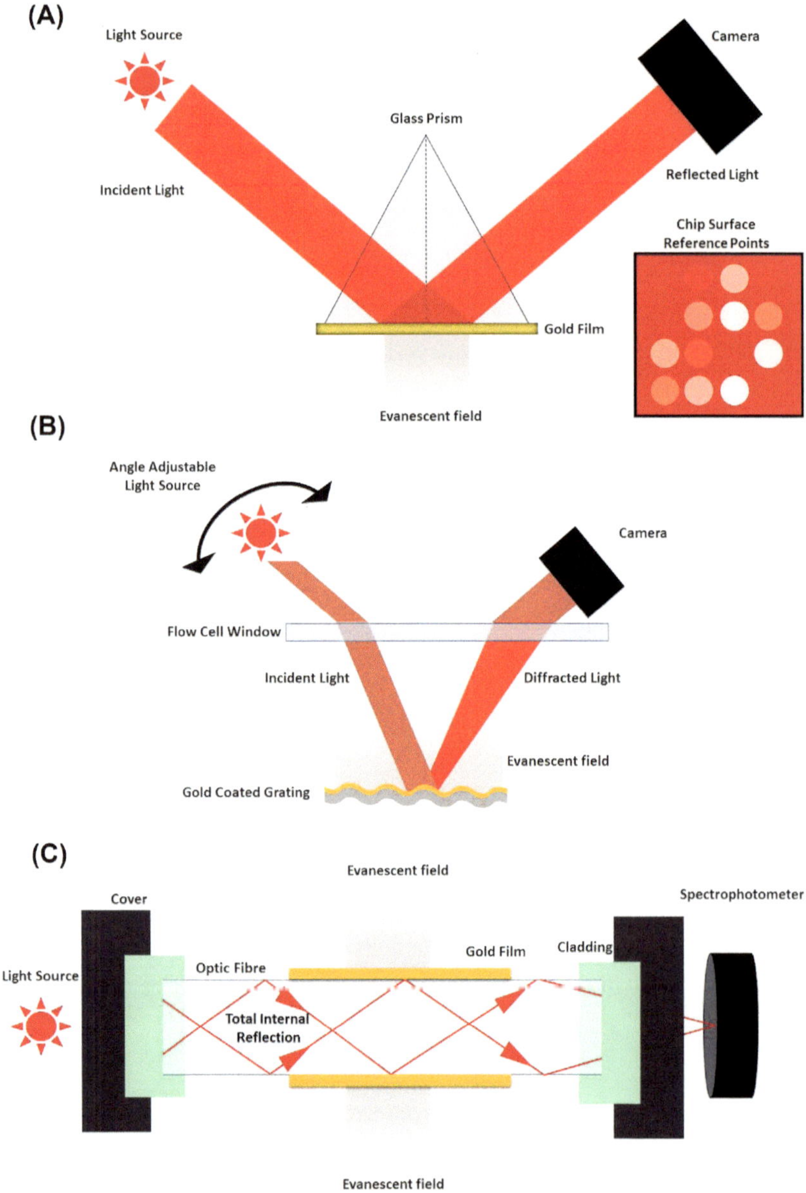

FIGURE 6.3 Surface plasmon resonance (SPR) imaging (A) uses a charge-coupled device camera for a broader incident light beam to monitor the reflectivity of multiple points on a chip surface simultaneously. Grating-coupled SPR (B) incorporates a gold film coated onto diffractive grating at which light is diffracted at multiple angles. Waveguide SPR (C) uses an optic fiber undergoing total internal reflection with a beam of white light and gold cladding to allow plasmon generation. A spectrophotometer monitors the intensity of the light across all wavelengths for detection.

target-containing solution must comply with specific optical properties (transparency, for example). Waveguide SPR employs an optic fiber undergoing TIR with a beam of white light. At the point of biosensing, the cladding of the fiber is replaced with a gold film for plasmon generation (Fig. 6.3C). Rather than monitoring a change in angle of resonance (which is not compatible with this configuration), a spectrophotometer is employed to monitor the intensity of the light across all wavelengths—with changes to the RI (caused by binding events) correlating to the wavelength at which surface plasmons are generated [13]. Due to the variation in SPR methodologies accessible for sensitive detection and characterization purposes, a range of SPR instruments have been introduced to the marketplace. Details of current commercially available SPR instruments and the SPR-based biosensor platform they employ are described in Table 6.1. Each SPR instrument has also been assigned a throughput value based on its capabilities of conducting simultaneous measurements.

2. SURFACE PLASMON RESONANCE–BASED IMMUNOASSAYS

2.1 Antibody Introduction

Antibodies, or immunoglobulins (IgG), in their native format are Y-shaped glycoproteins. They are a vital part of the adaptive immune system responsible for neutralizing pathogens such as bacteria, fungi, and viruses and marking them for destruction. Their value lies in their ability to identify and bind specifically to their respective target antigens. This antigen-selective trait is enabled through unique complementarity-determining regions located in the variable heavy and light chains of the antibody. In addition to the variable regions located at the ends of the two heavy and light polypeptide chains, standard immunoglobulins are made up of constant regions. The light chain has one constant fragment while the heavy chain has three (Fig. 6.4A). In addition to the standard immunoglobulin G (IgG) format, antibodies are present in IgD, IgE, IgM, and IgA formats in the body (see Ref. [15]). Immunoglobulins can be found in additional formats in nature, for example, IgY formats in birds [16] or the Ig new antigen receptor format present in sharks or nanobodies in camelids [17,18].

The biorecognition properties of antibodies make them extremely valuable both as therapeutic and diagnostic tools. Thus, antibodies have become an integral part of many therapies (see Ref. [19]) and biosensors (see Ref. [20]). They are produced as monoclonal (through hybridoma technology), polyclonal (through animal immunizations), or recombinant formats (through display technology) [21,22]. The advent of recombinant antibody generation technology has given rise to novel formats such as single chain fragment variables and antigen binding fragments, among others (Fig. 6.4B–D).

TABLE 6.1 Details of Surface Plasmon Resonance (SPR)-Based Biosensor Platforms Currently Available and Their Manufacturers. Only Instruments Capable of Exceeding Four Simultaneous Measurements Were Considered High-Throughput

Manufacturer	Instrument	Type	Throughput	Supplier Website
Accurion	Ellipsometric Platform EP^3	Fixed angle/ Imaging	Low	http://www.accurion.com
Affinité Instruments	P4SPR	Waveguide/ Imaging	Low	http://affiniteinstruments.com
Analytical micro-Systems	BIOSUPLAR 6	Fixed angle	Low	http://www.biosuplar.de
Biacore	Biacore 4000	Wedge beam	High	http://www.gelifesciences.com
	Biacore S200	Wedge beam	Low	
	Biacore 3000	Wedge beam	Low	
	Biacore C	Wedge beam	Low	
	Biacore T200	Wedge beam	Low	
	Flexchip	Grating-coupled	High	
	Biacore X100	Wedge beam	Low	
Bio-Rad Laboratories	ProteOn XPR36	Imaging	High	http://www.bio-rad.com
BioNavis	MP SPR Navi 420A ILVES	Angle-scanning	Low	http://www.bionavis.com
	MP SPR Navi 220A NAALI	Angle-scanning	Low	
	MP SPR Navi 210A VASA	Angle-scanning	Low	
	MP SPR Navi 200 OTSO	Angle-scanning	Low	
BiOptix	BiOptix 404pi	Fixed angle	Low	http://www.bioptix.com
Biosensing Instrument	BI-4500	Wedge beam	Low	http://biosensingusa.com
	BI-2500	Wedge beam	Low	
GWC Instruments	SPR-100	Fixed angle/ wavelength	Low	http://www.gwctechnologies.com
	SPRimager II	Imaging	High	

TABLE 6.1 Details of Surface Plasmon Resonance (SPR)-Based Biosensor Platforms Currently Available and Their Manufacturers. Only Instruments Capable of Exceeding Four Simultaneous Measurements Were Considered High-Throughput—cont'd

Manufacturer	Instrument	Type	Throughput	Supplier Website
HORIBA Scientific	XelPleX	Imaging	High	http://www.horiba.com
IBIS Technologies BV	OpenPlex	Imaging	High	
	MX96	Imaging	High	http://www.ibis-spr.nl
K-MAC	SPRmicro	Fixed-angle	Low	http://www.kmac.com
	SPR i	Imaging	High	
NanoSPR Devices	NanoSPR6	Angle scanning	Low	http://nanospr.com
	NanoSPR8	Angle-scanning	High	
	NanoSPR9	Fixed angle	Low	
Nicoya Lifesciences	OpenSPR	Waveguide	Low	https://nicoyalife.com
	Open-SPR-XT	Waveguide	Low	
Optre	Multiskop	Fixed angle	Low	http://www.optrel.de
Plexera LLC	PlexArray HT	Imaging	High	http://www.plexera.com
Reichert Technologies	Reichert4SPR	Wedge beam	Low	http://www.reichertspr.com
	Reichert2SPR	Wedge beam	Low	
	SR7000DC	Wedge beam	Low	
Seattle Sensors	SPIRIT	Wedge beam	High	http://www.seattlesensors.com
SENSIA	Indicator	Fixed angle	Low	http://www.sensia.es
	Indicator-G	Fixed angle	Low	
SensiQ Technologies	Pioneer FE	Wedge beam	Low	https://www.sensiqtech.com
	Pioneer	Wedge beam	Low	
	SensiQ	Wedge beam	Low	
	Discovery	Wedge beam	Low	

Continued

TABLE 6.1 Details of Surface Plasmon Resonance (SPR)-Based Biosensor Platforms Currently Available and Their Manufacturers. Only Instruments Capable of Exceeding Four Simultaneous Measurements Were Considered High-Throughput—cont'd

Manufacturer	Instrument	Type	Throughput	Supplier Website
Sierra Sensors	SPR-2/4	Imaging	Low	http://www.sierrasensors.com
	MASS-1	Imaging	High	
	MASS-2	Imaging	High	

Table on current instruments was generated using Refs. [9] and [14]. Only platforms actively advertised on supplier websites were included in this table.

2.2 Antibody Screening Using Surface Plasmon Resonance

The applications for SPR-based technologies have expanded dramatically since their introduction. Nevertheless, examination of antibody and target-molecule interactions remains one of SPR's most common uses. During antibody development it is pertinent that the most favorable candidates are identified promptly. It is possible to identify high-affinity antibodies and also to adjust their biophysical properties for specified applications. Depending on the desired function, antibody and target-molecule interactions can be long-lived or short-lived and, therefore, particular attention is needed to identify antibodies with specific reaction-rate and affinity properties. Measurements based only on direct enzyme-linked immunosorbent assays (ELISA) signals are subject to discrepancies arising from disparities in antibody expression levels, functional stability, and valency and cannot relate to specific affinity requirements [23,24]. The capacity for "real-time" and "label-free" analysis with vast assay versatility and high-throughput marks SPR as a powerful technology for both the screening and characterization of antibody and target-molecule interactions for therapeutic and diagnostic uses [25]. These key features enable fast analysis, accurate kinetic measurements, and reduce the additional costs and possible heterogeneities or other complications associated with labeling.

Leonard et al. [26] developed a rapid and reliable assay for the screening and characterization of recombinant single chain variable fragment (scFv) antibody fragments, directly from crude bacterial lysates. Ninety-six scFv antibodies targeting C-reactive protein (CRP) were selected by phage display and evaluated by SPR analysis using the BIAcore A100 (4000) instrument. Approximately 2000 RU of biotinylated anti-HA antibody was immobilized by spotting on the biosensor surface to screen 8 scFv antibodies per cycle delivering a total analysis time of 5 h 50 min for 96 scFv antibody fragments.

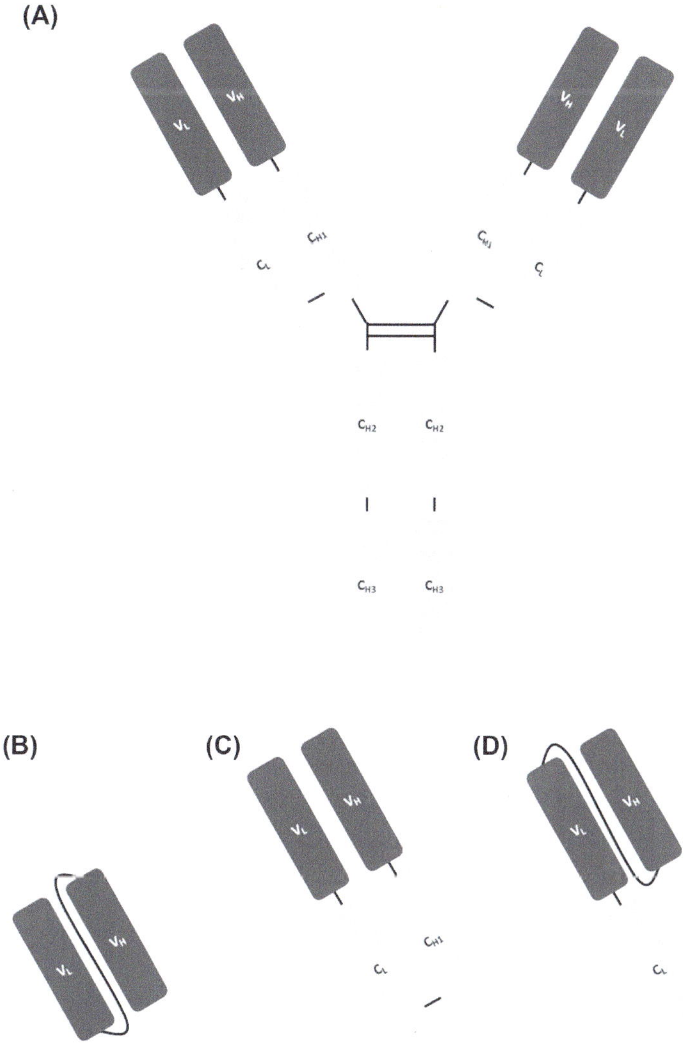

FIGURE 6.4 Antibody and recombinant antibody fragments. A typical immunoglobulin G antibody structure consists of two heavy chains and two light chains both of which contain variable and constant regions (A). A single heavy chain consists of one variable domain and three constant domains while a single light chain consists of one of each these domains. Together the variable domains of the heavy and light chains present the unique complementarity-determining region for target antigen binding. Three recombinant antibody fragments, a single chain variable fragment (scFv) (B), Fab antibody fragment (C), and a single chain antibody (scAb) (D), are also shown. *CH*, constant heavy chain; *CL*, constant light chain; *VH*, variable heavy chain; *VL*, variable light chain.

The antibody fragments were captured from crude bacterial extracts passed over the biosensor surface followed by 30 nM CRP. The anti-CRP scFv antibody fragments were ranked based on the percentage of the complex left after dissociation. Kinetic rate constants (k_a and k_d) and affinity (K_D) data were then obtained for six clones that bound monomeric CRP (50, 25, 12.5, 6.25, 3.125, 1.56, and 0 nM in HBS-EP) across a broad affinity range (2.54×10^{-8} to 3.53×10^{-10} M). Therefore, the antibody screening and characterization approach undertaken identified six high-affinity anti-CRP scFv antibody fragments useful for CRP detection. This SPR-based recombinant antibody screening method offered quicker analysis time and more detailed kinetic results for the 96 anti-CRP scFv clones analyzed than the traditionally used ELISA approach and negated expression level, valency, and stability issues. Furthermore, the authors concluded that combination of this screening approach with (1) the high sample throughput (~400 clones per 24 h using this assay format) and (2) the large sample storage capacity (10×96 well plates stored in a controlled environment within the instrument) of the Biacore A100 (4000) offers vast potential to both the biotechnology and pharmaceutical industries for screening extensive numbers of biological compounds [26].

2.3 Surface Plasmon Resonance Immunoassay Introduction

The detection of antibody and target-molecule interactions has been used in both basic research and real-world concerns in clinical, biotechnological, and environmental areas. Numerous companies such as Biacore, SensiQ Technologies, Windsor Scientific, Texas Instruments, Quantech, Bio-Rad, Ibis Technologies, Biosensing Instruments, BioNavis, and Nicoya Lifesciences have taken advantage of SPR transduction to develop and market commercial SPR instruments for "real-time" analysis of biomolecular interactions meeting the need for sensitive and versatile biosensing applications. The simplicity, speed, and direct ("label-free") nature of SPR biosensing coupled with the use of antibodies in SPR immunoassays have driven this technology beyond the scope of pure research and into the domain of routine analysis [27,28]. The development and application of SPR immunosensors exploiting the specificity of antibody and target-molecule immunoreactions and the high sensitivity of SPR signal transduction provide exceptional performance capabilities. Proteins, complex conjugates, toxins, drugs, and pesticides can be determined directly using SPR immunosensors [29]. Large molecule analysis has been extensively applied to SPR immunoassays using the high molecular weight of a molecule and sandwich assay formats to produce excellent sensitivities; however, small molecule immunosensing using SPR detection is more arduous. More recently, SPR immunoassays have overcome steric hindrance evident between small target-molecules and antibodies and issues associated with the labeling of small molecule, through careful considerations of biosensor interface design and signal enhancement [30]. Furthermore, the development of recombinant antibody phage display has allowed for the generation of highly specific antibody fragments to target small molecular weight target in immunoassays [31], making these biological recognition elements particularly amenable to SPR-based biosensing.

2.3.1 Small Molecular Weight Targets

Food safety issues are of major importance, and the challenges are significant due to the great diversity of foods and food components analyzed and rapidly changing food interests. Illnesses associated with food contaminants such as toxins, pesticides, antibiotics, and drugs pose a serious threat to human health. Mycotoxins are the toxic secondary metabolites produced by a wide range of mold and fungal species, which grow on food commodities and cause immunosuppression, nephrotoxicity, infertility issues, and hepatocellular carcinoma [6]. Recently in August 2016, it was reported that grain contamination with aflatoxin B_1 (AFB_1), a potent hepatocellular carcinogen, led to the death of 14 people who consumed contaminated material [32]. Additionally, poisoning and severe illnesses can be caused by phycotoxins and cyanotoxins produced by marine algae and cyanobacteria, which accumulate in the meat of frequently consumed shellfish such as mussels, clams, oysters, and scallops. SPR-based research has promoted detection and monitoring in food, marine, and environmental fields, where small target-molecules (>1000 Da) are of particular concern. A bifunctional protein cross-linker-based SPR biosensor for aflatoxin B_1 (AFB_1) detection was recently developed. A fusion protein was generated genetically by fusing a gold binding protein that strongly adheres gold surfaces to protein G, which can interact with the Fc portion of antibodies. The fusion protein was used as a bifunctional cross-linker for the rapid, self-orientated immobilization of anti-AFB_1 antibodies to a gold SPR biosensor surface. AFB_1 is a low molecular weight molecule, and the enhanced antibody immobilization strategy permitted the coating of 36% more anti-AFB_1 on the biosensor surface than a comparative surface lacking the gold binding protein. The resulting SPR immunoassay allowed the detection of 1 μg/mL of AFB_1 in buffer and in corn extracts. These results indicate that the gold surface used in SPR immunoassays can be optimized to detect very low concentrations of small molecular weight targets, which are traditionally difficult to measure [33]. Further examples of SPR immunoassays for the sensitive detection of small molecular weight targets in food, marine, and environmental analysis are described in Table 6.2.

2.3.2 Pathogens and Viruses

Foodborne diseases are spread by the consumption of foods or beverages contaminated with bacteria, viruses, and parasites resulting in ~48 million illnesses each year in the United States. Although few of these illnesses are recorded directly at the setting/vicinity of an outbreak of foodborne illness

TABLE 6.2 Examples of Surface Plasmon Resonance Immunoassays for Detection of Small Molecules in Food, Marine, and Environmental Fields

Target	Matrix	Biological Recognition Element	LoD	SPR Platforms	References
Food					
AFB_2	Almond extract	scFv	0.9 ng/mL	Biacore 3000	[34]
ZEN	Sorghum	scFv	7.8 ng/mL	Biacore 3000	[35]
DON	Wheat Breakfast cereal Baby food	mAb	57 μg/kg 9 μg/kg 6 μg/kg	Biacore Q	[36]
T-2 toxin and HT-2 toxin	Wheat Breakfast cereal Baby food	mAb	26 μg/kg 25 μg/kg 25 μg/kg	Biacore Q	[37]
Atrazine	Milk	mAb	5×10^{-8} M	BIOSUPLAR	[38]
Amoxicillin Ampicillin Benzylpenicillin Cloxacillin Cefroperazone Cephalexin	Milk	mAb	2 ng/mL 2 ng/mL 2 ng/mL 15 ng/mL 25 ng/mL 50 ng/mL	Biacore 3000	[39]
Sulfamethazine	Chicken sera	mAb	10 ng/mL	Biacore Q	[40]

Marine and Environmental					
Microcystin-LR	Freshwater	mAb	0.5 ng/mL 0.05 ng/mL	Biacore Q	[41]
Microcystin-LR	Buffer	scFv	1.7 ng/mL	Biacore 3000	[42]
OA	Mussels	mAb	0.24 μg/kg	Biacore 3000	[43]
DA	Clam extract	pAb	3 ppb (10 nM)	Spreeta 2000	[44]
Escherichia coli-K12	River/tap water	pAb	9.0×10^1 cfu/mL	Spreeta	[45]

AFB$_2$, aflatoxin B$_2$; *cfu*, colony forming units; *DA*, domoic acid; *DON*, deoxynivalenol; *LOD*, limit of detection; *mAb*, monoclonal antibody; *OA*, okadaic acid; *pAb*, polyclonal antibody; *ppb*, parts per billion; *scFv*, single chain variable fragment; *ZEN*, zearalenone.

(i.e., delis, cafes, restaurants, etc.), data collected during outbreak inquiries afford insights into the pathogens and foods that cause illness. The Centre for Disease Control and Prevention reported a total of 818 outbreaks, 13,360 affected patients, 1062 hospitalizations, 16 deaths, and 14 food recalls due to foodborne diseases in the United States between 2012 and 2013 [46]. Conventional methods for detecting pathogens include techniques based on microscopy, nucleic acids, and immunoassays. The microscopy-based methods require large sample quantities and tedious culture preparations. Nucleic acid—based polymerase chain reaction (PCR) methods are highly sensitive and selective; however, they require undamaged DNA [47]. Furthermore, where low levels of infection of the major pathogens occur, extremely sensitive detection platforms are necessary for pathogen identification and analysis [48]. Table 6.3 highlights SPR-based immunoassays developed for the highly sensitive and specific detection of common pathogens and viruses using minute sample quantities that do not require functional DNA and permit direct detection of the target of interest.

2.3.3 Disease Targets

Since the introduction of SPR and its advancement with surface plasmon resonance imaging (SPRi), the technology has emerged as a suitable and reliable platform in clinical analysis for "label-free," sensitive, and "real-time" detection of diseases described in Table 6.3 [59]. Present day disease diagnostics typically focus on ELISA, fluorescence-based ELISA, PCR, and imaging techniques. SPR-based methods are expanding at a rapid pace due to their ultrasensitive and specific detection capabilities, which are advantageous over, for example, ELISA-based methods. Further advancements in SPR could provide improved diagnostic platforms with important roles in sensitive disease detection, staging, diagnosis, prognosis, and monitoring [60]. Future research required in this area centers on the adaptability of SPR to provide methods for analysis of multiple targets and biomarkers in disease, as well as point of care (POC) applications, which would permit rapid and accurate disease detection in local general practitioner offices without the need for slow, centralized laboratory-based testing.

3. FUTURE TRENDS IN SURFACE PLASMON RESONANCE—BASED IMMUNOASSAYS

3.1 Surface Plasmon Resonance—Mass Spectrometry

SPR was first integrated with mass spectrometry (MS) in 1997 [61], to allow greater investigation into protein binding through characterization of biomolecular interactions between proteins and surface-immobilized targets by SPR, supplemented by determination of the identity of the bound proteins or peptides using MS [11]. SPR technology does not detect the identity of the

TABLE 6.3 Examples of Surface Plasmon Resonance (SPR) Immunoassays for Detection of Pathogens, Viruses, and Disease

Target	Matrix	Biological Recognition Element	LoD	SPR Platforms	References
Pathogens and Viruses					
Salmonella	Buffer	mAb and pAb	1×10^7 cfu/mL	Biacore 3000	[49]
Listeria monocytogenes	Buffer	pAb	1×10^5 cfu/mL	Biacore 3000	[50]
Cryptosporidium oocyst	Buffer	mAb	1×10^2 oocytes/mL	Biacore 2000	[51]
Influenza A	Buffer	mAb	1 ag or 0.2 fg/mm^2	BI-2000 SPR Instrument (Biosensing Instrument)	[52]
Feline calicivirus	Cell culture lysate	mAb and pAb	10^4 $TCID_{50}$ FCV/mL	Biacore T100	[53]
Disease Targets					
Prostate cancer • PSA	Buffer	single domain antigen binding fragment and mAb	1 ng/mL	Biacore 2000	[54]
Prostate cancer • tPSA	75% Serum	mAb	0.29 ng/mL	Biacore 3000	[55]

Continued

TABLE 6.3 Examples of Surface Plasmon Resonance (SPR) Immunoassays for Detection of Pathogens, Viruses, and Disease—cont'd

Target	Matrix	Biological Recognition Element	LoD	SPR Platforms	References
Prostate cancer • tPSA • fPSA	Buffer	mAb	0.4 ng/mL 0.005 ng/mL	BI-SPR 3000 system (Biosensing Instrument Inc., USA)	[56]
Oropharyngeal squamous cell carcinoma • IL8	Saliva	mAb	184 pM	Biacore X	[57]
Tumor marker • AFP	10% Serum	mAb and pAb	1 ng/mL	GWC SPRimager II	[58]

AFP, α-fetoprotein; *cfu*, colony forming units; *fPSA*, free prostate-specific antigen; *IL8*, interleukin 8; *LOD*, limit of detection; *mAb*, monoclonal antibody; *pAb*, polyclonal antibody; *PSA*, prostate-specific antigen; *scFv*, single chain variable fragment; *tPSA*, total prostate specific antigen.

individual cognate binding molecules involved in a biomolecular interaction; however, MS can be utilized for specific protein identification in concentrations comparable to that captured on a biosensor surface by SPR-based instruments. This process is based on intact proteins or peptides produced by digestion of proteins being transferred to gas phase using soft ionization methods in the ion source of mass spectrometers [62]. Matrix-assisted laser desorption/ionization time-of-flight mass spectrometry (MALDI-TOF-MS) has become an essential analytical tool in proteomics due to its ability to accurately and rapidly analyze multiple proteins to reveal intrinsic protein structural modifications, and its integration with SPR provides greater insights and understanding of protein interactions [63].

Kim et al. [64] described the use of an "ImmunoMALDI-TOF-MS" strategy involving a multiantibody array chip for MALDI-TOF-MS analysis of multiple captured proteins, which could be directly applied to an SPRi system to successfully monitor antibody and target binding. Each spot of the gold chip surface was immobilized with one of five different antibodies: nonspecific human IgG (control), antibodies for β2-microglobulin, human interleukin-6 (IL-6), CRP, or prostate-specific antigen (PSA). Complex protein mixture solutions including human serum were passed over the gold spots containing four target antigens (β2-microglobulin, IL-6, CRP, and PSA). The resulting mass spectrum of individual antibody spots showed highly specific target antigen peaks corresponding to the molecular weight of the target antigen in multiple charge states, without nonspecific binding in buffer solutions; however, peaks of serum components were nonspecifically observed in a mass range from 6000 to 9000 m/z. Furthermore, this system demonstrated that tryptic digests of captured PSA could be effectively detected by on-chip MALDI-TOF-MS [64]. This SPR-MS approach provides a basis for multiplexed protein analysis and the possibility for multibiomarker analysis. Methods for the determination of biomarkers in human samples are conducted in hospital settings or specialized laboratory facilities and use ELISA. Multiple cancer biomarkers often need to be examined and tested to establish an accurate and specific cancer diagnosis indicative of disease staging. Studies using human protein arrays have concluded that a range of 18 specific biomarkers tested against autoantibodies present in patient serum for colorectal cancer and four specific biomarkers tested against autoantibodies for the absence of colorectal cancer can deliver a highly sensitive disease staging, diagnosis, and prognosis [65]. Therefore, SPR-MS may be useful for detection of cancers where a range of biomarkers may need to be tested and identified to establish a proper diagnosis. In addition to difficulties associated with multimarker analysis, mapping protein networks and determining functional relationships between proteins and their cellular effects, as well as the characterization of protein modifications critical for the interaction, can pose a major challenge in proteomics. The combination of SPR with MS provides a highly compatible and adaptable tool for establishing new interacting protein partners and performing comprehensive characterization of their function and structure [66].

3.2 Surface Plasmon Resonance—Point-of-Care Applications

SPR is a proven versatile, "label-free" method for the observation of biomolecular interactions in "real-time." Recently, research efforts are focusing on the miniaturization of the biosensor system to allow for POC applications. The necessity for this development is driven by the underlying facts that SPR instruments are often large, expensive, and laboratory-based and require trained personnel for their use. The versatility and specificity of SPR detection makes it especially amenable for POC testing, allowing the monitoring and control of food and beverage safety, water examination, and environmental monitoring directly in the field, as well as disease testing in general practitioners' local offices. There are, however, difficulties associated with the integration of SPR with POC applications. Nonspecific binding of molecules other than the target in complex mixtures impinges on SPR signals; therefore, obtaining pure samples containing few other molecules, preferable for SPR biosensing, can prove difficult for POC situations. The size of an SPR-POC instrument may also be difficult to reduce with readout devices, optics, sensor chips, numerous tubes, pumps, and valves being required for simultaneous or multiple measurements. Texas instruments developed "Spreeta" the first commercially marketed, miniaturized SPR instrument [67]. However, a need remains for further research and delivery of portable SPR instruments capable of simultaneous analysis of multiple targets in complex samples on-site.

3.2.1 CD-Based Surface Plasmon Resonance

Microfluidic approaches have been exploited to achieve miniaturized SPR systems without the need for traditional large-scale SPR tubing, pumps, and valves. Hemmi et al. [68] developed a CD-based microfluidic platform using centrifugal force to drive sample solution toward a grating-based SPR biosensor. Reservoirs and microchannels fabricated in polydimethlylsiloxane to hold and move sample solution and reagents are fixed to a polycarbonate disk with a layer of gold-coated grating film for SPR detection purposes. The solutions in the reservoirs flow to the outermost reservoir by centrifugal force generated by the rotation of the microfluidic device at certain specific frequencies allowing an anti-IgA antibody solution, a blocking reagent, a IgA sample, and the washing solution to move. The solutions pass into the SPR detection chamber at the outermost reservoir, in sequence, with the SPR angle shift observed for each individual binding step [68]. The described CD-based SPR system does not require expensive prisms, pumps, valves, or tubing and would provide a suitable method for POC analysis. In addition, the use of centrifugal force could be exploited to separate complex matrices into purer fractions for easier detection that negates nonspecific binding. CD-based microfluidic centrifugal platforms have been successfully used for blood processing [69], and the ability to separate complex sample matrices offers a preprocessing step, which, when integrated with a CD-SPR biosensor, accomplishes an "all-in-one" SPR-POC detection device [70]. A SPR-based biosensor has been commercially developed by Biosurfit based on their unique technology spinit. This system permits all steps, including preprocessing, of a diagnostic test to be performed using a CD-SPR immunoassay. Our group engineered a recombinant antibody fragment specific for CRP, which was integrated with the spinit platform. The spinit CRP CD is used in conjunction with the spinit detection device to permit specific and quantitative measurement of CRP concentrations in whole blood, serum, and plasma samples. The CRP immunoassay is used for detection and evaluation of infection, tissue injuries, inflammatory responses, and associated diseases. The test requires only 5 μL of patient blood and has integrated positive and negative controls and a small readout device, therefore, presenting an excellent diagnostic device available for patient diagnosis directly in general practitioners' local offices and hospitals [71]. This device represents one of the best examples of POC platforms based on SPR developed to date and highlights the excellent potential within the field of SPR-POC biosensing.

3.2.2 Mobile Phone-Surface Plasmon Resonance

A major difficulty encountered during the development of POC devices is the miniaturization and combination of several components, including a fully integrated readout system. A new field of readout systems has emerged in the last decade with a growing trend toward the use of mobile phone–based devices in biotechnology. The common use of mobile phone devices with touch-screen displays, excellent Internet connectivity, high-resolution cameras, and high-performance central processing units (CPUs) has simplified the development of devices suited for POC applications. Digital microscopy, cytometry, colorimetric analysis, immunoassay/lateral flow readout, and healthcare monitoring have all been achieved by mobile phone devices without auxiliary instruments or laboratory conditions. Mobile phone devices represent an extremely apt system to underpin the development of next-generation POC applications for the monitoring of numerous fields, including clinical diagnosis of diseases, detection of pathogens, food safety, and environmental testing [72].

Fiber optic-SPR using a mobile phone device with a lightweight SPR adaptor was used to successfully evaluate the specific binding of Staphylococcal Protein A and bovine IgG in solution on a light-guiding silica capillary coated with a 50 nm gold film biosensing surface. Test, control, and reference channels were illuminated in the device by passing light through the "lead-in" fibers from the mobile phone device's LED flash, and the light passing from the "lead-out" fibers was then detected by the mobile phone device's camera. A smart application was developed to extract the light intensity from the camera images and RI changes. Comparable experiments were completed with a commercial SPR instrument (BIOSUPLAR 6) eliciting a limit of detection of 15.7 nM for biological IgG, while the mobile phone device SPR biosensor delivered a less sensitive limit of detection, however, in the same order of magnitude at 47.4 nM [73]. A SPRi platform has also been integrated with a mobile phone device to demonstrate an immunoassay for the capture of mouse IgG antibodies on an immobilized layer of rabbit antimouse IgG antibody with a nanomolar level limit of detection. "Real-time" bulk RI change measurements yielded noise equivalent RI changes as low as 4.12×10^{-5} RU, which are comparable with the detection performance of commercial instrument [74]. The performance of the mobile phone device SPR platform provide reliable, accurate, and repeatable detection abilities and suggests great promise for the future of SPR-POC biosensing. Additionally, integration of mobile phone devices with unmanned aerial vehicles (UAVs) has become a reality for sensitive diagnostics and fully "mobile healthcare." Recently Priye et al. [75] described a portable platform combined with a consumer-class quadcopter drone for on-board, time-resolved fluorescent detection and quantification of *Staphylococcus aureus* and λ-phage DNA using a mobile phone device camera and a smart image analysis application [75]. The UAV system is versatile and robust and permits the deployment of sophisticated diagnostics directly to the "point-of-need." An SPR biosensor was also incorporated into a UAV platform to collect and quantify the concentration of aerosolized biological agents. The aerial sampling and collection system was developed to collect and concentrate aerosolized particles, which are subsequently pressurized for detection at the SPR biosensor, all on-board a UAV. This new, drone-based, SPR detection system provides a new insight into how biological-threat agents could be assessed in the environment and shows the potential for the detection of a wide range of water, food, and environmental targets at the point-of-occurrence, without the need for direct manual testing [76].

4. CONCLUSIONS

In 1990, the fundamental technologies of surface chemistry, flow processes, and optical detection were integrated into one of the world's first SPR instrument. SPR systems allowed "label-free" detection of biomolecular interactions in their natural form in "real-time" and sparked the beginning of a new field in protein interaction characterization. Since then, numerous approaches have been adapted by researchers to achieve SPR detection. Fixed angle, angle scanning, wedge-shaped beam biosensors, and SPRi involving the incorporation of a CCD camera into the detection side of the typical SPR configuration, as well as grating-coupled SPR and waveguide SPR, which allow prism-free SPR configurations, have all been developed. These methods have permitted great expansion in the field of SPR detection, culminating in the availability of multiple commercial SPR instruments utilizing a range of SPR-based biosensor platforms and throughput capacities. Evidently, the SPR technique has proven to be a very versatile scaffold for biosensing use and provides attractive performance capabilities in terms of "label-free" detection, high specificity and selectivity, relatively fast preparation and testing proving "real-time" analysis, low limits of detection (nM–pM affinities), and excellent reproducibility. Accordingly, the coupling of SPR with the use of antibodies as a biological recognition element has resulted in the successful and sensitive detection of targets in disease and medicine, food safety, environment monitoring, and water analysis. Despite exquisite SPR characteristics for biosensing and proven success in immunological detection, POC devices utilizing SPR have not flourished to the same extent as other SPR-based platforms in the last two decades. Issues including difficulties in the miniaturization of the SPR fundamental components including readout systems, optics, biosensor chips, numerous tubes, pumps, and valves have hindered this process. Furthermore, pretreatment and separation of complex matrices is a necessity for SPR-POC testing, as nonspecific binding of molecules other than the target in complex mixtures may impinge on the SPR signal. Current research focusing on the use of centrifugal force for complex matrix processing in CD platforms and SPR detection have offered a solution to problematic separations, as well as negating the need for traditional SPR tubing, pumps, and valves, thus simplifying the miniaturization procedure. Furthermore, devices based on CD-SPR capable of blood processing for detection and evaluation of infection, tissue injuries, inflammatory responses, and associated diseases are now commercially available. The integration of SPR detection capacity with the use of mobile phone devices as readout systems is an exciting approach that has recently emerged in the biosensing field. These devices can function and permit detection levels comparable to that of traditional commercially available SPR instruments, indicating excellent promise for the future of next-generation SPR-based POC devices. Furthermore, SPR detection devices can now be integrated with UAVs to establish POC applications that can be completed in the absence of manual testing. In addition to the positive impact SPR-POC applications could impart, SPR-MS has been explored as a multitarget detection method. The ability to detect numerous targets or biomarkers is of particular concern in disease diagnosis, and exploiting SPR for characterization of binding molecules, as well as MS for identification of the binding molecules, could assist with accurate disease detection, diagnosis, and prognosis for patients.

Research on the use of SPR immunoassays has established an excellent platform for the reliable and sensitive detection of targets across multiple fields. The emergence of creative and exciting SPR enhancements for POC applications and the realization of UAV-detection capabilities indicates SPR immunoassay–based systems will be at the center of diagnostic, detection, and analytical instruments well into the future.

ACKNOWLEDGMENTS

The authors of this work are supported by the Dublin City University Daniel O'Hare Research Scholarship Scheme, the EU-funded PHARMATLANTIC Knowledge Transfer Network for Prevention of Mental Diseases and Cancer in the Atlantic Area, the Dairy Processing Technology Centre, and the Irish Research Council and Science Foundation Ireland (14/1A/2646).

The authors have declared no conflicts of interest.

REFERENCES

[1] G.E. Healthcare, Biacore Life Sciences > History. Biacore.com, 2016. https://www.biacore.com/lifesciences/history/index.html.

[2] H. Nguyen, J. Park, S. Kang, M. Kim, H. Hiep, Surface plasmon resonance: a versatile technique for biosensor applications, Sensors 15 (2015) 10481–10510.

[3] J.G. Quinn, R. O'Kennedy, Biosensor-Based estimation of kinetic and equilibrium constants, Anal. Biochem. 290 (1) (2001) 36–46.

[4] G.E. Healthcare, Biacore Life Sciences > Technology > Introduction > Following an Interaction. Biacore.com, 2016. https://www.biacore.com/lifesciences/technology/introduction/following_interaction/index.html.

[5] N.M. Saucedo, A. Mulchandani, Sensing of biological contaminants: pathogens and toxins, in: V. Scognamiglio (Ed.), Biosensors for Sustainable Food - New Opportunities and Technical Challenges (Radarweg 29, PO Box 211, 1000 AE Amsterdam, Netherlands), Zoe Kruze, 2016, pp. 73–91.

[6] K.L.M. Moran, J. Fitzgerald, D. McPartlin, J.H. Loftus, R. O'Kennedy, Biosensor-based technologies for the detection of pathogens and toxins, in: V. Scognamiglio (Ed.), Biosensors for Sustainable Food - New Opportunities and Technical Challenges (Radarweg 29, PO Box 211, 1000 AE Amsterdam, Netherlands), Zoe Kruze, 2016, pp. 93–120.

[7] S.G. Patching, Surface plasmon resonance spectroscopy for characterisation of membrane protein-ligand interactions and its potential for drug discovery, Biochim. Biophys. Acta 1838 (1 Pt A) (2014) 43–55.

[8] A.P. Van Der Merwe, Surface plasmon resonance, in: S.E. Harding, B. Chowdhry (Eds.), Protein-Ligand Interactions: Hydrodynamics and Calorimetry. A Practical Approach, Oxford University Press, Oxford, London, 2001, pp. 137–187.

[9] R.B.M. Schasfoort, A.J. Tudos, Handbook of Surface Plasmon Resonance, Royal Society of Chemistry, Thomas Graham House, Cambridge Royal Society of Chemistry, Thomas Graham House (290), 2008. Available online from: http://www.lavoisier.fr/notice/fr421618.html http://ebook.rsc.org/?DOI=10.1039/9781847558220.

[10] H. Huang, Y. Chen, Label-free reading of microarray-based proteins with high throughput surface plasmon resonance imaging, Biosens. Bioelectron. 22 (5) (2006) 644–648.

[11] R.B.M. Schasfoort, P. Schuck, Future trends in SPR technology, in: R.B.M. Schasfoort, A.J. Tudos (Eds.), Handbook of Surface Plasmon Resonance, Royal Society of Chemistry, Thomas Graham House, Cambridge Royal Society of Chemistry, Thomas Graham House (290), Science Park, Milton Road, Cambridge, CB4 0WF, 2008, pp. p354–p394.

[12] J. Homola, Surface plasmon resonance sensors for detection of chemical and biological species, Chem. Rev. 108 (2) (2008) 462–493.

[13] B. Gupta, A. Shrivastav, S. Usha, Surface plasmon resonance-based fiber optic sensors utilizing molecular imprinting, Sensors 16 (9) (2016), 1381(1-33).

[14] A. Marquart, SPR Instruments, 2016. http://www.sprpages.nl/instruments.

[15] H.W. Schroeder, L. Cavacini, L. Cavacini, Structure and function of immunoglobulins, J. Allergy Clin. Immunol. 125 (2 Suppl. 2) (2010) S41–S52.

[16] W. Dias da Silva, D.V. Tambourgi, IgY: a promising antibody for use in immunodiagnostic and in immunotherapy, Vet. Immunol. Immunopathol. 135 (3–4) (2010) 173–180.

[17] S. Krah, C. Schröter, S. Zielonka, M. Empting, B. Valldorf, H. Kolmar, Single-domain antibodies for biomedical applications, Immunopharmacol. Immunotoxicol. 38 (1) (2016) 21–28.

[18] S. Steeland, R.E. Vandenbroucke, C. Libert, Nanobodies as therapeutics: big opportunities for small antibodies, Drug Discov. Today 21 (7) (2016) 1076–1113.

[19] G.J. Weiner, Building better monoclonal antibody-based therapeutics, Nat. Rev. Cancer 15 (6) (2015) 361–370.

[20] A.D. Sheehan, J. Quinn, S. Daly, P. Dillon, R. O'Kennedy, The development of novel miniaturized immuno-sensing devices: a review of a small technology with a large future, Anal. Lett. 36 (3) (2003) 511–537.

[21] B. Byrne, E. Stack, N. Gilmartin, R. O'Kennedy, Antibody-based sensors: principles, problems and potential for detection of pathogens and associated toxins, Sensors (Basel) 9 (6) (2009) 4407–4445.

[22] P.J. Conroy, S. Hearty, P. Leonard, R.J. O'Kennedy, Antibody production, design and use for biosensor-based applications, Semin. Cell Dev. Biol. 20 (2009) 10–26.

[23] G.P. Rai, K.S. Venkateswaran, Limitations and practical problems in enzyme-linked immunosorbent assays, Def. Sci. J. 2 (1992) 71–84.

[24] S. Hearty, P. Leonard, R. O'Kennedy, Measuring antibody–antigen binding kinetics using surface plasmon resonance, Methods Mol. Biol. 907 (2012) 411–442.

[25] P.S. Katsamba, I. Navratilova, M. Calderon-Cacia, L. Fan, K. Thornton, M. Zhu, et al., Kinetic analysis of a high-affinity antibody/antigen interaction performed by multiple Biacore users, Anal. Biochem. 352 (2006) 208–221.

[26] P. Leonard, P. Säfsten, S. Hearty, B. McDonnell, W. Finlay, R. O'Kennedy, High throughput ranking of recombinant avian scFv antibody fragments from crude lysates using the Biacore A100, J. Immunol. Methods 323 (2007) 172–179.

[27] W.M. Mullett, E.P.C. Lai, J.M. Yeung, Surface plasmon resonance-based immunoassays, Methods 22 (2000) 77–91.

[28] D. Lemass, S. Hearty, R. O'Kennedy, Surface plasmon resonance-based immunoassays, in: H. Koistinen, U.H. Stenman (Eds.), Novel Approaches in Immunoassays, Future Medicine Ltd., Unitec House, 2 Albert Place, London N3 1QB, UK, 2014, pp. 20–40.

[29] D.R. Shankaran, V.K. Gobi, N. Miura, Recent advancements in surface plasmon resonance immunosensors for detection of small molecules of biomedical, food and environmental interest, Sens. Actuator B-Chem. 121 (2007) 158–177.

[30] J. Mitchell, Small molecule immunosensing using surface plasmon resonance, Sensors 10 (2010) 7323–7346.

[31] T. Fodey, P. Leonard, J. O'Mahony, R. O'Kennedy, M. Danaher, Developments in the production of biological and synthetic binders for immunoassay and sensor-based detection of small molecules, Trends Anal. Chem. 30 (2) (2011) 254–269.

[32] S. Buguzi, Tanzania: New Questions on Food Poison, News, 2016. http://allafrica.com/stories/201608020624.html.

[33] J.H. Park, Y.-P. Kim, I.-H. Kim, S. Ko, Rapid detection of aflatoxin B1 by a bifunctional protein crosslinker-based surface plasmon resonance biosensor, Food Control 36 (2014) 183–190.

[34] S.R. Edupuganti, O.P. Edupuganti, R.J. O'Kennedy, Biological and synthetic binders for immunoassay and sensor-based detection: generation and characterisation of and anti-AFB2 single-chain variable fragment (scFv), World Mycotoxin J. 6 (3) (2013) 273–280.

[35] S.R. Edupuganti, O.P. Edupuganti, R.J. O'Kennefy, Generation of anti-zearalenone scFv and its incorporation into surface plasmon resonance-based assay for the detection of zearalenone in sorghum, Food Control 34 (2) (2013) 668–671.

[36] J. Meneely, T. Fodey, L. Armstrong, M. Sulyok, R. Krska, C. Elliott, Rapid surface plasmon resonance immunoassay for the determination of deoxynivalenol in wheat, wheat products, and maize-based baby food, Agric. Food Chem. 58 (2010) 8936–8941.

[37] J.P. Meneely, M. Sulyok, S. Baumgartner, R. Krska, C.T. Elliott, A rapid optical immunoassay for the screening of T-2 and HT-2 toxin in cereals and maize-based baby food, Talanta 81 (1–2) (2010) 630–636.

[38] M. Tomassetti, E. Martini, L. Campanella, G. Favero, G. Sanzó, F. Mazzei, A new surface plasmon resonance immunosensor for triazine pesticide determination in bovine milk: a comparison with conventional amperometric and screen-printed immunodevices, Sensors 15 (5) (2015) 10255–10270.

[39] G. Cacciatore, M. Petz, S. Rachid, R. Hakenbeck, A. Bergwerff, Development of an optical biosensor assay for detection of beta-lactam antibiotics in milk using the penicillin-binding protein 2x, Anal. Chim. Acta 520 (1–2) (2004) 105–115.

[40] W. Haasnoot, M. Bienenmann-Ploum, F. Kohen, Biosensor immunoassay for the detection of eight sulfonamides in chicken serum, Anal. Chim. Acta 483 (1–2) (2003) 171–180. https://doi.org/10.1016/S0003-2670(02)01545-3.

[41] S. Devlin, J. Meneely, B. Greer, K. Campbell, V. Vasconcelos, C. Elliott, Production of a broad specificity antibody for the development and validation of an optical SPR screening method for free and intracellular microcystins and nodularin in cyanobacteria cultures, Talanta 122 (2014) 8–15.

[42] E. Stack, Studies on the Immunosuppressive Effects and Detection of Naturally-occurring Toxins, 2012.

[43] B. Prieto-Simón, H. Miyachi, I. Karube, H. Saiki, High-sensitive flow-based kinetic exclusion assay for okadaic acid assessment in shellfish samples, Biosens Bioelectron. 25 (6) (2016) 1395–1401.

[44] R.C. Stevens, S.D. Soelberg, B.-T.L. Eberhart, S. Spencer, J.C. Wekell, T.M. Chinowsky, et al., Detection of the toxin domoic acid from clam extracts using a portable surface plasmon resonance biosensor, Harmful Algae 6 (2) (2007) 166–174.

[45] F.C. Dudak, H.I. Boyaci, Development of an immunosensor based on surface plasmon resonance for enumeration of *Escherichia coli* in water samples, Food Res. Int. 40 (2007) 803–807.

[46] Centres for Disease Control and Prevention (CDC), Surveillance for Foodborne Disease Outbreaks, United States, 2013, Annual Report, CDC, Atlanta, Georgia: US Department of Health and Human Services, 2015. Available online from: http://www.cdc.gov/foodsafety/pdfs/foodborne-disease-outbreaks-annual-report-2013-508c.pdf.

[47] S.B. Inbaraj, B.H. Chen, Nanomaterial-based sensors for detection of foodborne bacterial pathogens and toxins as well as pork adulteration in meat products, J. Food Drug Anal. 24 (1) (2016) 15–28.

[48] N. Gilmartin, R. O'Kennedy, Nanobiotechnologies for the detection and reduction of pathogens, Enzyme Microb. Technol. 50 (2) (2012) 87–95.

[49] G.C. Bokken, R.J. Corbee, F. Knapen van, A.A. Bergwerff, Immunochemical detection of salmonella group B, D and E using an optical surface plasmon resonance biosensor, FEMS Microbiol. Lett. 222 (1) (2003) 75–82.

[50] P. Leonard, S. Hearty, J. Quinn, R. O'Kennedy, A generic approach for the detection of whole *Listeria monocytogenes* cells in contaminated samples using surface plasmon resonance, Biosens. Bioelectron. 19 (10) (2004) 1331–1335.

[51] C.D. Kang, S.W. Lee, T.H. Park, S.J. Sim, Performance enhancement of real-time detection of protozoan parasite, *Cryptosporidium* oocyst by a modified surface plasmon resonance (SPR) biosensor, Enzyme Microb. Technol. 39 (3) (2006) 387–390.

[52] S. Wang, X. Shan, U. Patel, X. Huang, J. Lu, J. Li, et al., Label-free imaging, detection, and mass measurement of single viruses by surface plasmon resonance, PNAS U.S.A. 107 (37) (2010) 16028–16032.

[53] B.J. Yakes, E. Papafragkou, S.M. Conrad, J.D. Neill, J.F. Ridpath, W. Burkhardt, et al., Surface plasmon resonance biosensor for detection of feline calicivirus, a surrogate for norovirus, Int. J. Food Microbiol. 162 (2) (2013) 152–158.

[54] L. Huang, G. Reekmans, D. Saerens, J.-M. Friedt, F. Frederix, L. Francis, et al., Prostate-specific antigen immunosensing based on mixed self-assembled monolayers, camel antibodies and colloidal gold enhanced sandwich assays, Biosens. Bioelectron. 21 (2005) 483–490.

[55] Y. Uludag, I.E. Tothill, Cancer biomarker detection in serum samples using surface plasmon resonance and quartz crystal microbalance sensors with nanoparticle signal amplification, Anal. Chem. 84 (2012) 5898–5904.

[56] Z. Jiang, Y. Qin, Z. Peng, S. Chen, C. Deng, J. Xiang, The simultaneous detection of free and total prostate antigen in serum samples with high sensitivity and specificity by using the dual-channel surface plasmon resonance, Biosens. Bioelectron. 62 (2014) 268–273.

[57] C. Yang, E. Brooks, Y. Li, P. Denny, C. Ho, F. Qi, et al., Detection of picomolar levels of interleukin-8 in human saliva by SPR, Lab Chip 5 (10) (2005) 1017–1023.

[58] W. Hu, H. Chen, Z. Shi, L. Yu, Dual signal amplification of surface plasmon resonance imaging for sensitive immunoassay of tumor marker, Anal. Biochem. 453 (2014) 16–21.

[59] S. Mariani, M. Minunni, Surface plasmon resonance applications in clinical analysis, Anal. Bioanal. Chem. 406 (9) (2014) 2303–2323.

[60] S. Barizuddin, S. Bok, S. Gangopadhyay, Plasmonic sensors for disease detection - a review, J. Nanomed. Nanotechnol. 7 (3) (2016) 1–10. https://doi.org/10.4172/2157-7439.1000373.

[61] J.R. Krone, R.W. Nelson, D. Dogruel, P. Williams, R. Granzow, BIA/MS: interfacing biomolecular interaction analysis with mass spectrometry, Anal. Biochem. 244 (1) (1997) 124–132.

[62] J. Borch, P. Roepstorff, SPR/MS: recovery from sensorchips for protein identification by MALDI-TOF mass spectrometry, in: N.J. Mol, M.J. Fischer (Eds.), Methods in Molecular Biology, vol. 627, Humana Press, Clifton, N.J., 2010, pp. 269–281.

[63] D. Nedelkov, Integration of SPR biosensors with mass spectrometry (SPR-MS), in: N.J. Mol, M.J.E. Fischer (Eds.), Methods in Molecular Biology vol. 627, Humana Press, Clifton, N.J., 2010, pp. 261–268.

[64] Y. Kim, S. Yi, C. Lee, Y. Jung, B. Chung, Gold patterned biochips for on-chip immuno-MALDI-TOF MS: SPR imaging coupled multi-protein MS analysis, Analyst 137 (2011) 386–392.

[65] G. Kijanka, S. Hector, E.W. Way, F. Murray, R. Cummins, D. Murphy, et al., Human IgG antibody profiles differentiate between symptomatic patients with and without colorectal cancer, Gut 59 (2010) 69–78.

[66] J. Buijs, G. Franklin, SPR-MS in functional proteomics, Brief. Funct. Genomics 4 (1) (2005) 39–47.

[67] M. Couture, S.S. Zhou, J.F. Masson, Modern surface plasmon resonance for bioanalytics and biophysics, Phys. Chem. Chem. Phys. 15 (2013) 11190–11216.

[68] A. Hemmi, T. Usui, A. Moto, T. Tobita, N. Soh, K. Nakano, et al., A surface plasmon resonance sensor on a compact disk-type microfluidic device, J. Sep. Sci. 34 (2011) 2913–2919.

[69] D.J. Kinahan, S.M. Kearney, M.T. Glynn, J. Ducrée, *Spira mirabilis* enhanced whole blood processing in a lab-on-a-disk, Sens. Actuators A-Phys. 215 (2014) 71–76.

[70] D. Wang, S. Fan, Microfluidic surface plasmon resonance sensors: from principles to point-of-care applications, Sensors (Basel) 16 (8) (2016), 1175(1–18).

[71] Biosurfit, Biosurfit > Available Tests. Biosurfit.com, 2016. http://biosurfit.com/spinit-available-tests/.

[72] S. Vashist, O. Mudanyali, E. Schneider, R. Zengerle, A. Ozcan, Cellphone-based devices for bioanalytical sciences, Anal. Bioanal. Chem. 406 (14) (2013) 3263–3277.

[73] Y. Liu, Q. Liu, S. Chen, F. Cheng, H. Wang, W. Peng, Surface plasmon resonance biosensor based on smart phone platforms, Sci. Rep. 5 (2015) 1–9.

[74] H. Guner, E. Ozgur, G. Kokturk, M. Celik, E. Esen, A.E. Topal, et al., A smartphone-based surface plasmon resonance imaging (SPRi) platform for on-site biodetection, Sens. Actuators B-Chem. 239 (2017) 571–577.

[75] A. Priye, S. Wong, Y. Bi, M. Carpio, J. Chang, M. Coen, et al., Lab-on-a-drone: toward pinpoint deployment of smartphone-enabled nucleic acid-based diagnostics for mobile health care, Anal. Chem. 88 (9) (2016) 4651–4660.

[76] M.C. Palframan, H.A. Gruszewski, D.G. Schmale, C.A. Woolsey, Design and implementation of an aerobiological sampling system for real-time characterization with a surface plasmon resonance equipped UAV, AIAA Infotech@Aerospace (I@A) Conference, American Institute of Aeronautics and Astronautics (AIAA), 2013.

Chapter 7

Lateral Flow Immunoassays

Aart van Amerongen[1], Jeroen Veen[2], Hugo A. Arends[2], Marjo Koets[1]
[1]*Wageningen University & Research, Wageningen, The Netherlands;* [2]*HAN University of Applied Sciences, Arnhem, The Netherlands*

1. INTRODUCTION

1.1 Lateral Flow Immunoassays

1.1.1 History of the Technology

The lateral flow (immuno)assay is a relatively old diagnostic platform that was developed in the 1980s. Preceding research has focused on the development of nanoparticles that were used in, e.g., agglutination assays (gold; [1]) and on the design of new assay principles, the flow through and the lateral flow technologies. The first lateral flow test that entered the market was the Clearblue 1-step test in 1988 (Unipath) [2] as a home test for the detection of the pregnancy hormone. Blue latex particles were used as signal labels. Soon other companies followed with other nanoparticle-labels (gold; Organon Teknika) or adapted devices. The lateral flow test was the first diagnostic platform that could be performed by laypersons, although the rates of false-negative and false-positive test results were too high in those years. Over the almost 30 years since, a very diverse panel of lateral flow immunoassays has been developed and marketed. The platform is used for the detection of targets such as microorganisms (infectious agents, pathogens), contaminants such as antibiotics and mycotoxins, proteins (biomarkers, allergens), pesticides, toxins, carbohydrates, and specific DNA/RNA sequences [3–13]. Increasingly, lateral flow immunoassays are being used as human diagnostics to rapidly assess infection-causative agents, disease-related biomarkers, and allergy responses [14–21]. Obviously, the old platform can be transformed into a modern diagnostic tool that can be integrated with nowadays technologies such as optical and electrochemical transducers and wireless data transfer. Nevertheless, being a point-of-care (PoC) technology, lateral flow assays will preferably be staying Affordable, Sensitive, Specific, User-friendly, Rapid and Robust, Equipment-free and Delivered to end users (ASSURED), although equipment will be needed for recent platform developments into multianalyte assays [22].

Handbook of Immunoassay Technologies. https://doi.org/10.1016/B978-0-12-811762-0.00007-4

1.1.2 Basic Technology

The principle of the lateral flow technology has been described many times. To our knowledge, the first review of this technology Posthuma-Trumpie and coauthors explain the lateral flow platform and provide an SWOT analysis (Strengths, Weaknesses, Opportunities, and Threats) [23]. As a short introduction, a description of a sandwich lateral flow assay will be given: a lateral flow test consists of a set of membranes that are mounted (glued) on a plastic card for support. The first membrane is the sample pad that will collect the sample and, if necessary, filter particular components from the sample such as blood cells that would clog the test membrane if allowed to flow in (see Fig. 7.1 for a schematic overview of a lateral flow test). The sample pad overlays a so-called conjugate pad by 1−2 mm such that the sample will flow from the sample pad to the conjugate pad by means of capillary force. In the conjugate pad the signal label, often colored nanoparticles, is present in a dry form. The signal label is bound to binding molecules such as antibodies that are specific for the target analyte(s) (the detection molecules). In most cases, conjugate pads are made of glass fiber or polyester. On addition of sample and buffer (if buffer components are not dried into the sample pad), the label will be released from the pad and will flow into the next membrane, a nitrocellulose test membrane that is overlayed by the conjugate pad (1−2 mm). Over the years, characteristics and properties of nitrocellulose membranes have improved substantially, and this is the main reason that nitrocellulose is the primary membrane used in rapid and simple diagnostics [24]. In the nitrocellulose membrane, often having dimensions of 2.5 by 0.5 cm, binding molecules, specific for the same target-analyte(s), have been immobilized in lines or spots (the capture molecules). Finally, at the distal side, the nitrocellulose membrane is overlaid by an absorbance pad that guarantees that the fluid will flow through the nitrocellulose membrane into the absorbance pad. Through all membranes the fluid flows by capillary force; no external pump or other means to move the fluid is needed. If the target analyte is present, it will be sandwiched between the detection and capture molecules leading to immobilization of the signal label at the line or spot position of the capture molecules. If sufficient label is immobilized, a color or other detection signal is visible or detectable, respectively.

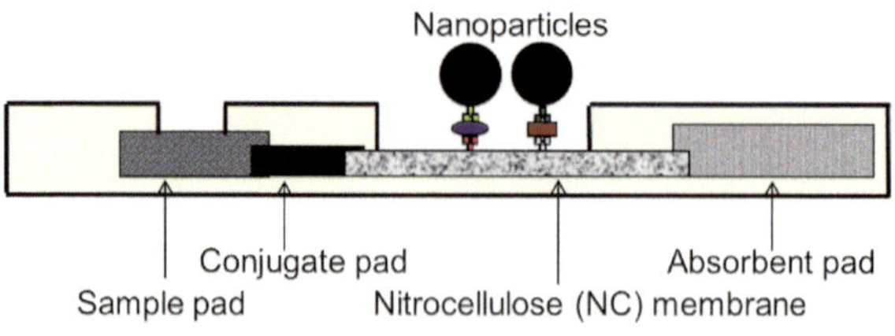

FIGURE 7.1 Side view of a lateral flow immunoassay showing a sandwich-type of assay.

In most cases, the set of membranes is packaged in a plastic device that has a sample application and a reading window. In particular cases, other windows may be present, for example, to add sample and running buffer at different positions. The positioning and the way the set of membranes is secured are crucial to the reproducibility and performance of the test. Pressing membranes, especially the nitrocellulose one, will damage the membrane pore structure and, consequently, its flow characteristics.

1.1.3 Recognition Elements

Recognition elements, often binding molecules, are an essential part of the lateral flow platform. Specificity and sensitivity are being determined by these molecules and, therefore, choosing the right molecules is very important. In most cases, antibodies are used to specifically bind target analytes. The origin of these antibodies is mouse, rat, rabbit, guinea pig, and human if monoclonal antibodies are being used, or rabbit, goat, sheep, donkey, and some other animals if polyclonal antibodies are being used. Other binding molecules are oligonucleotides, aptamers, peptides, and enzymes. Oligonucleotides are being used in several ways: to interact specifically with a complementary nucleotide sequence from the target organism/cell, or as unique addresses in the assay to distinguish various targets in multianalyte assays. Increasingly, aptamers are being applied in diagnostic assays. These RNA, DNA, or peptidic molecules are versatile tools, not only in diagnostics but also as therapeutics. A recent review on aptamers has been published by Vorobyeva et al. [25]. Enzymes have been essential reagents in immunoassays for decades. The possibility to amplify signals has been a great benefit to the diagnostic community; it is one of the reasons that diagnostic assays are being applied in such large numbers. In addition to being a reagent to amplify signals, enzymes can also be used as specific recognition molecules, for example, to convert a target analyte to a derivative that can be measured in one way or another. This may be very attractive in cases where the development of a specific binding molecule is difficult or impossible, e.g., if molecules that resemble the target analyte in large detail would generate cross-reactive signals in immunoassays. In such cases, the application of enzymes could be of help in discriminating between closely related analytes.

As compared to end-point assays, such as ELISAs, lateral flow assays require binding molecules, especially capture molecules immobilized on the nitrocellulose membrane, which have a fast k_{on} rate [26]. Target analytes in the sample will pass the test zone in few seconds, whereas incubations in an end-point assay have a duration of over 30 min. Therefore, choosing the right binding molecules is of utmost importance; evaluation of suitable binding molecule pairs should be done in an assay that has similar affinity requirements as the lateral flow assay format.

1.1.4 Signal Labels

Signal labels are key to lateral flow immunoassays. In the majority of commercially available lateral flow tests nanoparticles are being used for this purpose. These particles not only have a color or other detection principle but are also used to couple binding/detection molecules to their surface and to provide a high potential for binding target analytes. Colored latex and especially gold nanoparticles have been used most often in these assays. Other labels such as upconverting phosphors, magnetic particles, carbon, quantum dots, cellulose nanobeads, and silica have been reviewed recently [27–29]. Alternatively, chemical labels are used that have an intrinsic signal such as fluorescence or near-infrared [30,31]. The advantage of chemical labels as compared to (nano)particles, which are in the range of 10 nm (quantum dots) to micrometers, is their small size (<1 nm). Binding molecules, antibodies have dimensions of 5–15 nm, with such small labels are being released from the conjugate pad much easier than the complex conjugate of binding molecules immobilized on the surface of nanoparticles [32].

An example of a multicolor test as developed by BioSensing & Diagnostics, Wageningen University & Research, is shown in Fig. 7.2. Silica nanoparticles (approximately 50 nm in diameter) were produced from scratch,

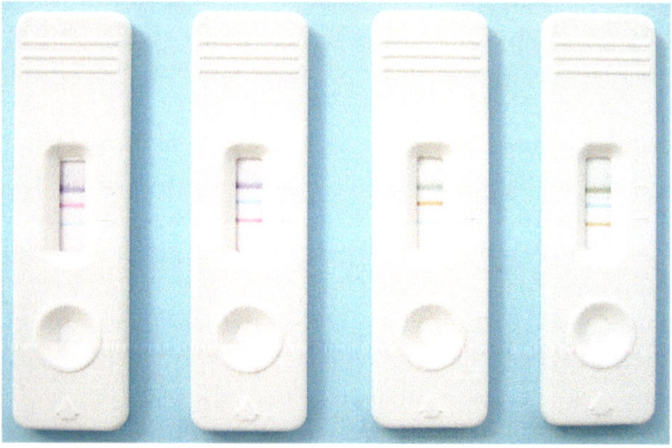

FIGURE 7.2 Multicolor test with silica nanoparticles (approximately 50 nm in diameter) that were produced from scratch. Three colored batches were prepared with internally and covalently bound dyes: phthalocyanine (PC; blue), sulfo-rhodamine (SR; pink), or dabsyl chloride (DC; yellow-orange). Half of each batch was labeled with a monoclonal antibody (MAb) against human serum albumin (HSA) and the other half with a rabbit IgG fraction. On the nitrocellulose membranes, the lower line was sprayed with HSA, the second line with goat antirabbit IgG (PAb), and the third line with a mix of these capture molecules. The silica batch colored with PC and conjugated to the anti-HSA MAb was combined with the silica batch labeled with SR and DC, respectively, both conjugated to the rabbit IgG (Fig. 7.2, first and third device). The silica batch colored with PC and conjugated to rabbit IgG was combined with the silica batch labeled with SR and DC, respectively, both conjugated to the MAb anti-HSA (Fig. 7.2, second and fourth device). Developed by Jan Wichers, BioSensing & Diagnostics, Wageningen University & Research.

and various dyes were covalently bound inside the particles: phthalocyanine (PC; blue), sulforhodamine (SR; pink), or dabsyl chloride (DC; yellow-orange) was incorporated. Half of each batch of colored silica nanoparticles were labeled with a monoclonal antibody (MAb) against human serum albumin (HSA) and the other half with a rabbit IgG fraction. On the nitrocellulose membranes, the lower line was sprayed with HSA, the second line with goat antirabbit IgG (PAb), and the third line with a mix of these capture molecules. The silica batch colored with PC and conjugated to the anti-HSA MAb was combined with the silica batch labeled with SR and DC, respectively, both conjugated to the rabbit IgG (Fig. 7.2, first and third device). The silica batch colored with PC and conjugated to rabbit IgG was combined with the silica batch labeled with SR and DC, respectively, both conjugated to the MAb anti-HSA (Fig. 7.2, second and fourth device). As can be seen the binding to the single antigen lines resulted in the color as expected, whereas the third line, having both capture molecules in a mix, shows the mixed color of the silica batches combined, i.e., purple in the case of silica-PC and silica-SR and green in the case of silica-PC and silica-DC. Interestingly, this setup can also be used as a "positive hapten" test, the hapten being a low molecular weight antigen such as an antibiotic or a mycotoxin. In a conventional competitive lateral flow assay, the color of the specific test line disappears on adding increasing concentrations of the hapten. The multicolor approach could be used to induce a change in color instead of the disappearance of a single-color line. Alternatively, sandwich assays could be developed with one, although multispecific line. In the present example in Fig. 7.2 the maximum wavelength of the color observed at the specific line could be assigned to capture molecule 1 or 2 or to combinations of these molecules in particular ratios.

7.7.5 Storage of Lateral Flow Devices

Following the application of bioreagents on the membrane(s), solvent fluid is removed in a drying step, often at elevated temperature (e.g., 37°C). To keep the bioreagents in their native, and thus functional, conformation surfactants and/or sugars such as trehalose and sucrose are added, especially to the detection molecules. Subsequently, the membrane set is assembled in a plastic cassette and packaged in an aluminum pouch with a silica desiccant bag inside to keep the humidity low. On the average lateral flow, devices can be stored for at least 1 year. On opening the pouch to run the test the dry reagents and particles should redissolve following the addition of sample (and buffer). Especially the release of the (nano)particles from the conjugate pad is not straightforward; it appears to be difficult to reach a high reproducibility for this process [32]. As a remedy, some producers lyophilize the signal label (binding molecules immobilized on the (nano)particles) and additives (surfactants, sugars) to a pellet that is added to a plastic tube. On running the test, the sample and buffer are used to redissolve the pellet into suspension and then, the membrane strip is placed in the tube. In this way, a better assay reproducibility is reached. Another possibility is to pipet the label in the tube and to simply evaporate the fluid by drying at 37°C (BioSensing & Diagnostics, Wageningen University & Research). The tube with the dried conjugate is then packaged in an aluminum pouch together with a ready-to-use nitrocellulose membrane that has been assembled on a plastic backing with an absorbance pad on the distal end. No conjugate pad has been added, and a sample pad is only applied if the sample contains material, e.g., cells or particulate matter, that would disturb a correct performance of the test. Again, sample and buffer are used to redissolve the dried label by gently shaking the tube. Next, the assembled membrane is put in the tube to start the run (Fig. 7.3). The reproducibility of this method appeared to be high as was shown in two studies in which fungal α-amylase was detected in bakeries [12,33] and malaria-specific amplicons in a field study in Kenya [34]. Especially given the increased interest in quantitative lateral flow assays, the reproducibility of each of the production and performance steps should preferably be optimal.

2. ADVANCES IN LATERAL FLOW IMMUNOASSAYS

2.1 Coupling to a Range of Detection Principles

Traditionally, lateral flow immunoassays are read by naked eye; there is a line or no line (qualitative assay). However, if a faint line is the result of the assay, unequivocal judgment appears to be difficult, leading to false-negative,

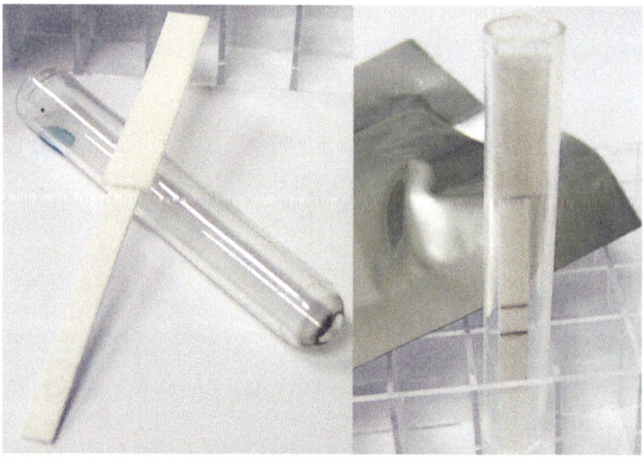

FIGURE 7.3 Tube format of a lateral flow immunoassay. Signal label conjugated to detection binding molecules is pipetted into a plastic tube and dried at 37°C (left panel) and then packaged in an aluminum pouch together with a ready-to-use nitrocellulose membrane that has been assembled on a plastic backing with an absorbance pad on the distal end. Sample and buffer are used to redissolve the dried label by gently shaking the tube. Next, the assembled membrane is put in the tube to start the run (right panel).

false-positive, or questionable results. Therefore, an independent and objective way to read a line will lead to fewer interpretation errors. This explains the introduction of lateral flow readers some 15 years ago. In addition to reading colorimetrically, readers enabled the introduction of other, more sensitive detection principles such as fluorescence, upconverting phosphor, magnetism, electrochemistry, chemiluminescence, surface-enhanced Raman spectroscopy (SERS), and near-infrared [15,30,35–48]. Interestingly, reading a lateral flow test by Google glass was reported as well [49]. In the meantime reading lateral flow tests has become possible using smartphones that are broadly available. The new detection principles also led to increased sensitivity and reproducibility of these assays. In addition, readers also enabled the quantification of lateral flow assays.

A quite recent development in the lateral flow immunoassay technology, i.e., the multianalyte or multiplex format, will further promote the application of readers. If an array of dots will be developed on a lateral flow membrane, the interpretation by the naked eye would fail, and a reader would be necessary. For this format, the readers that have been developed to read lines should be transformed into an apparatus that recognizes the array grid and processes the subimages, the individual spots, to a final result. In the next section, the multianalyte lateral flow assays will be further introduced.

2.2 Multianalyte and Quantitative Lateral Flow Immunoassays

The way forward for the lateral flow immunoassay technology is the development of sensitive, multianalyte, and quantitative assays that can compete with some of the diagnostics now used in the laboratory. The strong points already available (one-step, fast, simple, low cost, PoC/on-site, long shelf life) would be extended to meet the desire of many potential users. Nowadays, quantitative lateral flow tests for single or a low number of analytes are already on the market supported by dedicated readers, but %CV and reproducibility are still insufficient and should be improved substantially. This would require improvements at various levels, from sample pretreatment or integrated pretreatment modules, membranes, signal labels, the release of dry labels, and up to detection principles [32]. Preferably, such improvements would be part of one device that would enable a "sample in, result out" approach. The trend for multianalyte assays is partly based on the fact that a lot of diagnostic questions require the detection of 5–10 different analytes to explain up to 90% of the particular problem. Especially, for screening human diseases, multianalyte tests would revolutionize the diagnostic practice; the rapid and multiple answers could lead to timely medication and/or treatment.

In developing multianalyte lateral flow assays, there is much to be learned from the development of multianalyte planar and suspension immunoassay arrays that started some 30 and 20 years ago, respectively [50]. Development items such as the application of analytes in arrays, reading arrays, and software to process results could be transferred without substantial changes to the multianalyte lateral flow technology. The successful integration of available knowledge would speed up this development to a great extent. In literature, some tens of patents and articles have been published on lateral flow microarray assays [51–56], but only some of these have been marketed recently (February 2017).

Successful development of lateral flow microarray assays requires solutions to some problems that result from the increase in the number of different target analytes combined with some well-known characteristics of lateral flow membranes and their performance. In lateral flow assays, the flow speed is, for example, not constant over the length of the membrane. Due to issues such as slower flow speeds in regions more downstream from the sample application area (more interaction time for binding molecule and analyte!), evaporation of fluid, and the characteristics of the membrane and the absorbance pad, interaction time between binding molecules (capture and detection) and the analytes may differ. This means that the location of a spot with reference to the beginning of the strip is partly determining the maximal signal that can be reached for that particular spot. Another point to consider is the arrangement of the spots in the array on the membrane. Spots may be arranged in a lattice structure (e.g., a hexagonal lattice), in a rectangular array, or in any other format, and the array may include several rows of spots.

At BioSensing & Diagnostics (Wageningen University & Research) we carried out an experiment to study the influence of microarray pattern and spot position on the final assay signal (Fig. 7.4). Five different microarray patterns were printed on lateral flow membranes each having 12 identical spots (biotinylated goat antimouse IgG). The strips were run with a conjugate of carbon nanoparticles and neutravidin. The average spot intensity per row was normalized against the average of row 1 (100%). As can be seen in the lower panel of Fig. 7.4, the overall trend is a lower intensity for spots more distal from the beginning of the strip (decrease in carbon nanoparticles concentration due to binding to spots in more proximal rows). However, differences can also be observed for spots in the same row but in different patterns. The position of the spot and the arrangement obviously play a role in the concentration of the reagents that flow through the membrane [19]. If the arrangement is rectangular, a spot in row 2 will be directly "in the shadow" of the spot in the same column but in row 1. Depending on the pitch distance between row 1 and 2, the mixing of reagents following the flow around and through the spot in row 1 may be restored in time or still be suboptimal. The latter would lead to less reagent passing the spot in row 2, and this would be even worse for spots in the same column but in rows at a greater distance from the beginning of the strip. In addition, and even if the mixing of reagents would have been optimal, specific binding of signal label to the spot in row 1 would lead to a slightly decreased concentration of label and target analyte on passing the spot in row 2, etc. Again, this would be worse for spots in the same column but more

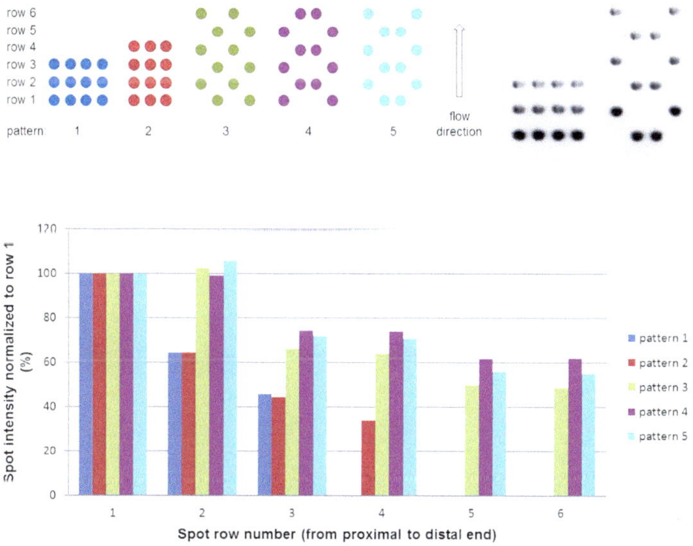

FIGURE 7.4 Influence of spot pattern and position on final signals. Five different microarray patterns were printed on lateral flow membranes each having 12 identical spots of biotinylated goat antimouse IgG (upper left panel). The strips were run with a conjugate of carbon nanoparticles and neutravidin. Two examples of frames recorded with a video camera are shown in the upper right panel. The average spot intensity per row was normalized against the average of row 1 as 100% (lower panel). Experiments were performed by Leander van Dijk at BioSensing & Diagnostics, Wageningen University & Research.

downstream from the sample application window (see Fig. 7.4). As is also clear from this figure, one way to improve this situation is to arrange spots in other formats, for example, in a lattice structure. Another option would be to increase the number of immobilized capture molecules to compensate for the decreased concentration of reagents in the fluid, but this approach will be limited by the binding capacity of the membrane. To make lateral flow microarray immunoassays fully quantitative, the above issues should be solved, although the introduction of positive control spots for each row may be an option as well. Such spots could be comparable to the control line in present lateral flow assays and could be used to normalize the results per row.

Even if the quantification problem would be solved, some other issues still delay the market introduction of these assays, all related to commercial production:

- The large-scale production of arrays on lateral flow membranes cannot be done with the present production equipment that applies lines of binding molecules on roles of nitrocellulose membranes. Scienion AG in Berlin has announced a new printing technology to speed up array production, Spot-on-the-Fly [57]. Potentially, this technology would enable the large-scale production of lateral flow microarray membranes. The much better reproducibility that is reached with inkjet printers as compared to equipment that is used to apply lines onto membranes would be a big advantage, especially in view of developing quantitative assays.

- Cutting lateral flow membranes is done by, for example, rotary slitting and guillotine shearing. This equipment can be programmed to cut the nitrocellulose membrane at every programmed width. However, cutting nitrocellulose membranes having arrays of spots cannot be done at random; a single membrane strip should contain a full array of spots. Recently, equipment was developed with an onboard camera to cut the membrane guided by vision technology (Kinematic; information through personal communication). Therefore, an inert dye is added during printing, which can be used to guide cutting, but also as a quality assurance that spots indeed have been printed.
- Microarray lateral flow assays cannot be read by the naked eye anymore. A dedicated reader will be needed to profit from the additional value of multianalyte assays. The authors of this chapter are involved in some developments that may lead to systems that can be used broadly to read lateral flow microarray tests. In the next section, two developments will be explained in more detail.

2.3 Reading MultiSpot Lateral Flow Assays

2.3.1 Lateral Flow Reader for Microarrays

The company Axxin (Melbourne, Australia) is developing next-generation diagnostic instruments for microarray lateral flow applications. The Axxin AX-2X Array Instrument is a portable, highly flexible, and easy-to-use rapid testing instrument platform designed to provide qualitative and quantitative results for visual colorimetric and fluorescent immunoassays. Initiated by Scienion (Berlin, Germany) and in collaboration with BioSensing & Diagnostics from Wageningen University & Research, initial experiments were performed to transform an existing lateral flow line reader into a reader for microarray applications [58]. In Fig. 7.5, left upper panel shows the layout of a lateral flow assay for the detection of specific amplicons that encode virulence factors of *Escherichia coli* O157 [59]. Following a 30 min PCR to amplify the gene coding for the *ehxA* virulence factor (enterohaemolysin), 1 μL of the PCR solution and four 5-times dilutions up to a factor 625 was added to lateral flow microarray membranes. A picture of the developed membranes is shown in Fig. 7.5 as well (left lower panel). Membranes were read in the reader adapted to recognize the array grid automatically and to return background-subtracted pixel gray values (PGV) of each of the spots. The right panel of this figure shows the typical sigmoidal curve that was fitted by using the average PGV of the triplicate spots per dilution. The adapted reader is as user-friendly as the original line-reader, "device in, results out,"

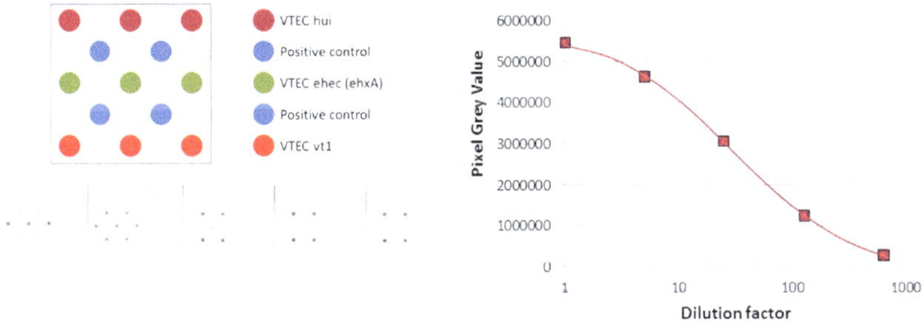

FIGURE 7.5 Lateral flow reader adapted for reading a microarray of spots (Axxin, Melbourne, Australia in collaboration with BioSensing & Diagnostics from Wageningen University & Research). A lateral flow microarray assay for the detection of specific amplicons that encode virulence factors of *Escherichia coli* O157 was used [59]. The layout is depicted in the upper left panel. Following a 30 min PCR to amplify the gene coding for the *ehxA* virulence factor (enterohaemolysin), 1 μL of the PCR solution and four 5-times dilutions up to a factor 625 was added to lateral flow microarray membranes. Pictures of the test zones are shown in the lower left panel. Membranes were read in the reader adapted to automatically recognize the array grid and to return background-subtracted pixel gray values (PGV) of each of the spots. The right panel of this figure shows the typical sigmoidal curve that was fitted by using the average PGV of the triplicate spots per dilution.

and enables the application of microarray lateral flow assays without having difficulties in reading and interpreting the individual spots.

2.3.2 Real-Time Video Reader

The HAN University of Applied Sciences and the group BioSensing & Diagnostics from Wageningen University & Research are developing a video reader that acquires and processes real-time images (frames) of the developing spots in a lateral flow microarray immunoassay. The reader consists of an inexpensive USB microscope camera and an LED-based illumination ring around the lens. This setup is assembled in a 3D-printed housing (see Fig. 7.6). The camera is connected to a laptop computer having dedicated software for image acquisition and processing. The full array is captured, and the software enables each spot to be addressed separately. During processing, a sequence of frames is acquired (typically 1000 frames in 10–15 min), wherein each spot is a region of interest. Features are calculated for each region of interest, resulting in a feature vector as a function of frame and spot. The frames sequence is indicative of the optical signals from the black carbon nanoparticles, used here as signal labels, at different points in time during the lateral flow. Additional information as to the evolution of binding reactions at each spot can be acquired and may allow detecting or quantifying the analytes more rapidly; for example, even before the lateral flow terminates, the liquid phase evaporates or asymptotically approaches an equilibrium state. Moreover, the additional information may allow detecting or quantifying the analytes

FIGURE 7.6 Video reader for the real-time recording of developing lateral flow microarray assays. The figure shows the 3D-printed housing as produced at the HAN University of Applied Sciences.

more accurately, especially since a number of spots can be used for a better statistical significance of the optical signals. It is also possible to compensate for differences in spot intensity that occur due to a different location with respect to the beginning of the strip. For compensation also positive reference spots can be used, which will be helpful in the normalization of data. Background values are defined as separate empty spots or are based on the average of the edge of the area surrounding each spot.

The output of the reader may be a semiquantitative signal that indicates that a particular antigen is below or above the threshold level, or, alternatively, is a quantitative result based on an internal software standard curve, or on the spot intensities of a series of specific antibody dilutions in the lateral flow microarray immunoassay. In the last option, quantitative data will be derived from a profile of the intensities of these spots. Frames are being acquired from the start of the experiments, and results can be output directly or as soon as sufficient significant data have been recorded. This means that the assay may be terminated well before the endpoint of the assay has been reached. Signal features may represent statistics (e.g., moments, cumulants, an average, and/or a variance) of pixels in the corresponding spot and may be based on local intensity and/or color of the pixels. Alternatively, the signal models may be time-dependent such as an extrapolated trend, e.g., the slope of the curve made up of the intensities of a particular spot over time, a base line and/or an asymptote of one or more of the signal features as a function of the points in time.

As an example, a lateral flow microarray immunoassay was developed for the detection of rat and mouse urinary antigens [60]. The antibodies were

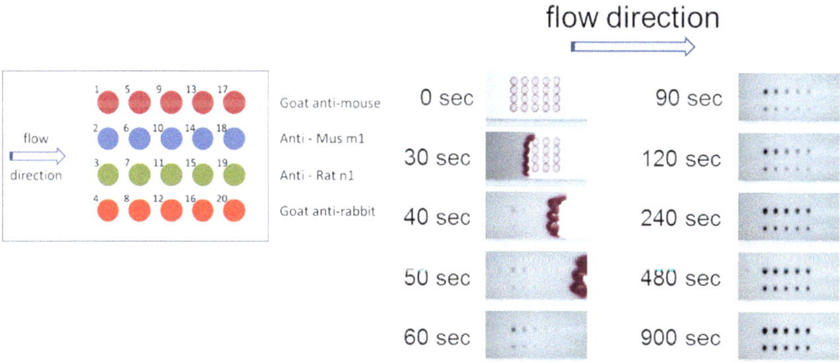

FIGURE 7.7 Lateral flow microarray immunoassay for the detection of rat and mouse urinary antigens [33]. Specific antibodies were developed and provided by Dr. Anne Renström from the Karolinska Institutet in Stockholm, Sweden. The layout of the array is shown in the left panel. The membrane had five identical spots per column as indicated. A carbon nanoparticles conjugate with anti-Mus m1 was run and video frames from the run with 1 μg/mL Mus m1 at indicated time points are shown in the right panel.

developed and provided by Dr. Anne Renström from the Karolinska Institutet in Stockholm, Sweden. In Fig. 7.7 the layout of the array on the lateral flow membrane is shown with five identical spots per column: one column with a monoclonal antibody MAb-6 recognizing the rat urinary antigen (Rat n1), a column with a polyclonal antibody (PAb) recognizing the mouse urinary antigen (Mus m1), and two control columns with goat antimouse and goat antirat, respectively. Two carbon nanoparticles conjugates were prepared: one with MAb-1 (anti-Rat n1) and the other one with the anti-Mus m1 PAb. In the example, three dilutions of Mus m1 (10, 100, and 1000 ng/mL) were run, and some of the frames from the recording of the Mus m1 1000 ng/mL experiments are shown in Fig. 7.7. It is obvious that the inert printing dye is removed on running the sample in the buffer and that the spots start appearing within 40 s. The sum of the local intensity differences of the first and the fifth spot of the Mus m1 concentrations are depicted in Fig. 7.8. It can be seen that in this particular setup the sum of local intensity differences is indeed larger for the first as compared to the fifth spot. In addition and as expected, per row, an increase is seen from 10 to 1000 ng/mL Mus m1. The final goal of this development is the printing of a dilution range of specific antibodies, one dilution per row. Such a range could increase the dynamic range of the assay and would also offer the possibility to cope with high-dose antigen effects (Hook effect). Furthermore, by performing dose−response experiments, it will be investigated if an algorithm can be found that can be used to return quantitative data on running unknown samples. Interestingly, from the slopes of the various curves, it should be possible to extract quantitative data information. If so, this would enable the very rapid

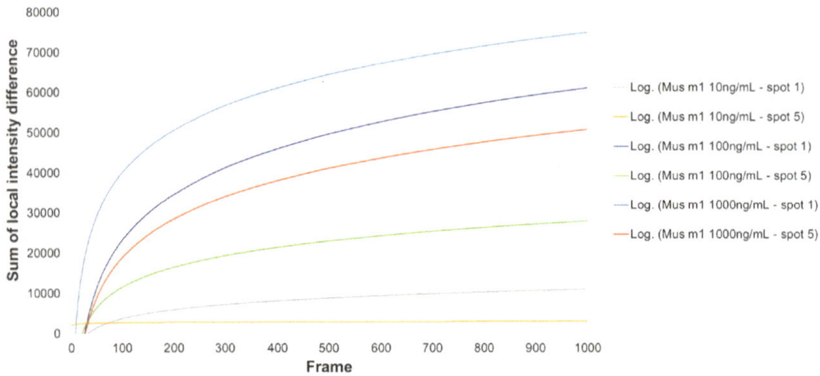

FIGURE 7.8 With reference to the legend of Fig. 7.7, the sum of the local intensity differences of the first and the fifth spot of the Mus m1 concentrations are depicted for 1000 frames as recorded in 15 min.

assessment of quantitative data, even far before the plateau phase of the assay would be reached.

2.3.3 Reading Arrays by a Smartphone Application

A recent trend in reading lateral flow assays is the use of smartphones [61–64]. Smartphones are worldwide available and are equipped with a camera and applications can be easily added. Some (potential) problems exist: there are several operating systems (Android, Apple, Microsoft) that would make it impossible to offer an all-in-one-solution. In addition, the quality of the camera (resolution of the chip, color/white balance, color temperature of the flash light, etc.) will influence the quality of the digital image and, most probably, the processing of the picture to final results. Adaptors have been introduced to standardize lighting conditions but are often dedicated to particular smartphones and, again, not an overall solution to the problems. An interesting option is the Cube-Reader from opTricon (Berlin, Germany) [65]. This very small (4 × 4 × 4 cm) and lightweight (40 g) reader is put over the reading window of the lateral flow cassette and returns results in seconds. Data can be transferred to PC or a laptop in seconds. Coupling to a smartphone would even further generalize this reading option. In that case, the smartphone would only be used as a data receiver (bluetooth or NFC) and transmission station, for example, for the (wireless) transfer of data to the doctor's office, or a centralized server. Subsequently, the data can be analyzed by an expert (doctor or production line manager) who could take immediate action to remedy a potential problem. Such scenarios are already available for some applications in which smartphones are being used as image recorders. It is very likely that this development will find widespread use in the near future, including Cube-Reader-like solutions.

3. CHALLENGES AND FUTURE DIRECTIONS

3.1 Updated SWOT Analysis

In a literature survey on lateral flow (immuno)assays by Posthuma-Trumpie et al. [23] a SWOT analysis was presented that characterized these assays at the time of the survey, i.e., 2009. As compared to that analysis, the Strength item does not need an update, but a number of points at the other items should be updated to the 2017 situation.

3.1.1 Weaknesses

"Only qualitative or semi-quantitative assays": Quantitative lateral flow assays have become possible and will expand in number in the near future.

"Usually designed for individual tests, not for high-throughput screening": Multianalyte or multiplex lateral flow assays have been described in patents and literature, and some have reached the market, although not yet as a traditional on-site/field test.

3.1.2 Opportunities

"New applications at point of care/need": Lateral flow immunoassays are already applied at the point of care, and many more will follow in the coming decade.

"Application on other biomatrices: tears, saliva, sweat": Assays that detect target analytes in these matrices have been described. Especially, since the introduction of readers and other signal detection principles allow for higher sensitivities, a new generation of tests may be expected. Analytes that are normally detected in the blood can often be measured in tears, saliva, and/or sweat as well. As compared to blood, a great advantage of these matrices is their noninvasive nature.

3.1.3 Threats

"Automated enzyme immunoassays": The introduction of readers initiated the automation of lateral flow assays. In the meantime, fully integrated systems have been developed, for example, by combining a DNA amplification to detection by lateral flow technology.

"Microparticle immunoassays (e.g., Luminex beads)": High-throughput and multianalyte bead technology (such as from Luminex and BD Biosciences) target other market segments than lateral flow assays. The latter are low-throughput, preferably applied in nonlaboratory/point-of-care/on-site settings and primarily meant for execution by laypersons. For both technologies, a market will exist.

"Lab-on-a-Chip technology": Despite great promise and a plethora of scientific publications, the commercial introduction of lab-on-a-chip devices is not in line with these expectations. The translation of excellent science on the

lab bench to robust and user-friendly devices appears to be difficult. The well-established lateral flow technology, perhaps rather a millifluidic than a microfluidic technology, is very well prepared for advanced future applications.

3.2 Combination With Amplification Procedures

A great number of lateral flow publications focus on combinations with DNA/RNA amplification strategies. One of the first examples was reported by Fong et al. [66]. The cycling probe technology was used to amplify specific DNA from the methicillin-resistant *Staphylococcus aureus* (MRSA), followed by the detection of amplified material using a lateral flow assay. A multiparameter oligonucleotide lateral-flow immunoassay for detection of antigens and antibodies was developed by Oku et al. [67]. Nucleic acid lateral flow immunoassays (NALFIAs) were presented by Koets et al. [68] and Van Amerongen and Koets [69]. Tagged-primers were used in PCR; all reverse primers were labeled with biotin and the forward primers with a discriminating label such as FITC and digoxigenin. Streptavidin was immobilized on carbon nanoparticles, and antibodies against the discriminating tags were immobilized on the nitrocellulose membrane. Examples shown were the detection of a synthase gene from *Pseudomonas putida*, a *Salmonella*-specific gene to detect this food pathogen, a wild-type gene and a GMO-insert gene in soy (Roundup Ready), and a dual-analyte format specific genes for *Escherichia coli* and *Bacillus cereus*, respectively. Very likely due to the future perspective that this combination of nucleic acid amplification and lateral flow detection will eventually lead to field applications, several studies show possibilities to detect tropical pathogens such as for sleeping sickness and tuberculosis [70,71]. A series of publications was published on the detection of the malaria-causing parasite *Plasmodium* [34,72–75]. For diagnosis of bacterial infection in arthroplasty, a PCR was combined with a lateral flow test in which specific amplicons were detected by a sandwich between a biotin-labeled oligonucleotide coupled to immobilized streptavidin in the membrane and an oligonucleotide probe conjugated to gold nanoparticles [76]. Genes coding for virulence factors of verotoxigenic *E. coli* and a NALFIA with two specificity lines for *Listeria* spp. and *Listeria monocytogenes*, respectively, are examples of rapid detection of food pathogens. In the *E. coli* example a multiplex PCR was developed that, without preceding DNA isolation, gave results for the five tested genes in 30 min. This was followed by the NALFIA detection and resulted in specific lines within 5 min [59]. In the case of *Listeria*, the development of the test was followed by an initial validation with milk samples and resulted in identical results as obtained with the microbiological standard method [11]. More recently applications were reported in which an internal amplification control was introduced [77], which combined the lateral flow test with loop-mediated isothermal amplification (LAMP) [78], detected DNA methylation sites [79], or made use of quantum dots as signal label with a low fmol-detection limit [5].

The combination of nucleic acid amplification followed by lateral flow detection of specific amplicons will only be successful if real on-site/field applications can be developed. Other technologies such as qPCR and next-generation sequencing (NGS) have been or are being introduced in diagnostic laboratories and are by far more suited for laboratory environments for reasons such as high-throughput, automation, and reproducibility. NALFIA-like applications will be commercially attractive if hurdles that exist for other technologies to be applied on-site/in the field can be taken. Several attempts have been published in which a cartridge is presented that enables both the amplification of nucleic acids and the lateral flow detection of specific amplicons [80–82]. To make such devices user-friendly, other points should be addressed such as (integrated) sample pretreatment [83], easy application of sample to the device, storage of dry amplification reagents, a cheap and user-friendly apparatus that controls amplification (e.g., temperature) and detection (e.g., imaging of the results), and a contamination-free device. The last point is necessary to prevent aerosols with high amplicon loading to contaminate the area where the test is executed. Integration of the various technologies that will be needed to enable commercial application of nucleic acid amplification combined with lateral flow detection may be expected within the next decade, provided the devices will be affordable.

3.3 Integration of Lateral Flow Immunoassays With Paper Diagnostics

Over the last decade, a new area of rapid diagnostics has evolved, i.e., paper diagnostics [84]. In paper diagnostics, paper-based materials are being used in a much broader sense than in lateral flow diagnostics that is almost exclusively restricted to nitrocellulose [85]. In this respect, lateral flow assays may be regarded as a subset of paper diagnostics. Several reviews on paper diagnostics excellently describe the field and show that the possibilities in paper diagnostics are much broader as compared to lateral flow technology [86–88]. Paper diagnostics are much more versatile in that they allow three-dimensional networks to be formed, easy compartmentalization, and integration of a vast number of other technologies such as electronics and microfluidics on paper [89]. To illustrate the diversity of assays that can be designed, some publications will be mentioned. Liver function with respect to alkaline phosphatase, aspartate aminotransferase, and total serum protein was measured with a vertical flow paper device [90]. A scanner, which was used to digitize the results, was the only piece of equipment applied in the execution of the test. It was noted that the scanner could be replaced by a smartphone to enable the equipment-free measurement of these liver functions. Two other studies show the added value of paper diagnostics in combination with the lateral flow technology [82,91]. Sample-to-result/-answer is presented in which a paper-based part is applied for amplification of nucleic acids and a lateral flow membrane for identification of specific amplicons. As an example of a simplified ELISA, a paper-based device was described for automating the sequential multistep procedures of a sandwich-type ELISA to measure the human pregnancy hormone [92]. This device only requires a single-step application of the sample solution and was based on a piece of nitrocellulose membrane with specially designed channels, where all the reagents are applied at different locations to control the fluid travel to the detection region. The inkjet printing method, a simple and low-cost process, was used to create the flow channel and device barrier patterns. The fabricated barrier was found to be an efficient boundary for the liquid along the printed design in the NC membrane, enabling direct control of the reagent flow time. Although this study uses lateral flow technology and a patterned nitrocellulose membrane, the absence of the words "lateral flow" in the text of this publication illustrates the strong overlap in technology and application fields between lateral flow technology and paper diagnostics. It is anticipated that further integration of these technologies will be seen in the near future.

4. BIBLIOGRAPHIC AND COMMERCIAL DATA

The lateral flow diagnostic platform is gaining increased interest both from the commercial and the scientific community. In the scientific literature, a substantial rise in the number of publications on lateral flow diagnostics can be observed. In a recent (January 2017) search in the Scopus database for "lateral flow" or immunochromatography-related keywords in article titles, an exponential growth in the number of articles could be seen, especially in the years 2015 and 2016 to over 300 articles per year (Fig. 7.9). In a total of almost 2400 articles, over 500 were published by Chinese researchers, approximately 450 by US researchers, and some 250 by Japanese groups. However, the actual number of publications on lateral flow diagnostics is much larger. When searching in article title, abstract, and keywords, a number of over 7000 articles was returned. In addition, "dipstick" was not a keyword for this search, although it is occasionally and erroneously used as a synonym to a lateral flow test.

Also from a commercial point of view the lateral flow immunoassay technology is big business. In a recent report by marketsandmarket.com (March 2016) the global lateral flow assay market is projected to reach USD 6.78 billion by 2020 [93]. The summary of the report states that the market was valued at USD 4.56 billion in 2015 and expected to grow at a CAGR (compound annual growth rate) of 8.3% from 2015 to 2020. Over the last 5 years, substantial growth and rapid technological advancements have been witnessed in this market that is mainly driven by the rising geriatric population, high prevalence of infectious diseases, rising usage of home-based

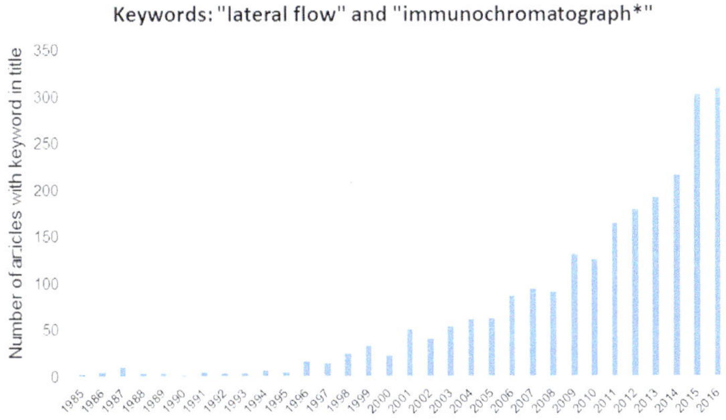

FIGURE 7.9 Search (January 2017) in the Scopus database for "lateral flow" or immunochromatography-related keywords in article titles. The figure shows the numbers per year from 1985 to 2016.

lateral flow assay tests, improved focus on decentralized or point-of-care diagnostic testing, and increasing application of lateral flow assay techniques in clinical diagnostics. However, the report also notes that the lateral flow market has difficulty in obtaining regulatory approvals due to the constantly changing standards. Also, variance in test results and insufficient reimbursements for lateral flow assay procedures are posing several challenges to the overall growth of the market.

The trend in both the scientific literature and the commercial market shows the strong interest in the lateral flow diagnostic platform. In the past particularly seen as a simple test to confirm a pregnancy, this platform is being transformed into a competitive diagnostic by the combination with and the incorporation of advanced technologies.

5. CONCLUSIONS

The lateral flow (immuno)assay is a relatively old diagnostic platform that was developed in the 1980s. Over the almost 30 years since, a very diverse panel of lateral flow immunoassays has been developed and marketed. The platform is used for the detection of targets such as microorganisms, contaminants such as antibiotics and mycotoxins, proteins (biomarkers, allergens), carbohydrates, and specific DNA/RNA sequences. At the moment the platform is being transformed into a modern diagnostic tool that can be integrated with nowadays technologies such as optical and electrochemical transducers and wireless data transfer. In addition, more sensitive detection principles have become available such as fluorescence, upconverting phosphor, magnetism, electrochemistry, chemiluminescence, SERS, and near-infrared.

The way forward for the lateral flow immunoassay technology is the development of sensitive, multianalyte, and quantitative assays that can compete with some of the diagnostics now used in the laboratory. The strong points already available (one-step, fast, simple, low cost, PoC/on-site, long shelf life) will be extended to meet the desire of many potential users. Especially, for screening human diseases, multianalyte tests will revolutionize the diagnostic practice; the rapid and multiple answers can lead to timely medication and/or treatment. Successful development of lateral flow microarray assays requires solutions to some problems that result from the increase in the number of different target analytes combined with some well-known characteristics of lateral flow membranes and their performance: the flow speed is not constant over the length of the membrane, the location of a spot with reference to the origin, and the arrangement of the spots in the array on the membrane influences the final results. Even if the quantification problem will be solved, a number of other issues still delays the market introduction of these assays, all related to commercial production: large scale production (being solved recently), cutting lateral flow membranes (also solved), and a dedicated reader. An adapted lateral flow reader, a real-time video reader, and reading by smartphone applications or by smartphone as an intermediate are being developed.

One of the challenges and future directions is the combination with nucleic acid amplification procedures. Integration of various technologies will be needed to enable commercial application of nucleic acid amplification combined with lateral flow detection. Another future direction is the integration of lateral flow immunoassays with paper diagnostics. In paper diagnostics, paper-based materials are being used in a much broader sense than in lateral flow diagnostics that is almost exclusively restricted to nitrocellulose. In this respect, lateral flow assays may be regarded as a subset of paper diagnostics. It is anticipated that further integration of these technologies will be seen in the near future.

The trend in both the scientific literature and the commercial market shows the strong interest in the lateral flow diagnostic platform. In the past particularly seen as a simple test to confirm a pregnancy, this platform is being transformed into a competitive diagnostic by the combination with and the incorporation of advanced technologies.

REFERENCES

[1] J. Leuvering, P. Thal, M. van der Waart, A. Schuurs, Sol particle immunoassay (SPIA), J. Immunoass. 1 (1980) 77–91.

[2] P. Kennedy, Who Made that Home Pregnancy Test_Kennedy_120727, 2012.

[3] B. Ngom, Y. Guo, X. Wang, D. Bi, Development and application of lateral flow test strip technology for detection of infectious agents and chemical contaminants: a review, Anal. Bioanal. Chem. 397 (3) (2010) 1113–1135.

[4] B.B. Dzantiev, N.A. Byzova, A.E. Urusov, A.V. Zherdev, Immunochromatographic methods in food analysis, TrAC Trends Anal. Chem. 55 (2014) 81–93.

[5] E. Sapountzi, S. Tragoulias, D. Kalogianni, P. Ioannou, T. Christopoulos, Lateral flow devices for nucleic acid analysis exploiting quantum dots as reporters, Anal. Chim. Acta 864 (2015) 48–54.

[6] W. Chen, R. Rahman, R. Antora, N. Saqib, P. Hossain, J. Zhang, et al., Lateral flow test strip approaches for rapid detection of food contaminants, residues, and chemical constituents of concern, Am. Res. Thoughts 1 (2015) 1222–1248.

[7] L. Anfossi, C. Baggiani, C. Giovannoli, G. D'Arco, G. Giraudi, Lateral-flow immunoassays for mycotoxins and phycotoxins: a review, Anal. Bioanal. Chem. 405 (2–3) (2013) 467–480.

[8] Y. Chen, K. Wang, Z. Liu, R. Sun, D. Cui, J. He, Rapid detection and quantification of tumor marker carbohydrate antigen 72-4 (CA72-4) using a superparamagnetic immunochromatographic strip, Anal. Bioanal. Chem. 408 (9) (2016) 2319–2327.

[9] K. Campbell, T. Fodey, J. Flint, C. Danks, M. Danaher, M. O'Keeffe, et al., Development and validation of a lateral flow device for the detection of nicarbazin contamination in poultry feeds, J. Agric. Food Chem. 55 (2007) 2497–2503.

[10] T. Klewitz, F. Gessler, H. Beer, K. Pflanz, T. Scheper, Immunochromatographic assay for determination of botulinum neurotoxin type D, Sens. Actuators B Chem. 113 (2) (2006) 582–589.

[11] M. Blažková, M. Koets, P. Rauch, A. van Amerongen, Development of a nucleic acid lateral flow immunoassay for simultaneous detection of *Listeria* spp. and *Listeria monocytogenes* in food, Eur. Food Res. Technol. 229 (6) (2009) 867–874.

[12] J. Bogdanovic, M. Koets, I. Sander, I. Wouters, T. Meijster, D. Heederik, et al., Rapid detection of fungal alpha-amylase in the work environment with a lateral flow immunoassay, J. Allergy Clin. Immunol. 118 (5) (2006) 1157–1163.

[13] A. van Amerongen, D. van Loon, L. Berendsen, J. Wichers, Quantitative computer image analysis of hCG colloidal carbon dipstick assay, Clin. Chim. Acta 229 (1994) 67–75.

[14] P.L. Corstjens, H.H. Fidder, K.C. Wiesmeijer, C.J. de Dood, T. Rispens, G.J. Wolbink, et al., A rapid assay for on-site monitoring of infliximab trough levels: a feasibility study, Anal. Bioanal. Chem. 405 (23) (2013) 7367–7375.

[15] M. Zangheri, F. Di Nardo, M. Mirasoli, L. Anfossi, A. Nascetti, D. Caputo, et al., Chemiluminescence lateral flow immunoassay cartridge with integrated amorphous silicon photosensors array for human serum albumin detection in urine samples, Anal. Bioanal. Chem. 408 (30) (2016) 8869–8879.

[16] J. Li, M. Zou, Y. Chen, Q. Xue, F. Zhang, B. Li, et al., Gold immunochromatographic strips for enhanced detection of avian influenza and Newcastle disease viruses, Anal. Chim. Acta 782 (2013) 54–58.

[17] C.M. Mugasa, T. Laurent, G.J. Schoone, P.A. Kager, G.W. Lubega, H.D. Schallig, Nucleic acid sequence-based amplification with oligochromatography for detection of *Trypanosoma brucei* in clinical samples, J. Clin. Microbiol. 47 (3) (2009) 630–635.

[18] Z. Parpia, R. Elghanian, A. Nabatiyan, D. Hardie, D. Kelso, p24 antigen rapid test for diagnosis of acute pediatric HIV infection, J. Acquir. Immune Defic. Syndr. 55 (2010) 413–419.

[19] S. Hong, Y. Park, Y. Jang, B.-H. Min, H. Yoon, Quantitative lateral-flow immunoassay for the assessment of the cartilage oligomeric matrix protein as a marker of osteoarthritis, BioChip J. 6 (3) (2012) 213–220.

[20] M.A. Dineva, D. Candotti, F. Fletcher-Brown, J.P. Allain, H. Lee, Simultaneous visual detection of multiple viral amplicons by dipstick assay, J. Clin. Microbiol. 43 (8) (2005) 4015–4021.

[21] M. Lonnberg, M. Drevin, J. Carlsson, Ultra-sensitive immunochromatographic assay for quantitative determination of erythropoietin, J. Immunol. Methods 339 (2) (2008) 236–244.

[22] G. Wu, M. Zaman, Low-Cost Tools for Diagnosing and Monitoring HIV Infection in Low-Resource Settings, Bulletin of the World Health Organization: World Health Organisation (WHO), 2012, pp. 914–920.

[23] G. Posthuma-Trumpie, J. Korf, A. van Amerongen, Lateral flow (immuno)assay: its strengths, weaknesses, opportunities and threats. A literature survey, Anal. Bioanal. Chem. 393 (2008) 569–582.

[24] G. Fridley, C. Holstein, S. Oza, P. Yager, The evolution of nitrocellulose as a material for bioassays, MRS Bull. 38 (4) (2013) 326–330.

[25] M. Vorobyeva, P. Vorobjev, A. Venyaminova, Multivalent aptamers: versatile tools for diagnostic and therapeutic applications, Molecules 21 (12) (2016).

[26] M. Brown, Antibodies: key to a robust lateral flow immunoassay, in: R. Wong, H. Tse (Eds.), Lateral Flow Immunoassay, Humana Press, 2009, pp. 59–74.

[27] I.Y. Goryacheva, P. Lenain, S. De Saeger, Nanosized labels for rapid immunotests, TrAC Trends Anal. Chem. 46 (2013) 30–43.

[28] D. Quesada-Gonzalez, A. Merkoci, Nanoparticle-based lateral flow biosensors, Biosens. Bioelectron. 73 (2015) 47–63.

[29] X. Huang, Z.P. Aguilar, H. Xu, W. Lai, Y. Xiong, Membrane-based lateral flow immunochromatographic strip with nanoparticles as reporters for detection: a review, Biosens. Bioelectron. 75 (2016) 166–180.

[30] L.G. Lee, E.S. Nordman, M.D. Johnson, M.F. Oldham, A low-cost, high-performance system for fluorescence lateral flow assays, Biosensors 3 (4) (2013) 360–373.

[31] C. Swanson, A. D'Andrea, Lateral flow assay with near-infrared dye for multiplex detection, Clin. Chem. 59 (4) (2013) 641–648.

[32] B. O'Farrell, Lateral flow immunoassay systems, in: W. DG (Ed.), The Immunoassay Handbook, Elsevier Ltd, 2013, pp. 89–107.

[33] M. Koets, I. Sander, J. Bogdanovic, G. Doekes, A. van Amerongen, A rapid lateral flow immunoassay for the detection of fungal alpha-amylase at the workplace, J. Environ. Monit. 8 (9) (2006) 942–946.

[34] P. Mens, A. van Amerongen, P. Sawa, P. Kager, H. Schallig, Molecular diagnosis of malaria in the field: development of a novel 1-step nucleic acid lateral flow immunoassay for the detection of all 4 human *Plasmodium* spp. and its evaluation in Mbita, Kenya, Diagn. Microbiol. Infect. Dis. 61 (4) (2008) 421–427.

[35] K. Faulstich, R. Gruler, M. Eberhard, D. Lentzsch, K. Haberstroh, Handheld and portable reader devices for lateral flow immunoassays, in: W. RC, T. HY (Eds.), Lateral Flow Immunoassay, Humana Press, New York, 2009, pp. 157–183.

[36] P.L. Corstjens, C.J. de Dood, J.W. Priest, H.J. Tanke, S. Handali, Cysticercosis Working Group in Peru, Feasibility of a lateral flow test for neurocysticercosis using novel up-converting nanomaterials and a lightweight strip analyzer, PLoS Neglected Trop. Dis. 8 (7) (2014) e2944.

[37] V.G. Panferov, I.V. Safenkova, A.V. Zherdev, B.B. Dzantiev, Setting up the cut-off level of a sensitive barcode lateral flow assay with magnetic nanoparticles, Talanta 164 (2017) 69–76.

[38] D.B. Wang, B. Tian, Z.P. Zhang, X.Y. Wang, J. Fleming, L.J. Bi, et al., Detection of Bacillus anthracis spores by super-paramagnetic lateral-flow immunoassays based on "Road Closure", Biosens. Bioelectron. 67 (2015) 608–614.

[39] H.A. Joung, Y.K. Oh, M.G. Kim, An automatic enzyme immunoassay based on a chemiluminescent lateral flow immunosensor, Biosens. Bioelectron. 53 (2014) 330–335.

[40] J. Li, K. Zhao, X. Hong, H. Yuan, L. Ma, J. Li, et al., Prototype of immunochromatographic assay strips using colloidal CdTe nanocrystals as biological luminescent label, Colloids Surfaces B Biointerfaces 40 (3–4) (2005) 179–182.

[41] J. Shen, Y. Zhou, F. Fu, H. Xu, J. Lv, Y. Xiong, et al., Immunochromatographic assay for quantitative and sensitive detection of hepatitis B virus surface antigen using highly luminescent quantum dot-beads, Talanta 142 (2015) 145–149.

[42] X. Song, M. Knotts, Time-resolved luminescent lateral flow assay technology, Anal. Chim. Acta 626 (2) (2008) 186–192.

[43] M. Li, H. Yang, S. Li, C. Liu, K. Zhao, J. Li, et al., An ultrasensitive competitive immunochromatographic assay (ICA) based on surface-enhanced Raman scattering (SERS) for direct detection of 3-amino-5-methylmorpholino-2-oxazolidinone (AMOZ) in tissue and urine samples, Sens. Actuators B Chem. 211 (2015) 551–558.

[44] C. Muehlethaler, M. Leona, J.R. Lombardi, Review of surface enhanced Raman scattering applications in forensic science, Anal. Chem. 88 (1) (2016) 152–169.

[45] Y. Xie, H. Chang, K. Zhao, J. Li, H. Yang, L. Mei, et al., A novel immunochromatographic assay (ICA) based on surface-enhanced Raman scattering for the sensitive and quantitative determination of clenbuterol, Anal. Methods 7 (2) (2015) 513–520.

[46] S. Bamrungsap, C. Apiwat, W. Chantima, T. Dharakul, N. Wiriyachaiporn, Rapid and sensitive lateral flow immunoassay for influenza antigen using fluorescently-doped silica nanoparticles, Microchim. Acta 181 (1–2) (2013) 223–230.

[47] C.Y. Shi, N. Deng, J.J. Liang, K.N. Zhou, Q.Q. Fu, Y. Tang, A fluorescent polymer dots positive readout fluorescent quenching lateral flow sensor for ractopamine rapid detection, Anal. Chim. Acta 854 (2015) 202–208.

[48] V. Venkatraman, R. Liedert, K. Kozak, A.J. Steckl, Integrated NFC power source for zero on-board power in fluorescent paper-based lateral flow immunoassays, Flexible Printed Electron. 1 (4) (2016) 044001.

[49] S. Feng, R. Caire, B. Cortazar, M. Turan, A. Wong, A. Ozcan, Google glass based immunochromatographic diagnostic test analysis, in: G. Coté (Ed.), Optical Diagnostics and Sensing XV – Toward Point-of-Care Diagnostics. Proceedings of SPIE, 2015, p. 93320M.

[50] P.J. Tighe, R.R. Ryder, I. Todd, L.C. Fairclough, ELISA in the multiplex era: potentials and pitfalls, Proteomics Clin. Appl. 9 (3–4) (2015) 406–422.

[51] D.J. Carter, R.B. Cary, Lateral flow microarrays: a novel platform for rapid nucleic acid detection based on miniaturized lateral flow chromatography, Nucleic Acids Res. 35 (10) (2007) e74.

[52] T. Chinnasamy, L.I. Segerink, M. Nystrand, J. Gantelius, H.A. Svahn, A lateral flow paper microarray for rapid allergy point of care diagnostics, Analyst 139 (10) (2014) 2348–2354.

[53] J. Gantelius, T. Bass, R. Sjoberg, P. Nilsson, H. Andersson-Svahn, A lateral flow protein microarray for rapid and sensitive antibody assays, Int. J. Mol. Sci. 12 (11) (2011) 7748–7759.

[54] J. Gantelius, C. Hamsten, M. Neiman, J.M. Schwenk, A. Persson, H. Andersson-Svahn, A lateral flow protein microarray for rapid determination of contagious bovine pleuropneumonia status in bovine serum, J. Microbiol. Methods 82 (1) (2010) 11–18.

[55] N.A. Taranova, N.A. Byzova, V.V. Zaiko, T.A. Starovoitova, Y.Y. Vengerov, A.V. Zherdev, et al., Integration of lateral flow and microarray technologies for multiplex immunoassay: application to the determination of drugs of abuse, Microchim. Acta 180 (11–12) (2013) 1165–1172.

[56] G.A. Posthuma-Trumpie, J.H. Wichers, M. Koets, L.B. Berendsen, A. van Amerongen, Amorphous carbon nanoparticles: a versatile label for rapid diagnostic (immuno)assays, Anal. Bioanal. Chem. 402 (2) (2012) 593–600.

[57] Spot-f (Spot-on-the-fly) – Spotting Technology to Speed up Array Production [Press Release], 2015.

[58] SCIENION and Axxin Jointly Collaborate on an Instrument Reader System for Multiplexed Lateral Flow Microarrays [Press Release], 2015.

[59] P. Noguera, G.A. Posthuma-Trumpie, M. van Tuil, F.J. van der Wal, A. de Boer, A.P. Moers, et al., Carbon nanoparticles in lateral flow methods to detect genes encoding virulence factors of Shiga toxin-producing *Escherichia coli*, Anal. Bioanal. Chem. 399 (2) (2011) 831–838.

[60] M. Koets, A. Renstrom, E. Zahradnik, J. Bogdanovic, I. Wouters, A. van Amerongen, Rapid one-step assays for on-site monitoring of mouse and rat urinary allergens, J. Environ. Monit. 13 (12) (2011) 3475–3480.

[61] D. Zhang, Q. Liu, Biosensors and bioelectronics on smartphone for portable biochemical detection, Biosens. Bioelectron. 75 (2016) 273–284.

[62] A. Roda, E. Michelini, M. Zangheri, M. Di Fusco, D. Calabria, P. Simoni, Smartphone-based biosensors: a critical review and perspectives, TrAC Trends Anal. Chem. 79 (2016) 317–325.

[63] W.C. Mak, V. Beni, A.P.F. Turner, Lateral-flow technology: from visual to instrumental, TrAC Trends Anal. Chem. 79 (2016) 297–305.

[64] E. Eltzov, S. Guttel, A. Low Yuen Kei, P.D. Sinawang, R.E. Ionescu, R.S. Marks, Lateral flow immunoassays - from paper strip to smartphone technology, Electroanalysis 27 (9) (2015) 2116–2130.

[65] opTricon, Cube-Reader [Internet], 2017. Available from: http://www.optricon.de/cube-monitoring-device-english.php.

[66] W. Fong, Z. Modrusan, J. Mcnevin, J. Marostenmaki, B. Zin, F. Bekkaoui, Rapid solid-phase immunoassay for detection of methicillin-resistant *Staphylococcus aureus* using cycling probe technology, J. Clin. Microbiol. 38 (2000) 2525–2529.

[67] Y. Oku, K. Kamiya, H. Kamiya, Y. Shibahara, T. Ii, Y. Uesaka, Development of oligonucleotide lateral-flow immunoassay for multi-parameter detection, J. Immunol. Methods 258 (2001) 73–84.

[68] Rapid and simple one-step and mini-array methods to detect and quantify amplified DNA and proteins using ligand-labeled colloidal particles, in: M. Koets, N. Barbier, E. Wolbert, H. Mooibroek, A. van Amerongen (Eds.), EURO FOOD CHEM XII – Strategies for Safe Food: Analytical, Industrial and Legal Aspects: Challenges in Organisation and Communication, Koninklijke Ylaamse Chemische Vereniging, Sectie Voeding, Bruges, Belgium, 2003.

[69] A. van Amerongen, M. Koets, Simple and rapid bacterial protein and DNA diagnostic methods based on signal generation with colloidal carbon particles, in: A. van Amerongen, D. Barug, M. Louwaars (Eds.), Rapid Methods for Biological and Chemical Contaminants in Food and Feed, Wageningen Academic Publishers, Wageningen, 2005, pp. 105–126.

[70] S. Deborggraeve, F. Claes, T. Laurent, P. Mertens, T. Leclipteux, J. Dujardin, et al., Molecular dipstick test for diagnosis of sleeping sickness, J. Clin. Microbiol. 44 (8) (2006) 2884–2889.

[71] P. Soo, Y. Horng, P. Hsueh, B. Shen, J. Wang, H. Tu, et al., Direct and simultaneous identification of *Mycobacterium tuberculosis* complex (MTBC) and *Mycobacterium tuberculosis* (MTB) by rapid multiplex nested PCR-ICT assay, J. Microbiol. Methods 66 (3) (2006) 440–448.

[72] P. Mens, A. Moers, L. de Bes, J. Flint, J. Sak, L. Keereecharoen, et al., Development, validation and evaluation of a rapid PCR-nucleic acid lateral flow immuno-assay for the detection of *Plasmodium* and the differentiation between falciparum and vivax, Malar. J. 11–279 (2012) 10.

[73] P. Mens, H. de Bes, P. Sondo, N. Laochan, L. Keereecharoen, A. van Amerongen, et al., Direct blood PCR in combination with nucleic acid lateral flow immunoassay for detection of *Plasmodium* species in settings where malaria is endemic, J. Clin. Microbiol. 50 (11) (2012) 3520–3525.

[74] A. Moers, R. Hallett, R. Burrow, H. Schallig, C. Sutherland, A. van Amerongen, Detection of single-nucleotide polymorphisms in *Plasmodium falciparum* by PCR primer extension and lateral flow immunoassay, Antimicrob. Agents Chemother. 59 (1) (2015) 365–371.

[75] S. Yongkiettrakul, J. Kampeera, W. Chareanchim, R. Rattanajak, W. Pornthanakasem, W. Kiatpathomchai, et al., Simple detection of single nucleotide polymorphism in Plasmodium falciparum by SNP-LAMP assay combined with lateral flow dipstick, Parasitol. Int. 66 (1) (2017) 964–971.

[76] D.P. Kalogianni, S. Goura, A.J. Aletras, T.K. Christopoulos, M.G. Chanos, M. Christofidou, et al., Dry reagent dipstick test combined with 23S rRNA PCR for molecular diagnosis of bacterial infection in arthroplasty, Anal. Biochem. 361 (2) (2007) 169–175.

[77] H. Huang, L. Jin, X. Yang, Q. Song, B. Zou, S. Jiang, et al., An internal amplification control for quantitative nucleic acid analysis using nanoparticle-based dipstick biosensors, Biosens. Bioelectron. 42 (2013) 261–266.

[78] S. Yongkiettrakul, W. Jaroenram, N. Arunrut, W. Chareanchim, S. Pannengpetch, R. Suebsing, et al., Application of loop-mediated isothermal amplification assay combined with lateral flow dipstick for detection of *Plasmodium falciparum* and *Plasmodium vivax*, Parasitol. Int. 63 (6) (2014) 777–784.

[79] W. Xu, N. Cheng, K. Huang, Y. Lin, C. Wang, Y. Xu, et al., Accurate and easy-to-use assessment of contiguous DNA methylation sites based on proportion competitive quantitative-PCR and lateral flow nucleic acid biosensor, Biosens. Bioelectron. 80 (2016) 654–660.

[80] D. Chen, M. Mauk, X. Qiu, C. Liu, J. Kim, S. Ramprasad, et al., An integrated, self-contained microfluidic cassette for isolation, amplification, and detection of nucleic acids, Biomed. Microdev. 12 (4) (2010) 705–719.

[81] X. Qiu, D. Chen, C. Liu, M. Mauk, T. Kientz, H. Bau, A portable, integrated analyzer for microfluidic – based molecular analysis, Biomed. Microdev. 13 (5) (2011) 809–817.

[82] L. Lafleur, J. Bishop, E. Heiniger, R. Gallagher, M. Wheeler, P. Kauffman, et al., A rapid, instrument-free, sample-to-result nucleic acid amplification test, Lab Chip 16 (19) (2016) 3777–3787.

[83] B. Park, S. Oh, J. Jung, G. Choi, J. Seo, D. Kim, et al., An integrated rotary microfluidic system with DNA extraction, loop-mediated isothermal amplification, and lateral flow strip based detection for point-of-care pathogen diagnostics, Biosens. Bioelectron. 91 (2017) 334–340.

[84] A. Martinez, S. Phillips, M. Butte, G. Whitesides, Patterned paper as a platform for inexpensive, low-volume, portable bioassays, Angew. Chem. 46 (8) (2007) 1318–1320.

[85] T. Lappalainen, T. Teerinen, P. Vento, L. Hakalahti, T. Erho, Cellulose as a novel substrate for lateral flow assay, Nordic Pulp Pap. Res. J. 25 (4) (2010) 536–550.

[86] J. Hu, S. Wang, L. Wang, F. Li, B. Pingguan-Murphy, T.J. Lu, et al., Advances in paper-based point-of-care diagnostics, Biosens. Bioelectron. 54 (2014) 585–597.

[87] M.N. Costa, B. Veigas, J.M. Jacob, D.S. Santos, J. Gomes, P.V. Baptista, et al., A low cost, safe, disposable, rapid and self-sustainable paper-based platform for diagnostic testing: lab-on-paper, Nanotechnology 25 (9) (2014) 094006.

[88] D.M. Cate, J.A. Adkins, J. Mettakoonpitak, C.S. Henry, Recent developments in paper-based microfluidic devices, Anal. Chem. 87 (1) (2015) 19–41.

[89] M.M. Hamedi, A. Ainla, F. Guder, D.C. Christodouleas, M.T. Fernandez-Abedul, G.M. Whitesides, Integrating electronics and microfluidics on paper, Adv. Mater. 28 (25) (2016) 5054–5063.

[90] S.J. Vella, P. Beattie, R. Cademartiri, A. Laromaine, A.W. Martinez, S.T. Phillips, et al., Measuring markers of liver function using a micropatterned paper device designed for blood from a fingerstick, Anal. Chem. 84 (6) (2012) 2883–2891.

[91] J.R. Choi, J. Hu, R. Tang, Y. Gong, S. Feng, H. Ren, et al., An integrated paper-based sample-to-answer biosensor for nucleic acid testing at the point of care, Lab Chip 16 (3) (2016) 611–621.

[92] A. Apilux, Y. Ukita, M. Chikae, O. Chailapakul, Y. Takamura, Development of automated paper-based devices for sequential multistep sandwich enzyme-linked immunosorbent assays using inkjet printing, Lab Chip 13 (1) (2013) 126–135.

[93] marketsandmarketscom, Lateral Flow Assay Market by Product (Reader, Kits) Application (Clinical Testing (Pregnancy, Infectious Disease, Cholestrol, Cardiac Marker), Veterinary, Drug Development) Technique (Sandwich, Competitive, Multiplex) End User – Global Forecast to 2020, March 2016.

Chapter 8

Paper-Based Immunoassays

Zhi Liu, Jie Hu, Zhiguo Qu, Feng Xu
Xi'an Jiaotong University, Xi'an, P.R. China

Immunoassays are a kind of biochemical tests that detect the presence or the concentration of antigen, antibody, protein, or excretive cytokine in solution [1]. Generally, these immunoassays recognize the specific target analyte based on the reaction between antibody and its antigen. The antibodies, including monoclonal and polyclonal antibodies, are regarded as the biorecognition elements encapsulated in the capture zone, where they can bind the target antigen and immunocomplex target-labels to present the detection signal [2]. Although the conventional immunoassay tests have a widespread range of applications in many different ways and perform high sensitivity, they still require multiple operation steps, expensive equipment, skillful technician, and long analysis time [3,4]. Therefore, paper-based immunoassays have been developed to address the aforementioned needs and they have gained the attention of many researchers.

As the main material to construct paper-based immunoassay devices, paper is claimed to be inexpensive, ubiquitous, disposable, easy to manipulate, and equipment-free, which is considered to be one of the attractive materials for preparing immunoassay devices [5]. Paper shows excellent physical properties (e.g., pliability, large surface area, various porosity, and durability coexist [6,7]) and good chemical properties (e.g., biocompatibility and hydrophilic [1]), which make it stand out in manufacturing microfluidic analytical devices. Due to its fibrous porous structure, paper can drive aqueous liquid flows by capillary force and provides a large surface for reagents immobilization and a big zone for reagents reactions. Recently, tremendous manufacturing technologies (e.g., cutting [8–10], wax printing [11–14], and photolithography [15,16]) are used for preparing the paper-based devices, resulting in a prosperous market in immunoassay applications.

In fact, paper has already played a vital role on their development history. Dipsticks, which were developed for detection of diabetes in the 1950s, are considered to be the forerunner of paper-based assay technologies [17]. After two decades of development, various paper-based assay devices with different structures and functions were developed; even some commercialized devices

Handbook of Immunoassay Technologies. https://doi.org/10.1016/B978-0-12-811762-0.00008-6

were emerged [1]. Afterward, the first immunoassay system driven by capillary force was reported by Glad et al. in 1978 [18]. With the development of the radioimmunoassay in the measurement of human chorionic gonadotropin (hCG) [19], another great milestone-style immunoassay device made a hit in the market for home pregnancy test in the 1980s [20], which was the first lateral flow assay (LFA) to detect hCG. A new paper-based immunoassay device, microfluidic paper-based assay device (μPAD), was reported by Martinez's research group for protein detection in 2007 [15]. Nowadays, various commercial paper-based immunoassay devices have been expanded to a widespread range of applications in clinical diagnostics, biodefense, food safety, and environmental monitoring [21].

In the current chapter, the strategies of paper-based immunoassay devices based on different technologies (e.g., colorimetric, thermal, electrochemical, and magnetic methods) are summarized as following. We also reviewed the development of paper-based immunoassays devices, including the methods to improve the sensitivity and automatically detect multitargets as well as to realize semiquantification or quantification detections. Finally, we conclude the future tendency of paper-based immunoassay devices.

1. PAPER-BASED IMMUNOASSAYS: STRATEGIES AND DETECTION PRINCIPLES

With the prosperous development of the economy and people's living level, paper-based immunoassays have paved a great step in biomedical and biochemical analyses. Tremendous paper-based immunoassays devices have been reported and can be categorized into three common formats, including paper-based enzyme-linked immunosorbent assay (P-ELISA), LFAs, and μPADs. The P-ELISA uses the paper microzone plate as an alliterative to the 96-microzone plastic plate in the traditional ELISA, resulting in low-cost, sample-saving immunoassays [22]. The analytical principle of P-ELISA is same with the traditional ELISA, which was introduced in the reported papers [23,24]. Since LFAs were commercialized in pregnancy tests with straight integrated strips, they have shown considerable advantages and excellent perspectives in a broad range of application fields [25]. Based on the different working methods of the antibodies and their corresponding antigens, the LFAs are always divided into two typical forms, such as sandwich LFAs and competitive LFAs [2,26,27]. Conventional LFAs are inexpensive, time-saving, easy-to-operate, and presenting qualitative answers in yes/no type. The key of μPADs is designing fluid flowing channel by using hydrophobic material (e.g., wax [11] and fluorosilanization [28]) to block the redundant fibrous porous structures. Both two-dimensional and three-dimensional μPADs without fixed structures are designed to control the liquid migration and to realize functional analysis [12,29]. Compared with the above paper-based devices, μPADs are rapid, flexible in design, and available for reagent storing.

All the aforementioned paper-based immunoassays transform the presence or concentration of target analytes by using labels. Different labels (e.g., colorimetric, thermal, electronic, and magnetic nanoparticles) are employed, showing a very different detection principle. Herein, we will review the advantages and drawbacks of these strategies for better understanding the application of paper-based immunoassays.

1.1 Colorimetric Method

The colorimetric detection method is known as the most popular and widespread method in immunoassays, mainly due to the considerable sensitivity and stability, especially the conveniently readable result by the naked eye without using extra expensive equipment [30]. Since the colorimetric nanoparticles, which are regarded as a tracer, play crucial roles in the immunoassays analysis, different strategies are achieved to obtain a convenient manipulation and an improved sensitivity by using varying colorimetric NPs (e.g., AuNPs, enzymes, carbon NPs, and magnetic NPs in Fig. 8.1) [2]. Hence, we try to summarize the impacts of these functional NPs on the immunoassays analysis.

1.1.1 AuNPs

Undoubtedly, AuNPs have been the most promising nanoparticles employed as the optical label in immunoassay technologies because of their ruby red color, low cost, ease of synthesis and manipulation, excellent biocompatibility, and good chemical stability (Fig. 8.1A) [26,31]. Generally, the functional AuNP conjugates are synthetized by the colloidal gold and detector probe (e.g., antibody) based on electrostatic effect in immunoassay detections [32]. It is worth noting that reasonable AuNPs size, the concentration of detector probe, and the conjugation pH should be explored in preparation AuNP conjugates because they greatly affect the immunoassay sensitivity [33]. Then, the prepared AuNP conjugates will be added in the paper-based devices (e.g., lateral flow immunoassays [LFIAs] [34] or μPADs [15]) for the following diagnosis. Despite the excellent properties of AuNPs, the immunoassay device that used conventional AuNPs often suffered from poor sensitivity mainly due to the insufficient signal intensity of signal label. Therefore, various composite nanomaterial (e.g., AuNPs decorated with enzymes [35], dual AuNPs conjugates [36], silica nanorods loaded with AuNPs [37], magnetic nanoparticle decorated with AuNPs [38], and silver nanoparticle covered with AuNPs [39]) are developed to amplify the color of the individual composite label on the test zone to decrease the detection limit. This will be discussed in detail in Section 2.1.

1.1.2 Enzymes

Enzymes are a kind of macromolecules with catalytic activities. The enzyme and its substrate are commonly used as indicator in ELISA test. As illustrated

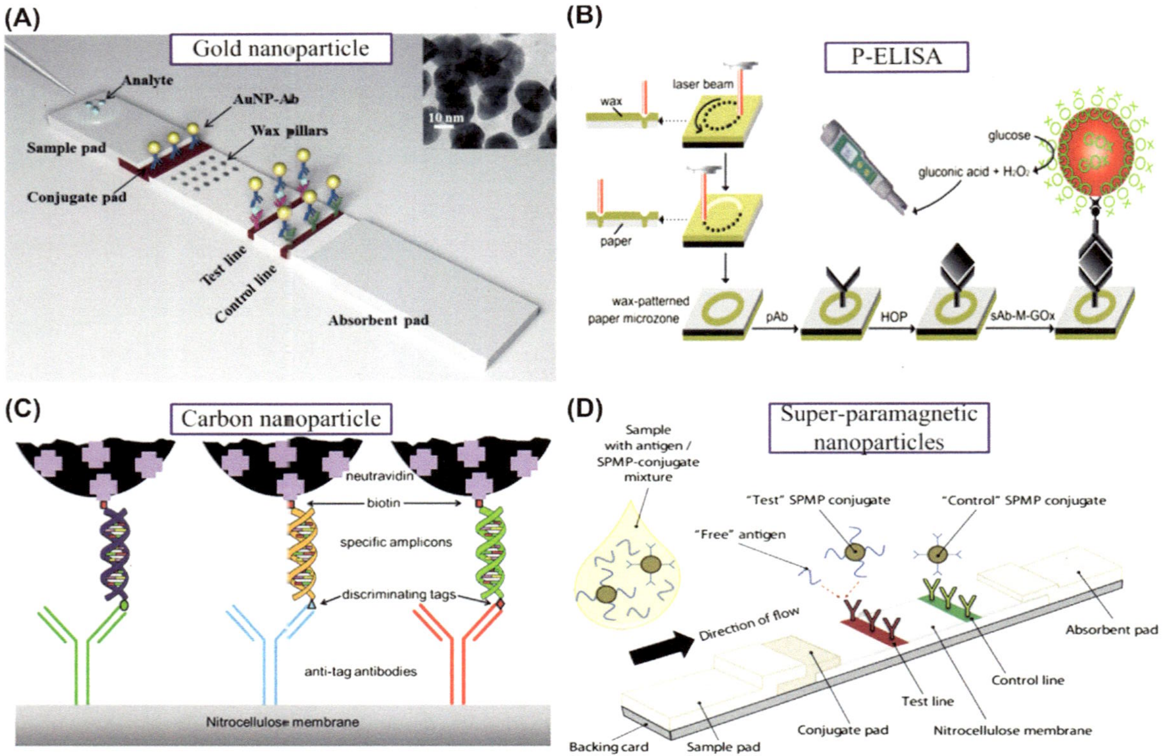

FIGURE 8.1 Colorimetric detection methoc. (A) AuNPs used in LFAs [34]; (B) enzyme employed in P-ELISA [21]; (C) carbon NPs used in LFAs [43]; (D) magnetic NPs used in LFAs [51].

previously, ELISA tests are limited to apply in large hospitals and experiment laboratories mainly because they require large sample volume, expensive equipment, skillful technicians, and long time to detect. To address this, P-ELISA platform, which is low-cost, sample-saving, and rapid compared with the traditional ELISA, has attracted significant attention in the immunoassays (Fig. 8.1B) [22,40,41]. Therefore, several strategies have been applied to develop the P-ELISA devices. For instance, paper devices (e.g., wax patterned microzone plate made by paper) were developed as a cheap, rapid, and sample-saving protocol for multisteps immunoassays analysis such as ELISA [21,22,40]. However, P-ELISA still requires multiple manipulation steps and depends on the environmental conditions (e.g., evaporation, relative humidity, and temperature). Alternatively, automatic or semiautomatic ELISA formats would meet the growing demand for simple and rapid immunoassays analysis with minimum manual operations. To reduce the manipulation steps on paper-based device, automated P-ELISA devices were developed by magnetic control technology [42]. Alternatively, fluidic technology also can achieve the automatic immunoassays by varying the pattern design and the reagent location on the channel of a nitrocellulose (NC) membrane [41].

1.1.3 Carbon Nanoparticles

The carbon NPs, also called black carbon, can be prepared by various physical and chemical methods, which are discussed in the reported review (Fig. 8.1C) [43]. Their promising properties (e.g., inexpensive, ease of preparation, nontoxicity, excellent stability, good optical and electrical features) make them widely applicable in the immunoassay detection [44]. Besides, the carbon NP also shows an intense black color, which presents a significant contrast against the white background of the detection pad (NC membrane), resulting in a higher signal intensity than that of AuNPs on the paper-based immunoassays device [45]. For instance, Linares et al. used the carbon NP as label to detect the nonstructural glycoprotein of dengue virus and exhibited a 10 times lower detection limit than that using AuNPs [46]. In another technology, the carbon tube, an allotrope of carbon, has also attracted attention toward immunoassay tests [43]. Apart from the aforementioned properties of carbon NPs, the carbon tube has a large surface area and can provide more binding sites, which make itself to be promising and highly efficient for capturing the antigen in immunoassays detections [31].

1.1.4 Magnetic Nanoparticles

In addition to AuNPs, enzymes, and carbon NPs, new labeling materials, such as magnetic NPs, are also applied in colorimetric immunoassay detection. They have strong brown color, low background noise, good stable surface carboxyl groups, excellent stability for storage, and magnetic properties [47]. These properties result in a high sensitivity in detection and ensure that they are promising materials to construct the paper-based immunoassay devices. The traditional preparation method of magnetic NPs is via the solvothermal reaction, which is quick and efficient to manipulate [48]. Like the carbon NPs, many of studies have used different types of magnetic NPs (e.g., Fe_3O_4 [49] and superparamagnetic iron oxide particles [50]) to make immunochromatography assays and presented high sensitivity (Fig. 8.1D) [51]. With the improved requirement for low limit of detection, composite nanobeads have also been proposed to construct the paper-based immunoassays devices on account of their vivid color and high magnetism. For instance, Liu et al. synthesized Fe_3O_4 NP aggregates and achieved a 40 times lower limit of detection by colorimetric method than that using individual Fe_3O_4 NP [52]. Yan et al. used the Fe_3O_4@silica composite nanobeads to detect the *Vibrio parahaemolyticus* and found that small particle size and excellent colloidal stability have a significant influence on the detection performance [53].

1.2 Thermal Method

Since the photothermal effects of metal NPs (e.g., AuNPs) is very obvious in nanoscale [54,55], these properties provide a thermal method in the immunoassay tests to achieve a good performance. Generally, after the common detection of the traditional paper-based immunoassay device, a laser is used to heat the test zone. Finally, the thermal signal is recorded and analyzed by the thermal contrast reader after 1 min irradiation by the laser. In contrast to the colorimetric method, the thermal method using the thermal contrast reader can produce a great performance in the immunoassay detection [56]. For example, Qin et al. applied the laser beam on the clinically used LFAs and achieved a 32-fold sensitivity enhancement by thermal contrast reader than the direct observation method by the naked eye (Fig. 8.2A) [57]. Recently, their research group developed a thermal contrast amplification reader and enhanced the sensitivity of the commercial LFAs to eightfold (Fig. 8.2B) [58]. Although the thermal contrast method exhibits a good sensitivity enhancement in paper-based immunoassays, it still has many drawbacks, such as needing additional instruments, complex operation steps, and skillful persons.

1.3 Electrochemical Method

Nowadays, electrochemical method, which is transforming the chemical reaction to an electronic signal, is also a good choice for immunoassay detection on account of its low background noise, controllable electrode potential, large produce response, and high-sensitivity performance [31,59–61]. With the development of industry technologies, many strategies have been developed to fabricate electrodes, such as sputtering, coating, printing (including screen printing and ink-jet printing), and handwriting techniques, making the electrodes printing precise and manipulation more easy in fabrication paper-based devices [62–65]. Much of recent research work regarding the paper-based devices using electrochemical method is focusing on environment monitoring, food safety, and medical diagnostics [66]. For instance, the paper-based electrochemical immunodevices are developed to detect cancer biomarkers by

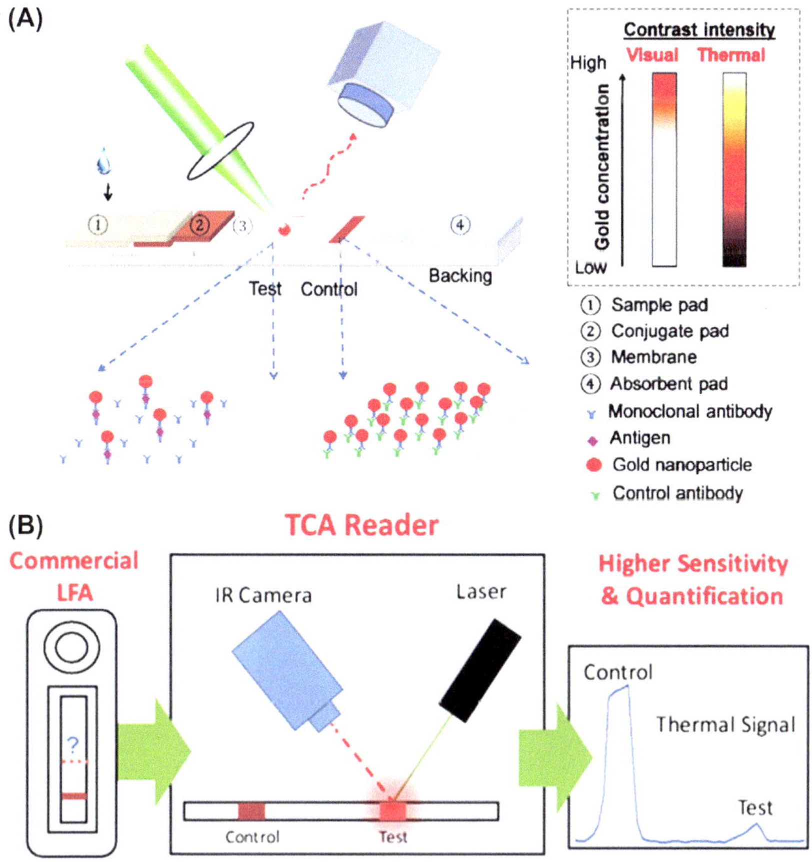

FIGURE 8.2 Thermal method. (A) Schematic of thermal contrast method for LFAs [57]; (B) thermal contrast amplification principle [58].

analyzing the potential curves from two-dimensional to the three-dimensional platform (Fig. 8.3) [67–70]. To end the market demand, some paper-based electrochemical immunodevices can be used for the ultrasensitive multiplexed detection, allowing a stable, reproducible, and accurate detections [71,72]. However, its development still suffers from the expensive and large equipment, complex manipulation steps, and technical staff.

1.4 Magnetic Method

In recent decades, the magnetic method becomes an attractive approach used in paper-based devices, mainly according to the stable magnetic signal of magnetic NPs [26]. Since the LFAs that used colorimetric method can only emit colored signal from top 10 μm layer of NC membrane, the magnetic field from magnetic NPs in the whole detection zone can be recorded by a magnetic assay reader without any loss, resulting in a dramatic sensitivity enhancement [50]. In the detection process, the magnetic NPs excited by the magnetic field

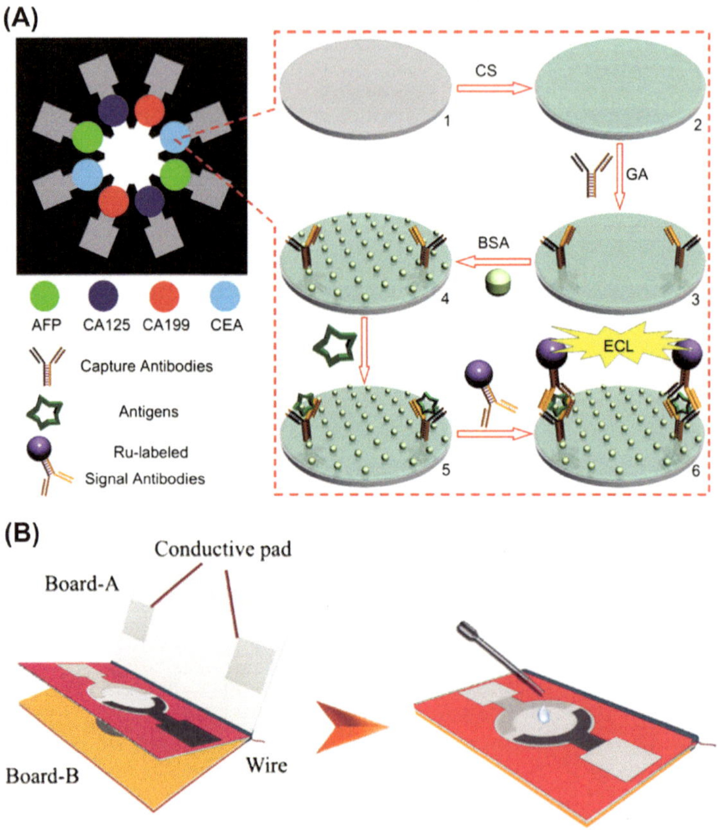

FIGURE 8.3 Electrochemical method. (A) Schematic of preparation and assay process for paper-based electrochemical device [68]; (B) a three-dimensional origami electrochemical immunoassay [70].

from the magnetic reader will generate a magnetic field simultaneously; then the magnetic signal from the magnetic NPs will be recorded and analyzed (Fig. 8.4) [31,73]. Therefore, thanks to the excellent properties of magnetic NPs as the previous introduction, the magnetic method presents a good performance in paper-based immunoassays. Additionally, a small size of magnetic NPs will decrease the whole immunoassay time and provide a weak magnetic signal; a high content of magnetite in nanoparticle will produce a strong magnetic signal [50]. Although enormous efforts have been made, there still exist sustainable challenges (including miniaturization or handheld of magnetic assay reader) in the application and design of magnetic immunoassay methods.

2. DEVELOPMENT OF THE PAPER-BASED IMMUNOASSAYS DEVICES

As previously discussed, most of the traditional paper-based immunoassays devices are limited by their drawbacks, including low sensitivity, challenges

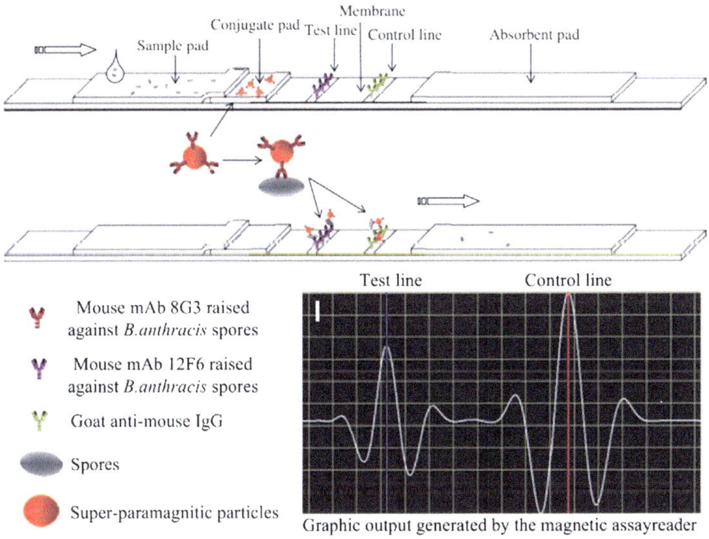

FIGURE 8.4 Schematic of the magnetic LFA system [73].

for automatic detection, and semiquantification detection. To end these needs, various strategies have been developed to enhance the performance of paper-based immunoassay devices.

2.1 Sensitivity Improvement

To decrease the detection limits of the paper-based immunoassays, many novel strategies have been reported. For instance, to optimize the properties of reagents in the preparation of paper based device is a direct method to obtain a better performance. As discussed in Section 1.1.1, composite labels (Fig. 8.5A), which can provide better characteristics, including high color signal, much binding site, and strong ability to capture the target, are excellent labels for signal enhancement in immunoassays [74]. Alternatively, an optimized concentration of label and capture probe is beneficial to obtain a higher sensitivity and to decrease the cost [33,75].

In another approach for sensitivity enhancement, microfluidic technologies are introduced to the paper-based immunoassays. Since the fluid flow behavior plays significant roles on the reagents mixing and reaction [76], microfluidic technologies have attracted the interest of many researchers in clinical or nonclinical applications. For example, architecture modifications in paper-based devices, such as varying the strip width [77] and adding a pad [78], provide an easy way to control the liquid migration behavior, resulting in a signal amplification in the detections (Fig. 8.5B). Besides the architecture modification, cutting boundaries [9,10] and hydrophobic boundaries (e.g., wax [11], photolithography [15]) are straightforward methods to construct the

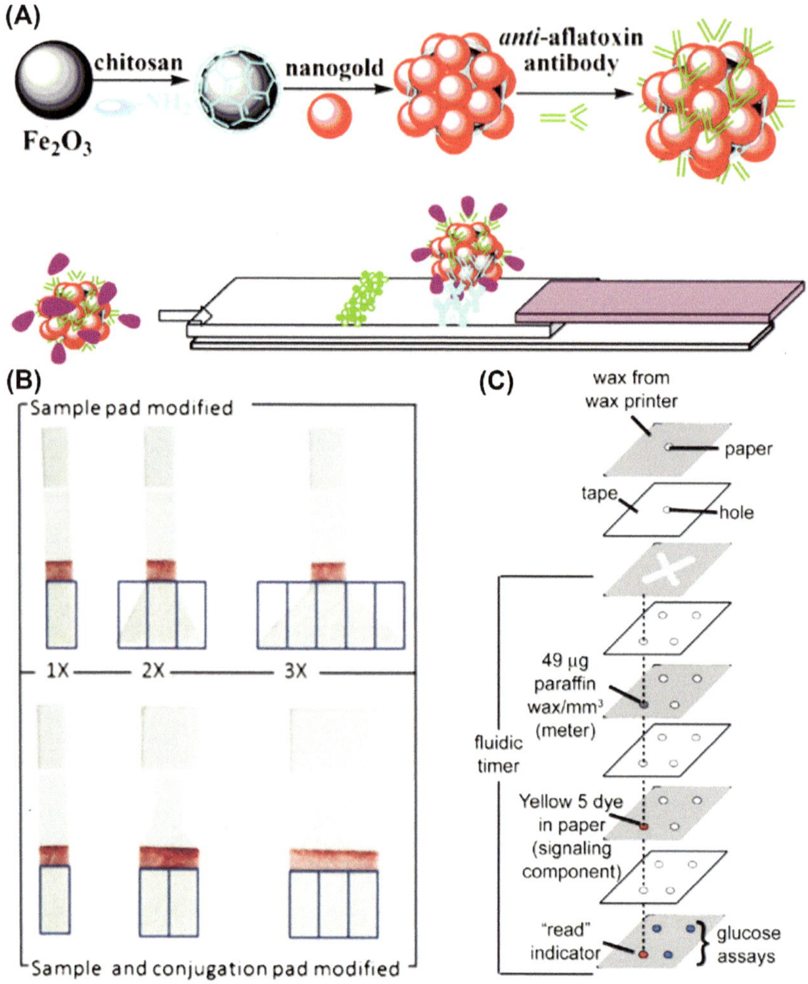

FIGURE 8.5 Methods of sensitivity improvement. (A) Composite labels (e.g., magnetic nanoparticle decorated with AuNPs) [38]; (B) microfluidic technology using architecture modifications [77]; (C) microfluidic technology using hydrophobic boundaries in µPAD [11].

microfluidic channel and to design the variation of liquid flow (Fig. 8.5C). Additionally, some strategies use barriers (e.g., trehalose [79], wax pillars [34]) to decrease the flow velocity to increase the reagent reaction time.

2.2 Automatic Detections

Many traditional paper-based microfluidic assays, such as ELISA, always limit by the complicated multioperation steps and multiple analytes detection

[80,81]. Thus, automatic multistep assays or one-step detections of multiple analytes are the future tendency for point-of-care tests. It is worth noting that microfluidic technologies are also useful strategies to control the multistep assay processes in sample pretreatment, sequential reactions, and detections [82], on account of the capillary action in paper and various two- or three-dimensional architecture. Yang et al. utilized the inherent characteristics of fibrous paper to separate the plasma by filtering the whole blood and performed a one-step assay [83]. On the other side, a dissolvable film [84] and flap value [85] regarded as a time shutoff can provide a reasonable fluid flow control approach (Fig. 8.6A), resulting in a controlled reagents release in fast and slow channels, separately.

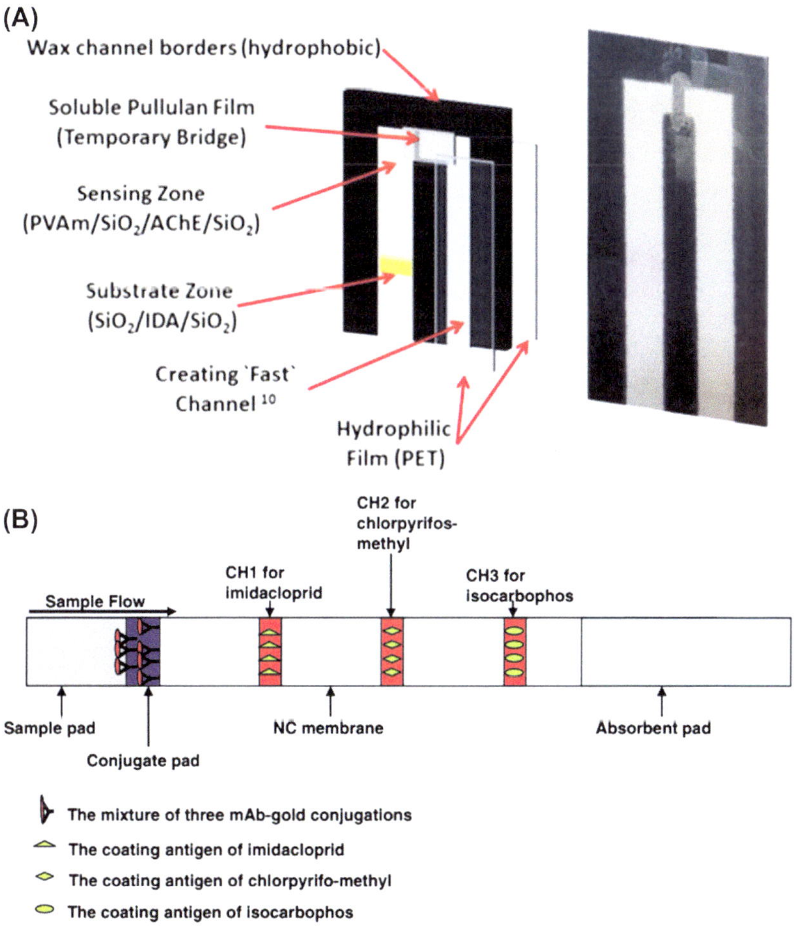

FIGURE 8.6 Automatic detections. (A) Automatic multistep assays in μPADs [84]; (B) automatic detection of multiple targets in LFAs [86].

Nowadays, the clinical sample or other samples often contain more than one disease target, which can provide many indexes to effectively evaluate the degree of the disease. The traditional paper-based device, which can detect only one target, can hardly meet this requirement in one-step operation. Thus, Wang et al. reported a bar-code LFA based on the competitive format to achieve the detection of three pesticides in one-step operation (Fig. 8.6B) [86]. However, the cross-reactivity is the vital challenge to the bar-code LFA in some detection of multiple targets. Therefore, a multiplexed LFA format was developed by integrating several LFA strips into a device, where each LFA strip can detect only one target without cross-reactivity [1].

2.3 Semiquantification Detection and Quantification Detection

Most paper-based diagnostics are easy and rapid to provide the qualitative detection results in the absence or presence of target analyte by using naked eyes, which is hardly to meet the demand of semiquantification detection and quantification detection. Recently, some novel strategies are reported to end this need. For instance, a bar-code-style immunoassay strip presents a considerable method to achieve the semiquantification assays [87,88]. Most of the bar-code LFIAs are based on the competition theory, where the targets and labels are competitive to immobilized in the test lines without interactions (Fig. 8.7A) [89−91]. Thus, a large dose of target analytes will produce a small number of test lines. In another strategy, Oh et al. presented a three-line (e.g., test line, antigen line and control line) immunoassays strip based on sandwich format immunoassays [92]. According to the Hook effect, it is found that the predicted C-reactive protein concentration obtained by comparing the signal of antigen line and control line has a linear relationship with the experimental measurement value, which can be used to evaluate the target concentration range.

Bar-code-style strategy measured by naked eyes can only provide semiquantification immunoassays to a certain degree, where it is hard to achieve the quantification detection and meet the requirement of clinical diagnosis. To achieve quantitative measurements of immune biomarkers, extra instruments, including image reader and analysis system, are used to collect and analyze the image information of the experimental results. As discussed previously, the thermal contrast reader, electrochemical workstation, and magnetic assay reader are used in different detection methods to record an accurate thermal, electronic signal and magnetic field for further analysis, but they suffer from some limitations, including high-cost, lager volume, complex operation steps, and the need for skillful persons. For the colorimetric detection method, digital camera and strip scanner are often used to take the pictures of result signal [56,59]. With the rapid development of

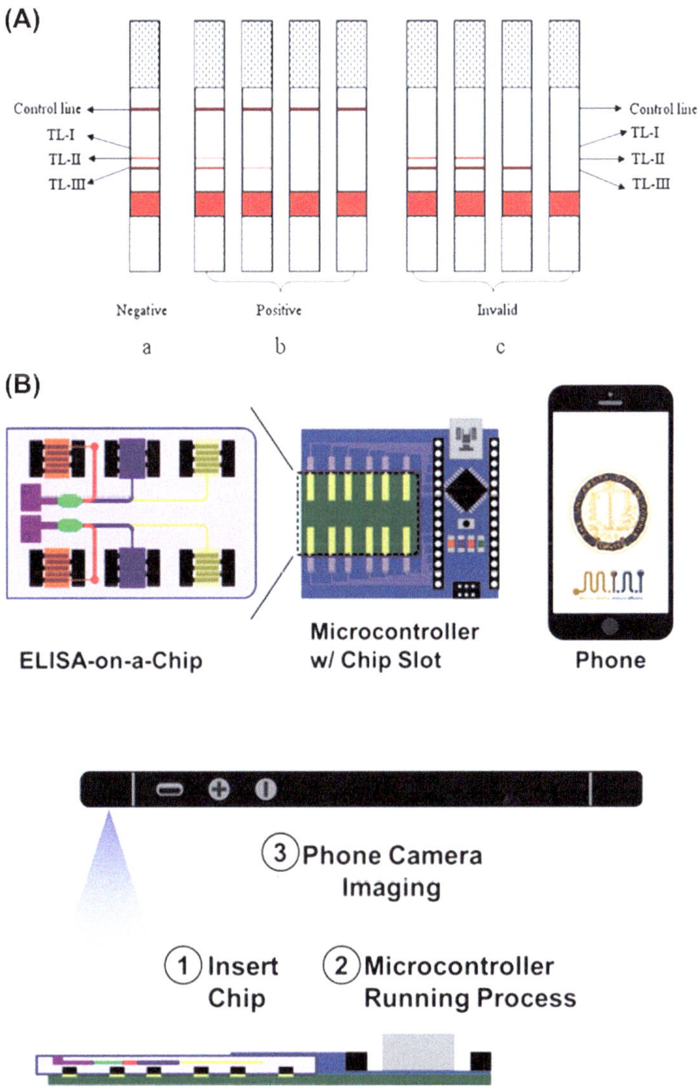

FIGURE 8.7 Semiquantification detection or quantification detection. (A) Bar-code lateral flow immunoassays [89]; (B) smartphone used in immunoassay diagnosis [93].

science and technology, smartphones, an integration of digital camera and analysis system, are portable, real-time, and convenient for immunoassay detection (Fig. 8.7B) [93–95]. Smartphones are often assembled with a microscopic imaging system to amplify the result signal, resulting in an accurate detection [96,97].

3. CONCLUSIONS

This chapter has summarized different strategies of paper-based immunoassays using various labels and has presented developments of paper-based devices in recent decades. Although enormous efforts have been made to improve the performance of paper-based devices, paper-based immunoassays diagnostics still require further investigation to satisfy the growing demand for clinical diagnosis, agriculture application, food safety, and environmental monitoring. The microfluidics technologies presented a considerable way to achieve an enhancement sensitivity and multifunctional diagnosis for paper-based immunoassay devices [98]. The future tendency toward extending the availability and application of microfluidics technology is to improve the microfluidic theory in various situations and establish available standard criteria to prepare the paper-based devices [99]. For colorimetric detection method, cameras and scanners provide an efficient strategy to record and recognize the result images, resulting in an enhancement performance of the paper-based immunoassay diagnostics. Although the data-acquisition attachments (including color reader, thermal contrast reader, electrochemical workstation, and magnetic assay reader) present a considerable performance in immunoassays, they still require great efforts to miniaturize their size and to reduce the cost for applications in developing countries [100]. The smartphone is a portable, handheld, and smart integration of result signal reader and analysis system, which provides widespread applications in various detection methods by attaching other instruments. As such, future trends should focus on new technologies to enhance assay sensitivity and to make the paper-based device more miniature, inexpensive, smart, and automatic.

REFERENCES

[1] A.K. Yetisen, M.S. Akram, C.R. Lowe, Paper-based microfluidic point-of-care diagnostic devices, Lab Chip 13 (12) (2013) 2210–2251.

[2] E.B. Bahadır, M.K. Sezgintürk, Lateral flow assays: principles, designs and labels, TrAC Trends Anal. Chem. 82 (2016) 286–306.

[3] T. Mazzu-Nascimento, et al., Development and statistical assessment of a paper-based immunoassay for detection of tumor markers, Anal. Chim. Acta 950 (2017) 156–161.

[4] B.D. Grant, et al., A paper based immunoassay to determine HPV vaccination status at the point-of-care, Vaccine 34 (46) (2016) 5656–5663.

[5] X. Li, D.R. Ballerini, W. Shen, A perspective on paper-based microfluidics: current status and future trends, Biomicrofluidics 6 (1) (2012) 011301.

[6] S. Jaganathan, H. Vahedi Tafreshi, B. Pourdeyhimi, Modeling liquid porosimetry in modeled and imaged 3-D fibrous microstructures, J. Colloid Interf. Sci. 326 (1) (2008) 166–175.

[7] N. Pan, On uniqueness of fibrous materials, in: Design & Nature II, vol. 6, WIT Press, Boston, 2004, p. 10.

[8] J. Hu, et al., Oligonucleotide-linked gold nanoparticle aggregates for enhanced sensitivity in lateral flow assays, Lab Chip 13 (22) (2013) 4352–4357.

[9] E. Fu, et al., Transport in two-dimensional paper networks, Microfluid. Nanofluid. 10 (1) (2011) 29–35.

[10] M.-B. Song, et al., Tear-off patterning: simple method for patterning nitrocellulose membranes to improve performance of point-of-care diagnostic biosensors, Lab Chip 15 (2015) 3006–3012.

[11] H. Noh, S.T. Phillips, Fluidic timers for time-dependent, point-of-care assays on paper, Anal. Chem. 82 (19) (2010) 8071–8078.

[12] E. Carrilho, A.W. Martinez, G.M. Whitesides, Understanding wax printing: a simple micropatterning process for paper-based microfluidics, Anal. Chem. 81 (16) (2009) 7091–7095.

[13] Z.W. Zhong, Z.P. Wang, G.X.D. Huang, Investigation of wax and paper materials for the fabrication of paper-based microfluidic devices, Microsyst. Technol. 18 (5) (2012) 649–659.

[14] Y. Lu, et al., Fabrication and characterization of paper-based microfluidics prepared in nitrocellulose membrane by wax printing, Anal. Chem. 82 (1) (2010) 329–335.

[15] A.W. Martinez, et al., Patterned paper as a platform for inexpensive, low volume, portable bioassays, Angew. Chem. Int. Ed. Engl. 46 (8) (2007) 1318–1320.

[16] A.W. Martinez, et al., Diagnostics for the developing world: microfluidic paper-based analytical devices, Anal. Chem. 82 (1) (2009) 3–10.

[17] A.H. Free, et al., Simple specific test for urine glucose, Clin. Chem. 3 (3) (1957) 163–168.

[18] C. Glad, A.O. Grubb, Immunocapillarymigration—a new method for immunochemical quantitation, Anal. Biochem. 85 (1) (1978) 180–187.

[19] J.L. Vaitukaitis, G.D. Braunstein, G.T. Ross, A radioimmunoassay which specifically measures human chorionic gonadotropin in the presence of human luteinizing hormone, Am. J. Obstet. Gynecol. 113 (6) (1972) 751–758.

[20] R. Hawkes, E. Niday, J. Gordon, A dot-immunobinding assay for monoclonal and other antibodies, Anal. Biochem. 119 (1) (1982) 142–147.

[21] S. Le, et al., Fabrication of paper devices via laser-heating-wax-printing for high-tech enzyme-linked immunosorbent assays with low-tech pen-type pH meter readout, Analyst 142 (3) (2017) 511–516.

[22] C.M. Cheng, et al., Paper-based ELISA, Angew. Chem. Int. Ed. 49 (28) (2010) 4771–4774.

[23] S. Avrameas, T. Ternynck, Enzyme-linked immunosorbent assay (ELISA) A2, in: P.J. Delves (Ed.), Encyclopedia of Immunology, second ed., Elsevier, Oxford, 1998, pp. 816–819.

[24] E.F. O'Connor, S. Paterson, R. de la Rica, Naked-eye detection as a universal approach to lower the limit of detection of enzyme-linked immunoassays, Anal. Bioanal. Chem. 408 (13) (2016) 3389–3393.

[25] K.M. Koczula, A. Gallotta, Lateral flow assays, Essays Biochem. 60 (1) (2016) 111–120.

[26] M. Sajid, A.-N. Kawde, M. Daud, Designs, formats and applications of lateral flow assay: a literature review, J. Saudi Chem. Soc. 19 (6) (2015) 689–705.

[27] C. Parolo, A. Merkoci, Paper-based nanobiosensors for diagnostics, Chem. Soc. Rev. 42 (2) (2013) 450–457.

[28] W. Jung, et al., Point-of-care testing (POCT) diagnostic systems using microfluidic lab-on-a-chip technologies, Microelectron. Eng. 132 (0) (2015) 46–57.

[29] C. Renault, et al., Three-dimensional wax patterning of paper fluidic devices, Langmuir 30 (23) (2014) 7030–7036.

[30] X. Huang, et al., Membrane-based lateral flow immunochromatographic strip with nano-particles as reporters for detection: a review, Biosens. Bioelectron. 75 (15) (2015) 166–180.

[31] D. Quesada-González, A. Merkoçi, Nanoparticle-based lateral flow biosensors, Biosens. Bioelectron. 73 (2015) 47–63.

[32] W. Zhou, et al., Gold nanoparticles for in vitro diagnostics, Chem. Rev. 115 (19) (2015) 10575–10636.

[33] I. Safenkova, A. Zherdev, B. Dzantiev, Factors influencing the detection limit of the lateral-flow sandwich immunoassay: a case study with potato virus X, Anal. Bioanal. Chem. 403 (6) (2012) 1595–1605.

[34] L. Rivas, et al., Improving sensitivity of gold nanoparticles-based lateral flow assays by using wax-printed pillars as delay barriers of microfluidics, Lab Chip 14 (2014) 4406–4414.

[35] C. Parolo, A. de la Escosura-Muñiz, A. Merkoçi, Enhanced lateral flow immunoassay using gold nanoparticles loaded with enzymes, Biosens. Bioelectron. 40 (1) (2013) 412–416.

[36] G. Shen, S. Zhang, X. Hu, Signal enhancement in a lateral flow immunoassay based on dual gold nanoparticle conjugates, Clin. Biochem. 46 (16–17) (2013) 1734–1738.

[37] H. Xu, et al., Gold-nanoparticle-decorated silica nanorods for sensitive visual detection of proteins, Anal. Chem. 86 (15) (2014) 7351–7359.

[38] D. Tang, et al., Magnetic nanogold microspheres-based lateral-flow immunodipstick for rapid detection of aflatoxin B2 in food, Biosens. Bioelectron. 25 (2) (2009) 514–518.

[39] Q. Fu, et al., Rough surface Au@Ag core–shell nanoparticles to fabricating high sensitivity SERS immunochromatographic sensors, J. Nanobiotechnol. 13 (1) (2015) 1–9.

[40] S. Wang, et al., Paper-based chemiluminescence ELISA: lab-on-paper based on chitosan modified paper device and wax-screen-printing, Biosens. Bioelectron. 31 (1) (2012) 212–218.

[41] A. Apilux, et al., Development of automated paper-based devices for sequential multistep sandwich enzyme-linked immunosorbent assays using inkjet printing, Lab Chip 13 (1) (2013) 126–135.

[42] X. Li, P. Zwanenburg, X. Liu, Magnetic timing valves for fluid control in paper-based microfluidics, Lab Chip 13 (13) (2013) 2609–2614.

[43] G.A. Posthuma-Trumpie, et al., Amorphous carbon nanoparticles: a versatile label for rapid diagnostic (immuno)assays, Anal. Bioanal. Chem. 402 (2) (2012) 593–600.

[44] C. Suárez-Pantaleón, et al., Development of an immunochromatographic assay based on carbon nanoparticles for the determination of the phytoregulator forchlorfenuron, Biosens. Bioelectron. 42 (2013) 170–176.

[45] M. Blažková, P. Rauch, L. Fukal, Strip-based immunoassay for rapid detection of thiabendazole, Biosens. Bioelectron. 25 (9) (2010) 2122–2128.

[46] E.M. Linares, et al., Enhancement of the detection limit for lateral flow immunoassays: evaluation and comparison of bioconjugates, J. Immunol. Methods 375 (1–2) (2012) 264–270.

[47] Y. Tang, et al., Preparation of functional magnetic nanoparticles mediated with PEG-4000 and application in *Pseudomonas aeruginosa* rapid detection, J. Biomed. Nanotechnol. 9 (2) (2013) 312–317.

[48] X. Xu, et al., Synthesis of magnetic microspheres with immobilized metal ions for enrichment and direct determination of phosphopeptides by matrix-assisted laser desorption ionization mass spectrometry, Adv. Mater. 18 (24) (2006) 3289–3293.

[49] Y.-M. Huang, et al., Rapid detection of aflatoxin M1 by immunochromatography combined with enrichment based on immunomagnetic nanobead, Chin. J. Anal. Chem. 42 (5) (2014) 654–659.

[50] Y. Wang, et al., Study of superparamagnetic nanoparticles as labels in the quantitative lateral flow immunoassay, Mater. Sci. Eng. C 29 (3) (2009) 714–718.

[51] R. Connolly, R. O'Kennedy, Magnetic lateral flow immunoassay test strip development – considerations for proof of concept evaluation, Methods 116 (2017) 132–140.

[52] C. Liu, et al., Lateral flow immunochromatographic assay for sensitive pesticide detection by using Fe_3O_4 nanoparticle aggregates as color reagents, Anal. Chem. 83 (17) (2011) 6778–6784.

[53] J. Yan, et al., Effect of physiochemical property of Fe_3O_4 particle on magnetic lateral flow immunochromatographic assay, Sensors Actuators B Chem. 197 (2014) 129–136.

[54] S. Shen, et al., Polyethylene nanofibres with very high thermal conductivities, Nat. Nano 5 (4) (2010) 251–255.

[55] A.O. Govorov, H.H. Richardson, Generating heat with metal nanoparticles, Nano Today 2 (1) (2007) 30–38.

[56] J. Hu, et al., Advances in paper-based point-of-care diagnostics, Biosens. Bioelectron. 54 (2014) 585–597.

[57] Z. Qin, et al., Significantly improved analytical sensitivity of lateral flow immunoassays by using thermal contrast, Angew. Chem. Int. Ed. 51 (18) (2012) 4358–4361.

[58] Y. Wang, et al., Thermal contrast amplification reader yielding 8-fold analytical improvement for disease detection with lateral flow assays, Anal. Chem. 88 (23) (2016) 11774–11782.

[59] D.M. Cate, et al., Recent developments in paper-based microfluidic devices, Anal. Chem. 87 (1) (2015) 19–41.

[60] S. Luo, et al., Ultrasensitive detection of pentachlorophenol based on enhanced electrochemiluminescence of Au nanoclusters/graphene hybrids, Sensors Actuators B Chem. 194 (2014) 325–331.

[61] Y. Xu, et al., Paper-based solid-state electrochemiluminescence sensor using poly(sodium 4-styrenesulfonate) functionalized graphene/nafion composite film, Anal. Chim. Acta 763 (2013) 20–27.

[62] Z. Li, et al., Recent advances in pen-based writing electronics and their emerging applications, Adv. Funct. Mater. 26 (2) (2016) 165–180.

[63] A.C. Siegel, et al., Thin, lightweight, foldable thermochromic displays on paper, Lab Chip 9 (19) (2009) 2775–2781.

[64] Y. Matsuda, et al., Electric conductive pattern element fabricated using commercial inkjet printer for paper-based analytical devices, Anal. Chem. 87 (11) (2015) 5762–5765.

[65] Z. Li, et al., Direct writing electrodes using a ball pen for paper-based point-of-care testing, Analyst 140 (16) (2015) 5526–5535.

[66] Z. Li, et al., Recent developments of three-dimensional paper-based electrochemical devices for cancer cell detection and anticancer drug screening, Curr. Pharm. Biotechnol. 17 (9) (2016) 802–809.

[67] P. Wang, et al., Paper-based three-dimensional electrochemical immunodevice based on multi-walled carbon nanotubes functionalized paper for sensitive point-of-care testing, Biosens. Bioelectron. 32 (1) (2012) 238–243.

[68] L. Ge, et al., Three-dimensional paper-based electrochemiluminescence immunodevice for multiplexed measurement of biomarkers and point-of-care testing, Biomaterials 33 (4) (2012) 1024–1031.

[69] W. Li, et al., Development of a 3D origami multiplex electrochemical immunodevice using a nanoporous silver-paper electrode and metal ion functionalized nanoporous gold-chitosan, Chem. Commun. (Camb) 49 (83) (2013) 9540–9542.

[70] C. Ma, et al., 3D origami electrochemical immunodevice for sensitive point-of-care testing based on dual-signal amplification strategy, Biosens. Bioelectron. 63 (2015) 7–13.

[71] Y. Wu, et al., Paper-based microfluidic electrochemical immunodevice integrated with nanobioprobes onto graphene film for ultrasensitive multiplexed detection of cancer biomarkers, Anal. Chem. 85 (18) (2013) 8661–8668.

[72] Y. Wu, et al., A paper-based microfluidic electrochemical immunodevice integrated with amplification-by-polymerization for the ultrasensitive multiplexed detection of cancer biomarkers, Biosens. Bioelectron. 52 (2014) 180–187.

[73] D.-B. Wang, et al., Rapid detection of *Bacillus anthracis* spores using a super-paramagnetic lateral-flow immunological detectionsystem, Biosens. Bioelectron. 42 (2013) 661–667.

[74] S. Shan, et al., Novel strategies to enhance lateral flow immunoassay sensitivity for detecting foodborne pathogens, J. Agric. Food Chem. 63 (3) (2015) 745–753.

[75] S. Qian, H.H. Bau, A mathematical model of lateral flow bioreactions applied to sandwich assays, Anal. Biochem. 322 (1) (2003) 89–98.

[76] Z. Liu, et al., Experimental and numerical studies on liquid wicking into filter papers for paper-based diagnostics, Appl. Therm. Eng. 88 (2015) 280–287.

[77] C. Parolo, et al., Simple paper architecture modifications lead to enhanced sensitivity in nanoparticle based lateral flow immunoassays, Lab Chip 13 (3) (2013) 386–390.

[78] B.J. Toley, et al., Tunable-delay shunts for paper microfluidic devices, Anal. Chem. 85 (23) (2013) 11545–11552.

[79] K.M. Schilling, et al., Fully enclosed microfluidic paper-based analytical devices, Anal. Chem. 84 (3) (2012) 1579–1585.

[80] B. Lutz, et al., Dissolvable fluidic time delays for programming multi-step assays in instrument-free paper diagnostics, Lab Chip 13 (14) (2013) 2840–2847.

[81] G.M. Whitesides, Viewpoint on "Dissolvable fluidic time delays for programming multi-step assays in instrument-free paper diagnostics", Lab Chip 13 (20) (2013) 4004–4005.

[82] C. Li, et al., Paper-based surfaces with extreme wettabilities for novel, open-channel microfluidic devices, Adv. Funct. Mater. 26 (66) (2016) 6121–6131.

[83] Y. Sun, A. Kharaghani, E. Tsotsas, Micro-model experiments and pore network simulations of liquid imbibition in porous media, Chem. Eng. Sci. 150 (2016) 41–53.

[84] S. Jahanshahi-Anbuhi, et al., Paper-based microfluidics with an erodible polymeric bridge giving controlled release and timed flow shutoff, Lab Chip 14 (1) (2014) 229–236.

[85] S. Jahanshahi-Anbuhi, et al., Creating fast flow channels in paper fluidic devices to control timing of sequential reactions, Lab Chip 12 (23) (2012) 5079–5085.

[86] L. Wang, et al., A bare-eye-based lateral flow immunoassay based on the use of gold nanoparticles for simultaneous detection of three pesticides, Microchim. Acta 181 (13) (2014) 1565–1572.

[87] J.-H. Cho, S.-H. Paek, Semiquantitative, bar code version of immunochromatographic assay system for human serum albumin as model analyte, Biotechnol. Bioeng. 75 (6) (2001) 725–732.

[88] H.-Y. Yin, et al., Development of a barcode-style lateral flow immunoassay for the rapid semi-quantification of gliadin in foods, Food Chem. 192 (2016) 934–942.

[89] D. Zhang, et al., A naked-eye based strategy for semiquantitative immunochromatographic assay, Anal. Chim. Acta 740 (2012) 74–79.

[90] D. Zhang, et al., Development of a detector-free semiquantitative immunochromatographic assay with major aflatoxins as target analytes, Sensors Actuators B Chem. 185 (2013) 432–437.

[91] Q. Fang, et al., A bare-eye based one-step signal amplified semiquantitative immunochromatographic assay for the detection of imidacloprid in Chinese cabbage samples, Anal. Chim. Acta 881 (2015) 82–89.

[92] Y.K. Oh, et al., A three-line lateral flow assay strip for the measurement of C-reactive protein covering a broad physiological concentration range in human sera, Biosens. Bioelectron. 61 (0) (2014) 285–289.

[93] A. Chen, et al., Smartphone-interfaced lab-on-a-chip devices for field-deployable enzyme-linked immunosorbent assay, Biomicrofluidics 8 (6) (2014) 064101.

[94] B. Li, et al., A smartphone controlled handheld microfluidic liquid handling system, Lab Chip 14 (20) (2014) 4085–4092.

[95] H. Huang, M. Sukop, X. Lu, Multiphase lattice Boltzmann methods: theory and application, in: H. Huang, M. Sukop, X. Lu (Eds.), Multiphase Lattice Boltzmann Methods: Theory and Application, Wiley Blackwell, New Jersey, 2015.

[96] D. Zhang, Q. Liu, Biosensors and bioelectronics on smartphone for portable biochemical detection, Biosens. Bioelectron. 75 (2016) 273–284.

[97] D.N. Breslauer, et al., Mobile phone based clinical microscopy for global health applications, PLoS One 4 (7) (2009) e6320.

[98] C.-Y. Wu, et al., Research highlights: increasing paper possibilities, Lab Chip 14 (17) (2014) 3258–3261.

[99] W.G. Lee, et al., Nano/microfluidics for diagnosis of infectious diseases in developing countries, Adv. Drug Deliv. Rev. 62 (4–5) (2010) 449–457.

[100] T.H. Schulte, R.L. Bardell, B.H. Weigl, Microfluidic technologies in clinical diagnostics, Clin. Chim. Acta 321 (1–2) (2002) 1–10.

Chapter 9

Acoustic Wave–Based Immunoassays

Aristea Grammoustianou[1,2], Electra Gizeli[1,2]
[1]*University of Crete, Heraklion, Greece;* [2]*Institute of Molecular Biology and Biotechnology, Foundation for Research and Technology-Hellas, Heraklion, Greece*

1. INTRODUCTION

In the last decade, the use of acoustic wave biosensors has increased significantly so that today these are fully established as a standard biophysical and bioanalytical tool. Some of the reasons for their wide implementation in research laboratories are their high sensitivity to surface-binding events, simplicity of the experimental procedures, and large range of materials that are compatible with the propagation of an acoustic wave and, thus, can be employed on the surface of a device. Moreover, the ability to monitor two signals, namely, the velocity and energy of the acoustic wave, and derive complementary information on the surface-binding events and interface properties, is a unique feature of acoustic wave sensors resulting in a plethora of information not accessible by other biosensing methods. Last but not least, a significant contribution to the rise in usage of acoustic wave technologies has been the recent availability of several commercial systems, all of which provide the means for simple and well-controlled real-time biomolecular interaction analysis. Indeed, it is now possible to have access to commercial acoustic sensor chips coated with several materials (Au, silica, titanium, hydroxyapatite, dextran, etc.) and/or prefunctionalized with a specific chemistry (biotin, carboxyl-groups, etc.) for the subsequent immobilization of a biorecognition molecule [1]. In addition, developments in the fluidics at both the milli- and microscales allow for precise control of flow rates, use of small volumes, parallel analysis of four or more samples, and fitting of experimental data with mathematical models for deriving quantitative results and physicochemical parameters.

Acoustic wave devices exploit the piezoelectric (PZ) effect that was discovered in 1880 by Jacques and Pierre Curie to generate an acoustic wave in a PZ substrate. The main types of acoustic wave devices used in biological and clinical applications are the quartz crystal microbalance (QCM) or

Handbook of Immunoassay Technologies. https://doi.org/10.1016/B978-0-12-811762-0.00009-8

thickness shear mode (TSM) resonator and the surface acoustic wave (SAW) device [2]. The former, operating at a fundamental frequency of 5 MHz and at harmonics going up to 65 MHz, are simple to fabricate and control electronically and for this reason, constitute the largest part of the commercially available systems. The SAW device production is more challenging since it relies on the use of photolithographically or ion-beam produced electrodes (interdigitated transducers) but operate at higher frequencies, i.e., from 100 to 500 MHz; the latter is desirable since higher operating frequencies are associated with a higher mass sensitivity, at least based on the well-known Sauerbrey equation [3]. SAW devices employed in biological sensing fall primarily in two main categories both of which support a shear-horizontal (SH) acoustic wave: the Love wave SAW (LW-SAW) device, a geometry relying on a two-layer structure where a film of polymer, silica, or other material is deposited on top of a quartz-based SH-SAW device and a lithium tantalate–based SAW SH device, which does not require an extra layer for waveguiding.

Regardless of the type of acoustic wave device used, the binding event is normally followed by monitoring the velocity and/or energy of the acoustic wave in real time. Measurements related to the former are normally recorded as frequency (F) or phase (Ph or φ) change, while dissipation (D) or amplitude (A) is the reported signal related to the acoustic wave energy. It is now well established that frequency/phase measurements can be directly associated with the amount of mass deposited on the device surface [3,4]; the phase signal can also be used to discriminate the molecular weight (Mw) of immobilized proteins [5]. Dissipation/amplitude changes are indicative of the viscosity or viscoelastic changes occurring at the device/liquid interface [2]. Moreover, recent advances in the field have shown that the combination of the two measurements (i.e., D/F) can be used to derive specific information on the conformation of discretely bound biomolecules [6–10] or the size of particles attached to the surface [11–13]. More information on the different types of acoustic wave devices and corresponding biophysical principles of acoustic wave/matter interaction can be found in several review articles [14–18] or book chapters [19]; of interest are also research papers comparing the different devices performance toward liquid sensing [20].

Both QCM and SAW sensors have been extensively studied for their use as immunosensors. Acoustic wave–based immunosensors rely on the specific biochemical interaction between an antibody and an analyte, typically an antigen. The type of analyte differs and can be a small molecule, an antibody or another protein. Label-free and real-time detection protocols as well as fast-response times are the main characteristics of acoustic wave immunosensors. This chapter discusses the most important research attempts to develop reliable acoustic wave–based immunosensors/immunoassays. For ease of reading, a classification based on the type of target detected, while within each category QCM and SAW immunosensors are reviewed separately under each category.

Moreover, they were classified based on the type of assay that was employed in direct, indirect and sandwich-amplified systems. Direct immunosensors use a single biorecognition element (i.e., antibody, aptamer) against the target of interest and provide a signal corresponding to the binding of the analyte. The indirect acoustic wave immunosensors detect the presence of the analyte of interest through a secondary recognition process, which is specifically related to the presence of the target analyte. Last, the sandwich-amplified immunosensors refer to cases where a secondary antibody, a nanoparticle, or liposome is employed for the enhancement of the acoustic signal and the increase of the assay sensitivity. It should be noted that in the vast majority of the reported acoustic immunosensors the frequency/phase signal is employed as the main means for analyte detection. This is because this signal, being directly related to the amount of deposited mass, has always shown a good correlation with the analyte's concentration. In addition, changes in the conformation of the analyte or viscoelastic properties of the interface are less pronounced in such assays making the energy measurement less sensitive to the binding event.

2. CLINICAL DIAGNOSTICS

The detection and quantification of disease-related biomarkers in biological samples (blood, urine, saliva, etc.) is of great importance for diagnosis and management of different types of cancers and chronic diseases such as Alzheimer's disease and Rheumatoid arthritis. The development of fast, noninvasive, sensitive, and cost-effective diagnostic platforms that will allow the detection of diseases at an early stage will change the future of clinical diagnostics [21–23]. Acoustic wave–based immunosensors have been extensively studied for their application in clinical diagnostics (Table 9.1).

2.1 Quartz Crystal Microbalance Immunosensors

2.1.1 Direct Immunosensors

Direct immunosensors use a single biorecognition element (i.e., antibody, aptamer) against the target of interest.

The first report of an immunosensor was produced by Shons et al. in 1972 [24]. They managed to use a 5 MHz quartz crystal microbalance for a rapid immunoassay by which they detected anti bovine serum albumin (BSA) antibodies that were bound to previously immobilized BSA. The sensitivity of the assay was similar to that obtained with traditional passive agglutination techniques; the assay time, however, was much faster. Since then, a wide variety of QCM immunosensors for clinical diagnosis of biomarkers implicated in pathological conditions such as cancer, heart diseases, and inflammation have been developed. A number of QCM immunosensors were designed for the detection of the C-reactive protein (CRP). CRP is an acute-phase protein and a biomarker of inflammation, cardiovascular diseases, and

TABLE 9.1 Acoustic Wave–Based Immunosensors for the Detection of Clinical Biomarkers

Marker	Biosensor Type	Limit of Detection (LOD)	Clinically Relevant LOD	Assay Time	Signal	Sample	References
C-reactive protein (CRP)	QCM	0.001 μg/dL	3 μg/mL	–	ΔF (Hz)	Human serum	Kurosawa et al. [25]
	QCM	0.130 ng/mL	3 μg/mL	–	ΔF (Hz)	Buffer	Kim N. et al. [59]
	QCM	–	3 μg/mL		ΔF (Hz)	Human serum	Aizawa et al. [60]
IgE	QCM	–	150–300 UI/mL	15 min	ΔF (Hz)	Buffer	Yao et al. [27]
	QCM	2.5 μg/L	150–300 UI/mL	15 min	ΔF (Hz)	Human serum	Yao et al. [28]
	QCM	5 UI/mL	150–300 UI/mL	–	ΔF (Hz)	Buffer	Su and Zhang [29]
IA-2 autoantibodies	QCM	–	0.02 nmol/L	–	ΔF (Hz), ΔD (10^{-6})	Human serum	Ayela et al. [31]
APP_{770}	QCM	–	N/A	–	ΔF (Hz)	Complete medium	Mustafa et al. [32]
Human ferritin	QCM	0.1 ng/mL	90–200 ng/mL	–	ΔF (Hz)	Buffer and mouse serum	Chou et al. [33]

α-Fetoprotein (AFP)	QCM	0.1 μg/L	<20 μg/L	–	ΔF (Hz)	Buffer and human serum	Chou et al. [34]
	QCM	50 ng/mL	<20 μg/L	–	ΔF (Hz)	Clinical patient sera	Tsai et al. [35]
Mesothelin	QCM	range of ng	N/A	–	ΔF (Hz)	Supernatant from two cell lines	Corso et al. [36]
Carcinoembryonic antigen (CEA)	QCM	1.5 ng/mL	≤3 ng/mL	–	ΔF (Hz)	Human serum	Tang et al. [37]
	QCM	0.5 ng/mL	≤3 ng/mL	–	ΔF (Hz)	Clinical serum specimens	Chen and Tang [38]
	QCM	–	≤3 ng/mL	–	ΔF (Hz)	Buffer	Shen and Lu [39]
CYP1B1	QCM	2.2 ± 0.9 nM	N/A	–	ΔF (Hz)	Normal and malignant cell lysates	Shen et al. [40]
Cancer antigen 15.3 (CA 15.3)	QCM	0.5 U/mL	<25 U/mL	<10 s	ΔF (Hz)	Buffer	Wang et al. [41]
GTPase RhoA	QCM	–	N/A	–	ΔF (Hz)	Buffer	Martínez-Rivas et al. [42]
Parathyroid hormone-related peptide (PTHrP)	Electromagnetic QCM	61 ng/mL	<2 pmol/L	–	ΔF (Hz)	Buffer	Crivianu-Gaita et al. [43]
rhIL-1ra	QCM-D	25 ng/mL	N/A	10 min	ΔF (Hz)	human serum	Carrigan et al. [44]
RhIL-18BPa	QCM-D	400 ng/mL	N/A	–	ΔF (Hz)	Buffer	Carrigan et al. [45]

Continued

TABLE 9.1 Acoustic Wave–Based Immunosensors for the Detection of Clinical Biomarkers—cont'd

Marker	Biosensor Type	Limit of Detection (LOD)	Clinically Relevant LOD	Assay Time	Signal	Sample	References
Rheumatoid arthritis (RA)-anti-CCP	QCM	Femtomol range	<20 U/mL	–	ΔF (Hz)	Human serum	Drouvalakis et al. [46]
Human serum albumin (HSA)	QCM	20 μg/mL	35–50 g/L	–	ΔF (Hz)	Urine	Sakti et al. [47]
HSA	Monilithic QCM (MQCM)	–	35–50 g/L	–	ΔF (Hz)	Buffer	Jaruwongrungsee et al. [48]
Insulin	QCM	1 ng/mL	<174 pmol/L (fasting)	–	ΔF (Hz)	Buffer	Saha et al. [49]
Cortisol	QCM	5 pg/mL	30–140 ng/mL	10 min	ΔF (Hz)	Buffer	Ito et al. [50]
Bone morphogenetic protein-2 (BMP-2)	QCM	0.5 μg/mL	N/A	–	ΔF (Hz)	Buffer	Michalzik et al. [51]
Matrix metalloproteinase-1 (MMP-1)	QCM	–	N/A	–	ΔF (Hz)	Buffer	Soumetz et al. [52]
Ceruplasmin (CP)	QCM	0.15 μg/mL	210–530 mg/L	–	ΔF (Hz)	Clinical samples	Wang H. et al. [53]

Cardiac Troponin T (cTnT)	QCM	5 ng/mL	0.05 ng/mL	–	ΔF (Hz)	Buffer	Wong-Ek et al. [54]
Urinary proteins (diabetes nephropathy)	QCM	0.01 mg/L	30–300 mg	15 min	ΔF (Hz)	Urine samples	Luo et al. [55]
Blood coagulation factor VIII	QCM	–	N/A	10 min	ΔF (Hz)	Patient sera	Yao et al. [56]
Human Immunoglobulin G (hIgG)	PQC	50 ng/mL	231–1600 mg/dL	–	ΔF (Hz)	hIgG sera	Chen H. et al. [58]
Fibrin degradation products (FDP)	QCM	<10 μg/mL	<10 μg/mL	10 min	ΔF (Hz)	Human serum	Aizawa et al. [61]
α-Amylase	QCM	0.08 U/mL	40–140 U/L	–	QCM measurements	Human serum	Gibbs et al. [63]
Prostate specific antigen (PSA)	QCM-1	0.29 ng/mL	4 ng/mL	–	ΔF (Hz)	human serum (75%)	Uludăg and Tothill [64]
Antihemoglobin (anti-Hb) and antimyoglobin (anti-Mb) antibodies	SAW	0.32 μg/mL and 0.35 μg/mL respectively	N/A	–	Frequency shift (Δf)	Buffer with biospecies of human blood (urea, ascorbic acid etc.)	Chang and Shih [69]
Interleukin-6 (IL-6)	LW-SAW	–	5–15 pg/mL	–	Phase shift (Δφ)	Human serum	Krishnamoorthy et al. [70]

LW-SAW, love wave surface acoustic wave; *QCM*, quartz crystal microbalance; *SAW*, surface accustic wave.

cancer. Kurosawa et al. [25] employed plasma polymerized films (pp-films) on the sensor surface to immobilize both anti-CRP monoclonal antibodies and anti-CRP $F(ab')_2$ fragments of these antibodies in an oriented way. Both capturing agents were able to detect CRP successfully. Acoustic wave–based immunoassays have been designed for the detection of antibodies produced by the immune system. Su et al. [26] detected the IgE produced in cases of allergy through binding of the antibody to anti-IgE monoclonal antibodies. Yao et al. [27,28] did the same but used biotinylated aptamers as the recognition element. Su and Zhang [29] compared QCM for their ability to detect the interaction of human IgE with surface-immobilized anti-IgE monoclonal antibodies and concluded that the QCM was simpler and less expensive compared to the SPR, although not as sensitive. IgM (produced in high levels in cases of infection) was acoustically monitored by Chu et al. [30] who constructed a PZ quartz crystal array system (composed of 5 MHz crystals) for simultaneous detection of two analytes (IgM and CRP). Ayela et al. [31] also combined the use of QCM and SPR for the detection of a synthetic diabetes-linked autoantigen (IA-2 peptide) in human serum. Another Alzheimer's disease protein, the APP_{770} was detected in complete medium (that is a complex medium containing other proteins, salts, and amino acids) when Mustafa et al. [32] employed QCM-1 sensors that carried two sensing spots on the same quartz crystal and coupled them to a total internal reflection ellipsometry (TIRE) method. Conjoined use of the two techniques allowed the determination of the unknown concentration of APP_{770} analyte.

A significant amount of literature on the acoustic detection of cancer biomarkers is available. Chou et al. [33,34] managed to detect ferritin (nonspecific tumor marker) in mouse serum and α-fetoprotein (AFP) (an early biomarker of hepatocellular carcinoma and teratoblastoma) in human serum. In both cases, they immobilized specific antibodies on self-assembled monolayers (SAMs) of cystamine and measured the frequency shifts of the QCM sensors. They achieved to detect AFP at concentrations of 0.1–100 μg/L. AFP in human serum has also been monitored by Tsai and Liu [35]. They employed 10 MHz amine-coated sensors on which they immobilized anti-AFP antibodies via glutaraldehyde (GA) cross-linking and they monitored the frequency change when 50 ng/mL of AFP were added on the sensor surface. Corso et al. [36] reported on an acoustic wave immunosensor for the real-time and less invasive detection of mesothelin, which is a marker of pancreatic and ovarian cancers. They used a QCM sensor and tested the supernatant of two different pancreatic cell lines in which they identified mesothelin at nanogram concentrations. Carcinoembryonic antigen (CEA), a tumor marker for clinical diagnosis of colorectal, pancreatic, gastric, and cervical cancer, has been acoustically identified as well [37–39]. Chen and Tang [38], in particular, employed a magnetic $CoFe_2O_4/SiO_2$ composite nanoparticle-functionalized interface and monitored the presence of CEA in clinical serum. Shen and Lu, in turn, used gold nanoparticles (AuNPs) coupled with mixed SAM to

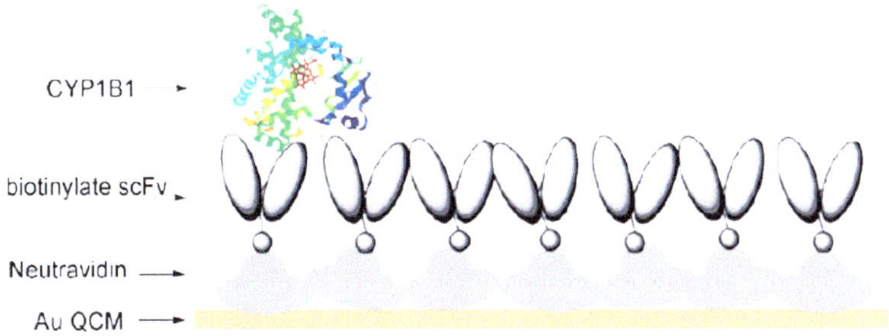

FIGURE 9.1 Representation of an antibody-based quartz crystal microbalance (QCM) immunosensor for detection of the cytochrome CYP1B1 [40].

increase the efficiency of the interaction between anti-CEA antibodies and CEA and to minimize nonspecific binding. Universal tumor marker cytochrome P450 1B1 has also been acoustically detected in normal and malignant cell lysates [40] (Fig. 9.1). One of the most important breast cancer markers the cancer antigen 15.3 (CA 15.3) was identified in an acoustic wave immunoassay that used QCM sensors modified with ZnO nanorods. The response time of these sensors was reported to be less than 10 s [41]. Martínez-Rivas et al. [42] presented a QCM sensor surface modified with nickel nanoislands of ~5 nm, which permitted detection and discrimination between the active and inactive form of the cancer biomarker GTPase RhoA. In 2016, Crivianu-Gaita et al. [43] reported for the first time on the detection the breast and prostate cancer metastasis biomarker parathyroid hormone-related peptide (PTHrP). They employed quartz discs for the fabrication of an electromagnetic PZ acoustic device analysis; the PZ disc operating as an electrode-free sensor was stimulated by an external magnetic field and achieved detection of PTHrP at a concentration of 61 ng/mL when whole and nonfragmented anti-PTHrP antibodies were surface immobilized.

Other clinical biomarkers detected with QCM immunoassays are sepsis biomarkers [44,45], rheumatoid arthritis-autoantibodies (Drouvalakis et al. [46]: the first report ever on the use of functionalized single-walled carbon nanotubes for surface modification), human serum albumin (HSA) [47,48], insulin [49], cortisol [50], recombinant human bone morphogenetic protein-2 [51], matrix metalloproteinase-1 [52], ceruloplasmin [53], cardiac troponin T [54], urinary proteins [55], and blood coagulation factor VIII [56].

2.1.2 Indirect Immunosensors

Miura et al. [57] used a PZ crystal resonator to detect methamphetamine (MA) in human urine. They designed an indirect immunoassay in which MA was conjugated to BSA and immobilized on the sensor surface. Then anti-MA antibodies previously mixed with MA in a buffer solution resulted in a shift in the resonant frequency and in the detection of 0.02–100 ppm of MA in approximately 7 min. Chen et al. [58] used liposomes, the PZ immunoagglutination assay (PEIA) and a 9 MHz PZ quartz crystal to detect human immunoglobulin G (hIgG). Anti-hIgG functionalized liposomes were immobilized on a BSA monolayer (used to reduce background interference) previously deposited on the sensor surface. The addition of hIgG serum caused liposomes agglutination and frequency change. Kim et al. [59] constructed a QCM immunosensor for detection of CRP. This time CRP was firstly immobilized on the sensor surface followed by the addition of anti-CRP antibodies mixed with free CRP. When the mixture of the anti-CRP antibody and free CRP reached the surface, a competitive immune reaction took place, where immobilized CRP and free CRP competed against the antibody. This interaction was acoustically monitored through changes in the frequency and the detection limit achieved was 0.130 ng/mL. Aizawa et al. [60] combined a QCM with the agglutination reaction of immunized latex beads creating the latex PZ immunoassay (LPEIA). Latex beads were functionalized with anti-CRP antibodies and adsorbed on the sensor surface. Subsequent addition of CRP serum led to the beads agglutination and to the change of the resonant frequency of the sensor. This method avoids the surface immobilization of antigen or antibody and thus, it is faster and less laborious. The whole assay was integrated into a laboratory-manufactured sensor, the CRP II, and measured the CRP concentration in human serum. The same LPEIA assay was employed by Aizawa et al. [61] for the detection of fibrinogen and fibrin degradation products (FDP). High levels of fibrinogen are linked to cardiovascular diseases. The assay enabled the detection of FDP in human serum, in 10 min and led to the fabrication of smart FDP-portable sensors [62]. α-Amylase is used in the diagnosis of pancreatitis. Gibbs et al. [63] constructed a QCM immunosensor that was based on oligosaccharide hydrogel films (such as glycogen and amylopectin) and detected 0.08 U/mL of α-amylase in human serum through degradation of the films than occurred with the addition of the analyte.

2.1.3 Sandwich-Amplified Immunosensors

Uludag and Tothill [64] reported on the development of a QCM immunoassay with AuNPs amplification for detection of cancer biomarkers (prostate-specific antigen [PSA] and PSA-alpha 1-antichymotrypsin [ACT] complex). Apart from proteins/cancer biomarkers, QCM immunoassays have been developed for the detection of whole cancer cells. Ma et al. [65] used a QCM sensor where human lung carcinoma cells (LCC) were captured by anti-LCC antibodies immobilized on the surface. The whole assay was amplified by the addition of anti-LCC modified AuNPs that reduced the limit of detection (LOD) to 100 cells/mL. Similarly, Shan et al. [66] detected 1160 leukemia cells/mL in a sensitive and selective manner. Leukemia cells were captured by surface-immobilized aptamers, then aminophenylboronic acid-modified AuNPs (APBA-AuNPs) were used for cells labeling, followed by silver enhancement that led to frequency signal enhancement.

2.2 Surface Acoustic Wave Immunosensors

SAW immunosensors have been less studied compared with QCM immunosensors for clinical diagnosis, probably because commercial SAW sensors and measuring electronics are not widely available yet.

2.2.1 Direct Immunosensors

Welsch et al. [67] were the first to construct a shear horizontal polarized surface acoustic wave (SH-SAW) for monitoring of immunochemical reactions. IgG immunoglobulin was captured by specific antibodies on the surface, and the reaction was monitored through shifts in the phase signal ($\Delta\varphi$). Campitelli et al. [68] also investigated the interaction of IgG with protein A using a portable SH-SAW sensor array. Chang and Shih [69] identified the existence of antihemoglobin (anti-Hb) and antimyoglobin (anti-Mb) antibodies in human blood using a SAW immunosensor. Krishnamoorthy et al. [70] fabricated a high-sensitivity $ZnO/SiO_2/Si$ LW-SAW immunosensor able to detect the glycoprotein interleukin-6 (IL-6) in human serum through monitoring of the frequency shift.

3. DETECTION OF MICROBIAL PATHOGENS AND TOXINS

Bacterial infections are a common phenomenon both in developed and underdeveloped world. The World Health Organization ranks communicable infections caused by microbial pathogens among the top 10 causes of death in 2015 (http://www.who.int/gho/mortality_burden_disease/causes_death/top_10/en/). Thus, there is an urgent need for rapid and accurate detection of bacterial pathogens and their toxins. The food industry, public health sectors, and biosafety agencies are three of the many fields concerned with the detection of bacterial pathogens. There has been a considerable amount of work done on the development of acoustic wave immunosensors for bacterial detection [71]. These sensors aim to replace the traditional, time-consuming methods such as cell culture and colony counting and the laborious and expensive polymerase chain reaction.

3.1 Quartz Crystal Microbalance Immunosensors

3.1.1 Direct Immunosensors

The first QCM immunosensor for microbial pathogens detection was presented by Muramatsu et al. [72]. They achieved detection of *Candida albicans* cells captured by anti–*Candida albicans* antibodies that had been immobilized on PZ crystals. Detection was carried out via monitoring of changes in the frequency shift at cell concentrations raging from 10^6 to 5×10^8 cells/cm^{-3}. Many research groups have investigated *Escherichia coli* detection in buffer and food samples with QCM immunosensors due to its implications in gastroenteritis disease [73–79]. Poitras and Tufenkji [80] studied the detection of *E. coli* following the change in the dissipation signal (ΔD) using a QCM-D instrument. Detection was carried out in less than 5 min, and no other study reports on the direct detection of any other bacterial pathogen through dissipation shift alone. Zhang et al. [81] used a hierarchically molecularly imprinted polymer QCM immunosensor to detect *E. coli.* Molecularly imprinted polymers (MIPs) are a template-based synthesis of cavities on polymer films. The cavities are complementary in size, shape, and interaction points to the template. In other words, MIPs are mimics of antibodies but more stable and more robust.

Salmonella species, which are major food pathogens, have repeatedly been detected by QCM immunoassays in buffer, milk, and food samples. Prusak-Sochaczewski and Luong [82] employed 9 MHz PZ crystals that were dipped in reagents solution to detect *Salmonella* Typhimurium (10^5–10^9 cells/mL). Similar approaches with small differences in the surface chemistry have also been reported [83–90]. In 2006 Olsen et al. [91] reported on the use of a filamentous phage as a specific and selective probe for *Salmonella* Typhimurium. Bacteriophage interaction resulted in resonance frequency changes and the detection of 10^2 cells/mL of *Salmonella* in less than 180 s.

Other bacterial pathogens probed involve *Mycobacterium tuberculosis* [92,93], *Treponema pallidum* [94], *Vibrio* species [95,96], *Pseudomonas aeruginosa* [97], *Bacillus* species [98–100], *Francisella tularensis* [101], *Listeria monocytogenes* [102], *Campylobacter jejuni* [103], and the marine pathogen *Edwardsiella tarda* [104].

Apart from bacterial pathogens, QCM immunosensors have been employed for the identification of bacterial and fungal toxins that act mostly as food and water contaminants. Such QCM immunoassays are most frequently based on antibody–toxin interaction. In 1997, Harteveld et al. [105] described a PZ crystal immunosensor for detection of the food poisoning Staphylococcal enterotoxin B. The ganglioside GM1 was employed by Spangler et al. [106] to bind *E. coli* heat-labile enterotoxin that is responsible for traveler's diarrhea. Ochratoxin A was also studied [107].

3.1.2 Indirect Immunosensors

Shen et al. [108] suggested the use of lectins (concanavalin A) as surface recognition elements for *E. coli* identification instead of the use of antibodies because they were shown to promote the strong adhesion of *E. coli* on the sensor surface. Wan et al. [109] managed to detect the marine pathogenic sulfate-reducing bacterium. In particular, they distinguished between the vancomycin-sensitive bacterium *D. desulfotomaculum* and the vancomycin-resistant *V. anguillarum* employing vancomycin-functionalized magnetic nanoparticles. Guntupalli et al. [110] presented a QCM-D immunosensor able to discriminate between methicillin-resistant (MRSA) and methicillin-sensitive (MSSA) strains of *Staphylococcus aureus*. Lytic bacteriophage spheroids were surface-immobilized and targeted the bacterial cells. Following the cells binding that changed the frequency and dissipation signal, the sensors were tested with penicillin-binding protein (PBP2) specific antibody conjugated latex beads. Sensors with MRSA captured responded to PBP2 antibody beads, while sensors with MSSA did not.

3.1.3 Sandwich-Amplified Immunosensors

Initial QCM/SAW applications involved the measurement of Δf in air, i.e., after drying the crystal; PZ crystals that had previously been coated with anti−*E. coli* antibodies were immersed in *E. coli* serum by Muramatsu et al. [111]. The immersion resulted in a shift of the resonant frequency of the crystal. Injection of polystyrene particles modified with anti−*E. coli* antibodies led to further increase of the resonant frequency shift and to *E. coli* detection with an LOD of 10^5 cells/cm^{-3}. Ben-Dov and Willner [112] demonstrated the detection of *Chlamydia trachomatis* in urine samples using 9 MHz QCM crystals coated with anti−*C. trachomatis* LPS antibodies. *C. trachomatis* was added on the sensor and then another step of anti−*C. trachomatis* LPS antibodies addition was used for amplification and confirmation. Alfonta and Willner [113] identified the cholera toxin produced by *Vibrio cholerae* using a QCM 9 MHz immunosensor and enzyme-modified (HRP)-ganglioside (GM1)-functionalized liposomes as frequency signal amplifiers. Su and Li [114] reported on the detection of *Helicobacter pylori* with a sandwiched and enzymatically amplified PZ immunosensor, while Kim G.H. et al. [115] created a 5 MHz dual QCM sensor that could detect *Salmonella* Typhimurium cells that had been previously captured by anti−*S. typhimurium*−coated paramagnetic beads. The complexes were magnetically transferred to the sensor surface and captured by surface-immobilized antibodies. *Salmonella typhimurium* in food samples was probed by Su and Li [116] who used a QCM immunosensor with simultaneous monitoring of the resonant frequency (ΔF) and motional resistance (ΔR). Both ΔF and ΔD achieved direct *Salmonella typhimurium* detection in chicken meat sample. When anti-*Salmonella* magnetic beads were used as signal amplifiers, ΔR proved to be more sensitive and reliable, reaching an LOD of 10^2 cells/mL. Jin X. et al. [117] detected Aflatoxin B (a fungal mycotoxin) with an ultrasensitive PZ technique and biocatalyzed deposition amplification. Aflatoxin B1 was identified in artificially contaminated milk samples at a concentration of 0.01 ng/mL. Jiang et al. [118] investigated the use of micro/nanobeads of various sizes and materials (i.e., magnetic/silica/polymer beads with a diameter varying from 30 to 970 nm) as QCM signal amplifiers for the detection of *E. coli* O157: H7. They showed that the largest frequency signal amplification was obtained by magnetic beads with a diameter of 350 nm. Guo et al. [119] went a step further and proceeded to the construction of a QCM immunosensor that simultaneously enriched and detected *E. coli* O157: H7. Enrichment was done in brain heart infusion (BHI) broth for 18 h and was followed by the detection with anti−*E. coli* antibody-functionalized AuNPs. The whole assay was completed within 24 h and achieved an LOD of 1−10 CFU/mL.

Ozalp et al. [120] in 2015 also reported on the detection of *S. typhimurium* using anti-*Salmonella* aptamers and magnetic particles.

3.2 Surface Acoustic Wave Immunosensors

3.2.1 Direct Immunosensors

Direct SAW immunoassays for bacterial detection are limited. This is probably due to the smaller penetration depth obtained with the high operating frequency of SAW devices when compared with the QCM ones. Berkenpas et al. [121] demonstrated the use of a langasite (LGS) SH-SAW sensor for *E. coli* recognition. *E. coli* detection was further studied by Dinh et al. [122] who employed an LW-SAW and concluded that the topology of the surface of waveguides was crucial for the specific detection of *E. coli*.

Reports on SAW detection of toxins are even fewer. Aflatoxin B conjugated with BSA was probed by Puiu et al. [123] with an LW-SAW in a study concerning the SAW detection of high molecular weight (MW) antigens.

3.2.2 Sandwich-Amplified Immunosensors

Pyun et al. [124] used an acoustic gravimetric flexural plate wave (FPW) transducer and microspheres of styrene—maleic—copolymerisate encapsulated magnetite (2 μm) functionalized with anti—*E. coli* antibodies to detect *E. coli* through changes in the resonance frequency of the PZ sensor.

4. DETECTION OF PARASITES

Parasites pose a threat to human health in the same way bacterial pathogens do. They too are related to foodborne and waterborne diseases. Malaria, a mosquito-borne disease is the most widely known infection caused by a parasite. However, the Centers for Disease Control and Prevention (CDC) reports on the reappearance of five neglected parasitic infections (toxoplasmosis, Chagas disease, cysticercosis, toxocariasis, and trichomoniasis) in the United States due to lacking surveillance and prevention (https://www.cdc.gov/parasites/npi/index.html). Accordingly, improved diagnostic assays are necessary.

4.1 Quartz Crystal Microbalance Immunosensors

4.1.1 Direct Immunosensors

Few QCM-based immunosensors for the detection of parasites are found in the literature. *Cryptosporidium parvum* associated with waterborne illnesses was identified by Poitras et al. [125] who used a QCM-D device and frequency and dissipation monitoring. *Schistosoma japonicum*, a major infectious agent of schistosomiasis that is a disease with high prevalence in Africa, Asia, and Latin America has been detected with QCM immunoassays [126,127].

4.1.2 Indirect Immunosensors

Toxoplasma gondii, responsible for toxoplasmosis, was detected with 9 MHz PZ quartz crystals by Wang et al. [128]. In this work they introduced, for the first time, the use of AuNPs as an alternative to latex particles, which were commonly used till then for LPEIAs. Likewise, Wang et al. [129] used a QCM sensor and submicrometer (280 nm in diameter) silica particles modified with *Schistosoma japonicum* antigens. Particles were agglutinated on anti–*Schistosoma japonicum* antibodies addition, resulting in detection of the parasite in human serum.

4.1.3 Sandwich-Amplified Immunosensors

Ding et al. [130] employed a QCM enzyme-catalyzed amplified immunoassay for detection of *Toxoplasma gondii.* Wu et al. [131] identified the presence of *Schistosoma japonicum* antibodies in patients' and normal human sera at concentrations of 5 ng/mL. A mercapto *Schistosoma japonicum* antigen was surface immobilized through self-assembly and followed by capturing of the *Schistosoma japonicum* antibodies. The horseradish peroxidase labeled protein A (HRP-PA) was used as "biocatalyst" that amplified the mass PZ immunoassay.

5. DETECTION OF VIRUSES

Viruses are another causative agent of communicable diseases and their rapid detection via reliable immunoassays is needed for preventing the spreading of the infections. Several acoustic wave devices have found extensive use in the detection of viruses with the help of specific antibodies (Table 9.2).

5.1 Quartz Crystal Microbalance Immunosensors

5.1.1 Direct Immunosensors

In 1994 Aberl et al. [132] achieved the detection of the human immunodeficiency virus (HIV) in human serum when they designed a flow-through system with an integrated QCM immunosensor. Anti-HIV antibodies in human serum were captured by specific surface-immobilized antigens, and the capturing was followed through monitoring of the changes of the resonance frequency of the crystal. König and Grätzel [133] detected five human herpes viruses with a 10 MHz AT-cut quartz piezoimmunosensor. The technique they developed was rapid ($\sim$90 min), sensitive, and selective. König and Grätzel [134], again, were the first to detect human hepatitis viruses type A and type B using PZ quartz immunosensors and achieving a limit of detection as good as the one achieved with the ELISA. Picornaviruses [135], bovine ephemeral fever (a viral disease of cattle) [136], the African swine fever virus [137] and H3N2 canine influenza virus [138] have also been detected by QCM immunosensors. Plant-infecting viruses such as tobacco mosaic virus (TMV) and orchid viruses

TABLE 9.2 Acoustic Wave–Based Immunosensors for the Detection of Viruses

Virus	Biosensor Type	Limit of Detection (LOD)	Assay Time	Signal	Sample	References
Picornaviruses: Human rhinovirus (HRV) and foot-and-mouth disease virus (FMDV)	MIP-QCM	–	–	ΔF (Hz)	Buffer	Jenik et al. [135]
Bovine ephemeral fever virus (BEFV)	QCM	5 μg/mL	–	ΔF (Hz)	Buffer	Lee and Chang [136]
H3N2 canine influenza virus (CIV)	QCM	2^2 HAU	–	ΔF (Hz)	Buffer	Kim et al. [138]
Tobacco mosaic virus (TMV)	MIP-QCM	100 ng/mL	–	ΔF (Hz)	Tobacco plant sap	Dickert et al. [139]
Orchid viruses: *Cymbidium mosaic virus* (CymMV) and Odontoglossum ringspot virus (ORSV)	QCM	1 ng of each	–	ΔF (Hz)	Crude sap of infected *Oncidium* orchid	Eun et al. [140]
Maize chlorotic mottle virus (MCMV)	QCM	250 ng/mL	–	ΔF (Hz)	Crude extractions of infected plants	Huang X. et al. [141]
Dengue virus	MIP-QCM	1 μg/mL	<1 h	ΔF (Hz)	Clinical samples	Tai et al. [142]
	QCM	–	~1 h	ΔF (Hz)	Buffer	Su et al. [143]
	QCM	1.727 ± 0.291 mg/mL	–	ΔF (Hz)	Clinical samples	Wu et al. [144]

Avian influenza virus (AIV) H5N1	QCM	2^{-4} HAU/50 μL	10 min	ΔF (Hz)	Chicken tracheal swab samples	Wang R. et al. [145]
	QCM	0.0128 HAU	30 min	ΔF (Hz)	Buffer	Wang and Li [149]
	QCM	0.0128 HAU	2 h	ΔF (Hz)	Chicken tracheal swab samples	Li et al. [152]
	QCM	1 HAU	1 h	ΔF (Hz)		Brockman et al. [150]
M13 bacteriophage	Bond-rupture QCM	20 bacteriophages	–	acoustic energy emission (integrated signal, A.U.)	Buffer	Dultsev et al. [146]
	QCM	–	–	ΔF (Hz)	Buffer	Uttenhaler et al. [147]
M13 bacteriophage particles	QCM	–	–	ΔF (Hz), Δmass	Buffer	Yang L.M.C. et al. [148]
Influenza A and B	QCM	1×10^3 PFU/mL	–	ΔF (Hz)	Clinical samples	Peduru Hewa et al. [151]
Coxsakie B4 and Sin Nombre Virus	SH-SAW	–	within min	ΔPhase	Complex solutions (bodies of water of river, sewage effluent)	Bisoffi et al. [153]

Continued

TABLE 9.2 Acoustic Wave–Based Immunosensors for the Detection of Viruses—cont'd

Virus	Biosensor Type	Limit of Detection (LOD)	Assay Time	Signal	Sample	References
Hepatitis B virus	LW-SAW	10 pg/μL	–	Frequency shift	Whole blood samples	Lee et al. [154]
HIV-1 and HIV-2	SAW	12 $TCID_{50}$s for HIV-1 and 87 $TCID_{50}$s for HIV-2	5 min	Phase shift ($\Delta\varphi$)	Human serum	Bisoffi et al. [155]
Ebola Zaire virus	SAW	1.9×10^4 PFU/mL	–	Phase shift ($\Delta\varphi$)	Buffer	Baca et al. [156]

LW-SAW, love wave surface acoustic wave; *MIP-QCM*, Molecularly imprinted polymer-quartz crystal microbalance; *QCM*, quartz crystal microbalance; *SAW*, surface acoustic wave.

have been identified. Dickert et al. [139] demonstrated the detection of TMV in tobacco plant sap using a MIP-QCM immunosensor and Eun et al. [140] described a QCM immunoassay that identified the existence of two orchid viruses in the crude sap of infected flowers and at concentrations comparable to those detected by ELISA. Huang X. et al. [141], using an antibody-based QCM immunosensor, managed to detect *maize chlorotic mottle virus* at a concentration similar to that of the ELISA method (250 ng/mL). Tai et al. [142] were the first to demonstrate the use of a MIP-QCM immunosensor for the detection of Dengue virus in patients' sera. Su et al. [143] and Wu et al. [144] designed QCM immunosensors for the Dengue virus. They created "cocktail immunochips" on which two different anti-Dengue antibodies specific for different viral proteins were attached. This approach raised the signal level and increased the sensitivity. Wang et al. [145] detected the avian influenza (AIV) H5N1 virus when they made use of a nanowall-based QCM sensing surface on which they immobilized an anti-H5N1 aptamer for virus recognition. This approach achieved the detection of AIV H5N1 in 10 min. Validation of the assay was carried out in chicken tracheal swab samples. M13 bacteriophage has been probed for its significance in genetic engineering [146–148]. Dultsev et al. [146] notably approached M13 bacteriophage detection differently from the approaches used in the studies mentioned so far. They employed a bond-rupture QCM; they sought to identify a genetically modified M13 phage displaying a maltose-binding protein. The phage was surface immobilized through a potato starch layer that contained branched polymers of maltose. The QCM crystal was oscillated at increasing amplitude until bond rupture occurred, i.e., until all of the phage–surface interactions were broken and the phage detached. By measuring the amount of acoustic energy emission generated on phage dissociation, it was possible to detect down to 20 phages.

5.1.2 Indirect Immunosensors

Wang and Li [149] employed a QCM aptasensor that was based on a hydrogel formed by a single-stranded DNA cross-linked to a specific aptamer against AIV H5N1 virus for AIV H5N1 detection. The hydrogel was not responsive in the absence of the virus. However, on exposure to the H5N1 virus, the cross linking between the single-stranded DNA and the aptamer ruptured, and hydrogel swelled. They created, in this way, a "smart" AIV-responsive hydrogel for rapid and specific virus detection. The assay time reported was 30 min, far less than the 1 h presented by Brockman et al. [150].

5.1.3 Sandwich-Amplified Immunosensors

Influenza A and B viruses in nasal washes of hospitalized patients were probed by Hewa et al. [151] who employed a QCM immunosensor in combination with AuNPs modified with antibodies specific for influenza A and B and detected 1×10^3 PFU/mL of each virus. In 2011, Li et al. [152] fabricated an antibody-based QCM immunosensor for the detection of avian influenza (AIV) H5N1 virus. The immunosensor was amplified by the final addition of magnetic nanobeads (30 nm) modified with anti-H5N1 antibodies. The time of detection was determined to be 2 h, and the sensor was used for virus detection in chicken tracheal swab sample. Brockman et al. [150] reduced the detection time of AIV H5N1 virus to 1 h. They employed an aptamer-based QCM sensor that specifically identified the virus. They also used magnetic nanobeads (150 nm) as frequency signal amplifiers.

5.2 Surface Acoustic Wave Immunosensors

In 2008, Bisoffi et al. [153] explored, for the first time, the construction of an antibody-based SAW immunosensor for viral detection. They designed a SAW-based platform able to detect *Coxsakie B4* and Sin *Nombre virus* (which is a potential bioterrorism agent). Their platform, allowing the fast (within seconds) and selective detection of viral agents, was further tested in a real-world scenario (i.e., detection in water from the sewage system and a river in an urban area) (Fig. 9.2). Lee et al. [154] carried out the detection of Hepatitis B virus in whole blood samples using an LW-SAW immunosensor

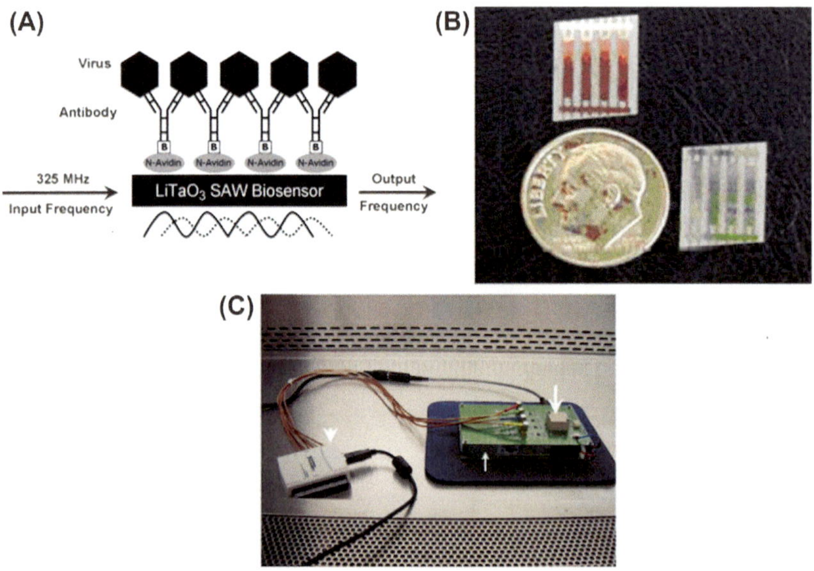

FIGURE 9.2 (A and B) Bisoffi et al. [153] developed a lithium tantalite ($LiTaO_3$) Surface acoustic wave (SAW)-based platform for direct detection of the Coxsakie B4 and the Sin Nombre virus. (C) The main components of the platform are depicted: the SAW detection board (*thin arrow*), the fluidic casing (*thick arrow*) and the output interface device (*arrowhead*).

without any sample pretreatment. They were followed by Bisoffi et al. [155] who detected and distinguished between HIV types 1 and 2 using a SAW immunosensor. Experiments were conducted in human serum and detection was complete 5 min after addition of the viral antigen through monitoring of the phase shift ($\Delta\varphi$). Baca et al. [156] presented a point-of-care SAW platform for the direct, rapid, and reagent-free detection of Ebola Zaire virus.

6. QUARTZ CRYSTAL MICROBALANCE AND SURFACE ACOUSTIC WAVE-BASED ELECTRONIC NOSES

This section refers to PZ electronic nose devices. Electronic noses respond to gaseous compounds and consist of three main parts: a sample delivery system, a detection system, and a computing system. Both QCM and SAW have been used for the development of electronic nose sensors that detected a wide range of analytes, from drugs to aerosolized biological agents. Owen et al. [157] employed a QCM electronic nose for the detection of aerosolized Influenza A viral particle. Aqueous solutions of the virus were aerosolized with a nebulizer and then captured by anti-Influenza A antibodies on the QCM sensor surface. The achieved limit of detection was 4 particles/mL. Lee et al. [158] detected airborne *Vaccinia* viruses when they combined an electrospray aerosol generator with a QCM device. The virus suspension was aerosolized and neutralized by the aerosol generator and transferred into a QCM chamber. Frequency shifts on virus surface capturing were proportional to the number of viruses. A real-time PZ electronic nose was constructed to monitor the exhaled breath of patients diagnosed with respiratory failure caused by severe pneumonia or other extrapulmonary diseases [159]. The core of the device was a PZ quartz crystal microbalance array of 24 transducers. Each transducer was coated with a different peptide, each one specific for one of these six bacterial pathogens: *Pseudomonas aeruginosa*, *Acinetobacter baumannii*, *Klebsiella pneumonia*, *Staphylococcus aureus*, and *Acinetobacter lwoffii*. Patients' breath was inserted into the QCM array via a catheter placed directly through their trachea; as it passed over the array, each transducer responded differently. Kovár et al. [160] presented the detection of aerosolized *E. coli* using a PZ sensor coated with anti–*E. coli* antibodies. *E. coli* cells were spread by an aerosol generator in a desktop bioaerosol chamber. *E. coli* particles were then collected with a cyclone SASS 2300, suspended in buffer and analyzed by the QCM immunosensor. The total assay time was 16 min, and the LOD was 1.45×10^4 CFU/mL. Stubbs et al. [161,162] coupled a two port SAW device operating at 250 MHz with a vapor generator for cocaine detection. They managed to detect a few picograms of cocaine molecules, suggesting that such an electronic nose could be used by low enforcement as a cost-effective alternative to detection dogs.

7. QUARTZ CRYSTAL MICROBALANCE AND SURFACE ACOUSTIC WAVE IMMUNOASSAYS IN ENVIRONMENTAL MONITORING AND ALLERGENS DETECTION

Environmental monitoring covers the areas of air quality, soil, and water quality monitoring and examines the risks posed by human activities on the natural environment. Most studies on the application of acoustic wave–based immunoassays for environmental monitoring reviewed here were focused on water quality monitoring. Continuous population growth, migration, and overconsumption are a few of the factors responsible for water contamination [163]. Human health and whole ecosystems sustainability are at risk because of water contamination that results from industrial practices, domestic sewage, and agricultural pesticides [164].

Comparable to water contamination, food, and airborne allergens are also a health problem, affecting a considerable number of people worldwide.

Yokohama et al. [165] investigated the existence of the herbicide atrazine using a PZ quartz crystal sandwiched immunosensor. Das et al. [166] proposed the use of an MIP-QCM for real-time detection of Hexachlorobenzene (an animal carcinogen and a probable human carcinogen) in water. The assay was selective, sensitive (10^{-12} mol/L), and with an extremely rapid response time (<10 s). Zhang et al. [167] discussed an indirect PZ immunoassay for monitoring environmental wastewater. They detected the presence of penicillin in wastewater through the detection of the growth of *Staphylococcus aureus*; the sensor was immersed in a solution containing culture medium, bacteria, and penicillin. The growth of bacteria was inhibited in the presence of penicillin affecting the chemical and physical characteristics of the culture system. This culture inhibition, and subsequently the existence of penicillin in wastewater from a hospital, was monitored through PZ impedance analysis for a week. Oshima et al. [168] designed a QCM immunosensor for the identification of plasma Vitellogenin. Vitellogenin is a female-fish biomarker used to monitor environmental-endocrine interference caused by anthropogenic and natural compounds. The sensor responded to plasma from *Cyprinus carpio* and reached an LOD of 100 μg/mL.

Sun et al. [169] were the first to use a QCM sensor for the detection of shrimp allergen in food. The sensor they presented showed good performance and a short analysis time (<10 min). Toma et al. [170] combined an SH-SAW chip with a sandwich immunoassay, generating a SAW immunosensor for the detection of airborne house dust mite (HDM) allergens. In particular, they detected the HDM *Dermatophagoides farina* group 1 (Der f 1) associated with severe allergenicities. The LOD of the sensor was 6.1 ng/mL, lower than that set by the World Health Organization, while detection was performed within 24 min.

8. INTEGRATED ACOUSTIC WAVE IMMUNOSENSORS FOR POINT OF CARE

This review section is dedicated to point-of-care (PoC) diagnostic acoustic wave immunosensors developed by research groups for the detection of clinical biomarkers, pathogenic microorganisms, viruses, etc. An essential part in the development of PoC acoustic platforms is the integration of the immunosensor with a microfluidic system; for this reason, the PoC immunosensors presented in this section also involve the use of microfluidics coupled with acoustic devices.

Godber et al. [171] designed a microfluidic QCM immunosensor using automated 2-channel or 4-channel RAP♦*id* instruments (Akubio Ltd, Cambridge, UK). This system evolved around a QCM (16.5 MHz) crystal that resonates on voltage application and measure the resonant frequency shifts in real time. The immunosensor was able to detect myoglobin between 1 and 100 nM in less than 6 min. Determination of enzyme cofactor specificities was also achieved in less than 10 min. Frisk et al. [172] fabricated an integrated QCM-based electronics narcotic system. Fluidic connections and electronics were incorporated in a single MEMS device. The MEMS system absorbed airborne narcotics molecules, and a subsequent competitive immunoassay took place on the QCM surface. Cocaine and ecstasy measurements were carried out with an LOD of 100 and 200 ng, respectively, in less than 30 s.

Choi et al. [173] developed a fully automated SAW immunosensor system with a disposable cartridge for PoC simultaneous testing of three cardiac markers of acute myocardial infarction in human blood. The SAW system completed all the steps needed, i.e., plasma separation, solution mixing, air bubble removal, AuNPs-based sandwich immunoassay, signal enhancement by gold staining, washing and antigen detection through frequency shift in 12 min. Lee et al. [174] have also reported on SAW immunodetection of cardiac markers. Mitsakakis and Gizeli [175] were the first to combine microfluidics on a SAW sensor, resulting in the compartmentalization of the surface into several subareas for the semiparallel detection of several analytes with a single sensor. This construction, referred to as "microfluidics-on-SAW (μF-on-SAW)," was used to detect a group of four cardiac markers that had never been explored before. In their work, a parallel-channel microfluidic module mounted on a 155 MHz LW SAW device allowed for the multiplex and semiparallel detection of all four markers on the same sensor (Fig. 9.3A). Goto et al. [176] described the fabrication of a POC disposable SH-SAW immunosensor operating at 250 MHz for the detection of HSA via antibody–antigen interaction. Matatagui et al. [177] combined for the first time an LW-SAW operating at 163 MHz with a PDMS microfluidics chip that could work as a PoC system for the detection of M13 bacteriophage. The immunoassay per se was a sandwich-amplified one, where anti-M13 antibodies were surface immobilized and followed by injection of the M13 phage.

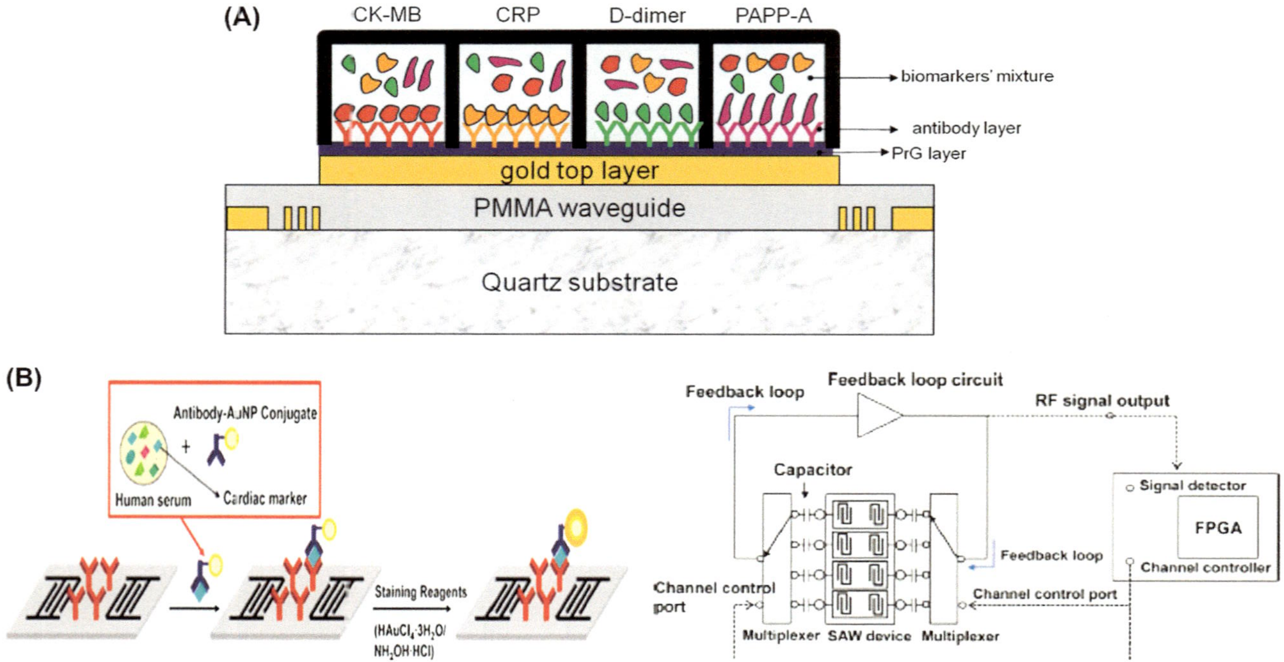

FIGURE 9.3 (A) Mitsakakis and Gizeli [175] detected four cardiac markers with an integrated point-of-care surface acoustic wave (SAW) system; they combined a parallel-channel microfluidic module with Love wave-SAW device; (B) Lee et al. [174] used antibody-functionalized AuNPs as signal amplifiers to detect cTnT. On the left panel the experimental setup regarding the immunoassay is shown; on the right panel a diagram of the measuring system is depicted.

Subsequent addition of secondary anti-M13 antibodies led to a bigger shift in the frequency and velocity signal. The sandwich assay achieved detection of the M13 phage at a concentration of 5×10^7 PFU/mL. Lee et al. [178] presented a fully automated centrifugal PoC system based on a SAW immunosensor for sensitive detection of cardiac troponin I (cTnTI) in buffer and whole blood. The immunoassay developed relied on AuNPs amplification and consecutive gold staining. The SAW immunosensor was contained in a disposable microfluidic cartridge and achieved detection of the cTnTI biomarker in buffer with a detection limit of 6.7 pg/mL. This detection limit was comparable with the one achieved for cTnTI detection in whole blood (Fig. 9.3B). Zhang et al. [179] built a miniaturized platform for CEA detection using an LW-SAW immunosensor integrated with a PDMS microfluidic chip and chip cartridge. The CEA detection was based on a sandwich-amplified assay in which capture anti-CEA antibodies were surface immobilized followed by injection of CEA, which had been previously incubated with AuNPs (30 nm). A final gold staining step was carried out. The LOD achieved was 1.26 ng/mL, comparable with the sensitivity of a commercial chemiluminescence immunoassay (Abbott, architect i2000).

9. COMMERCIAL ACOUSTIC WAVE IMMUNOSENSORS

The largest share of acoustic wave immunosensor market belongs to laboratory-based systems for research purposes. Several acoustic sensor platforms and chips (both QCM and SAW), which can be used to develop an immunoassay, are available in the biosensors market. Biolin Scientific (formerly Q-Sense) was the first company to launch a commercial acoustic sensor in Sweden, the QCM-D, which is a system that provides simultaneously two measurements (frequency and dissipation). AWSensors Scientific Technology, a recently established Spanish company, designs, fabricates, and sells platforms for acoustic bio-applications. Among their innovations is the development of 150 MHz high fundamental frequency QCM sensor (HFF-AWS), which has been shown to exhibit a very high sensitivity and resolution among the existing commercial instruments [180]. Attana is another Sweden-based company that offers innovative platforms and surface technologies for acoustic monitoring of molecular interactions in crude matrices (serum, periplasmic extracts, cell culture supernatants, and lysates) with no need for sample pretreatment. Initium, a Japan based company, provides the "Affinix" QCM-based molecular interaction analyzer and the appropriate sensor chips. Sierra Sensors and TTP Labtech market similar QCM-based products. The SEISMOS (formerly S-Sens K5) is an LW-based biosensor developed by the CEASAR research institute in Bonn, Germany. Detection is carried out through changes in phase and amplitude signals. Certain studies reported on the use of SEISMOS [181–183].

In addition to the above commercially available acoustic platforms, which have been developed for a broad range of applications, some new products have emerged targeting the more specialized market sector of immunosensors. Atonomics A/S have developed a portable LW-SAW sensing (immunoassay) disposable cartridge, which is combined with a microfluidic separator and a micro PoC reader. The device can perform biomarkers' detection in blood. OJ-Bio LTD developed a hand-held device that uses an SH-SAW biochip [184]. The device was successfully used in the detection of CRP protein in serum; the company is now focusing on the remote diagnosis of HIV and other viruses in real samples.

10. MARKET POTENTIAL AND CONCLUSIONS

Currently, there is no doubt that acoustic wave immunosensors have been established as a reliable and sensitive tool for analyte detection. This is obvious from the large number of papers presented in this chapter and other review articles on immunosensors and immunoassays as one of the key detection platforms [185,186]. For acoustic immunosensors to reach the market, they should fulfill several criteria. Ideally, a reliable immunosensor needs to be sensitive, specific, rapid, accurate, robust, small in size, low cost, and easy to use by nonskilled personnel [172,187,188]. Most of the immunosensors reviewed in this chapter are shown to satisfy some of these prerequisites, and specifically, they offer fast, label-free and real-time detection. Overall, they still lag behind in sensitivity compared with established diagnostic techniques such as ELISA and are not portable. The employment of nanoparticles and carbon nanotubes that have been shown to increase the sensor sensitivity could lead to the generation of an acoustic immunosensor able to exceed the limits of detection achieved with conventional immunoassays. Also, the majority of them have not been tested in field applications and with real complex samples. Another issue of significance is the lifetime of the sensors, which should preferably be at least 1 month [189]. The majority of the acoustic immunosensors presented in this chapter could only be generated a few times due to the frail nature of the antibodies. The shift of the scientific interest toward MIP-QCM immunosensors, which are more robust, modifiable, and stable [190,191], could be the solution to this problem. Last, nowadays one of the great challenges for the development of successful, marketable immunoassays is their miniaturization and integration in automated, high-throughput, cost-effective diagnostic platforms for on-site applications and testing of crude samples (such as blood, urine, and saliva). The integration of a PZ device with microfluidics and wireless measurement and collection of data would be a critical step for the development of a desired acoustic wave–based immunosensor. The target market is there and is well defined: food and environmental industries, hospitals and diagnostic centers, biosafety agencies, and pharmaceutical industries that need biosensors for drug discovery and screening; what remains is for the right sensor to appear.

ACKNOWLEDGMENTS

We would like to thank our funding agent, the European Commission for the generous support of our research through the following grants: H2020-ICT-2015-GA687681 (Love-Food2Market); H2020-ICT-2015-GA687785 (LiquBiopSens); and H2020-FETOPEN-1-2016-2017-GA737212(CATCH-U-DNA).

REFERENCES

[1] C.I. Cheng, Y.P. Chang, Y.H. Chu, Biomolecular interactions and tools for their recognition: focus on the quartz crystal microbalance and its diverse surface chemistries and applications, Chem. Soc. Rev. 41 (2012) 1947–1971.

[2] D.S. Ballantine, R.M. White, S.J. Martin, A.J. Ricco, E.T. Zellers, G.C. Frye, et al., Acoustic Wave Sensors: Theory, Design, and Physico-chemical Applications, Academic Press, 1997.

[3] G. Sauerbrey, Verwendung von schwingquarzen zur wägung dünner schichten und zur mikrowägung, Z. Phys. 155 (1959) 206–222.

[4] J.E. Solma Solma, M.A. Cooper, Does the Sauerbrey equation hold true for binding of peptides and globular proteins to a QCM?: A systematic study of mass dependence of peptide and protein binding with a piezoelectric sensor, Sens. Biosens. Res. 11 (2016) 71–77.

[5] K. Mitsakakis, A. Tsortos, E. Gizeli, Quantitative determination of protein molecular weight with an acoustic sensor; significance of specific versus nonspecific binding, Analyst 139 (2014) 3918–3925.

[6] A. Tsortos, G. Papadakis, K. Mitsakakis, K. Melzak, E. Gizeli, Quantitative determination of size and shape of surface-bound DNA using an acoustic wave sensor, Biophys. J. 94 (2008) 2706–2715.

[7] A. Tsortos, G. Papadakis, E. Gizeli, Shear acoustic wave biosensor for detecting DNA intrinsic viscosity and conformation: a study with QCM-D, Biosens. Bioelectron. 24 (2008) 836–841.

[8] A. Tsortos, A. Grammoustianou, R. Lymbouridou, G. Papadakis, E. Gizeli, The detection of multiple DNA targets with a single probe using a conformation-sensitive acoustic sensor, Chem. Commun. 51 (2015) 11504–11507.

[9] P. Mateos-Gil, A. Tsortos, M. Velez, E. Gizeli, Monitoring structural changes in intrinsically disordered proteins using QCM-D: application to the bacterial cell division protein ZipA, Chem. Commun. 52 (2016) 6541–6544.

[10] A. Tsortos, G. Papadakis, E. Gizeli, On the hydrodynamic nature of DNA acoustic sensing, Anal. Chem. 12 (2016) 6472–6478.

[11] E. Tellechea, D. Johannsmann, N.F. Steinmetz, R.P. Richter, I. Reviakine, Model independent analysis of QCM data on colloidal particle adsorption, Langmuir 25 (2009) 5177–5184.

[12] I. Reviakine, M. Callego, D. Johannsmann, E. Tellechea, Adsorbed liposome deformation studied with quartz crystal microbalance, J. Chem. Phys. (2012) 136, 084702.

[13] A.L.J. Olsson, I.R. Quevedo, D. He, M. Basnet, N. Tufenkji, Using the quartz crystal microbalance with dissipation monitoring to evaluate the size of nanoparticles deposited on surfaces, ACS Nano 7 (2013) 7833–7843.

[14] K. Länge, B.E. Rapp, M. Rapp, Surface acoustic wave biosensors: a review, Anal. Bioanal. Chem. 391 (2008) 1509–1519.

[15] I. Reviakine, D. Johannsmann, R.P. Richter, Hearing what you cannot see and visualizing what you hear: interpreting quartz crystal microbalance data from solvated interfaces, Anal. Chem. 83 (2011) 8838–8848.
[16] G.N.M. Ferreira, A.C. da-Silva, B. Tome, Acoustic wave biosensors: physical models and biological applications of quartz crystal microbalance, Trends Biotechnol. 27 (2009) 689–697.
[17] B.A. Cavic, M. Thompson, G.L. Hayward, Acoustic waves and the study of biochemical macromolecules and cells at the sensor–liquid interface, Analyst 124 (1999) 1405–1420.
[18] A. Janshoff, H.J. Galla, C. Steinem, Piezoelectric mass-sensing devices as biosensors-an alternative to optical biosensors? Angew. Chem. Int. Ed. Engl. 39 (2000) 4004–4032.
[19] E. Gizeli, C.R. Lowe, Biomolecular Sensors, 2002.
[20] K. Mitsakakis, A. Tsortos, J. Kondoh, E. Gizeli, Parametric study of SH-SAW device response to various types of surface perturbations, Sens. Actuator B Chem. 138 (2009) 408–416.
[21] S.O. Kelley, What are clinically relevant levels of cellular and biomolecular analytes? ACS Sens. 2 (2016) 193–197.
[22] C. Rossetti, M.A. Świtnicka-Plak, T.G. Halvorsen, P.A.G. Cormack, B. Sellergren, L. Reubsaet, Automated protein biomarker analysis: on-line extraction of clinical samples by molecularly imprinted polymers, Sci. Rep. 7 (2017) 44298.
[23] S.O. Kelley, Advancing ultrasensitive molecular and cellular analysis methods to speed and simplify the diagnosis of disease, Acc. Chem. Res. 50 (2017) 503–507.
[24] A. Shons, F. Dorman, J. Najarian, An immunospecific microbalance, J. Biomed. Mater. Res. 6 (1972) 565–570.
[25] S. Kurosawa, M. Nakamura, J.W. Park, H. Aizawa, K. Yamada, M. Hirata, Evaluation of a high-affinity QCM immunosensor using antibody fragmentation and 2-methacryloyloxyethyl phosphorylcholine (MPC) polymer, Biosens. Bioelectron. 20 (2004) 1134–1139.
[26] X. Su, F.T. Chew, S.F.Y. Li, Self-assembled monolayer-based piezoelectric crystal immunosensor for the quantification of total human immunoglobulin E, Anal. Biochem. 273 (1999) 66–72.
[27] C. Yao, Y. Qi, Y. Zhao, Y. Xiang, Q. Chen, W. Fu, Aptamer-based piezoelectric quartz crystal microbalance biosensor array for the quantification of IgE, Biosens. Bioelectron. 24 (2009) 2499–2503.
[28] C. Yao, T. Zhu, Y. Qi, Y. Zhao, H. Xia, W. Fu, Development of a quartz crystal microbalance biosensor with aptamers as bio-recognition element, Sensors 10 (2010) 5859–5871.
[29] X. Su, J. Zhang, Comparison of surface plasmon resonance spectroscopy and quartz crystal microbalance for human IgE quantification, Sens. Actuator B Chem. 100 (2004) 309–314.
[30] X. Chu, J.H. Jiang, G.L. Shen, R.Q. Yu, Simultaneous immunoassay using piezoelectric immunosensor array and robust method, Anal. Chim. Acta 336 (1996) 185–193.
[31] C. Ayela, F. Roquet, L. Valera, C. Granier, L. Nicu, M. Pugniere, Antibody–antigenic peptide interactions monitored by SPR and QCM-D A model for SPR detection of IA-2 autoantibodies in human serum, Biosens. Bioelectron. 22 (2007) 3113–3119.
[32] M.K. Mustafa, A. Nabok, D. Parkinson, I.E. Tothill, F. Salam, A. Tsargorodskaya, Detection of -amyloid peptide (1–16) and amyloid precursor protein (APP770) using spectroscopic ellipsometry and QCM techniques: a step forward towards Alzheimers disease diagnostics, Biosens. Bioelectron. 26 (2010) 1332–1336.

[33] S.F. Chou, W.L. Hsu, J.M. Hwang, C.Y. Chen, Development of an immunosensor for human ferritin, a nonspecific tumor marker, based on a quartz crystal microbalance, Anal. Chim. Acta 453 (2002) 181–189.

[34] S.F. Chou, W.L. Hsu, J.M. Hwang, C.Y. Chen, Determination of α-φetoprotein in ηuman σerum by a quartz crystal microbalance-based immunosensor, Clin. Chem. 48 (2002) 913–918.

[35] W.C. Tsai, I.C. Lin, Development of a piezoelectric immunosensor for the detection of alpha-fetoprotein, Sens. Actuator B Chem. 106 (2005) 455–460.

[36] C.D. Corso, D.D. Stubbs, S.H. Lee, M. Goggins, R.H. Hruban, W.D. Hunt, Real time detection of mesothelin in pancreatic cancer cell line supernatant using an acoustic wave immunosensor, Cancer Detect Prev. 30 (2006) 180–187.

[37] D.Q. Tang, D.J. Zhang, D.Y. Tang, H. Ai, Amplification of the antigen–antibody interaction from quartz crystal microbalance immunosensors via back-filling immobilization of nanogold on biorecognition surface, J. Immunol. Methods 316 (2006) 144–152.

[38] Z.G. Chen, D.Y. Tang, Antigen–antibody interaction from quartz crystal microbalance immunosensors based on magnetic CoFe2O4/SiO2 composite nanoparticle-functionalized biomimetic interface, Bioproc. Biosyst. Eng. 30 (2007) 243–249.

[39] G. Shen, J. Lu, Piezoelectric immunosensor based on gold nanoparticles capped with mixed self-assembled monolayer for detection of carcinoembryonic antigen, Thin Solid Films 518 (2010) 5010–5013.

[40] Z. Shen, H. Yan, F.F. Parl, R.L. Mernaugh, X. Zeng, Recombinant antibody piezoimmunosensors for the detection of cytochrome P450 1B1, Anal. Chem. 79 (2007) 1283–1289.

[41] X. Wang, H. Yu, D. Lu, J. Zhang, W. Deng, Label free detection of the breast cancer biomarker CA15.3 using ZnOnanorods coated quartz crystal microbalance, Sens. Actuator B Chem. 195 (2014) 630–634.

[42] A. Martínez-Rivas, P. Chinestra, G. Favre, S. Pinaud, C. Séverac, J.C. Faye, et al., Detection of label-free cancer biomarkers using nickel nanoislands and quartz crystal microbalance, Int. J. Nanomed. 5 (2010) 661–668.

[43] V. Crivianu-Gaita, M. Aamer, R.T. Posaratnanathan, A. Romaschin, M. Thompson, Acoustic wave biosensor for the detection of the breast and prostate cancer metastasis biomarker protein PTHrP, Biosens. Bioelectron. 78 (2016) 92–99.

[44] S.D. Carrigan, G. Scott, M. Tabrizian, Real-Time QCM-D immunoassay through oriented antibody immobilization using cross-linked hydrogel biointerfaces, Langmuir 21 (2005) 5966–5973.

[45] S.D. Carrigan, G. Scott, M. Tabrizian, Rapid three-dimensional biointerfaces for real-time immunoassay using hIL-18BPa as a model antigen, Biomaterials 26 (2005) 7514–7523.

[46] K.A. Drouvalakis, S. Bangsaruntip, W. Hueber, L.G. Kozar, P.J. Utz, H. Dai, Peptide-coated nanotube-based biosensor for the detection of disease-specific autoantibodies in human serum, Biosens. Bioelectron. 23 (2008) 1413–1421.

[47] S.P. Sakti, P. Hauptmann, B. Zimmermann, F. Buhling, S. Ansorge, Disposable HSA QCM-immunosensor for practical measurement in liquid, Sens. Actuator B Chem. 78 (2001) 257–262.

[48] K. Jaruwongrungsee, U. Waiwijit, A. Wisitsoraat, M. Sangworasil, C. Pintavirooj, A. Tuantranont, Real-time multianalyte biosensors based on interference-free multichannel monolithic quartz crystal microbalance, Biosens. Bioelectron. 67 (2015) 576–581.

[49] S. Saha, M. Raje, R. Suri, Sandwich microgravimetric immunoassay: sensitive and specific detection of low molecular weight analytes using piezoelectric quartz crystal, Biotechnol. Lett. 24 (2002) 711–716.

[50] T. Ito, N. Aoki, S. Kaneko, K. Suzuki, Highly sensitive and rapid sequential cortisol detection using twin sensor QCM, Anal. Methods 6 (2014) 7469–7474.
[51] M. Michalzik, J. Wendler, J. Rabe, S. Buttgenbach, U. Bilitewski, Development and application of a miniaturised quartz crystal microbalance (QCM) as immunosensor for bone morphogenetic protein-2, Sens. Actuator B Chem. 105 (2005) 508–515.
[52] F.C. Soumetz, L. Pastorino, C. Ruggiero, Development of a piezoelectric immunosensor for matrix metalloproteinase-1 detection, Conf. Proc. IEEE Eng. Med. Biol. Soc. (2009) 2775–2778.
[53] H. Wang, D. Li, Z. Wu, G. Shen, R. Yu, A reusable piezo-immunosensor with amplified sensitivity for ceruloplasmin based on plasma-polymerized film, Talanta 62 (2004) 201–208.
[54] K. Wong-ek, O. Chailapakul, N. Nuntawong, K. Jaruwongrungsee, A. Tuantranont, Cardiac troponin T detection using polymers coated quartz crystal microbalance as a cost-effective immunosensor, Biomed. Tech. 55 (2010) 279–284.
[55] Y. Luo, M. Chen, Q. Wen, M. Zhao, B. Zhang, X. Li, et al., Rapid and simultaneous quantification of 4 urinary proteins by piezoelectric quartz crystal microbalance immunosensor array, Clin. Chem. 52 (2006) 2273–2280.
[56] C. Yao, L. Qu, W. Fu, Detection of fibrinogen and coagulation factor VIII in plasma by a quartz crystal microbalance biosensor, Sensors 13 (2013) 6946–6956.
[57] N. Miura, H. Higobashi, G. Sakai, A. Takeyasu, T. Uda, N. Yamazoe, Piezoelectric crystal immunosensor for sensitive detection of methamphetamine (stimulant drug) in human urine, Sens. Actuator B Chem. 13–14 (1993) 188–191.
[58] H. Chen, J.H. Jiang, Y.F. Li, T. Deng, G.L. Shen, R.Q. Yu, A novel piezoelectric immunoagglutination assay technique with antibody-modified liposome, Biosens. Bioelectron. 22 (2007) 993–999.
[59] N. Kim, D.K. Kim, Y.J. Cho, Development of indirect-competitive quartz crystal microbalance immunosensor for C-reactive protein, Sens. Actuator B Chem. 143 (2009) 444–448.
[60] H. Aizawa, S. Kurosawa, K. Ogawa, M. Yoshimoto, J. Miyake, H. Tanaka, Conventional diagnosis of C-reactive protein in serum using latex piezoelectric immunoassay, Sens. Actuator B Chem. 76 (2001) 173–176.
[61] H. Aizawa, S. Kurosawa, M. Tozuka, J.W. Park, K. Kobayashi, H. Tanaka, Conventional detection method of fibrinogen and fibrin degradation products using latex piezoelectric immunoassay, Biosens. Bioelectron. 18 (2003) 765–771.
[62] H. Aizawa, S. Kurosawa, M. Tozuka, J.W. Park, K. Kobayashi, Rapid detection of fibrinogen and fibrin degradation products using a smart QCM-sensor, Sens. Actuator B Chem. 101 (2004) 150–154.
[63] M.J. Gibbs, A. Biela, S. Krause, α-Amylase sensor based on the degradation of oligosaccharide hydrogel films monitored with a quartz crystal sensor, Biosens. Bioelectron. 67 (2015) 540–545.
[64] Y. Uludag, I.E. Tothill, Development of a sensitive detection method of cancer biomarkers in human serum (75%) using a quartz crystal microbalance sensor and nanoparticles amplification system, Talanta 82 (2010) 277–282.
[65] Z. Ma, J. Wu, T. Zhou, Z. Chen, Y. Dong, J. Tang, et al., Detection of human lung carcinoma cell using quartz crystal microbalance amplified by enlarging Au nanoparticles in vitro, New J. Chem. 26 (2002) 1795–1798.
[66] W. Shan, Y. Pan, H. Fang, M. Guo, Z. Nie, Y. Huang, et al., An aptamer-based quartz crystal microbalance biosensor for sensitive and selective detection of leukemia cells using silver-enhanced gold nanoparticle label, Talanta 126 (2014) 130–135.

[67] W. Welsch, C. Klein, M. von Schickfus, S. Hunklinger, Development of a surface acoustic wave immunosensor, Anal. Chem. 68 (1996) 2000–2004.

[68] A. Campitelli, W. Wlodarski, M. Hoummady, W. Sawyer, Shear horizontal surface acoustic wave based immunosensing system, Solid State Sensors Actuators, 1997 Transducers '97 Chic. 1997 Int. Conf. 2 (1997) 187–190.

[69] H.W. Chang, J.S. Shih, Surface acoustic wave immunosensors based on immobilized C60-proteins, Sens. Actuator B Chem. 121 (2007) 522–529.

[70] S. Krishnamoorthy, A.A. Iliadis, T. Bei, G.P. Chrousos, An interleukin-6 ZnO/SiO2/Si surface acoustic wave biosensor, Biosens. Bioelectron. 24 (2008) 313–318.

[71] A.K. Deisingh, M. Thompson, Biosensors for the detection of bacteria, Can. J. Microbiol. 50 (2004) 69–77.

[72] H. Muramatsu, K. Kajiwara, E. Tamiya, I. Karube, Piezoelectric immunosensor for the detection of *Candida albicans* microbes, Anal. Chim. Acta 188 (1986) 257–261.

[73] M. Plomer, G.G. Guilbault, B. Hock, Development of a piezoelectric immunosensor for the detection of enterobacteria, Enzyme Microb. Technol. 14 (1992) 230–235.

[74] N. Kim, I.S. Park, Application of a flow-type antibody sensor to the detection of *Escherichia coli* in various foods, Biosens. Bioelectron. 18 (2003) 1101–1107.

[75] X.L. Su, Y. Li, A self-assembled monolayer-based piezoelectric immunosensor for rapid detection of *Escherichia coli* O157:H7, Biosens. Bioelectron. 19 (2004) 563–574.

[76] N. Adanyi, M. Varadi, N. Kim, I. Szendro, Development of new immunosensors for determination of contaminants in food, Curr. Appl. Phys. 6 (2006) 279–286.

[77] B. Serra, M. Gamella, A.J. Reviejo, J.M. Pingarrón, Lectin-modified piezoelectric biosensors for bacteria recognition and quantification, Anal. Bioanal. Chem. 391 (2008) 1853–1860.

[78] L.J. Wang, C.S. Wu, Z.Y. Hu, Y.F. Zhang, R. Li, P. Wang, Sensing *Escherichia coli* O157:H7 via frequency shift through a self-assembled monolayer based QCM immunosensor, J. Zhejiang Univ. Sci. B 9 (2008) 121–131.

[79] H.C. Han, Y.R. Chang, W.L. Hsu, C.Y. Chen, Application of parylene-coated quartz crystal microbalance for on-line real-time detection of microbial populations, Biosens. Bioelectron. 24 (2009) 1543–1549.

[80] C. Poitras, N. Tufenkji, A QCM-D-based biosensor for *E. coli* O157:H7 highlighting the relevance of the dissipation slope as a transduction signal, Biosens. Bioelectron. 24 (2009) 2137–2142.

[81] F. Zhang, H. Li, X. Wang, H.Y. Low, X. Li, Hierarchically imprinted polymer substrates for enhanced attachment of *Escherichia coli*, J. Colloid Interf. Sci. 343 (2010) 109–114.

[82] E. Prusak-Sochaczewski, J.H.T. Luong, G.G. Guilbault, Development of a piezoelectric immunosensor for the detection of *Salmonella typhimurium*, Enzyme Microb. Technol. 12 (1990) 173–177.

[83] I.S. Park, N. Kim, Thiolated *Salmonella* antibody immobilization onto the gold surface of piezoelectric quartz crystal, Biosens. Bioelectron. 13 (1998) 1091–1097.

[84] I.S. Park, W.Y. Kim, N. Kim, Operational characteristics of an antibody-immobilized QCM system detecting *Salmonella* spp, Biosens. Bioelectron. 15 (2000) 167–172.

[85] S.T. Pathirana, J. Barbaree, B.A. Chin, M.G. Hartell, W.C. Neely, V. Vodyanoy, Rapid and sensitive biosensor for *Salmonella*, Biosens. Bioelectron. 15 (2000) 135–141.

[86] Y.S. Fung, Y.Y. Wong, Self-assembled monolayers as the coating in a quartz piezoelectric crystal Immunosensor to detect *Salmonella* in aqueous solution, Anal. Chem. 73 (2001) 5302–5309.

[87] S.H. Si, X. Li, Y.S. Fung, D.R. Zhu, Rapid detection of *Salmonella enteritidis* by piezoelectric immunosensor, Microchem J. 68 (2001) 21–27.

[88] S. Babacan, P. Pivarnik, S. Letcher, A. Rand, Piezoelectric flow injection analysis biosensor for the detection of *Salmonella typhimurium*, J. Food Sci. 67 (2002) 314–320.

[89] Y.Y. Wong, S.P. Ng, M.H. Ng, S.H. Si, S.Z. Yao, Y.S. Fung, Immunosensor for the differentiation and detection of *Salmonella* species based on a quartz crystal microbalance, Biosens. Bioelectron. 17 (2002) 676–684.

[90] E.V. Olsen, S.T. Pathirana, A.M. Samoylov, J.M. Barbaree, B.A. Chin, W.C. Neely, et al., Specific and selective biosensor for *Salmonella* and its detection in the environment, J. Microbiol. Methods 53 (2003) 273–285.

[91] E.V. Olsen, I.B. Sorokulova, V.A. Petrenko, I.H. Chen, J.M. Barbaree, V.J. Vodyanoy, Affinity-selected filamentous bacteriophage as a probe for acoustic wave biodetectors of *Salmonella typhimurium*, Biosens. Bioelectron. 21 (2006) 1434–1442.

[92] F. He, L. Zhang, Rapid diagnosis of *M. tubercylosis* using a piezoelectric immunosensor, Anal. Sci. 18 (2002) 397–401.

[93] F. He, L. Zhang, J. Zhao, B. Hu, J. Lei, A TSM immunosensor for detection of *M. tuberculosis* with a new membrane material, Sens. Actuator B Chem. 85 (2002) 284–290.

[94] H. Aizawa, S. Kurosawa, H. Tanaka, S.I. Wakida, Z.A. Talib, J.W. Park, et al., Conventional diagnosis of *Treponema pallidum* in serum using latex piezoelectric immunoassay, Mater. Sci. Eng. C 17 (2001) 127–132.

[95] R.M. Carter, J.J. Mekalanos, M.B. Jacobs, G.J. Lubrano, G.G. Guilbault, Quartz crystal microbalance detection of *Vibrio cholerae* O139 serotype, J. Immunol. Methods 187 (1995) 121–125.

[96] S. Buchatip, C. Ananthanawat, P. Sithigorngul, P. Sangvanich, S. Rengpipat, V.P. Hoven, Detection of the shrimp pathogenic bacteria, *Vibrio harveyi*, by a quartz crystal microbalance-specific antibody based sensor, Sens. Actuator B Chem. 145 (2010) 259–264.

[97] N. Kim, I.S. Park, D.K. Kim, Characteristics of a label-free piezoelectric immunosensor detecting *Pseudomonas aeruginosa*, Sens. Actuator B Chem. 100 (2004) 432–438.

[98] R.D. Vaughan, R.M. Carter, C.K. O'Sullivan, G.G. Guilbault, A quartz crystal microbalance (QCM) sensor for the detection of *Bacillus cereus*, Anal. Lett. 36 (2003) 731–747.

[99] S.H. Lee, D.D. Stubbs, J. Cairney, W.D. Hunt, Rapid detection of bacterial spores using a quartz crystal microbalance (QCM) immunoassay, IEEE Sens. J. 5 (2005) 737–743.

[100] R. Hao, D. Wang, X. Zhang, G. Zuo, H. Wei, R. Yang, et al., Rapid detection of *Bacillus anthracis* using monoclonal antibody functionalized QCM sensor, Biosens. Bioelectron. 24 (2009) 1330–1335.

[101] M. Pohanka, P. Skladal, Piezoelectric immunosensor for *Francisella tularensis* detection using Immunoglobulin M in a limiting dilution, Anal. Lett. 38 (2005) 411–422.

[102] R.D. Vaughan, C.K. O'Sullivan, G.G. Guilbault, Development of a quartz crystal microbalance (QCM) immunosensor for the detection of *Listeria monocytogenes*, Enzyme Microb. Technol. 29 (2001) 635–638.

[103] M.E. Yakovleva, A.P. Moran, G.R. Safina, T. Wadström, B. Danielsson, Lectin typing of *Campylobacter jejuni* using a novel quartz crystal microbalance technique, Anal. Chim. Acta 694 (2011) 1–5.

[104] S.R. Hong, S.J. Choi, H.D. Jeong, S. Hong, Development of QCM biosensor to detect a marine derived pathogenic bacteria *Edwardsiella tarda* using a novel immobilisation method, Biosens. Bioelectron. 24 (2009) 1635–1640.

[105] J.L.N. Harteveld, M.S. Nieuwenhuizen, E.R.J. Wils, Detection of Staphylococcal Enterotoxin B employing a piezoelectric crystal immunosensor, Biosens. Bioelectron. 12 (1997) 661–667.

[106] B.D. Spangler, E.A. Wilkinson, J.T. Murphy, B.J. Tyler, Comparison of the Spreeta® surface plasmon resonance sensor and a quartz crystal microbalance for detection of *Escherichia coli* heat-labile enterotoxin, Anal. Chim. Acta 444 (2001) 149–161.

[107] J.C. Vidal, P. Duato, L. Bonel, J.R. Castillo, Use of polyclonal antibodies to ochratoxin A with a quartz–crystal microbalance for developing real-time mycotoxin piezoelectric immunosensors, Anal. Bioanal. Chem. 394 (2009) 575–582.

[108] Z. Shen, M. Huang, C. Xiao, Y. Zhang, X. Zeng, W.G. Wang, Non-labeled QCM biosensor for bacterial detection using carbohydrate and lectin recognitions, Anal. Chem. 79 (2007) 2312–2319.

[109] Y. Wan, D. Zhang, B. Hou, Determination of sulphate-reducing bacteria based on vancomycin-functionalised magnetic nanoparticles using a modification-free quartz crystal microbalance, Biosens. Bioelectron. 25 (2010) 1874–1950.

[110] R. Guntupalli, I. Sorokulova, E. Olsen, L. Globa, O. Pustovyy, T. Moore, et al., Detection and identification of methicillin resistant and sensitive strains of *Staphylococcus aureus* using tandem measurements, J. Microbiol. Methods 90 (2012) 182–190.

[111] H. Muramatsu, Y. Watanabe, M. Hikuma, T. Ataka, I. Kubo, E. Tamiya, et al., Piezoelectric crystal biosensor system for detection of *Escherichia coli*, Anal. Lett. 22 (1989) 2155–2166.

[112] I. Ben-Dov, I. Willner, Piezoelectric immunosensors for urine specimens of *Chlamydia trachomatis* employing quartz crystal microbalance microgravimetric analyses, Anal. Chem. 69 (1997) 3506–3512.

[113] L. Alfonta, I. Willner, Electrochemical and quartz crystal microbalance detection of the Cholera toxin employing horseradish peroxidase and GM1-functionalized liposomes, Anal. Chem. 73 (2001) 5287–5295.

[114] X. Su, S.F.Y. Li, Serological determination of *Helicobacter pylori* infection using sandwiched and enzymatically amplified piezoelectric biosensor, Anal. Chim. Acta 429 (2001) 27–36.

[115] G.H. Kim, A.G. Rand, S.V. Letcher, Impedance characterization of a piezoelectric immunosensor part II: *Salmonella typhimurium* detection using magnetic enhancement, Biosens. Bioelectron. (2003) 91–99.

[116] X.L. Su, Y. Li, A QCM immunosensor for *Salmonella* detection with simultaneous measurements of resonant frequency and motional resistance, Biosens. Bioelectron. 21 (2005) 840–848.

[117] X. Jin, X. Jin, X. Liu, L. Chen, J. Jiang, G. Shen, et al., Biocatalyzed deposition amplification for detection of aflatoxin B1 based on quartz crystal microbalance, Anal. Chim. Acta 645 (2009) 92–97.

[118] X. Jiang, R. Wang, Y. Wang, X. Su, Y. Ying, J. Wang, et al., Evaluation of different micro/nanobeads used as amplifiers in QCM immunosensor for more sensitive detection of *E. coli* O157:H7, Biosens. Bioelectron. 29 (2011) 23–28.

[119] X. Guo, C.S. Lin, S.H. Chen, R. Ye, V.C.H. Wu, A piezoelectric immunosensor for specific capture and enrichment of viable pathogens by quartz crystal microbalance sensor, followed by detection with antibody functionalized gold nanoparticles, Biosens. Bioelectron. 38 (2012) 177–183.

[120] V.C. Ozalp, G. Bayramoglu, Z. Erdem, M.Y. Arica, Pathogen detection in complex samples by quartz crystal microbalance sensor coupled to aptamer functionalized core–shell type magnetic separation, Anal. Chim. Acta 853 (2015) 533–540.

[121] E. Berkenpas, P. Millard, M. Pereira da Cunhaa, Detection of *Escherichia coli* O157:H7 with langasite pure shear horizontal surface acoustic wave sensors, Biosens. Bioelectron. 21 (2006) 2255–2262.

[122] D.H. Dinh, E. Pascal, L. Vellutini, B. Bennetau, D. Rebière, C. Dejous, et al., Novel optimized biofunctional surfaces for Love mode surface acoustic wave based immunosensors, Sens. Actuator B Chem. 146 (2010) 289–296.

[123] M. Puiu, A.M. Gurban, L. Rotariu, S. Brajnicov, C. Viespe, C. Bala, Enhanced sensitive love wave surface acoustic wave sensor designed for immunoassay formats, Sensors 15 (2015) 10511–10525.

[124] J.C. Pyun, H. Beutel, J.U. Meyer, H.H. Ruf, Development of a biosensor for *E. coli* based on a flexural plate wave (FPW) transducer, Biosens. Bioelectron. 13 (1998) 839–845.

[125] C. Poitras, J. Fatisson, N. Tufenkji, Real-time microgravimetric quantification of *Cryptosporidium parvum* in the presence of potential interferents, Water Res. 43 (2009) 2631–2638.

[126] C. Wang, H. Wang, Z. Wu, G. Shen, R. Yu, A piezoelectric immunoassay based on self-assembled monolayers of cystamine and polystyrene sulfonate for determination of *Schistosoma japonicum* antibodies, Anal. Bioanal. Chem. 373 (2002) 803–809.

[127] Z.Y. Wu, G.L. Shen, S.P. Wang, R.Q. Yu, Quartz-crystal microbalance immunosensor for *Schistosoma japonicum*-infected rabbit serum, Anal. Sci. 19 (2003) 437–440.

[128] H. Wang, C. Lei, J. Li, Z. Wu, G. Shen, R. Yu, A piezoelectric immunoagglutination assay for *Toxoplasma gondii* antibodies using gold nanoparticles, Biosens. Bioelectron. 19 (2004) 701–709.

[129] H. Wang, Y. Zhang, B. Yan, L. Liu, S. Wang, G. Shen, et al., Rapid, simple, and sensitive immunoagglutination assay with SiO_2 particles and quartz crystal microbalance for quantifying *Schistosoma japonicum* antibodies, Clin. Chem. 52 (2006) 2065–2071.

[130] Y. Ding, H. Wang, G. Shen, R. Yu, Enzyme-catalyzed amplified immunoassay for the detection of *Toxoplasma gondii* -specific IgG using Faradaic impedance spectroscopy, CV and QCM, Anal. Bioanal. Chem. 382 (2005) 1491–1499.

[131] Z. Wu, J. Wu, S. Wang, G. Shen, R. Yu, An amplified mass piezoelectric immunosensor for *Schistosoma japonicum*, Biosens. Bioelectron. 22 (2006) 207–212.

[132] F. Aberl, H. Wolf, C. Koblinger, S. Drost, P. Woias, S. Koch, HIV serology using piezoelectric immunosensors, Sens. Actuator B Chem. 18–19 (1994) 271–275.

[133] B. Konig, M. Gratzel, A novel immunosensor for Herpes viruses, Anal. Chem. 66 (1994) 341–344.

[134] B. Konig, M. Gratzel, A piezoelectric immunosensor for hepatitis viruses, Anal. Chim. Acta 309 (1995) 19–25.

[135] M. Jenik, R. Schirhagl, C. Schirk, O. Hayden, P. Lieberzeit, D. Blaas, et al., Sensing picornaviruses using molecular imprinting techniques on a quartz crystal microbalance, Anal. Chem. 81 (2009) 5320–5326.

[136] Y.G. Lee, K.S. Chang, Application of a flow type quartz crystal microbalance immunosensor for real time determination of cattle bovine ephemeral fever virus in liquid, Talanta 65 (2005) 1335–1342.

[137] E. Uttenthaler, C. Koßlinger, S. Drost, Characterization of immobilization methods for African swine fever virus protein and antibodies with a piezoelectric immunosensor, Biosens. Bioelectron. 13 (1998) 1279–1286.

[138] Y.K. Kim, S.I. Lim, Y.Y. Cho, S. Choi, J.Y. Song, Detection of H3N2 canine influenza virus using a quartz crystal microbalance, J. Virol. Methods 208 (2014) 16–20.

[139] F.L. Dickert, O. Hayden, R. Bindeus, K.J. Mann, D. Blaas, E. Waigmann, Bioimprinted QCM sensors for virus detection – screening of plant sap, Anal. Bioanal. Chem. 378 (2004) 1929–1934.

[140] A.J.C. Eun, L. Huang, F.T. Chew, S.F.Y. Li, S.M. Wong, Detection of two orchid viruses using quartz crystal microbalance (QCM) immunosensors, J. Virol. Methods 99 (2002) 71–79.

[141] X. Huang, J. Xu, H.F. Ji, G. Li, H. Chen, Quartz crystal microbalance based biosensor for rapid and sensitive detection of maize chlorotic mottle virus, Anal. Methods 6 (2014) 4530–4536.

[142] D.F. Tai, C.Y. Lin, T.Z. Wu, J.S. Huang, P.Y. Shu, Artificial receptors in serologic tests for the early diagnosis of dengue virus infection, Clin. Chem. 52 (2006) 1486–1491.

[143] C.C. Su, T.Z. Wu, L.K. Chen, H.H. Yang, D.F. Tai, Development of immunochips for the detection of dengue viral antigens, Anal. Chim. Acta 479 (2003) 117–123.

[144] T.Z. Wu, C.C. Su, L.K. Chen, H.H. Yang, D.F. Tai, K.C. Peng, Piezoelectric immunochip for the detection of dengue fever in viremia phase, Biosens. Bioelectron. 21 (2005) 689–695.

[145] R. Wang, L. Wang, Z.T. Callaway, H. Lu, T.J. Huang, Y. Li, A nanowell-based QCM aptasensor for rapid and sensitive detection ofavian influenza virus, Sens. Actuator B Chem. 240 (2017) 934–940.

[146] F.N. Dultsev, R.E. Speight, M.T. Fiorini, J.M. Blackburn, C. Abell, V.P. Ostanin, et al., Direct and quantitative detection of bacteriophage by "hearing" surface detachment using a quartz crystal microbalance, Anal. Chem. 73 (2001) 3935–3939.

[147] E. Uttenthaler, M. Schraml, J. Mandel, S. Drost, Ultrasensitive quartz crystal microbalance sensors for detection of M13-Phages in liquids, Biosens. Bioelectron. 16 (2001) 735–743.

[148] L.M.C. Yang, J.E. Diaz, T.M. McIntire, G.A. Weiss, R.M. Penner, Covalent virus layer for mass-based biosensing, Anal. Chem. 80 (2008) 933–943.

[149] R. Wang, Y. Li, Hydrogel based QCM aptasensor for detection of avian influenza virus, Biosens. Bioelectron. 42 (2013) 148–155.

[150] L. Brockman, R. Wang, J. Lum, Y. Li, QCM aptasensor for rapid and specific detection of avian influenza virus, OJAB 2 (2013) 97–103.

[151] T.M.P. Hewa, G.A. Tannock, D.E. Mainwaring, S. Harrison, J.V. Fecondo, The detection of influenza A and B viruses in clinical specimens using a quartz crystal microbalance, J. Virol. Methods 162 (2009) 14–21.

[152] D. Li, J. Wang, R. Wang, Y. Li, D. Abi-Ghanem, L. Berghman, et al., A nanobeads amplified QCM immunosensor for the detection of avian influenza virus H5N1, Biosens. Bioelectron. 26 (2011) 4146–4154.

[153] M. Bisoffi, B. Hjelle, D.C. Brown, D.W. Branch, T.L. Edwards, S.M. Brozik, et al., Detection of viral bioagents using a shear horizontal surface acoustic wave biosensor, Biosens. Bioelectron. 23 (2008) 1397–1403.

[154] H.J. Lee, K. Namkoong, E.C. Cho, C. Ko, J.C. Park, S.S. Lee, Surface acoustic wave immunosensor for real-time detection of hepatitis B surface antibodies in whole blood samples, Biosens. Bioelectron. 24 (2009) 3120–3125.

[155] M. Bisoffi, V. Severns, D.W. Branch, T.L. Edwards, R.S. Larson, Rapid detection of human immunodeficiency virus Types 1 and 2 by use of an improved piezoelectric biosensor, J. Clin. Microbiol. 51 (2013) 1685–1691.

[156] J.T. Baca, V. Severns, D. Lovato, D.W. Branch, R.S. Larson, Rapid detection of Ebola virus with a reagent-free, point-of-care biosensor, Sensors 15 (2015) 605–8614.

[157] T.W. Owen, R.O. Al-Kaysi, C.J. Bardeen, Q. Cheng, Microgravimetric immunosensor for direct detection of aerosolized influenza A virus particles, Sens. Actuator B Chem. 126 (2007) 691–699.

[158] J. Lee, J. Jang, D. Akin, C.A. Savran, R. Bashir, Real-time detection of airborne viruses on a mass-sensitive device, Appl. Phys. Lett. 93 (2008) 13901.

[159] C.H. Shih, Y.J. Lin, K.F. Lee, P.Y. Chien, P. Drake, Real-time electronic nose based pathogen detection for respiratory intensive care patients, Sens. Actuator B Chem. 148 (148) (2010) 153–157.

[160] D. Kovar, Z. Farka, P. Skladal, Detection of aerosolized biological agents using the piezoelectric immunosensor, Anal. Chem. 86 (2014) 8680–8686.

[161] D.D. Stubbs, S.H. Lee, W.D. Hunt, Cocaine detection using surface acoustic wave immunoassay sensors, IEEE Int. Freq. Control Symp. PDA Exhib. (2002) 289–293.

[162] D.D. Stubbs, S.H. Lee, W.D. Hunt, Investigation of cocaine plumes using surface acoustic wave immunoassay sensors, Anal. Chem. 75 (2003) 6231–6235.

[163] S. Dailianis, Environmental impact of anthropogenic activities: the use of mussels as a reliable tool for monitoring marine pollution, in: Mussels: Anatomy, Habitat and Environmental Impact, © 2010, Nova Science Publishers, Inc., 2010.

[164] T. Moore, M.H. Depledge, L. Fleming, P. Hess, D. Lees, P. Leonard, et al., Oceans and human health (OHH): a European perspective from the marine board of the European science foundation (marine board-ESF), Microb. Ecol. 65 (2013) 889–900.

[165] K. Yokoyama, K. Ikebukuro, E. Tamiya, I. Karube, N. Ichiki, Y. Arikawa, Highly sensitive quartz crystal immunosensors for multisample detection of herbicides, Anal. Chim. Acta 304 (1995) 139–145.

[166] K. Das, J. Penelle, V.M. Rotello, Selective picomolar detection of hexachlorobenzene in water using a Quartz crystal microbalance coated with a molecularly imprinted polymer thin film, Langmuir 19 (2003) 3921–3925.

[167] S. Zhang, W. Wei, J. Zhang, Y. Mao, L. Deng, Monitoring environmental wastewater using a piezoelectric impedance microbial sensing technique, Intern. J. Environ. Anal. Chem. 82 (2002) 113–122.

[168] K. Oshima, H. Nakajima, S. Takahashi, Y. Kera, M. Shimomura, S. Miyauchi, Quartz crystal microbalance assay for determination of plasma vitellogenin, Sens. Actuator B Chem. 105 (2005) 473–478.

[169] S. Xiulan, Z. Yinzhi, S. Jingdong, S. Liyan, Q. He, Z. Weijuan, A quartz crystal microbalance-based immunosensor for shrimp allergen determination in food, Eur. Food Res. Technol. 231 (2010) 563–570.

[170] K. Toma, D. Miki, C. Kishikawa, N. Yoshimura, K. Miyajima, T. Arakawa, et al., Repetitive immunoassay with a surface acoustic wave device and a highly stable protein monolayer for on-site monitoring of airborne dust mite allergens, Anal. Chem. 87 (2015) 10470–10474.

[171] B. Godber, M. Frogley, M. Rehak, A. Sleptsov, K.S.J. Thompson, Y. Uludag, et al., Profiling of molecular interactions in real time using acoustic detection, Biosens. Bioelectron. 22 (2007) 2382–2386.

[172] T. Frisk, N. Sandstrom, L. Eng, W. van der Wijngaart, P. Mansson, G. Stemme, An integrated QCM-based narcotics sensing microsystem, Lab. Chip 8 (2008) 1648–1657.

[173] Y.S. Choi, J.P. Do, H.J. Lee, S.S. Lee, J. Lee, Y.H. Lee, et al., A fully-automated surface acoustic wave immunosensing system for the detection of cardiac markers in whole blood, in: 15th International Conference on Miniaturized Systems for Chemistry and Life Sciences October 2–6, 2011, Seattle, Washington, USA, 2011, pp. 1855–1857.

[174] J. Lee, Y.S. Choi, Y. Lee, H.J. Lee, J.N. Lee, S.K. Kim, et al., Sensitive and simultaneous detection of cardiac markers in human serum using surface acoustic wave immunosensor, Anal. Chem. 83 (2011) 8629–8635.

[175] K. Mitsakakis, E. Gizeli, Detection of multiple cardiac markers with an integrated acoustic platform for cardiovascular risk assessment, Anal. Chim. Acta 699 (2011) 1–5.

[176] M. Goto, O. Iijima, T. Kogai, H. Yatsuda, Point-of-Care SH-SAW biosensor, Ultrason. Symp. (IUS), 2010 IEEE (2010) 736–739.

[177] D. Matatagui, D. Moynet, M.J. Fernández, J. Fontecha, J.P. Esquivel, I. Gràcia, et al., Detection of bacteriophages in dynamic mode using a Love-wave immunosensor with microfluidics technology, Sens. Actuator B Chem. 185 (2013) 218–224.

[178] W. Lee, J. Jung, Y.K. Hahn, S.K. Kim, Y. Lee, J. Lee, et al., A centrifugally actuated point-of-care testing system for the surface acoustic wave immunosensing of cardiac troponin, Analyst 138 (2013) 2558–2566.

[179] X. Zhang, Y. Zou, C. An, K. Ying, X. Chen, P. Wang, A miniaturized immunosensor platform for automatic detection of carcinoembryonic antigen in EBC, Sens. Actuator B Chem. 205 (2014) 94–101.

[180] C. March, J.V. Garcia, A. Sanchez, A. Arnau, Y. Jimenez, P. Garcia, et al., High-frequency phase shift measurement greatly enhances the sensitivity of QCM immunosensors, Biosens. Bioelectron. 65 (2015) 1–8.

[181] S. Joseph, T.M.A. Gronewold, M.D. Schlensog, K. Olbrich, E. Quandt, M. Famulok, et al., Specific targeting of ultrasound contrast agent (USCA) for diagnostic application: an in vitro feasibility study based on SAW biosensor, Biosens. Bioelectron. 20 (2005) 1829–1835.

[182] T.M.A. Gronewold, S. Glass, E. Quandt, M. Famulok, Monitoring complex formation in the blood-coagulation cascade using aptamer-coated SAW sensors, Biosens. Bioelectron. 20 (2005) 2044–2052.

[183] M. Perpeet, S. Glass, T. Gronewold, A. Kiwitz, S.I. Malave a, et al., SAW sensor system for marker-free molecular interaction analysis, Anal. Lett. 39 (2006) 1747–1757.

[184] H. Yatsuda, T. Kogai, M. Goto, N. Yoshimura, Shear-horizontal surface acoustic wave biosensors for POCT, Freq. Control Symp. (FCS), 2014 IEEE Int. (2014).

[185] P. Xiomei, B. Zhnag, J. Tang, B. Liu, W. Lai, D.Q. Tang, Sandwich type immunosensors and immunoassays exploiting nanostructure labels: a review, Anal. Chim. Acta 758 (2013) 1–18.

[186] M.I. Mohammed, M.P.Y. Desmulliez, Lab-on-a-chip based immunosensor principles and technologies for the detection of cardiac biomarkers: a review, Lab. Chip 11 (2011) 569–595.

[187] G.L. Cote, R.M. Lec, M.V. Pishko, Emerging biomedical sensing technologies and their applications, Ieee Sens. J. 3 (3) (2003) 251–266.

[188] H. Sharma, R. Mutharasan, Review of biosensors for foodborne pathogens and toxins, Sens. Actuator B Chem. 183 (2013) 535–549.

[189] J.H.T. Luong, K.B. Male, J.D. Glennon, Biosensor technology: technology push versus market pull, Biotechnol. Adv. 26 (2008) 492–500.

[190] A.L. Hillberg, K.R. Brain, C.J. Allender, Molecular imprinted polymer sensors: implications for therapeutics, Adv. Drug Deliv. Rev. 57 (2005) 1875–1889.

[191] L. Uzun, A.P.F. Turner, Molecularly-imprinted polymer sensors: realising their potential, Biosens. Bioelectron. 76 (2016) 131–144.

Chapter 10

Interferometry-Based Immunoassays

Panagiota Petrou[1], Eleni Makarona[2], Sotirios Kakabakos[1], Georgios Koukouvinos[1], Konstantinos Misiakos[2], Ioannis Raptis[2]

[1]*Institute of Nuclear & Radiological Sciences & Technology, Energy & Safety, Athens, Greece;* [2]*Institute of Nanoscience and Nanotechnology, Athens, Greece*

1. INTRODUCTION: GENERAL CONTEXT

Science in the 21st century has been characterized by awe-inspiring breakthroughs in biology and the related disciplines that have already achieved a significant impact on our lives. Current and planned efforts for the immediate future by numerous research groups, funded by companies as well as the public sector, are focusing on the development of the next generation of technologies and sensing systems for a wide range of diverse applications, such as the analysis of minute concentrations of single or multiple biomarkers, discovery of more efficient drugs, food authenticity, safety and quality monitoring throughout the whole food supply chain, and environmental monitoring. These sensing elements are expected to be combined with other modules stemming from the digital era to establish smart system networks accessible through the worldwide net, setting a new framework for ameliorated quality of life and improvement of services for all citizens around the globe.

The core element of these sensing modules is the "biosensor" [1]. The operation of a biosensor can be analyzed in two distinct but closely interacting parts: the specific biomolecule that interacts (binds or recognizes) with the analyte under investigation and the transducer that translates the biomolecular interaction to a measurable signal (typically an electrical or optical one). The intense research efforts during the past 20 years have led to the development of a large variety of transducers relying on and categorized based on a wide range of transduction principles, such as optical [2–4], piezoelectric [5,6], or electrochemical [7,8].

The transducer's output, i.e., the biosensor response, is measured by an external electronic unit, usually named measuring apparatus or reader. This electronic unit includes all the necessary modules for the interface with the biosensor. In the most recent implementations, the reader is not just limited to

Handbook of Immunoassay Technologies. https://doi.org/10.1016/B978-0-12-811762-0.00010-4

measuring and recording the biosensor's response, but it also provides all the connections for the fluidic interfacing for the sample insertion and circulation, and it takes care of the data transmission to the data processing unit (usually a computer, and more recently and more desirably for point-of-need [PoN] applications, a smartphone or tablet). In particular, in the case of optical biosensors, the reader additionally accommodates external optical components, such as light sources and photodetectors or spectrum analyzers. The developed readers can be designed either for bench-top configurations more suitable for standard analytical laboratory determinations or for handheld embodiments for applications at the PoN, e.g., for healthcare, food safety, water quality monitoring, environmental monitoring, forensics.

The importance of biosensors and the wide spectrum of potential applications have attracted significant funding from both the public and private sector for numerous research projects. These are not only dedicated to the development of novel transduction principles and biosensor designs and configurations but also targeted at the realization of holistic systems of enhanced performance, improved analytical capabilities, and high modularity. Indicative of the importance of biosensors in modern life was the foundation of the Qualcomm Tricorder XPRIZE, a $10 million global competition to stimulate innovation and integration of precision diagnostic technologies and to tackle needed breakthroughs in sensing technologies, and the H2020 Food Scanner, a €1 million challenge to develop an affordable and noninvasive mobile solution enabling users to measure and analyze their food intake.

Both labeled and label-free sensing principles have been pursued and intensively explored. But the label-free ones appear to offer competitive advantages in terms of sample preparation, analysis duration, and real-time monitoring capabilities, which render them to the antagonistically preferred approaches for PoN/point-of-care (PoC) applications. Furthermore, the most recent label-free biosensors have been proved able to provide excellent bioanalytical performance in terms of the limit of detection (LOD), sensitivity, resolution, and assay time. Even though there are several transducers employing label-free bioanalytical protocols (e.g., quartz crystal microbalance or microcantilevers), the majority of them are optical ones. Optical label-free biosensors offer a long list of concrete advantages, the most important of which are as follows:

1. galvanic isolation between the excitation and detection electronics that allows for simultaneous, multianalyte determinations along with appropriate multiplexing functions,
2. the excellent bioanalytical performance (LOD, high dynamic range, high sensitivity, and resolution),
3. their fabrication with conventional microelectronic/micromachining processes that guarantee cost-efficient mass production at existing facilities.

To date, among optical label-free biosensors, surface plasmon resonance (SPR) has been the dominating player that has already reached the market with several vendors offering such systems. Alongside the SPR systems, which despite their success are bound to the confines of laboratories, numerous other optical transducers have been emerging targeting at the PoN/PoC domain that has not yet found suitable solutions. The majority of the emerging optical biosensors rely on planar waveguides. This particular type of biosensors has exhibited superior bioanalytical performance due to their ability to monitor minute refractive index changes over the appropriately functionalized sensor surface (due to the binding of the target molecules by the specific application bioreceptors) in real time. The fast signal acquisition and the use of very accurate optical modules with high sensitivity allow for the detection of infinitesimally thin biomolecular adlayers (on the order of a picometer) causing minuscule changes of either the intensity of the output signal or the spectral content. In addition, the detection of more than one analyte is achieved through multiplexing of individual biosensors densely integrated into an array format on the same substrate. Last but not least, the fabrication of the majority of planar waveguide-based optical label-free biosensors can be performed with conventional microelectronic/micromachining processes. This way the tremendous improvements of these technologies can be straightforwardly applied to optical biosensors improving their analytical performance, size, reproducibility, and cost. Moreover, thanks to the advances in the Si processing technologies, it is possible to massively and cost-efficiently fabricate at transducers that is hard to realize in high volumes with past technologies.

2. PRINCIPLES OF OPERATION

This section is dedicated to a brief, but concise, presentation of the operational principles of interferometric optoelectronic biosensors, commonly employed in immunoassay detection schemes. The section begins with a brief introduction to the underlying mechanisms of label-free optical sensing, common to most types of optoelectronic sensors, followed by subsections concentrating on the operating principles of the various interferometric sensors.

2.1 Label-Free Optical Sensing

In general, an optical biochemical sensor may be conceived as a device that modulates an input optical signal, when a biochemical event takes place. The modulation usually affects a property of the incoming light, such as its intensity, phase, or spectral content, while the degree of modulation depends on and can be directly correlated to the specifics of the biochemical event under investigation (e.g., the presence and/or the concentration of the targeted analyte). In other words, an optical sensor may be regarded as an optical modulator, where the interaction of molecules provides the signal modulation. Optical sensors and optical sensing techniques are divided, as already mentioned, into two major categories: labeled and label-free ones. In labeled techniques, the modulation is enhanced by the presence of molecules that may further interact with the incoming light, such as fluorescent molecules, metallic nanoparticles, or radioisotopes that are tethered to the analyte entities. Even though labeled techniques may in some cases offer enhanced sensitivity and/or lower LODs, the additional preparation steps required by the coupling chemistries add to the cost and complexity of the analytical method and prohibit real-time monitoring and studying of interaction kinetics. Hence, labeled techniques become unsuitable for PoC/PoN applications, where expeditious results are of paramount importance. For that reason, label-free techniques have emerged as the highly preferred option for the next generation of portable applications in immunosensing, since they allow for preservation of the binding affinity and kinetics, simpler analysis procedures, and reduction of analysis cost.

In light of the aforementioned, the basic layout of a label-free optical sensor may be conceived as consisting of three distinct parts, as shown schematically in Fig. 10.1:

1. *An input section*, where the light destined to perform the interrogation of the biological entities is introduced. This section is most commonly encountered in Si-based optical sensors as planar waveguides suitably designed to support specific modes of the incoming light. In the majority of cases, its geometrical characteristics and constituent materials are selected so that only the fundamental mode (monomodal waveguides) may be supported and propagated within the sensor. In very few cases (as in the case of bimodal waveguide interferometry), a selected number of modes is allowed as will be analyzed in the following subsections.
2. *The sensing area (optical transducer)* over which the selected biorecognition events take place. The sensing area is, in essence, the part of the optical sensor appropriately functionalized with capture molecules

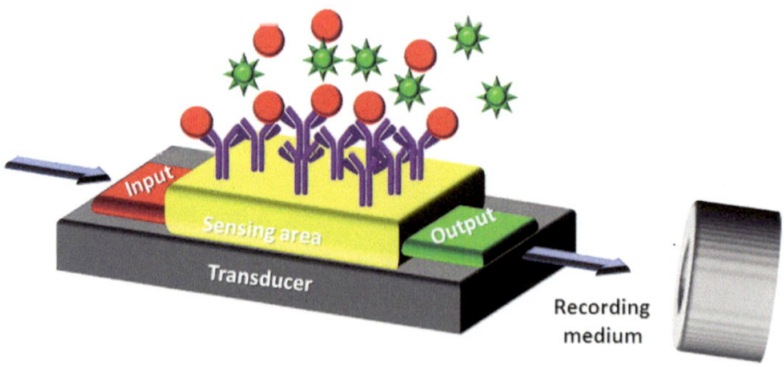

FIGURE 10.1 Schematic representation of an optical biosensor where its three main parts are shown, namely the input section, the "actual" transducer with the appropriately functionalized sensing area where the biorecognition event and optical modulation take place, and the detection part comprised of the output section of the sensor and the recording medium.

crafted to detect specific analytes. It is within this area that the guided light interacts via its evanescent field with the biological entities during a biochemical event and is modulated because of that event. When light propagates within a solid medium due to total internal reflection—as is the case for planar waveguides and integrated optical devices—although all of the incident energy is reflected at the boundary of the media forming the waveguide, an electromagnetic field with exponentially decreasing intensity propagates in the less dense medium (Fig. 10.2). This field is known as an evanescent wave and has a decay length or penetration depth, Δz, on the order of the wavelength of the incident light, λ, and is expressed as

$$\Delta z = \frac{\lambda}{2\pi}\sqrt{N^2 - n_c^2} \tag{10.1}$$

where N is the effective index of refraction of the guided mode, n_c is the refractive index of the covering medium over the sensing area, and λ is the wavelength of the guided mode. Typically, the penetration depth of the evanescent field for Si-based optical sensors is on the order of 100–200 nm. The confined nature of the evanescent wave close to the interface is ideal for studying surface and near-surface phenomena. In other words, it is the evanescent wave that "senses" the biochemical events as changes in the refractive index of the sensing area's surrounding material (Fig. 10.2). Mathematically speaking, these changes are expressed as changes in the

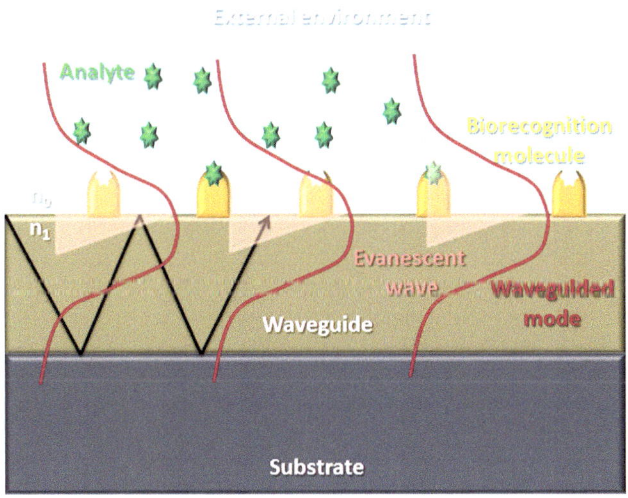

FIGURE 10.2 Schematic representation of the evanescent field probing the buildup of an adlayer on a biorecognition event on the sensor's sensing area.

complex phase velocity of the propagating beam and more specifically as changes of the real part of the complex effective index of refraction (the imaginary part is related to absorption and takes effect in labeled techniques, where luminescent molecules are employed).

During a biochemical event, two distinct effects occur that may induce changes in the refractive index of the surrounding medium and that can be "sensed" by the guided mode's evanescent field. The first mechanism involves the capturing event itself: on binding of the targeted (analyte) molecules on the probe molecules immobilized onto the sensing area surface, the thickness of the adlayer is increased, and its refractive index may also change. The second effect pertains to the refractive index of the carrier medium itself, which on introduction of the sample may change. Both these effects alter the effective index of refraction of the guided mode according to

$$\delta N = \left(\frac{\partial N}{\partial d}\right)\delta d + \left(\frac{\partial N}{\partial n_c}\right)\delta n_c \tag{10.2}$$

where d is the adlayer's thickness and n_c is the covering medium index of refraction (usually a buffer solution). It is this change in the effective index of refraction, δN, that is "perceived" through the evanescent field and is the driving force modulating the physical properties of the propagating beam. The property that is modulated depends on the architecture and operating principle of each optical sensor, but the underlying mechanism described above is common to all types of optical sensors.

3. After the sensing area, the guided light that has "probed" and thus "interrogated" the biochemical event is steered to the output section of the sensor, where it is coupled to *the detection part* of the system. In the majority of cases, the detection part consists of external recording media, such as photodiodes, CCD cameras, spectrometers, and optical parametric analyzers, which are capable of detecting and recording the modulations (in intensity or spectral content or fringe translation, etc., as will be analyzed in the following subsections) induced to the guided light by the biochemical event occurring over the sensing area. As soon as the signal is monitored and recorded, the biochemical detection is completed, and a new cycle of analysis may begin. Depending on the operating principle and the system development, the recorded signal may provide an immediate answer (as in the case of qualitative measurements, such as measurements that simply detect the presence of an analyte or warn about a certain entity surpassing a predetermined concentration level). However, in most cases, signal processing algorithms must be employed and embedded in the final system to obtain an accurate quantitative determination.

In light of the aforementioned, the sensitivity and analytical capability of an optical sensor—most commonly referred to as the LOD—is directly related to the minimum detectable effective refractive index change it can record, ΔN_{min}:

$$\Delta N_{min} = G(\lambda)\frac{\lambda}{L} \quad (10.3)$$

where λ is the vacuum wavelength of the input light, L is the length (size) of the sensing area, and $G(\lambda)$ is a "form factor" that is related to the design and nature of the sensor. Nonetheless, two equally important aspects about the analytical capabilities and performance of optical sensors, namely the resolution and the dynamic range, are often overlooked in general discussions and comparisons among various sensors and sensing principles. The former refers to the smallest detectable incremental change of the parameter under inspection that can be unambiguously detected in the output signal. In other words, it represents the ability of the sensor to distinguish between two measurements that are closely ranged. The term "resolution" is often erroneously used in place of sensitivity or LOD, but it is a different figure of merit and should not be confused. As far as the dynamic range is concerned, it describes the total range of input values that the sensor can respond to, from minimum to maximum. This again is an important feature of a sensor, since a device may be proven to be of high sensitivity but unable to perform measurements across the entire range of values that are relevant to a particular application. It is worth noting at this point that in the majority of reports found in literature the aspects of resolution and dynamic range are rarely discussed, and the performance of each introduced device is evaluated primarily in terms of LOD. This renders it tough to assess the various optical sensors to the full extent or to formulate a common frame of performance criteria within to place and compare the different devices.

2.2 Interferometric Sensors

Interferometric sensors have emerged during the last 15 years as one of the most powerful analytical tools exhibiting impressive LODs. The LODs have reached down to effective index of refraction changes of $\Delta N_{eff,min} = 9 \times 10^{-9}$ RIU (corresponding to a bulk refractive index change of $\Delta n = 4 \times 10^{-8}$ RIU) [9] or even the detection of *Listeria monocytogenes* down to a minimum detectable concentration of 2.8×10^5 CFU/mL, which is well below the infective dosage for humans as well as the LOD achieved by other more conventional detection methods [10]. They are mostly encountered in the form of Mach–Zehnder (MZIs) or Young interferometers (YIs), but lately, newly developed interferometric principles, such as the bimodal interferometry, broadband MZI (BB-MZI), or multiwavelength YI (MW-YI), have arisen offering competitive advantages compared with the more standard approaches. In a sense, interferometric sensors are the oldest amidst optical sensors, since their principles of operation date back to the early and late 19th century. This is evident in the pioneering works of Thomas Young and his famous double-slit experiment [11], followed by the refinement of Ludwig Mach [12] over an article of Ludwig Zehnder [13] concerning an apparatus capable of measuring phase shifts between two beams caused by a sample or a change in length of one of the paths.

The basic, underlying idea is relatively simple and is shared among all interferometric types of sensors, with the main difference being the output section and the signal recording method. An interferometric sensing scheme consists of an input monochromatic beam that is split into two beams of equal power, one to be used as the reference beam and the other to be used as the sensing/probing one. The latter is steered to travel through a specially prepared area (sensing area), where the biorecognition events are bound to take place and induce local changes of the refractive index. These changes cause—through the interaction of the evanescent field of the sensing beam with the biomolecular layers—a phase lag between the sensing and reference beams. At the end of the sensing area, the two beams may either be recombined in a common path, and one is left with an MZI (Fig. 10.3A), or they may be allowed to interfere in free space and recorded as moving fringes on a screen placed at a distance, and then one is left with an YI (Fig. 10.3B). This simplified description is mathematically expressed in its core through the following equation:

$$\Delta\varphi(\lambda) = \frac{2\pi L}{\lambda}[N_r(\lambda) - N_s(\lambda)] = \frac{2\pi L}{\lambda}\Delta N_{rs}(\lambda) \qquad (10.4)$$

where $\Delta\varphi(\lambda)$ is the phase lag for a given input wavelength λ between the reference and sensing beams due to the local effective refractive index differences $\Delta N_{rs}(\lambda)$ between the reference area (with effective refractive index $N_r(\lambda)$) and the sensing area of length L (with effective refractive index $N_s(\lambda)$). Any biorecognition binding event that can alter the effective refractive index of the sensing arm by $\delta N_s(\lambda)$ induces a new phase lag $\Delta\varphi'(\lambda)$ between the two beams:

$$\Delta\varphi'(\lambda) = \frac{2\pi L}{\lambda}[N_r(\lambda) - (N_s(\lambda) + \delta N_s(\lambda))] = \frac{2\pi L}{\lambda}\Delta N'_{rs}(\lambda) \qquad (10.4')$$

In the case of an MZI, the interference of the two beams into the common path—due to the wave nature of light—will result in a recombined beam of intensity I_{out} that will vary according to

$$I_{\text{out}} = \frac{I_{\text{in}}}{2}(1 + \cos\Delta\varphi(\lambda)) = \frac{I_{\text{in}}}{2}\left(1 + \cos\left(\frac{2\pi L}{\lambda}\Delta N_{rs}(\lambda)\right)\right) \qquad (10.5)$$

where I_{in} is the intensity of the unsplit original input beam. As it can be seen, the MZI biochemical sensing recording is, in essence, a light intensity recording, and in particular, one which is sinusoidally dependent on the effective refractive index changes resulting from the biochemical reactions.

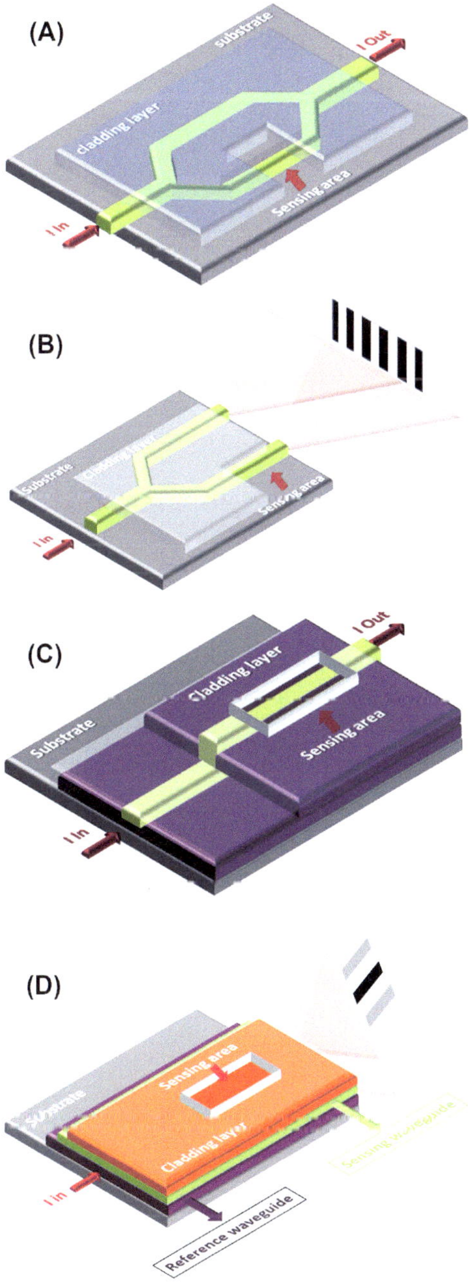

FIGURE 10.3 Schematic representation of a (A) Mach–Zehnder, (B) Young, (C) bimodal, and (D) dual-polarization interferometer.

In the case of the YI configuration, the two output beams are left to interfere in free space instead of recombining in a common path, resulting in a pattern of dark and bright fringes recorded onto a screen (most often a CCD camera), called an interferogram (Fig. 10.3B). The intensity distribution along the direction x of the recording screen is provided by

$$I_{\text{out}}(x,\lambda) = I_r + I_s + 2\sqrt{I_r I_s}\cos\left[\frac{2\pi b}{\lambda D}x + \Delta\varphi(\lambda)\right] \tag{10.6}$$

with I_r, I_s being the intensity of the reference and sensing arm, respectively, D is the distance of the YI from the screen, and b the distance between the two paths of the two beams. $\Delta\varphi(\lambda)$ represents the phase lag between the beams and is given by the same equation applying to an MZI, i.e., Eq. (10.4). In very much the same way, a biorecognition binding event alters the effective refractive index of the sensing arm by $\delta Ns(\lambda)$, resulting in a different phase lag $\Delta\varphi'(\lambda)$ between the two propagating beams, which, instead of the sinusoidally modified intensity of the recombined beam as in the case of the MZI, results in a spatial translation of the interference fringes along the detection screen by a distance dictated by Eq. (10.6). It is through this fringe spatial shift that a biochemical event is recorded and evaluated.

A critical prerequisite for an interferometric operation, not readily apparent through the above mathematical expressions, is the fact that the input light must be monomodal and of a single polarization. This prerequisite is essential as each supported mode for every polarization has a different effective refractive index, which makes it propagate with a different velocity and is affected in a different way by the biochemical interactions. Hence, should more than one mode coexist within the sensor, the output signal would be the overlap of several modes and would appear scrambled as the recording media cannot discern photons of the same wavelength and would record the arriving photons of any polarization and mode in the same manner.

Nonetheless, dual-mode coexistence within the sensor *can* be exploited in a different way, and this is exactly what was achieved with the recently developed bimodal interferometric sensors (Bi-MI) [14,15]. In Bi-MI, instead of splitting a monochromatic input beam into two beams of equal intensity propagating along two different paths, the input light is coupled into a planar ridge waveguide supporting a single transversal mode, which is the fundamental mode. After a certain distance, this waveguide is shaped in such a way that it can support two transversal modes: the fundamental and the first-order one (Fig. 10.3C). These two supported modes probe, each in its own way (as each one has its own propagation velocity/effective refractive index), the biochemical event along the sensing area. After that, they are left to propagate within the chip at the end of which they form an interference pattern to be recorded by an external two-section photodetector aligned against the chip.

For a Bi-MI the phase difference between the two modes is expressed in an analogous way to the classical interferometry via

$$\Delta\varphi(\lambda) = \frac{2\pi L}{\lambda}[N_1(\lambda) - N_0(\lambda)] = \frac{2\pi L}{\lambda}\Delta N_{1,0}(\lambda) \quad (10.7)$$

where N_0 and N_1 are the effective refractive indices of the two propagating modes, respectively, $\Delta N_{1,0}$ their difference, L the sensing area length, and λ the wavelength of the light coupled in the chip.

The recorded signal by the two-section photodetector is expressed through

$$S_R = \frac{I_{up} - I_{down}}{I_{up} + I_{down}} \propto V\cos(\Delta\varphi) \quad (10.8)$$

with I_{up} and I_{down} being the currents measured in the two sections of the photodiode and V is the so-called visibility factor, usually ranging from 0.5 to 0.7. S_R is normalized to the total optical power propagating in the BiMW structure.

In a slightly different twist, dual polarization interferometry (DP-I) exploits the two polarizations of the same input beam instead of the two first modes of the same polarization as Bi-MI does. The general idea of DP-I is schematically represented in Fig. 10.3D. A DP-I consists of two slab waveguides separated by an isolation layer. The bottom waveguide serves as the reference arm and the top, which can be exposed to the analytes, serves as the sensing one. The light beams traveling through the stacked arms interfere in free space and result in an interferogram in a similar way to an YI. By switching the polarization of the input light, the two polarization modes TE and TM of the waveguides are alternately excited. Measurement of the interferogram for both polarizations allows both the refractive index and the thickness of the adsorbed layer to be calculated [16,17]. The polarization can be switched rapidly, allowing real-time measurements of chemical reactions taking place on the sensing area.

Bi-MI and DP-I are not the only interferometric principles that have provided modifications over classical interferometry. Traditionally, interferometry has been conceptualized in strict relation to monochromatic sources. Very recently, broadband interferometry (BB-I) made its appearance. Not only has its feasibility been proven, it has also been established that it is capable of providing solutions to several inherent limitations of "classical," single-wavelength (monochromatic) interferometry (SW-I), such as phase ambiguity and signal fading (for a detailed discussion on the limitations see the review paper by E. Makarona et al. [18]). In brief, in BB-I under certain conditions that can be satisfied with the appropriate design of the constituent waveguides, the monochromatic input source may be substituted with a broadband source. That way a wide spectral range comes into play and the spectral content provides for additional observables to the intensity variation of a single wavelength as in SW-I.

In its earlier versions, BB-I was attempted in MZI configurations. In the 2003 report of Luo et al. [19], a white-light source was coupled to an SiO_2-based MZI, and the transmitted light was monitored with an external optical spectrum analyzer in the 1200−1700 nm spectral window. Even though the sensitivity of the particular device was rather low, it was the first indication that recording the interference intensity variations in a wide spectral range is the step toward resolving the phase ambiguity and signal fading issues. This demonstration was followed by the work of Li et al. [20] in 2010, where again broadband NIR light was transmitted through an MZI structure and recorded by an optical spectrum analyzer yielding moderate sensitivities, but exhibiting the great potential of spectral interrogation. During the same year, the first theoretical background and ramifications of BB-MZI started being formulated, and a theoretical LOD of an effective adlayer thickness of 2.5 pm with a sensing area as short as 600 microns was predicted [21].

A few years later, by taking advantage of a monolithic Si-based optocoupler (that incorporates on a single chip the emitting source, the interferometric transducer as well as the detector in the form of a photodiode), it was established that BB-MZI may be applied in two distinct operating regimes. In one embodiment, provided that the sensing and reference arms are suitably engineered to have a bell-shaped phase difference centered at the emission source spectrum, the BB-MZI may operate in an analogous way to its monochromatic counterpart even with a broadband input source. When operated as a refractometer, the LOD was established to be on the order of 1×10^{-5} with a sensing arm length of only 600 μm. The same device was tested with model immunoassays detecting the binding of 1 nM of streptavidin and 10 nM of antimouse IgG [22]. In the second embodiment, the same optocoupler was used, but the integrated photodiode was replaced with an external spectrophotometer to detect and analyze the transmission spectrum. In this case, the two MZI arms have to be photonically engineered for their phase difference to have a linear dependence on the wavelength [23]. When this condition holds, the transmission spectrum contains two characteristic frequencies that are recorded simultaneously, one for the TE and one for the TM, since both polarizations are excited at the input. Through signal splitting in the Fourier transform domain and peak argument tracking, the two polarizations can be deconvoluted from the composite signal and offer the ability to calculate independently—through a single measurement alleviated from the need of external polarizers—between cover refractive index changes and binding events leading to the formation of an adlayer. LODs of 10^{-6} RIU and a sensitivity of 20 μM/RIU were demonstrated limited by temperature fluctuations.

On par with BB-MZI, it was equally theorized that employing a broadband source in the YI format will enhance the analytical capabilities of the interferometer through the interrogation of multiple wavelengths. One of the earliest theoretical approaches of multiwavelength YI (MW-YI) made by

Mulder et al. [24] suggested the usage of three distinct wavelengths taking the dual-wavelength scheme proposed by Stamm et al. [25] or the dual-wavelength/polarization approach of Nellen and Lukosz [26] one step further. Mulder et al. theorized that since the evanescent field is smaller for shorter wavelengths and more expanded for longer ones according to Eq. (10.1), it is possible to discriminate between short molecules (such as proteins) and bulk RI changes or larger particles, offering thus the ability to distinguish specific binding from nonspecific binding and bulk RI changes. Their simulations for Si_3N_4-based waveguides predict an LOD of 10^{-6} RIU with the added benefit of the deconvolution of binding events from pure cover media refractometric changes through a single interferogram.

Recently, the feasibility of an YI operating with a wide input spectrum was demonstrated with an input source emitting in the VIS-NIR range (650−850 nm) [27]. Even though the entire spectrum was not exploited and an interference filter of appropriate bandwidth was employed (to redefine the spectral range of the interfering photons in a way to allow the interference fringes to be detectable), a multianalyte model experiment allowed for the determination of biotin−streptavidin and rabbit IgG−antirabbit IgG binding.

3. SENSOR SURFACE FUNCTIONALIZATION

The path from a transducer to a biosensor passes through two main milestones: the transducer surface activation and its functionalization with biomolecules. Although closely related to one another, the two procedures can be individually optimized since for the first one, the determining factor is the transducer material and for the other the molecule to be immobilized. Ultimately, the result should be a surface that contains a sufficient amount of properly working (with respect to their ability to recognize the targeted analyte) recognition molecules and does not interact with other components of the sample [28]. Talking about immunosensors, the biorecognition molecules are restricted to antibodies (or fragments of them) and antigens either their native forms or analogs of them. The latter could include peptides in the case of protein antigens or protein conjugates in the case of low molecular weight antigens. In antibody-modified transducer surfaces, the detection of analyte could be performed either directly (on its binding to the immobilized antibody) or after the implementation of a second antibody directed to a different epitope than that of the immobilized antibody (Fig. 4.4A). Transducers modified with antigens could be employed in competitive assays for the determination of the antigen concentration in the sample (Fig. 4.4B) or in indirect assays for the detection of antibodies against this antigen in the sample (Fig. 4.4C). Depending on the recognition molecule, the immobilization can be achieved by physical adsorption, covalent coupling, and/or affinity binding [18,29]. In all cases, the transducer surfaces should be properly activated to facilitate the recognition molecule immobilization. As it was mentioned previously, this is a

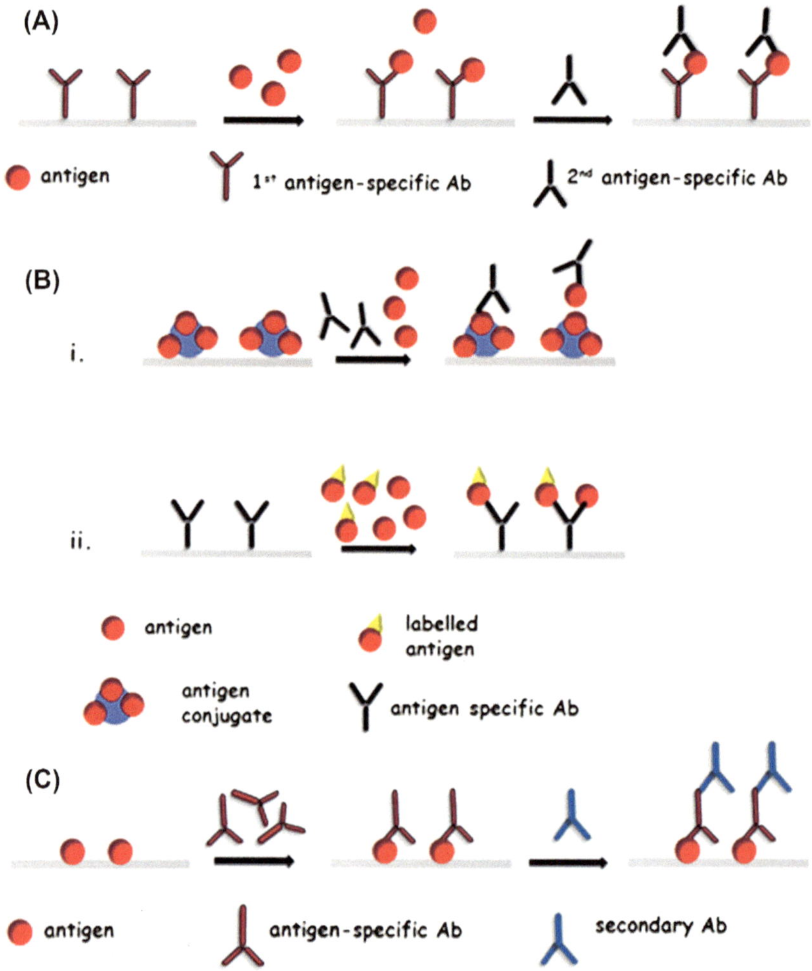

FIGURE 10.4 Schematic representation of (A) noncompetitive immunoassay, (B) competitive immunoassay for analyte determination through immobilization of analyte (i) or of antibody, and (C) indirect assay for the detection of antibodies against a particular antigen in the sample.

material-driven procedure. Therefore, we will focus on methods for activation/ biofunctionalization of silicon-based surfaces, since most of the interferometric transducers so far referenced are made of silicon oxide, silicon nitride, silicon-on-insulator (SOI), and less often of glass substrates, Ta_2O_5, or polymers. Nevertheless, the biofunctionalization approaches followed for glass- and Ta_2O_5-based sensors are similar to those applied for the silicon-based ones. Afterward, the most frequently used methods to introduce chemical groups on the sensor surface (sensor activation) will be described; then the biofunctionalization approaches will be presented and, finally, a brief description of the approaches used for the elimination of nonspecific binding will be mentioned.

3.1 Chemical Activation of Transducers

The chemical activation of a transducer's surface aims at the introduction of reactive groups that could be used in the subsequent step of covalent coupling of biorecognition molecules or at the adjustment of the surface wetting properties so as to allow for immobilization of biomolecules via physical adsorption. This is usually accomplished by deposition of a self-assembled alkyl-silane layer [28]. The effective creation of a silane layer relies on the condensation of siloxane groups on the silane molecules with hydroxyl groups that are part of the native silicon oxide layer, always present on silicon surfaces, or that have been created on the sensor surface through oxidation. Oxidation may be accomplished by surface pretreatment in oxygen plasma [22,30], piranha solutions prepared by mixing concentrated sulfuric acid (H_2SO_4) with a 30% (v/v) hydrogen peroxide solution (H_2O_2) at ratios from 3:1 to 7:3 [31], nitric acid [32], heated sulfuric−chromic acid solution [33], acetone [34], mixtures of ammonium hydroxide and hydrogen peroxide [9], or commercially available cleaning solutions such as Hellmanex [35] or Micro-90 solution [36]. It should be noted that the same procedure can be applied to metal oxide surfaces such as aluminum (Al_2O_3), titanium (TiO_2), or tantalum (Ta_2O_5), whereas oxygen plasma treatment has also been applied to clean/hydrophilized waveguides made of polyimide [37]. The oxidized surfaces are then either immersed in a silane solution (aqueous or organic depending on the silane used) or exposed to silane vapors generated by applying pressure reduction and/or heating to a silane solution. The desired outcome of the silanization procedure is the creation of a smooth and homogeneous silane monolayer [38,39]. This is not so straightforward as one would expect based on the many applications of this chemistry for modification of silicon-based surfaces. It seems that the outcome can be influenced by a large number of parameters including the solvent used—water [40,41] or anhydrous organic solvents [42,43]— the silane content, the duration, and the temperature of the treatment [28]. It is, therefore, recommended that researchers should perform some optimization of selected "literature" protocols to ensure that the results are the desired ones for the particular sensor and application they are working with. The silanes first used for modification of interferometric sensors were hydrophobic, e.g., hexamethyldisilane [4] or dichlorodimethylsilane [44], and were used to render the surface suitable for protein molecules immobilization through physical adsorption. Nonetheless, the majority of silanes used since then include amino-, epoxy-, mercapto-, or carboxy-terminated alkyl silanes, which are primarily implemented for covalent coupling of recognition biomolecules. 3-Aminopropyltriethoxysilane

(APTES) is the most popular modification molecule for silicon-based surfaces and leads to the introduction of reactive amine groups onto the surface. There are numerous available protocols in the literature for surface modification with APTES using aqueous or dry organic solutions or vapor deposition, followed in most cases by thermal curing to stabilize the silane layer through promotion of the condensation reaction with the surface hydroxyl groups [28]. APTES (or any other triethoxysilane) reacts more readily with the surface hydroxyl groups when water is present [28]; a notion that does not apply to other types of reactive alkyl silanes. Other amine-silanes that have been used for modification of interferometric transducers include 4-aminobutyldimethylethoxysilane [45] and N-[3-(trimethoxysilyl) propyl]-ethylene diamine. Epoxy-organosilanes, and in particular 3-glycidoxypropyltrimethoxysilane (GOPTS), have also been applied for the chemical modification of silicon-based interferometric sensors [9,33]. In contrast to APTES, anhydrous organic solvents (mainly toluene) are recommended for GOPTS deposition, and meticulous washing with organic solvents supplemented by ultrasonic treatment might be necessary to ensure the formation of a monolayer [39,46]. Concerning mercapto-alkyl silanes, the reagent almost exclusively used is 3-mercaptopropyltrimethoxysilane (MPTMS), although the conditions used for the modification differ between reports with respect to silane content, the solvent, and the duration of the procedure [37,47,48]. Carboxy-terminated silanes, and especially carboxyethylsilanetriol, have been successfully applied for the chemical activation of interferometric transducers made of silicon nitride [15,49,50]. This approach has the advantage that aqueous solutions are used for silane deposition avoiding any organic solvents. The chemical structures of the most frequently used silanes are provided in Fig. 10.5.

To avoid silanes, and the concomitant problems arising from the difficulty to completely control the silanization procedure, alternative modifications have been proposed. These methods rely on the etching of the native oxide layer (usually with concentrated HF) so as to create a surface rich in H-moieties. In a second step, these moieties are replaced by functional groups or organic molecules [51–53]. Until now, the application of this alternative procedure for the chemical activation of silicon-based transducers is rather limited [54–56]. This could be ascribed to the rather tedious conditions and dangerous chemicals used.

3.2 Immobilization of Recognition Molecules

The selection of the approach to be implemented for the immobilization of recognition molecules on the chemically activated surface of a biosensor depends on both the nature of the molecule to be attached and the reagent used for the said activation. In general, immobilization of recognition molecules can be accomplished by physical absorption, covalent coupling, or affinity binding.

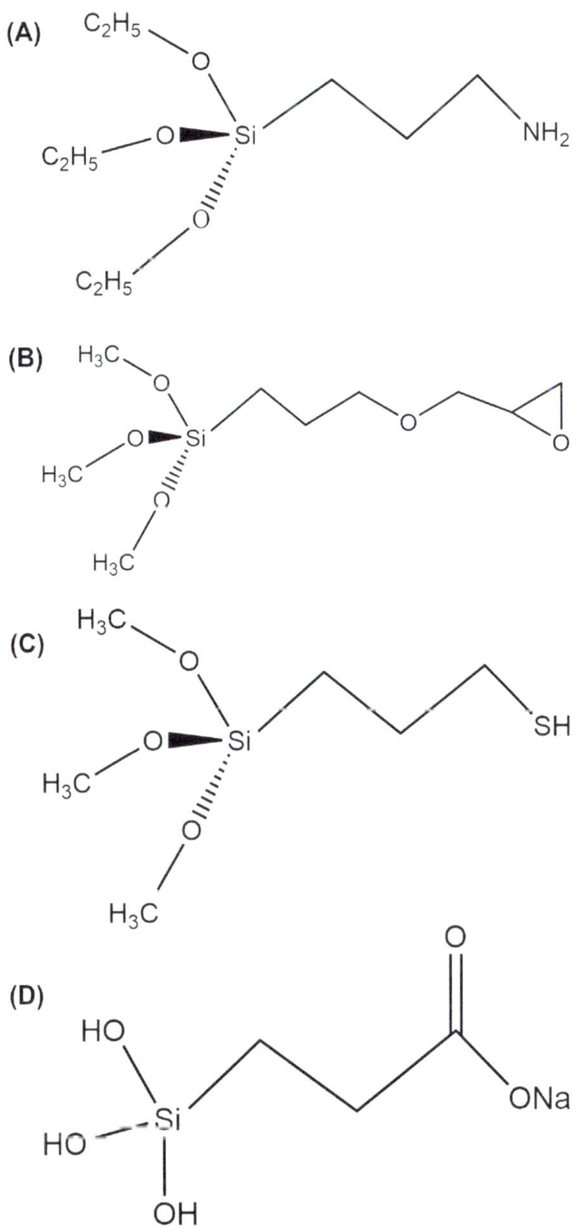

FIGURE 10.5 Chemical structures of (A) 3-aminopropyltriethoxysilane, (B) 3-glycidoxypropyltrimethoxysilane, (C) 3-mercaptopropyltrimethoxysilane, and (D) carboxyethylsilanetriol.

Physical adsorption is the simplest approach to immobilize molecules and especially proteins onto interferometric transducers. There are opposing opinions regarding the stability of the immobilized biomolecules, especially when the analysis is performed under flow conditions or when regeneration and reuse of the sensor surface is desirable [28]. The notion that physical adsorption does not result in stably attached molecules arises mainly from the fact that it is based on weak van der Waals interactions (interactions between induced, permanent, or transient electric dipoles) between the surface and the protein molecules. Although true, the high number of interactions created for a single molecule leads to its very strong and irreversible attachment onto the surface. Thus, physical adsorption is still employed for biofunctionalization of interferometric transducers [41,46,57]. Although in the early-reported interferometric sensors physical adsorption of recognition molecule was performed without any modification of the transducer surface [58,59], in more recent ones modification with hydrophobic silanes [4,44] or APTES has been adopted [41,46,60]. Physical adsorption is also the most straightforward procedure for the biofunctionalization of waveguides made of polymers [61] as most polymers readily adsorb protein molecules. The most significant limitation of physical adsorption is the fact that the attached biomolecules do not have any particular orientation, which in turn might mean that the immunoreactivity of the immobilized molecules is not optimal, a limitation that some of the covalent coupling strategies may overcome.

Covalent coupling offers a high diversity of immobilization strategies combined by stable bonding of immobilized molecules and has, therefore, been widely adopted for functionalization of interferometric transducers. In addition, it is more suitable for immobilization of small molecular weight molecules, e.g., antigens onto the transducer surface. The particular chemistry applied depends on the available reactive groups on the surface and the molecule. Thus, molecules having free carboxy groups can be attached to surfaces modified with amine-terminated alkyl silanes, through activation of these groups with carbodiimide (e.g., EDC), N-hydroxysuccinimide (NHS), or a combination of these two reagents [32,34,62,63]. Glutaraldehyde is a bifunctional reagent that has also been used to couple molecules onto aminosilane-modified surfaces. The reaction leads to the introduction of reactive aldehyde groups onto the transducer surfaces, which can in a second step react with amine groups onto biomolecules to form imines (Schiff bases). Reduction by a reagent, such as sodium cyanoborohydride, converts the imines to stable amine bonds [31,64]. Conversion of surface amine groups to carboxyl groups through reaction with succinic anhydride and subsequent activation by EDC/NHS is also a procedure that can be employed for coupling of molecules with free amine groups [45]. Alternatively, amine activation with cyanuric chloride and tribenzylamine [65] or p-phenylene diisothiocyanate [66] could provide the necessary reactivity for coupling of amine-bearing molecules. Transducers with epoxy-terminated silanes can be directly modified with biomolecules through the reaction of surface epoxy groups with their amine or sulfhydryl groups, or after the conversion of epoxy groups to other functional moieties, such as aldehyde groups, prepared via oxidation of epoxy groups with sodium periodate [33,67,68], or amine groups introduced via reaction with small bifunctional molecules, such as ethylenediamine [68], or polymers, such as amino-dextran [69]. Thiol-modified surfaces are suitable for coupling of molecules with thiol groups through the formation of disulfide bonds [47], or amine groups after reaction with heterobifunctional cross-linkers that have a maleimido-group [36,37,48]. A possible advantage of this coupling chemistry is that owing to the reversibility of the disulfide bond, the transducer surfaces can be easily regenerated by treatment with a thiol-containing reagent, such as dithiothreitol. Finally, the coupling of biomolecules on transducers modified with carboxylate-terminated silanes can be accomplished through reaction with free amine-groups after conversion of carboxy moieties to active esters, e.g., by reaction with EDC/NHS [49,50]. Activation and coupling should be performed in two steps to avoid the formation of protein–protein aggregates that may affect the performance of the bound biomolecules. In addition, careful handling is required to prevent hydrolysis of active ester groups before the addition of the molecule to be attached onto the surface.

Affinity binding may also be opted for biofunctionalization of interferometric transducers. In this case, the recognition molecules are coupled to the surface not directly, but through specific binding molecules. Since the recognition element does not interact directly with the surface, its functionality is not affected by changes in the structure or steric hindrance effects, and it is more efficiently preserved compared with direct physical adsorption or covalent coupling. The most popular immobilization scheme based on affinity binding is that exploiting the biotin–(strept)avidin reaction. This can be realized either by modification of the transducer surface with biotin moieties followed by sequential binding of streptavidin and biotinylated antianalyte-specific antibody [35] or directly by avidin [33] or streptavidin [70] analogously followed by binding of the biotinylated antianalyte-specific antibody. Another affinity binding approach uses bacterial proteins, protein A or protein G, which bind to the Fc fragment of immunoglobulins from different species for the immobilization of analyte-specific antibodies onto transducer surfaces [59,71].

Irrespective of the surface activation strategy followed, the procedure is usually applied on the chip level. However, it is highly desirable that the biofunctionalization is confined to a very specific area, e.g., on interferometric sensors, with both arms exposed to the analyte, only onto the sensing arm, or on arrays of sensors aiming to multianalyte determinations only on selected sensors dedicated to the specific analyte. For this purpose, specially designed cuvettes [72] or flow cells [15,49] have been used to apply one or more recognition molecules onto the sensing channel of a single or multiple sensors, respectively. Moreover, robotic spotters have also been implemented for the spatially selective deposition of different recognition molecules onto arrays of transducers realized on a single chip [32,41,57,60]. The latter procedure is more suitable for large-scale biofunctionalization of transducers.

3.3 Elimination of Nonspecific Binding

Nonspecific binding, i.e., the binding of molecules other than the ones specifically recognized from the immobilized onto to the sensor recognition molecule, can seriously affect the transducer performance and is one of the main reasons that analyte detection in real samples is often not as successful as detection in a buffer solution. To maintain the nonspecific binding as low as possible, a blocking step is implemented after the attachment of the recognition element onto the transducer surface. The blocking agents commonly employed are the same ones used in standard microtiter plate immunoassays and mainly include nonreactive proteins, such as bovine serum albumin (BSA), gelatine, and casein. [41,46,48,57,70], short chain hydrophilic polymers such as poly(ethyleneglycol) (PEG) or Tween 20 [67,68], or commercially available blocking buffers such as SuperBlock [36]. Transducer surfaces that have been chemically activated with an agent such as GOPTS can be chemically blocked using small molecular weight amine-containing compounds, such as ethanolamine [10,15,49,50]. Alternative approaches include coverage of the transducer with a nonfouling polymer, such as poly(carboxybetaine methacrylate), which also offers functional groups for the attachment of recognition biomolecules. Such a polymer, modified with 3,4-dihydroxy-L-phenylalanine (DOPA) moieties [73], has been employed for the functionalization of a silicon oxide–based SPR sensor but not in interferometric sensors. Thus, new strategies and materials that could reduce nonspecific binding in interferometric immunosensors are still in demand.

4. APPLICATION OF INTERFEROMETRIC IMMUNOSENSORS

Immunochemical reactions have been implemented to evaluate the analytical performance of interferometric sensors since the first attempts to evaluate them as potential candidates for diagnostics. Nowadays, and about 30 years after the development of the first interferometric immunosensor, the interferometric transducers that have been used for immunosensing schemes include MZI, YI, and bimodal interferometer. From the application point of view, most reports target at the determination of analytes related to clinical diagnostics, environmental monitoring, or food safety. In the following section, the immunosensing applications of each type of interferometric transducers will be presented.

4.1 Mach–Zehnder Interferometers

MZIs are the most developed and widely exploited category of interferometric transducers. The waveguide structures are made almost exclusively from silicon nitride (Si_3N_4) or silicon oxynitride (SiON), while the cladding layers are composed of silicon dioxide (SiO_2), allowing both immobilization of recognition molecules via physical adsorption or covalent coupling. Thus, the 1991 interferometric transducer developed by Heideman et al. [58] was evaluated with respect to the detection of human chorionic gonadotropin (hCG) at a concentration of 25 nM after modification with an antibody against the specific hormone. An improved version of the same sensor [4] allowed the detection of hCG binding to the anti-hCG antibody–modified transducer from a solution of 50 pM within 20 min. A few years later, Brosinger et al. employed an MZI-based on rib SiON waveguides [65] for the determination of 2,4-dichlorophenoxyacetic acid (2,4-D) at a concentration as low as 1 μg/mL following a competitive immunoassay format. The same year, Schipper et al. using thinner waveguides made of Si_3N_4 demonstrated the detection of hCG using transducers modified with an anti-hCG antibody at a concentration of 30 nM. Similarly, Drapp et al. using a 3 × 3 coupler at the MZI output [45] achieved an improvement of signal-to-noise ratio by a factor of 10. The sensor was used for simazine detection with a detection limit of 0.1 ng/mL. At the beginning of the new millennium, research efforts focused on the improvement of sensitivity of MZI sensors fabricated with silicon technology. Thus, Prieto et al. fabricated MZIs based on antiresonant reflecting optical waveguides (ARROW-based MZIs), which were evaluated as immunosensors through the detection of the insecticide carbaryl following a competitive immunoassay format [74]. Optimization of MZIs' performance was also attempted through redesign of the divisors that split and recombine the two traveling beams to minimize the optical transmission losses. Thus, implementing MZIs with optimized S-bends, Sarkar et al. detected *L. monocytogenes* at a concentration of 2.8×10^5 CFU/mL, which is well below the infective dosage for humans and the LOD achieved by most of the alternative detection methods [10]. At the same time, the concentration that provided maximum plateau signals was 2.8×10^{10} CFU/mL, providing for a dynamic range of 5 orders of magnitude. Asymmetric MZIs, i.e., MZIs with small path differences between the two arms, have been employed for the immunochemical detection of aflatoxin M1 (AFM1) in milk samples [75] via modification of the transducers with the $F(ab')_2$ fragments of a rabbit polyclonal antibody against AFM1. According to the operation principle of the asymmetric MZIs, the difference in path length between the two arms is expected to lead to a wavelength dependency of the output phase. Thus, by scanning the asymmetric MZI over a small bandwidth, a phase plot is obtained from which the phase shift due to biomolecular interactions can be monitored with high accuracy. The sensor provided an LOD for AFM1 of

3 μg/L, which is well above the maximum allowable limit set by EU (0.05 μg/mL). Thus preconcentration of the sample might be necessary.

All the aforementioned devices are fabricated on silicon materials by CMOS-compatible technology offering the advantage of reproducible mass production. Nevertheless, polymers have also been exploited for the fabrication of MZIs due to their lower cost compared with silicon-based ones, especially when single-use, disposable devices are desired. The polymer material for fabrication of the waveguides must have the highest possible refractive index difference from the cladding layer and be compatible with the available laser sources (either in visible or near infrared). These requirements are not so easy to fulfill with polymeric materials, and as a result successful polymeric MZIs have been realized almost exclusively on the well-known resist SU-8, with a few reports involving polyimide as the material of choice. From the so-far-published, all-plastic interferometric transducers [37,48,76,77], only a few of them have been tested as potential immunosensors. Regarding the detection of immunochemical interactions, the polyimide-based MZI transducer reported by Melnik et al. was applied for the determination of human IgG after modification with streptavidin for indirect immobilization of a biotinylated antihuman IgG antibody. Direct binding provided a detection limit of 3.1 nM, which was reduced to 30 pM when a second antihuman IgG antibody was used for signal enhancement [78]. This LOD makes the sensor appropriate for diagnostic applications and supports the implementation of plastic MZIs as low-cost, disposable immunosensors.

Most silicon- and polymer-based MZIs reported in the literature are single-wavelength devices, i.e., they use external lasers as light sources. However, single-wavelength interferometric biosensors have several limitations, as it was mentioned in the previous sections. Therefore, alternative solutions were sought out and transducers based on spectral interferometry were introduced after c.2000 [19–21,79]. Among them, the most revolutionary approach was that developed and exploited by the NCSR "Demokritos" group [21,23,60,79], which implements LEDs and silicon nitride planar waveguides monolithically integrated and optically coupled on the same silicon substrate. This integration of the active and passive optical components was achieved by using mainstream planar technology and self-alignment techniques, which enabled efficient light coupling between the LED and the silicon nitride waveguide [60]. The LEDs are silicon avalanche diodes that emit a broadband spectrum covering all the visible and near-infrared range and then biased beyond their breakdown point. The polychromatic nature of the integrated LED results in a new detection concept, that is, broadband Mach–Zehnder interferometry (BB-MZI), which is highly advantageous compared with single-wavelength MZI transducers, as explained in the previous section. Although two different transducer configurations were tested, the first including an integrated photodetector and termed as "fully integrated" and the second implementing an external spectrometer for output spectrum recording and termed as "semiintegrated," the latter was selected for applications in immunochemical detection of analytes owing to its higher sensitivity. In particular, the fully integrated sensor provided detection limits for the simultaneous detection of antimouse IgG and streptavidin of 10 nM and 1 nM, respectively[60], whereas the semiintegrated device provided, for the same binding assays, detection limits of 32 and 5 pM, respectively [46]. Thus, the analytical capabilities of the semiintegrated device were further demonstrated through the detection of bovine milk in goat milk within 10 min at levels as low as 0.04% (v/v) and with a dynamic range that varied between 0.1 and 1.0% (v/v), showing the potential of the device for application to goat milk adulteration control [41]. The device was also applied for the detection of the mycotoxin ochratoxin A in beer samples with a detection limit of 2.0 ng/mL and a dynamic range 4.0–100 ng/mL, demonstrating its versatility and great potential for food analysis [57]. In parallel, C-reactive protein (CRP), an inflammation marker, was determined in human serum samples within few minutes with a detection limit of 2.1 ng/mL and a quantification limit of 4.2 ng/mL [46]. Moreover, the CRP concentration values determined in human serum samples were in excellent agreement with those determined for the same samples by a standard clinical laboratory method run in a clinical analyzer, showing the applicability of this particular interferometric immunosensor in the clinical diagnostics field and potential PoC applications.

A system based on MZIs (MiniLab) and commercialized by Optiqua Technologies Pte Ltd. has been used for the detection of Bisphenol A, an endocrine disruptor, in drinking water samples implementing a competitive immunoassay [69]. The immunosensor's working range extended from 0.5 to 5 ng/mL. The sensor could be regenerated and reused for more than 200 times in a period of 6 months. For the analysis of water samples, a cutoff corresponding to 50% of the normalized sensor response was adopted as the limit between positive and negative samples. The high number of research references as well as the existence of commercially available systems is an indication that the MZI-based sensing systems are mature enough for "real-world" applications.

4.2 Young Interferometers

YIs have evolved almost in parallel with MZIs [80], but their development was not so widespread, despite the fact that they have a simpler readout than MZIs and are less sensitive to temperature and wavelength drifts. Brandenburg et al. have developed two different types of YI, the first based on silicon nitride strip waveguides fabricated on a silicon oxide substrate [81] and a second one based on monomodal Ta_2O_5 waveguides [9]. The first transducer was evaluated using model binding assays, biotin/streptavidin and protein G/IgG, with a detection limit of 390 fM (58.5 fg/mL) for the binding of IgG onto protein G. Furthermore, it was employed to detect serum proteins after modification with antibodies as well as pharmaceutical substances (methotrexate) in serum [81] at μg/mL concentration ranges. The second transducer was used to monitor the production of a recombinant protein in cell lysates [35] and to detect tuberculosis-specific antibodies in serum samples from infected patients [82]. For the first application, a detection limit of 160 ng/mL was reported, a value well below the required concentration of 1 μg/mL. In addition, this detection limit was significantly lower than that determined by an ELISA method or by a commercially available SPR biosensor. Regarding the second application, the performance of the YI was compared with a grating coupler and a commercially available reflectometric interference spectroscopy (RIfS) sensor, and its superior performance with respect to two other sensors was demonstrated.

Another research group fabricated and evaluated a four-channel YI that allowed separate measurements in each sensing area by the application of an appropriate fluidic module [83,84]. The particular transducer was assessed as an immunosensor for the detection of human serum albumin (HSA) after modification with an anti-HSA antibody providing a detection limit of 20 fg/mm^2 in terms of protein mass coverage, which was based on the response obtained by a 50 μg/mL HSA solution and the noise level. The particular transducer was also applied for the detection of herpes simplex virus type 1 (HSV-1) after modification with a specific antibody, and a detection limit of about 850 particles/mL was achieved in the buffer but the detection limit was higher when the analysis was performed in serum [85].

As with MZIs, most YIs are using single-wavelength light sources. Nonetheless, oligochromatic or polychromatic sources have also been suggested [24], but these setups have not been evaluated by analysis of real samples. The same applies for polymer-based YIs [86], which have been assessed with respect to refractive index sensitivity but not in terms of detection of biomolecular interactions and especially immunochemical reactions. The only polychromatic YI reported so far, but not strictly applied as such, was the BB-YI suggested by the Demokritos group [27]. This BB-YI had its own monolithically integrated broadband source and a broadband output spectrum that spun in the VIS-NIR part of the spectrum. However, only a selected narrower band of the output spectrum with respect to the wide emission spectrum of the integrated source was monitored and analyzed through the use of an appropriate interference filter. The BB-YI under this "quasi-polychromatic" format was evaluated as a potential immunosensor through the sequential determination of biotin-streptavidin and rabbit IgG—antirabbit IgG binding. Concentrations of 1 nM streptavidin and 10 nM anti-IgG were easily detected through peak shifts of more than one period, suggesting that the analytical capabilities of the BB-YI may well extend down to the sub-nM range rendering BB-YIs as potential competitive biosensors for both PoN and PoC applications.

4.3 Bimodal Interferometers

Bimodal interferometers are based on an entirely different design than the previously mentioned transducers [14]. They are also fabricated by standard microelectronic processes on silicon wafers, but they consist of a single silicon nitride waveguide that is patterned so as to have two different zones: the first one with single-mode behavior and the second one supporting two modes (zero- and first-order modes) propagating at different velocities depending on the refractive index of the silicon dioxide cladding layer. The setup employs monochromatic light sources and microscope objective lenses for coupling of light into the waveguide. A two-sectional photodiode is used to collect the interference pattern that changes when the refractive index at the sensing area changes as a result of a biomolecular interaction. An updated version of the sensor has been applied for the immunodetection of human thyroid stimulating hormone (hTSH) via an indirect immunoassay demonstrating a detection limit of 20 pM, which is within the clinically relevant range for testing of thyroid gland status [15]. The transducer was also evaluated with respect to direct bacteria detection. After its modification with specific antibodies, the detection of *Bacillus cereus* to a concentration down to 12 CFU/mL and of *Escherichia coli* down to 4 CFU/mL was achieved [87]. In general, the performance of this type of transducer is comparable to that of other interferometric transducers and, in particular, single-wavelength MZIs.

5. CONCLUSIONS AND FUTURE PERSPECTIVES

Label-free interferometric immunosensors have emerged as a powerful tool for a plethora of analytical applications in a variety of sectors, such as healthcare, food safety, and environmental monitoring. Their analytical performance—especially in terms of sensitivity and LODs—has already been proven to be capable of satisfying the needs of all these sensitive sectors related to our well-being, health, and safety. Furthermore, their fabrication procedures are in par with standard fabrication techniques applicable to large-scale production at a reasonable cost, allowing for their future development into analytical systems, especially for PoN/PoC applications. Finally, one of their most important attributes and competitive characteristics is their design versatility enabled not only by the flexibility of the available transducer architectures and principles of operation but equally by the versatility of surface activation strategies.

Nonetheless, despite their great potential, interferometric transducers are still to "escape" the strict confines of laboratory settings or simulated environments and find the means to evolve into viable products applied on the spot and/or actual settings. This bottleneck, which has been identified by all technology providers of label-free interferometric immunosensors, is mostly related to the systemic development and not the actual performance of the sensors per se. The main hurdles and open issues still faced are the appropriate strategies to render interferometric immunosensors into truly compact and moveable systems to become realistic solutions for PoN/PoC applications and the sample preparation steps, which have to be simplified per case avoiding complex and time-consuming steps.

Provided that the systemic challenges are addressed and solved, which is definitely the trend and focus in today's research on interferometric immunosensing, there is a great future for this highly sensitive and versatile class of optical biosensors. They will provide viable solutions for a broad range of sectors and will evolve into powerful tools for analytical determinations.

REFERENCES

[1] IUPAC Definition for Biosensor "A Device that Uses Specific Biochemical Reactions Mediated by Isolated Enzymes, Immunosystems, Tissues, Organelles or Whole Cells to Detect Chemical Compounds Usually by Electrical, Thermal or Optical Signals." https://goldbook.iupac.org/B00663.html.

[2] A. Brecht, J. Ingenhoff, G. Gauglitz, Direct monitoring of antigen-antibody interactions by spectral interferometry, Sens. Actuators B Chem. 6 (1992) 96–100.

[3] B. Liedberg, I. Lundstrom, E. Stenberg, Principles of biosensing with an extended coupling matrix and SPR, Sens. Actuators B Chem. 11 (1993) 63–72.

[4] R.G. Heideman, R.P.H. Kooyman, J. Greve, Performance of a highly sensitive optical waveguide Mach-Zehnder interferometer immunosensor, Sens. Actuators B Chem. 10 (1993) 209–217.

[5] P. Skládal, Piezoelectric biosensors, TrAC Trends Anal. Chem. 79 (2016) 127–133.

[6] G. Marrazza, Piezoelectric biosensors for organophosphate and carbamate pesticides: a review, Biosensors 4 (2014) 301–317.

[7] L.C. Clark Jr., C. Lyons, Electrode systems for continuous monitoring in cardiovascular surgery, Ann. NY Acad. Sci. 102 (1962) 29–45.

[8] D. Grieshaber, R. MacKenzie, J. Vörös, E. Reimhult, Electrochemical biosensors - sensor principles and architectures, Sensors 8 (2008) 1400–1458.

[9] K. Schmitt, B. Schirmer, C. Hoffmann, A. Brandenburg, P. Meyrueis, Interferometric biosensor based on planar optical waveguide sensor chips for label-free detection of surface bound bioreactions, Biosens. Bioelectron. 22 (2007) 2591–2597.

[10] D. Sarkar, N. Siva Kumar Gunda, I. Jamal, S.K. Mitra, Optical biosensors with an integrated Mach-Zehnder Interferometer for detection of *Listeria monocytogenes*, Biomed. Microdev. 16 (2014) 509–520.

[11] T. Young, Philos. Trans. R. Soc. Lond. 94 (1804) 1–16.

[12] L. Mach, Uber einen interferenzrefraktor. Zeitschrift für Instrumentenkunde, Z. für Instrumentenkd. 12 (1892) 89–93.

[13] L. Zehnder, Ein neuer interferenzrefraktor, Z. für Instrumentenkd. 11 (1891) 275–285.

[14] K.E. Zinoviev, A.B. González-Guerrero, C. Domínguez, L.M. Lechuga, Integrated bimodal waveguide interferometric biosensor for label-free analysis, J. Light. Technol. 29 (2011) 1926–1932.

[15] D. Duval, A.B. Gonzalez-Guerrero, S. Dante, J. Osmond, R. Monge, L.J. Fernandez, K.E. Zinoviev, C. Domınguez, L.M. Lechuga, Nanophotonic lab-on-a-chip platforms including novel bimodal interferometers, microfluidics and grating couplers, Lab Chip 12 (2012) 1987–1994.

[16] M.J. Swann, L.L. Peel, S. Carrington, N.L. Freeman, Dual-polarization interferometry: an analytical technique to measure changes in protein structure in real time, to determine the stoichiometry of binding events, and to differentiate between specific and nonspecific interactions, Anal. Biochem. 329 (2004) 190–198.

[17] J. Escorihuela, M.Á. González-Martínez, J.L. López-Paz, R. Puchades, Á. Maquieira, D. Gimenez-Romero, Dual-polarization interferometry: a novel technique to light up the nanomolecular world, Chem. Rev. 115 (2015) 265–294.

[18] E. Makarona, P. Petrou, S. Kakabakos, K. Misiakos, I. Raptis, Point-of-Need bioanalytics based on planar optical interferometry, Biotechnol. Adv. 34 (2016) 209–233.

[19] D.H. Luo, R.A. Levy, Y.F. Hor, J.F. Federici, R.M. Pafchek, An integrated photonic sensor for in situ monitoring of hazardous organics, Sens. Actuators B Chem. 92 (2003) 121–126.

[20] Y. Li, E. Harris, L. Chen, X. Bao, Application of spectrum differential integration method in an in-line fiber Mach-Zehnder refractive index sensor, Opt. Express 18 (2010) 8135–8143.

[21] M. Kitsara, M. Misiakos, I. Raptis, E. Makarona, Integrated optical frequency-resolved Mach-Zehnder interferometers for label-free affinity sensing, Opt. Express 18 (2010) 8193–8206.

[22] K. Misiakos, S.E. Kakabakos, P.S. Petro, H.H. Ruf, A monolithic silicon optoelectronic transducer as a real-time affinity biosensor, Anal. Chem. 76 (2004) 1366–1373.

[23] K. Misiakos, I. Raptis, A. Salapatas, E. Makarona, A. Botsialas, M. Hoekman, R. Stoffer, G. Jobst, Broad-band Mach-Zehnder interferometers as high performance refractive index sensors: theory and monolithic implementation, Opt. Express 22 (2014) 8856–8870.

[24] H.K.P. Mulder, A. Ymeti, V. Subramaniam, J.S. Kanger, Size-selective detection in integrated optical interferometric biosensors, Opt. Express 20 (2012) 20934–20950.

[25] C. Stamm, R. Dangel, W. Lukosz, Biosensing with the integrated-optical difference interferometer: dual wavelength operation, Opt. Commun. 153 (1998) 347–359.

[26] P.M. Nellen, W. Lukosz, Integrated optical input grating couplers as direct affinity sensors, Biosens. Bioelectron. 8 (1993) 129–147.

[27] E. Savra, A. Malainou, A. Salapatas, A. Botsialas, P. Petrou, I. Raptis, E. Makarona, S.E. Kakabakos, K. Misiakos, Monolithically-integrated Young interferometers for label-free and multiplexed detection of biomolecules, Proc. SPIE 9752 (2016) 97520N.

[28] M.-J. Banuls, R. Puchades, A. Maquieira, Modifications development of silicon based label free integrated optical (IO) biosensors: a review, Anal. Chim. Acta 777 (2013) 1–16.

[29] E. Mauriz, M.C. García-Fernández, L.M. Lechuga, Towards the design of universal immunosurfaces for SPR-based assays: a review, TrAC Trend Anal. Chem. 79 (2016) 191–198.

[30] L. Wu, L. Cai, A. Liu, W. Wang, Y. Yuan, Z. Li, Self-assembled monolayers of perfluoroalkylsilane on plasma-hydroxylated silicon substrates, Appl. Surf. Sci. 349 (2015) 683–694.

[31] P. Kozma, F. Kehl, E. Ehrentreich-Förster, C. Stamm, F.F. Bier, Integrated planar optical waveguide interferometer biosensors: a comparative review, Biosens. Bioelectron. 58 (2014) 287–307.

[32] A. Densmore, M. Vachon, D.X. Xu, S. Janz, R. Ma, Y.H. Li, G. Lopinski, A. Delâge, J. Lapointe, C.C. Luebbert, Q.Y. Liu, P. Cheben, J.H. Schmid, Silicon photonic wire biosensor array for multiplexed real-time and label-free molecular detection, Opt. Lett. 34 (2009) 3598–3600.

[33] B.H. Schneider, E.L. Dickinson, M.D. Vach, J.V. Hoijer, L.V. Howard, Highly sensitive optical chip immunoassays in human serum, Biosens. Bioelectron. 15 (2000) 13–22.

[34] J. Hong, J.S. Choi, G. Han, J.K. Kang, C.M. Kim, T.S. Kim, D. Sung Yoon, A Mach-Zehnder interferometer based on silicon oxides for biosensor applications, Anal. Chim. Acta 573–574 (2006) 97–103.

[35] C. Hoffmann, K. Schmitt, A. Brandenburg, S. Hartmann, Rapid protein expression analysis with an interferometric biosensor for monitoring protein production, Anal. Bioanal. Chem. 387 (2007) 1921−1932.

[36] J. Xu, D. Suarez, D.S. Gottfried, Detection of avian influenza virus using an interferometric biosensor, Anal. Bioanal. Chem. 389 (2007) 1193−1199.

[37] E. Melnik, R. Bruck, P. Muellner, R. Hainberger, M. Lämmerhofer, Multi-step surface functionalization of polyimide based evanescent wave photonic biosensors and application for DNA hybridization by Mach-Zehnder interferometer, Anal. Chim. Acta 699 (2011) 206−215.

[38] K. Awsiuk, A. Bernasik, M. Kitsara, A. Budkowski, P. Petrou, S. Kakabakos, S. Prauzner-Bechcicki, J. Rysz, I. Raptis, Spectroscopic and microscopic characterization of biosensor surfaces with protein/amino-organosilane/silicon structure, Colloid Surf. B Biointerfaces 90 (2012) 159−168.

[39] K. Awsiuk, A. Budkowski, P. Petrou, A. Bernasik, M.M. Marzec, S. Kakabakos, J. Rysz, I. Raptis, Model immunoassay on silicon surfaces: vertical and lateral nanostructure vs. protein coverage, Colloid Surf. B Biointerfaces 103 (2013) 253−260.

[40] N. Massad-Ivanir, G. Shtenberg, A. Tzur, M.A. Krepker, E. Segal, Engineering nanostructured porous SiO2 surfaces for bacteria detection via "direct cell capture", Anal. Chem. 83 (2011) 3282−3289.

[41] M. Angelopoulou, A. Botsialas, A. Salapatas, P.S. Petrou, W. Haasnoot, E. Makarona, G. Jobst, D. Goustouridis, A. Siafaka-Kapadai, I. Raptis, K. Misiakos, S.E. Kakabakos, Assessment of goat milk adulteration with a label-free monolithically integrated optoelectronic biosensor, Anal. Bioanal. Chem. 407 (2015) 3995−4004.

[42] H.K. Hunt, C. Soteropulos, A.M. Armani, Bioconjugation strategies for microtoroidal optical resonators, Sensor 10 (2010) 9317−9336.

[43] S. Lee, S.C. Eom, J.S. Chang, C. Huh, G.Y. Sung, J.H. Shin, Label-free optical biosensing using a horizontal air-slot SiNx microdisk resonator, Opt. Express 18 (2010) 20638−20644.

[44] E.F. Schipper, A.J.H. Bergevoet, R.P.H. Kooyman, J. Greve, New detection method for atrazine pesticides with the optical waveguide Mach-Zehnder immunosensors, Anal. Chim. Acta 341 (1997) 171−176.

[45] B. Drapp, J. Piehler, A. Brecht, G. Gauglitz, B.J. Luff, J.S. Wilkinson, J. Ingenhoff, Integrated optical Mach-Zehnder interferometers as simazine immunoprobes, Sens. Actuators B Chem. 38−39 (1997) 277−282.

[46] A. Psarouli, A. Botsialas, A. Salapatas, G. Stefanitsis, D. Nikita, G. Jobst, N. Chaniotakis, D. Goustouridis, E. Makarona, P.S. Petrou, I. Raptis, K. Misiakos, S.E. Kakabakos, Fast label-free detection of C-reactive protein using broad-band Mach-Zehnder interferometers integrated on silicon chips, Talanta 165 (2017) 458−465.

[47] B. Sepúlveda, J.S. Río, M.F. Moreno, F.J. Blanco, K. Mayora, C. Domínguez, L.M. Lechuga, Optical biosensor microsystems based on the integration of highly sensitive Mach−Zehnder interferometer devices, J. Opt. A Pure Appl. Opt. 8 (2006) S561−S566.

[48] R. Bruck, E. Melnik, P. Muellner, R. Hainberger, M. Lämmerhofer, Integrated polymer-based Mach-Zehnder interferometer label-free streptavidin biosensor compatible with injection molding, Biosens. Bioelectron. 26 (2011) 3832−3837.

[49] S. Dante, D. Duval, B. Sepúlveda, A.B. González-Guerrero, J. Ramón Sendra, L.M. Lechuga, All-optical phase modulation for integrated interferometric biosensors, Opt. Express 20 (2012) 7195−7205.

[50] S. Dante, D. Duval, D. Farina, A. Gonzalez-Guerrero, L.M. Lechuga, Linear readout of integrated interferometric biosensors using a periodic wavelength modulation, Laser Phot. Rev. 9 (2015) 248−255.

[51] A. Arafat, M. Giesbers, M. Rosso, E.J.R. Sudhölter, K. Schroën, R.G. White, L. Yang, M.R. Linford, H. Zuilhof, Covalent biofunctionalization of silicon nitride surfaces, Langmuir 23 (2007) 6233–6244.

[52] L. Scheres, A. Arafat, H. Zuilhof, Self-assembly of high-quality covalently bound organic monolayers onto silicon, Langmuir 23 (2007) 8343–8346.

[53] P. Thissen, O. Seitz, Y.J. Chabal, Wet chemical surface functionalization of oxide-free silicon, Prog. Surf. Sci. 87 (2012) 272–290.

[54] T. Bocking, K.A. Kilian, K. Gaus, J.J. Gooding, Modifying porous silicon with self-assembled monolayers for biomedical applications: the influence of surface coverage on stability and biomolecule coupling, Adv. Funct. Mater. 18 (2008) 3827–3833.

[55] M.-J. Banuls, V. González-Pedro, C.A. Barrios, R. Puchades, A. Maquieira, Selective chemical modification of silicon nitride/silicon oxide nanostructures to develop label-free biosensors, Biosens. Bioelectron. 25 (2010) 1460–1466.

[56] C.F. Carlborg, K.B. Gylfason, A. Kazmierczak, F. Dortu, M.J. Banuls, A. Maquieira Catala, G.M. Kresbach, H. Sohlstrom, T. Moh, L. Vivien, J. Popplewell, G. Ronan, C.A. Barrios, G. Stemme, W. van der Wijngaart, A packaged optical slot-waveguide ring resonator sensor array for multiplex label-free assays in labs-on-chips, Lab Chip 10 (2010) 281–290.

[57] V. Pagkali, P.S. Petrou, A. Salapatas, E. Makarona, J. Peters, W. Haasnoot, G. Jobst, A. Economou, K. Misiakos, I. Raptis, S.E. Kakabakos, Detection of ochratoxin A in beer samples with a label-free monolithically integrated optoelectronic biosensor, J. Hazard Mater. 323 (2017) 75–83.

[58] R.G. Heideman, R.P.H. Kooyman, J. Greve, Development of an optical waveguide interferometric immunosensor, Sens. Actuators B Chem. 4 (1991) 297–299.

[59] A. Ymeti, J.S. Kanger, J. Greve, G.A.J. Besselink, P.V. Lambeck, R. Wijn, R.G. Heideman, Integration of microfluidics with a four-channel integrated optical Young interferometer immunosensors, Biosens. Bioelectron. 20 (2005) 1417–1421.

[60] K. Misiakos, I. Raptis, E. Makarona, A. Botsialas, A. Salapatas, P. Oikonomou, A. Psarouli, P.S. Petrou, S.E. Kakabakos, K. Tukkiniemi, M. Sopanen, G. Jobst, All-silicon monolithic Mach-Zehnder interferometer as a refractive index and bio-chemical sensor, Opt. Express 22 (2014) 26803–26813.

[61] A. Der, S. Valkai, A. Mathesz, I. Andó, E.K. Wolff, P. Ormos, Protein-based all-optical sensor device, Sens. Actuators B Chem. 151 (2010) 26–29.

[62] Q. Liu, X. Tu, K. Woo Kim, J. Sheng Kee, Y. Shin, K. Han, Y.-J. Yoon, G.-Q. Lo, M. Kyoung Park, Highly sensitive Mach–Zehnder interferometer biosensor based on silicon nitride slot waveguide, Sens. Actuators B Chem. 188 (2013) 681–688.

[63] Q. Liu, K. Woo Kim, Z. Gu, J. Sheng Kee, M. Kyoung Park, Single-channel Mach-Zehnder interferometric biochemical sensor based on two-lateral-mode spiral waveguide, Opt. Express 22 (2014) 27910–27920.

[64] Q. Liu, S.Y. Lim, R.A. Soo, M.K. Park, Y. Shin, A rapid MZI-IDA sensor system for EGFR mutation testing in non-small cell lung cancer (NSCLC), Biosens. Bioelectron. 74 (2015) 865–871.

[65] F. Brosinger, H. Freimuth, M. Lacher, W. Ehrfeld, E. Gedig, A. Katerkamp, F. Spener, K. Cammann, A label-free affinity sensor with compensation of unspecific protein interaction by a highly sensitive integrated optical Mach–Zehnder interferometer on silicon, Sens. Actuators B Chem. 44 (1997) 350–355.

[66] R.P. Gandhiraman, V. Gubala, L.C.H. Nam, C. Volcke, C. Doyle, B. James, S. Daniels, D.E. Williams, Deposition of chemically reactive and repellent sites on biosensor chips for reduced non-specific binding, Colloid Surf. B 79 (2010) 270–275.

[67] B.H. Schneider, J.G. Edwards, N.F. Hartman, Hartman interferometer: versatile integrated optic sensor for label-free, real-time quantification of nucleic acids, proteins, and pathogens, Clin. Chem. 43 (1997) 1757–1763.

[68] B.H. Schneider, E.L. Dickinson, M.D. Vach, J.V. Hoijer, L.V. Howard, Optical chip immunoassay for hCG in human whole blood, Biosens. Bioelectron. 15 (2000) 597–604.

[69] K. Ma, T. Ekblad, M. Klein Koerkamp, H. Kelderman, M. van Wijlen, A. Duarah, J. Yue, L. Zhang, M. Vee-Meng Wong, M. Hoo Lim, Contaminant detection in treated water using Optiqua's MiniLab biosensing system: a case study for Bisphenol A, Intern. J. Environ. Anal. Chem. 95 (2015) 366–378.

[70] X. Jiang, Y. Chen, F. Yu, L. Tang, M. Li, J.J. He, High-sensitivity optical biosensor based on cascaded Mach–Zehnder interferometer and ring resonator using Vernier effect, Opt. Lett. 39 (2014) 6363–6366.

[71] D. Schlatter, R. Barner, C. Fattinger, W. Huber, J. Hiibscher, J. Hurst, H. Keller, C. Mangold, F. Miiller, The difference interferometer: application as a direct affinity sensor, Biosens. Bioelectron. 8 (1993) 109–116.

[72] A. Brandenburg, R. Krauter, C. Kunzel, M. Stefan, H. Schulte, Interferometric sensor for detection of surface-bound bioreactions, Appl. Opt. 34 (2000) 6396–6405.

[73] Y. Gao, Q. Gan, Z. Xin, X. Cheng, F.J. Bartol, Plasmonic Mach-Zehnder interferometer for ultrasensitive on-chip biosensing, ACS Nano 5 (2011) 9836–9844.

[74] F. Prieto, B. Sepulveda, A. Calle, A. Llobera, C. Domınguez, L.M. Lechuga, Integrated Mach–Zehnder interferometer based on ARROW structures for biosensor applications, Sens. Actuators B Chem. 92 (2003) 151–158.

[75] T. Chalyan, R. Guider, L. Pasquardini, M. Zanetti, F. Falke, E. Schreuder, R.G. Heideman, C. Pederzolli, L. Pavesi, Asymmetric Mach–Zehnder interferometer based biosensors for aflatoxin M1 detection, Biosensor 6 (2016) 1–10.

[76] B.Y. Shew, Y.C. Cheng, Y.H. Tsai, Monolithic SU-8 micro-interferometer for biochemical detections, Sens. Actuators A Phys. 141 (2008) 299–306.

[77] G. Calo, A. Farinola, V. Petruzzelli, Design and optimization of high sensitivity photonic interferometric biosensors on polymeriv waveguides, Prog. Electromagn. Res. Lett. 33 (2012) 151–166.

[78] E. Melnik, R. Bruck, P. Müellner, T. Schlederer, R. Hainberger, M. Lämmerhofer, Human IgG detection in serum on polymer based Mach-Zehnder interferometric biosensors, J. Biophot. 9 (2016) 218–223.

[79] K. Misiakos, E. Makarona, M. Kitsara, I. Raptis, Monolithic silicon optocoupler engineering based on tapered waveguides, Microelectron. Eng. 85 (2008) 1074–1076.

[80] A. Brandenburg, R. Henninger, Integrated optical Young interferometer, Appl. Opt. 25 (1994) 5941–5947.

[81] E. Brynda, M. Houska, A. Brandenburg, A. Wikerstal, Optical biosensors for real-time measurement of analytes in blood plasma, Biosens. Bioelectron. 17 (2002) 665–675.

[82] T. Nagel, E. Ehrentreich-Forster, M. Singh, K. Schmitt, A. Brandenburg, A. Berka, F.F. Bier, Direct detection of tuberculosis infection in blood-serum using three optical label-free approaches, Sens. Actuators B Chem. 129 (2008) 934–940.

[83] A. Ymeti, J.S. Kanger, J. Greve, P.V. Lambeck, R. Wijn, R.G. Heideman, Realization of a multichannel integrated Young interferometer chemical sensor, Appl. Opt. 42 (2003) 5649–5660.

[84] A. Ymeti, J. Greve, P.V. Lambeck, R. Wijn, R.G. Heideman, J.S. Kanger, Drift correction in a multichannel integrated optical Young interferometer, Appl. Opt. 44 (2005) 3409–3412.

[85] A. Ymeti, J. Greve, P.V. Lambeck, T. Wink, S.W.F.M. van Hovell, T.A.M. Beumer, R.R. Wijn, R.G. Heideman, V. Subramaniam, J.S. Kanger, Fast, ultrasensitive virus detection using a Young interferometer sensor, Nano Lett. 7 (2007) 394–397.

[86] M. Hiltunen, J. Hiltunen, P. Stenberg, S. Aikio, L. Kurki, P. Vahimaa, P. Karioja, Polymeric slot waveguide interferometer for sensor applications, Opt. Express 22 (2014) 7229.

[87] J. Maldonado, A.B. González-Guerrero, C. Domínguez, L.M. Lechuga, Label-free bimodal waveguide immunosensor for rapid diagnosis of bacterial infections in cirrhotic patients, Biosens. Bioelectron. 85 (2016) 310–316.

Chapter 11

Nanomaterial- and Micromaterial-Based Immunoassays

Minli You, Zedong Li, Penghui Zhang, Dan Bai, Min Lin, Feng Xu
Xi'an Jiaotong University, Xi'an, P.R. China

1. INTRODUCTION

Since the first immunoassay for insulin was described by Solomon Berson and Rosalyn S. Yalow in 1959, technological evolutions in life science were always accompanied with the development of immunoassays. Especially in the 1970s, several important milestones were established: the first homogenous immunoassay was developed by Rubenstein et al. [1], the method of monoclonal antibody production was described by Kohler and Milstein [2], and chemiluminescence was introduced into immunoassay as nonisotopic label [3], whereas immunoassay technologies still need advancements in detection sensitivity and multiplex analysis in the 21st century [4]. The development of microscopic materials (e.g., nanomaterials and micromaterials) provides new opportunity for the immunoassay. Nano- and micromaterials are beyond the limit of human eye observation, but they provide many unique properties, such as high surface-to-volume ratio, colloidal and diffusion properties, and other unique features (e.g., optical, electronic, or magnetic properties). The advantages brought by nano- and micromaterials to the immunoassays can be summarized by the following points: (1) faster kinetics of antibody-antigen reaction, shortening assay time; (2) higher detection sensitivity, critical in detection of disease biomarker with low abundance; (3) capability of multianalyte detection, allowing particular biomarker panel detection and decreasing cost; (4) less required sample volume, enabling more tests or less sample taken from patients, which is extremely important for neonates.

Nanomaterials describe atoms or molecules that agglomerate together to form objects in at least one dimension between 1 and 100 nm. Therefore, nanomaterials can be differentiated based on the number of dimensions on the nanoscale; for example, two-dimensional nanomaterials are featured in one

Handbook of Immunoassay Technologies. https://doi.org/10.1016/B978-0-12-811762-0.00011-6

dimension (e.g., the thickness of its surfaces) on the nanoscale; nanotubes show two nanoscale dimensions to the cross section, while its length is allowed to elongate greater; three dimensions of nanoparticles all belong to the nanoscale. In many materials, nanoscale causes many interesting properties in term of electronics, magnetics, optics, and mechanics, resulting in intensive research. Like nanomaterials, micromaterials deal with objects with any one dimension ranging from 1 to 1000 μm, whose microscale size also results in some unique characters. Based on the main matrix, we can classify micromaterials as metallic, polymeric, ceramic, and so on. Immunoassays are established on the specific biorecognition between antigen (Ag) and antibody (Ab), aimed at measuring the number of one of them. Currently, the commonly used immunoassay in clinics is enzyme-linked immunosorbent assay (ELISA), which conducts in a microplate and uses enzyme-catalyzed color change to measure the analyte. The ELISA method permits up to hundreds of tests, and simultaneously conducts and finally quantifies them. But it is still limited by complicated operation, long reaction time, unsatisfactory sensitivity, high volumes of sample and reagent requirement, and the need for skillful operators. Although radioisotope ELISA achieves improved sensitivity, it comes with radiation problems. For solving these problems, nano- and micromaterials are introduced to immunoassays. This critical chapter focuses on recent developments of nano- and micromaterials applied in immunoassays for improving sensitivity, decreasing cost and time consumption, simplifying operation steps, and reducing the reaction reagents.

2. MICROMATERIAL-BASED IMMUNOASSAY

2.1 Fluorescent Polystyrene Microsphere

Polystyrene (PS) microsphere is the most widely used micromaterial in immunoassays, which generally ranges from 1 to 100 μm. The small size and sphere structure feature great reaction area and fast-binding kinetics closing to solution-phase kinetics. Encoded PS microspheres allow high throughput detections in the immunoassay within a single sample. More accurate barcodes provide the possibility of detecting more target analytes at the same time. To this end, various encoding schemes have been utilized, such as spectrometric encoding [5,6], graphical encoding [7,8], chemical encoding [9,10], and physical encoding [11]. Additionally, the combination of different encoding schemes also can further increase the barcode number, but with a more complicated fabrication process and decode manner. Among these encoding technologies, graphical encoding relies on sophisticated equipment to synthesize and read out the encoded structure [8]. Chemical encoding suffers from high cost, and a complex and time-consuming decoding process. Physical encoding (e.g., size and refractive index) is limited by encoding capacity, which usually requires additional encoding dimensions of modern flow cytometers [12,13]. In comparison, spectrometric encoding possesses great encoding capacity, a simple fabrication method, and high decoding speed. Therefore, it is the most widely used encoding method, such as Luminex xMAP system, which has been widely applied in various clinical automatic biochemical analyzers. The Luminex xMAP system utilizes about 5.6 μm PS microspheres internally encoded with various ratios of two organic dyes with distinct emitting spectra. To precisely control amounts of these organic dyes, a library of barcodes is created consisting of more than 100 different microspheres with unique spectral emission. The surface of each microsphere can be modified with various recognition elements (e.g., nucleic acid, antibody). Because the microspheres can be identified with their unique emission, nearly 100 different targets can be measured simultaneously in a sample test, as shown in Fig. 11.1A. However, organic dye encoded PS microspheres suffer from several drawbacks [14]; for instance, multiple excitation lasers are generally needed for decoding organic days embedded in xMAP spheres with different excitation wavelengths, which increases the cost [15]. Additionally, organic dyes suffer from low photobleaching threshold and inevitable signal interference.

To solve these problems, various newly developed fluorescent nanoparticles are used to replace the organic dyes to embed in PS microspheres. Among them, quantum dots (QDs) [17,18], upconversion nanoparticles (UCNPs) [19,20], and noble metal particles [21] are most widespread. These fluorescent nanoparticles all possess a superior photostability and a wide range of emission wavelengths upon single excitation light. Therefore, the combination of them in preparing PS microspheres can generate far more barcodes than the Luminex xMAP system. QDs feature narrow-band and size-tunable emissions upon the same excitation wavelength, thus allowing higher encoding capacities (Fig. 11.1B and C). UCNPs are excited by near infrared light and also able to emit multiple color light [22]. Moreover, it can minimize the interference of the background fluorescence [23] and its tunable fluorescence lifetime could be another possible encoding dimension [24]. Noble metal particle-based surface-enhanced Raman scattering signals also can be utilized in encoding PS microspheres in order to realize sensitive and multiplex detection [25]. Spectrometric encoding PS microspheres not only can provide high array density, but also feature good repeatability, and high throughput and sensitivity. Therefore, the PS microspheres that involved new functional particles bring significant improvement, contributing to high throughput analysis.

2.2 Magnetic Microbeads

Magnetic microbeads (MMBs) used in immunoassays are generally coated with an organic shell to produce a core-shell structure. The common coating materials involve polystyrene, silica, dextran, and albumin, which could prevent magnetic aggregation and contribute to a final diameter size above 1 μm,

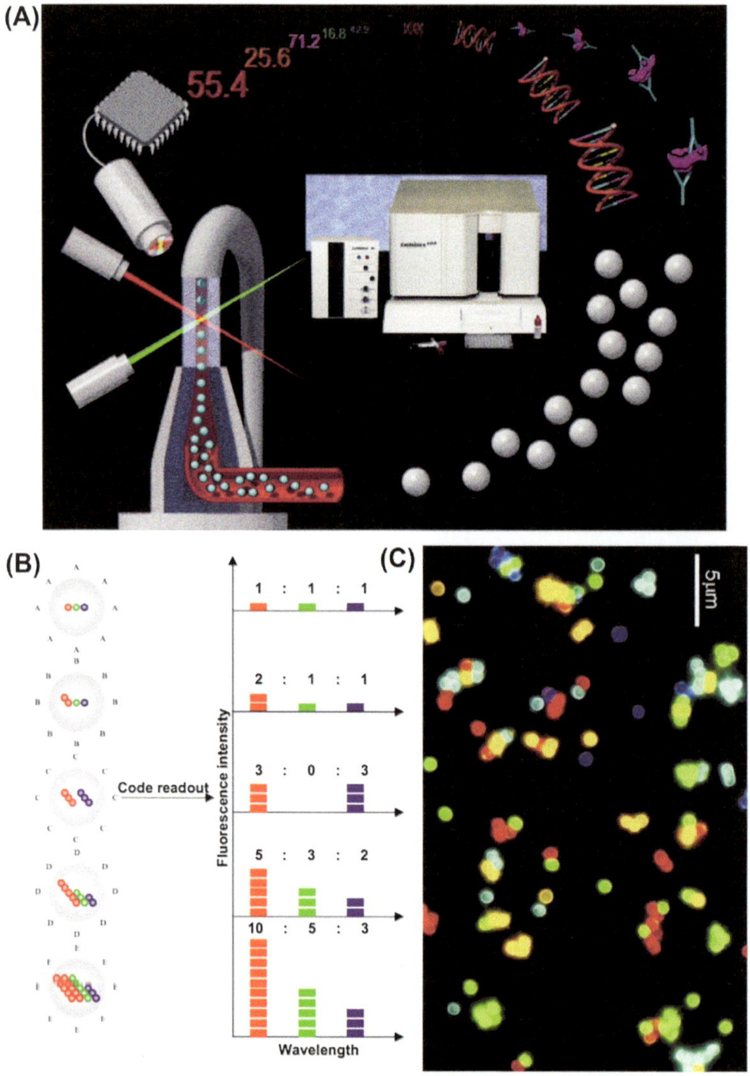

FIGURE 11.1 Fluorophore-encoded polystyrene microbeads. (A) Luminex xMAP system and components. (B) Schematic illustration and (C) fluorescence micrograph of multiplexing QDs encoded polystyrene microspheres [16,17].

which is easier to recover by the magnetism and overcome viscous resistance in solution. The active ligands outside the MMBs allow the further conjugation of biorecognition ligands. Various MMBs with different sizes, surrounding matrix, magnetic magnitude, or material composition have been commercialized now. The wide applications of MMBs in immunoassays benefit from huge surface area, good dispersibility, and magnetic controllability, which promotes the improvement of sensitivity and reaction rate and simplifies the sample pretreatment and purification steps, which are normally accompanied by traditional immunoassay methods [26].

In most related immunoassays, MMBs are mainly used as reaction support and to isolate target molecules. Compared with traditional immunoassay, the utilization of MMBs improves reaction efficiency and is able to isolate and enrich signal molecules. Generally, specific antibodies are preconjugated onto the MMBs to capture target molecules in a complex sample. Under guidance of magnetic field, MMBs could be concentrated on the surface of the reaction chamber to perform the following detection after several washing steps. According to the final signal detection modes, MMB-based immunoassays could be classified as fluorescent detection, electrochemical detection, colorimetric detection, and direct measurement of MMBs.

Fluorescent detection uses fluorophore to perform the signal transfer from biomolecular information to photonic signal. They are often conjugated with second antibodies and bond with target molecular and MMBs to form a sandwich structure, Fig. 11.2A. Commonly used fluorophores in this method [27,30,31], such as QDs, organic dyes, carbon dots, and UCNPs, offer higher sensitivity by virtue of fluorescence. Additionally, a few enzyme-based fluorescence immunoassays are also used with MMBs, similar to ELISA by just replacing the microtiter plate by MMBs as reaction support. Electrochemical reaction is the most utilized detection method in MMB-based immunoassay [32]. In this method, the magnetic field significantly improves the MMBs' concentration on the electrode surface, subsequently attracting lots of enzymes as tracers enriched on the electrode surface and resulting in a high electrochemical reaction, as shown in Fig. 11.2B. The use of MMBs removes tedious modification steps of recognition elements onto solid surface and provides the opportunity of electrode regeneration for electrochemical sensors. Last, some direct detection methods of MMBs are also developed on the basis of their optical and magnetic properties. The optical method measures the covered fraction of the surface by magnetic particles, and then calculates the amount of target analyte, Fig. 11.2C [33,34]. The magnetic method is to sense the variations of the magnetic field aroused by MMBs located on the sensor surface, Fig. 11.2D [29,35]. The magnetic method is very interesting; it allows magnetic sensors integrated into the chips without any interference to the detection signal. However, nonspecific surface adsorption of the MMBs is hard to totally avoid, which will cause undeserved background signal.

2.3 Nanomaterial-Based Immunoassay

With the enormous improvement of material science and nanotechnology, various nanomaterials can be synthesized to a designed structure with different superior properties, and delicate modification methods are developed to conjugate nanomaterials with various biomolecules. Utilizing different transforming manners

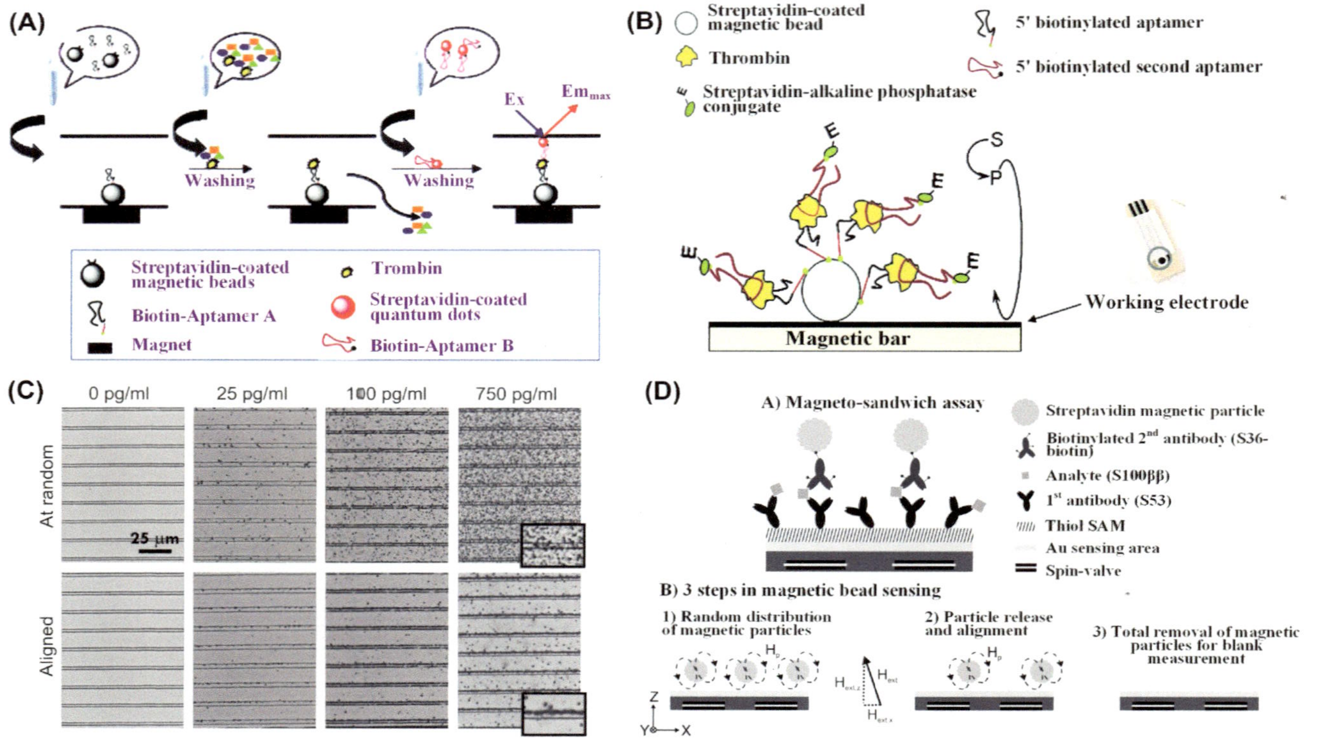

FIGURE 11.2 Magnetic microbead (MMB)-based immunoassays. (A) Quantum dot–assisted sandwich assay for thrombin detection employing MMBs. (B) Electrochemical sandwich assay coupled to MMBs. (C) Optical micrographs of MMBs on the sensor surface. (D) Magnetic sensor of MMB for protein detection [27–29].

(e.g., colorimetric, electrochemical, fluorescent, and magnetic techniques) of detection signal, the bioconjugated nanomaterials have remarkably boosted immunoassay performance, resulting in less reaction time, higher sensitivity and specificity, lower cost, and greater multiplexing capability. Based on the transducing techniques, nanomaterial-based immunoassays are classified into three categories (i.e., colorimetric, electrochemical, and fluorescent immunoassays); we focus on representing the different strategies of improving sensitivity, simplifying operation, and accelerating reaction for immunoassays.

3. COLORIMETRIC IMMUNOASSAY

3.1 Lateral Flow Assay

Colloidal suspension of gold nanoparticles (AuNPs) in a liquid, usually water, shows different color as different particle sizes (e.g., red color for <100 nm particles, blue color for >100 nm particles or particle aggregates). Therefore, AuNPs could be directly used as a colorimetric indictor in immunoassay, especially point-of-care (POC) applications. Lateral flow assay (LFA) is a simple device used to detect the presence of a target analyte in sample. Unlike ELISA, which needs multiple washing steps, the LFA performs a sample-to-answer test and removes the need for operation by professional persons and with specialized costly equipment. A successful and well-known commercial product is the home pregnancy test.

The principle of LFA is easy to understand. The device is composed of five components: sample pad, conjugate pad, nitrocellulose (NC) membrane, absorbent pad, and backing pad. The first four components are stacked orderly on the backing pad to form a table strip. The sample pad acts as a sponge and absorbs the to-be-detected sample. When the sample pad is totally wet, the sample fluid migrates to the conjugate pad, where AuNPs modified with specific antibodies were embedded in a salt-sugar matrix. In this position, the target antigens in sample fluid begin to react with the specific antibodies on the surface of AuNPs after the salt-sugar matrix is redissolved in sample fluid. Afterward, the complex of AuNP-antigens flows to the NC membrane with sample fluid driven by capillary force. Generally, there are two detection zones in NC membrane, namely test line and control line. Primary antibodies specific for target antigens are preembedded in the test line. When AuNP-antigen complex flows through a test line, it will be trapped by the antibodies on the test line and form a sandwich structure, while the AuNPs without binding with antigens could flow smoothly through the test line. The control line positioned behind the test line is used to capture the unbinding AuNPs by secondary antibodies. The final absorbent pad is used to store the reaction test fluid. Therefore, a normal positive detection of LFA will displays two red lines (i.e., test line and control line) and negative detection shows only one red line (i.e., control line). Here, the control line is used to confirm that the AuNP probes are excess, meaning the test line signal is valid. Based on this, the optical density of the test line could qualify or semiquantify the concentration of target antigens. This standard LFA has been used for various target detections, such as infection diseases [36], cardiovascular diseases [37], food safety [38], environment monitoring [39], antiterrorism [40], and so on.

Despite the advantages of equipment-free, operator-friendly, and naked-eye readout, LFA is not as sensitive as ELISA. For solving this problem, intensive efforts have been extended into the signal amplification of AuNP-based LFA [38,42,44]. A remarkable amplification strategy for AuNP-based LFA is silver enhancement. In this strategy, a silver enhancer and sodium thiosulfate are used to soak the test line of AuNP-based LFA to enhance the signal, Fig. 11.3A [41]. In a recent study, silver enhancement reaches two orders of magnitude improvement in sensitivity for ricin detection, from 50 ng/mL to 100 pg/mL [45]. However, this approach additionally asks for a liquid operation for silver enhancement, losing the superiority of the LFA method as a rapid one-step test. Therefore, another significant amplification strategy is designed by utilizing a second, larger AuNP to couple with the first one trapped in the test line, Fig. 11.3B [42]. This strategy is used to sensitively detect troponin I, whose clinical cutoff level is low to pg/mL.

To enable the naked-eye assay for troponin I, two conjugate pads are designed to hold first AuNP probes and second AuNP probes, respectively. The first AuNP probes are conjugated with antitroponin I antibody and blocked with bovine serum albumin (BSA). The second AuNP probes are conjugated with anti-BSA antibody and blocked with human serum albumin. When analyte solution is dropped into the sample pad and flows through second conjugate pad, the second AuNP probes will migrate with the fluid. When it arrives at the first conjugate pad, the second AuNP probes will react with the first AuNP probes and the latter will simultaneously react with analytes to form the complex, which subsequently binds with antibodies embedded in test line. In this design, the different combinations of two AuNPs with different sizes were systemically studied to improve their sensitivity. With the optimized combination, 100-fold increase of detection limit is achieved compared with traditional LFA, reaching a naked-eye detection with a limit as low as 0.01 ng/mL for troponin I in serum samples. Afterward, this strategy is also used for highly sensitive detection of nucleic acid [46] and *E. coli* [38]. Furthermore, the flow speed in LFA also causes an important effect in the test result. A reduced flow speed of reagent solution has been verified to be positive for improving sensitivity (Fig. 11.3C) [43,47], while causing a risk of specificity decrease.

In addition, the existing AuNP-based LFAs have a considerable drawback; they can only be used for qualitative (positive or negative) analysis and are unable to quantify the target concentrations. However, most cases in clinical investigation demand concentration information from specimens. To satisfy this need, multiple dots or lines in the detection zone are designed. In

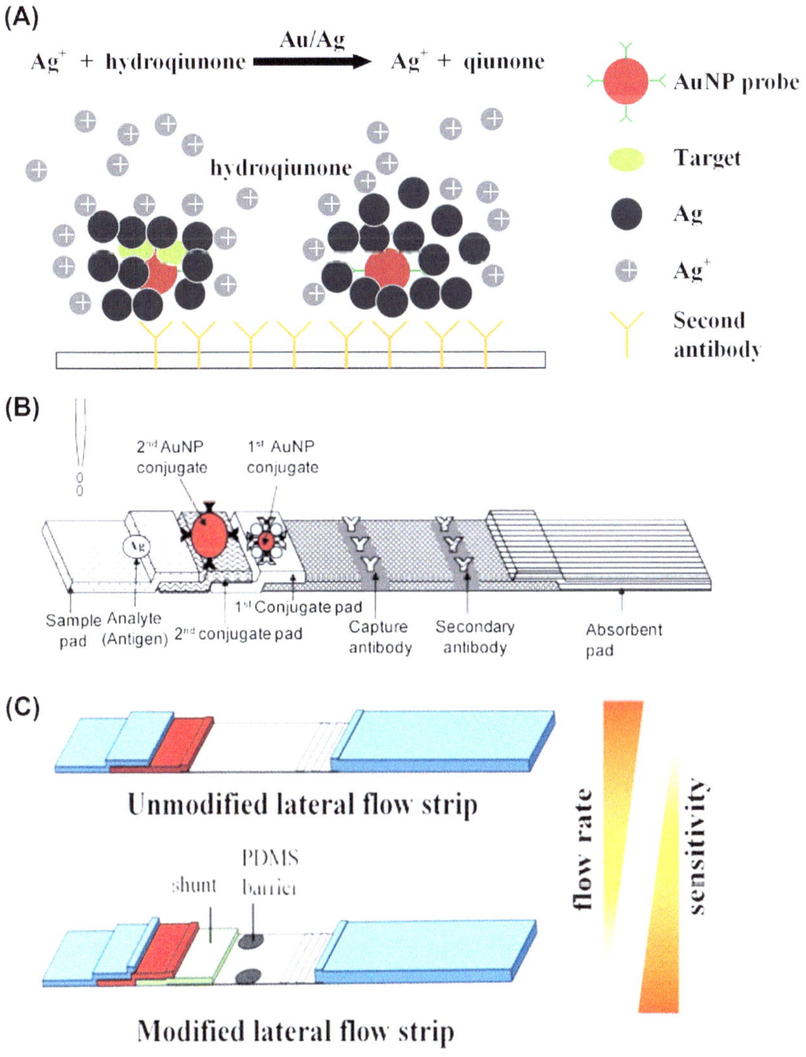

FIGURE 11.3 Sensitivity improvements of gold nanoparticle (AuNP)-based lateral flow assay (LFA). (A) The silver staining LFA. (B) Dual AuNP conjugate-based LFA method. (C) Polydimethylsiloxane barrier modified LFA [41–43].

immunoassays, the multiple test dots could output a semiquantification analysis of target molecules, simultaneously minimizing the hook effect, Fig. 11.4A [48]. Interestingly, Hu et al. [49] proposed a code method to determine the target analyte concentration, Fig. 11.4B. They defined the colorimetric pattern of four test lines representing a certain analyte

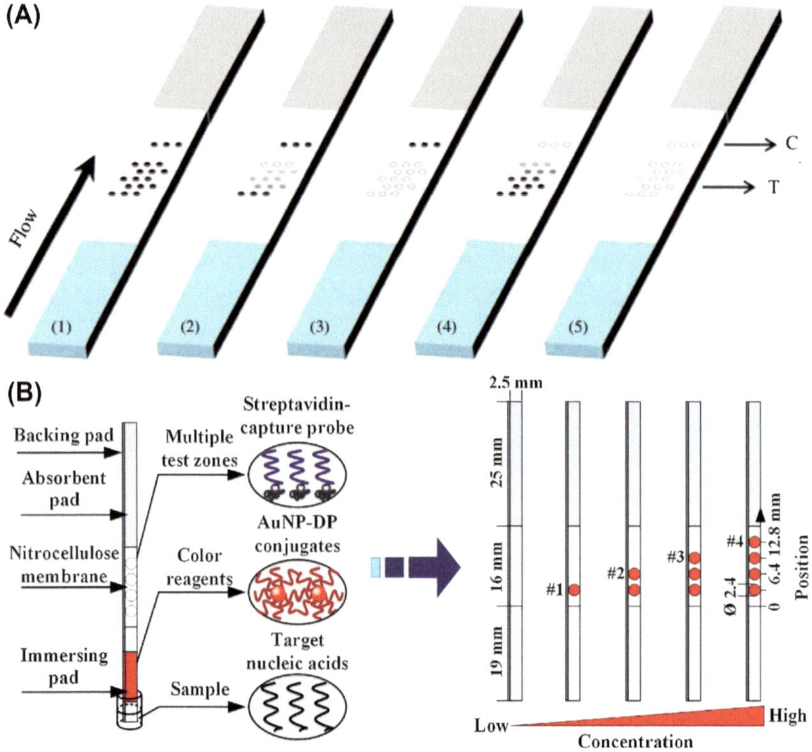

FIGURE 11.4 Semiquantitative detection of lateral flow assay using multiple dots for (A) interested protein or (B) nucleic acid [48,49].

concentration in 20 min after LFA detection, although these methods failed in accurate quantitation detection of target analyte. It is still a challenge to develop a naked-eye readable quantitative LFA.

3.2 Plate-Based Colorimetric Immunoassay

Noble metal nanoparticles (i.e., AuNPs and Ag nanoparticles) demonstrate an ultrahigh molar extinction coefficient, which is several orders of magnitude higher than that of common dyes. Additionally, these nanoscale metal-materials (e.g., nanofilm or nanoparticle) display a unique optical property called localized surface plasmon resonance (LSPR), arising from the interaction between the local plasmons across the surface of nanomaterials and incident light. The slight change of size and shape of nanostructures or their interparticle distances could induce a distinct shift in LSPR. Consequently, many studies have been conducted about the applications of LSPR spectrum shift of metal nanomaterials in plate-based immunoassays.

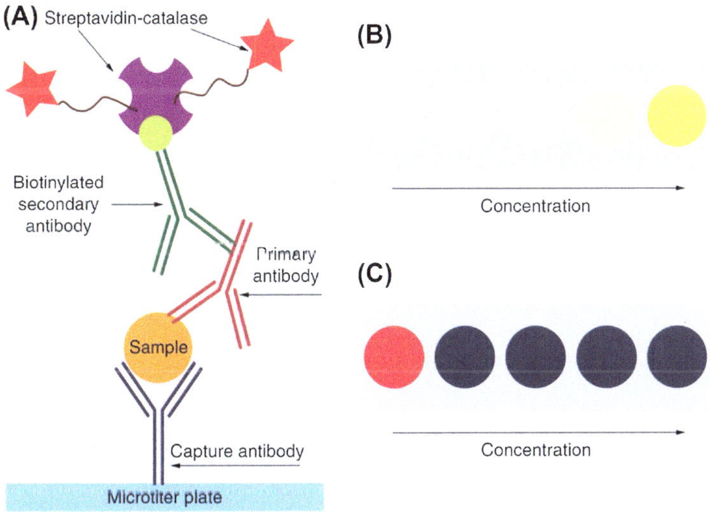

FIGURE 11.5 Plasmonic enzyme-linked immunosorbent assay (ELISA). (A) Schematic of classical sandwich immunoassay. Different visual detection manners of (B) conventional colorimetric ELISA and (C) plasmonic ELISA [53].

As we know, metal nanomaterials in solution display quite different colors depending on the different size or shape of nanomaterials. Thus, ultrasensitive colorimetric immunoassay is possible to be designed by utilizing enzymes to catalyze the shape change of the nanomaterials. Several reports have demonstrated the ultrasensitive immunoassays by using enzyme-controlled growth of metal nanoparticles [50–52]. As illustrated in Fig. 11.5, a plasmonic ELISA was proposed. In conventional colorimetric ELISA, the signal is produced by the colored substrate under catalysis of enzyme (Fig. 11.5A and B), and the color intensity could be quantified by a plate reader. However, we are still unable to determine the low concentration of analyte with confidence.

In plasmonic ELISA, a very low concentration of analyte could be readable by the naked eye, depending on hydrogen peroxide (H_2O_2) in solution to promote the reduction of gold ions to AuNPs. When H_2O_2 was at high concentration, the gold ions would be reduced to uniform small AuNPs at a fast rate, which results in a red color solution. When the concentration of H_2O_2 was low, the reduction rate of gold ions was slowed down and the obtained nanoparticles with strange morphology was aggregated, which displayed a blue color (Fig. 11.5C). To specifically detect the target analyte, a catalase was linked with a secondary antibody via the affinity between streptavidin and biotin. In the presence of the target analyte, a sandwich structure was formed, which dragged the catalase from solution, decreasing the H_2O_2 concentration by converting it to water and oxygen. Hence, decreased H_2O_2 concentration leads to nanoparticle aggregation and exhibits a blue color. In comparison, if the analyte is absent in solution, the hydrogen peroxide won't be consumed and finally a red color will be displayed. The red or blue color change is easy to be read by the naked eye, therefore this plasmonic ELISA is used for highly sensitive detection of prostate-specific antigen (PSA) and HIV-1 capsid antigen p24 with an ultralow concentration of 10^{-18} g/mL in whole serum, which is proper for use in resource-limited areas. In the same year, another plasmonic ELISA was reported, in which the reduction of Ag ions by hydrogen peroxide on the surface of AuNPs resulted in readable color change by eyes. Differently, this immunoassay indicated an inverse relationship between the LSPR shift and target analyte concentration [51], which means that lower analyte concentration leads to greater LSRP shift. Although the plasmonic ELISA methods possess ultrahigh sensitivity, they still need complicated operation steps and a long time for detection (almost 3 h), and they can't achieve multianalyte detection.

Another strategy of plate-based colorimetric immunoassay is based on the fact that optical properties of nanomaterials are dependent on interparticle distance. Nobel metal nanoparticles are very sensitive to the distance between neighboring particles, resulting in LSRP blue shift. Therefore, an abundance of research has been developed to detect the presence of the targets by using the naked eye to read the color change from red to blue/purple due to the aggregation between the noble metal nanoparticles. For example, a method for detecting Cu^{2+} ions by azide- and alkyne-terminal AuNPs was reported, Fig. 11.6A [53]. In this assay, the Cu^{+} ion was found able to catalyze the azides-alkynes cycloaddition, leading the azide- and terminal alkyne-functionalized AuNPs to aggregate, whereas the Cu^{+} ions could be derived from reduction of Cu^{2+} ions in the presence of sodium ascorbate. Therefore, when the Cu^{2+} ions sample is added in mixed solution of azide- and alkyne-modified AuNPs and sodium ascorbate, the original red color of the AuNP solution will change to blue. This cycloaddition reaction catalyzed by Cu^{+} ions is highly selective and tolerated in a wide range of solvents, temperatures, and pH values. Thus, the naked-eye readable colorimetric immunoassay is highly specific, while achieving ~50 μM of detection limit. Afterward, a universal AuNP-aggregation—based colorimetric immunoassay is developed.

This method uses acetylcholinesterase (AChE) to catalyze acetylthiocholine (ATC) to generate thiocholine, which further results in aggregation of citrate-capped AuNPs, Fig. 11.6B [54]. Via preconjugating AChE with secondary antibodies, a specific detection of target analyte can be performed. In the presence of analyte, the secondary antibodies and primary antibodies could specifically recognize target analytes and form a sandwich structure on the plate surface. Subsequently, the AChE conjugated with secondary antibodies will catalyze the ATC in concentration and induce AuNPs to aggregate. To achieve a lower detection limit, an improved method is proposed, which exhibits an ultrahigh sensitivity compared to the sensitivity of polymerase chain reaction [55]. In this assay, thousands of AChE (>30,000 AChE per magnetic bead) and secondary antibodies are both modified on the surface of magnetic beads. As a

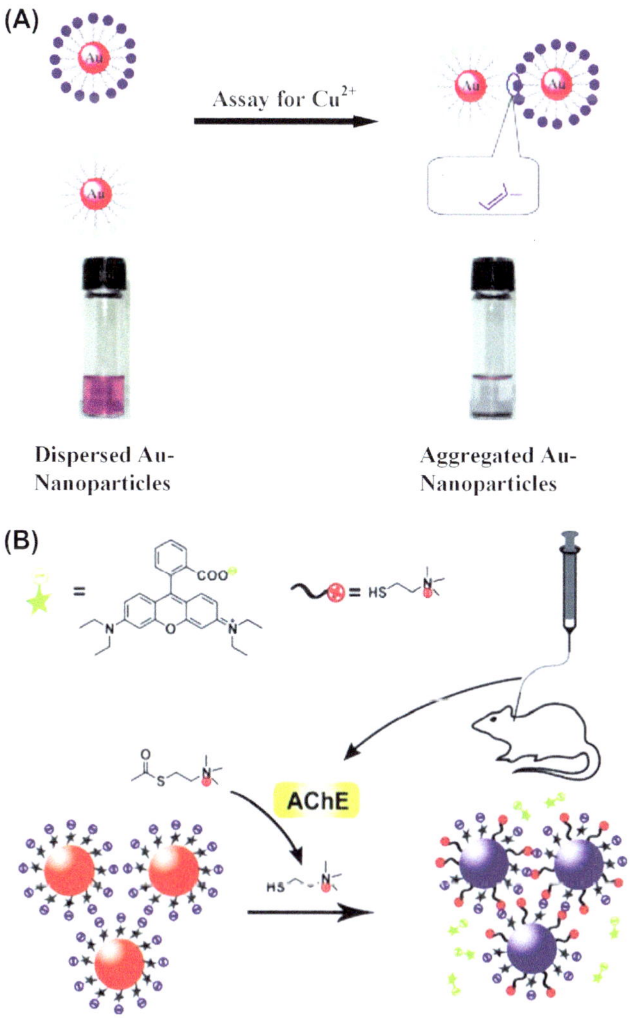

FIGURE 11.6 Gold nanoparticle (AuNP) aggregation-based homogeneous colorimetric immunoassay. (A) Cu^{2+} and sodium ascorbate-induced AuNPs aggregation for Cu^{2+} detection. (B) Electrostatic interaction-induced AuNP aggregation for acetylcholinesterase detection [53,54].

consequence, each target analyte can bring much more AChE to conduct catalytic reactions. This signal amplification strategy leads to four orders of magnitude in sensitivity improvement compared with traditional ELISA.

4. ELECTROCHEMICAL IMMUNOASSAY

Electrochemical immunoassays, based on antibody−antigen interactions on the electrochemical immunosensors, have increasingly gained attention recently, ascribed to their attractive merits of miniaturized instrumentation, high sensitivity, and specificity. An electrochemical immunosensor usually composed of three electrodes (i.e., working electrode, counter electrode, reference electrode) is mainly realized through direct or indirect modification of antibody on the working electrode. During the detection process, the antibody–antigen interaction occurs on the modified working electrode surface and produces electron transfer (in the form of current or potential), which is quantitatively monitored to reflect the target concentration. So far, the key orientations for pushing electrochemical immunoassays forward involve signal amplification, detection limits reduction, and high throughput detection [56]. For achieving these goals, great amounts of nanomaterials have been introduced on electrochemical immunosensors through either promoting electron transfer or facilitating antibody immobilization such as carbon-based nanomaterials and metal-based nanoparticles [56–58]. Generally, the nanomaterials introduced on electrochemical immunosensors are mainly based on two processes: (1) nanomaterials acting as electrode materials during fabrication; (2) modification of electrode surfaces using nanomaterials.

Traditionally, carbon paste is often employed serving as the working electrode material of electrochemical immunosensors due to advantages of low cost, easy accessibility, and chemical inertness. However, carbon paste electrodes are also subjected to relatively low electric conductivity. This intrinsic defect would drop electron transfer efficiency on the electrode surface and reduce the detection sensitivity of the immunosensor. To overcome this limitation, carbon nanotubes characterized by high electric conductivity and surface-to-volume ratio have been adopted as alternatives to conventional materials. For example, an electrochemical immunosensor was developed through fabricating electrodes with single-walled nanotubes [59]. The electrodes were prepared by adding the mixture of carbon nanotubes and mineral oil (with the ratio 60%: 40%) in a Teflon tube. Finally, these carbon nanotube–based electrodes have demonstrated better electrochemical properties than conventional ones, reflected as lower peak potential difference and higher anodic current.

On the other hand, the most popular way to introduce nanomaterials on electrochemical immunosensors is achieved through modifying electrode surfaces using nanomaterials. A great deal of nanomaterials classified by nonmetallic nanomaterials (e.g., carbon and silica) and metallic nanomaterials (e.g., silver and gold) are used to modify the electrode surface in order to amplify the detection signal for ultrasensitive detection, which is of great importance to immunoassays of some serious diseases (e.g., cancer) where the relevant species is at an extremely low level. For instance, multiwalled carbon nanotubes (MWCNTs) conjugated with immunoglobulin G (IgG) were introduced on an electrochemical immunosensor through modification for detecting clenbuterol (CLB) in swine urine [60]. As shown in Fig. 11.7, the working electrode of the immunosensor first was modified with a layer of

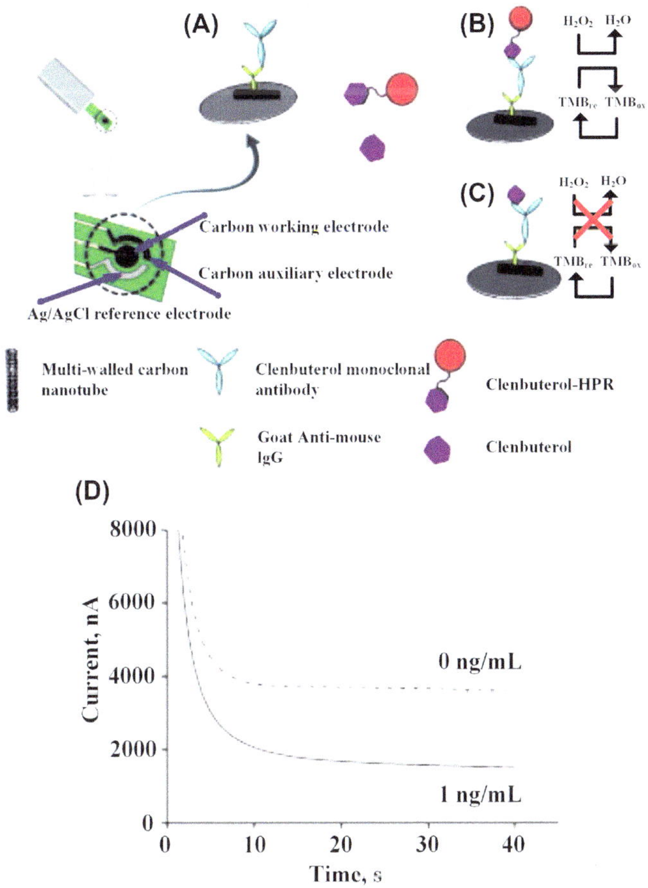

FIGURE 11.7 Preparing the electrochemical immunosensor using multiwalled carbon nanotubes s. (A) Modification of the working electrode. (B and C) Competitive immunoassay on the immunoelectrodes using clenbuterol (CLB)-horseradish peroxidase conjugate (B) and free CLB (C). (D) Monitored electrochemical current signal [60].

conjugated MWCNTs-IgG for capturing the CLB monoclonal antibodies. Then, the mixture of CLB-HRP (i.e., horseradish peroxidase) conjugate and free CLB solution were introduced to realize competitive immunoassay on this immunosensor, and TMB substrate was used to react with HRP, resulting in an electrochemical signal. This type of immunosensor was ultrasensitive with the detection limit of 0.1 ng/mL, the sensitivity of which is at least 10 times that of an immunosensor without carbon nanotubes. Further, a screen-printed electrochemical immunosensor assembled by using MWCNTs labeled with HRP and goat-rabbit IgG was developed to ultrasensitively monitor cancer biomarkers [61].

As demonstrations, the detection limits of PSA and interleukin 8 were as low as 5 and 8 pg/mL, respectively. Meanwhile, high-throughput electrochemical immunoassay has gained increasing attention with the peculiar capability of simultaneous monitoring of multiple targets, which can be realized through modifying the working electrode with different antibodies. As an example, an electrochemical immunosensor with eight separate working electrodes was developed to detect four cancer biomarkers [62]. In their work, the working electrode zone first was modified with MWCNTs to strengthen the electric conductivity and increase the effective reaction surface, and then coated with chitosan to immobilize antibodies. Finally, four capture antibodies corresponding to four cancer biomarkers (i.e., r-fetoprotein, carcinoma antigen 125, carcinoma antigen 199, and carcinoembryonic antigen (CEA)) were introduced onto the eight working electrodes for immunoassays, and the sensitive detection limits of 14.0 pg/mL, 11.0 mU/mL, 13.0 mU/mL, and 9.0 pg/mL in human serum samples were obtained, respectively.

In addition to carbon nanotubes, liposomes in diameters from nanometers to micrometers composed with a spherical aqueous cavity surrounded by a phospholipid bilayer are widely used for immunoassays [63,64]. Thanks to its high-volume inner cavity and large area of outside surface, liposomes hold the capability to encapsulate electroactive markers or enzymes in the cavity and bind antibodies on the outside surface. A sandwiched electrochemical immunosensor based on liposomes with diameter of 238 nm (functioning as a signal amplifier) and carbon nanotubes (modifying electrodes for antibodies immobilization) was developed to detect cholera toxin [63]. The liposomes first were modified with ganglioside on their surface for specific recognition of cholera toxin and then loaded with potassium ferrocyanide into their cavity for electrochemical analysis. The modified liposomes were then added to bond to cholera toxin, which was captured by antibodies on the electrode, forming the sandwiched immunoassay (Fig. 11.8). Finally, the electrochemical analysis was performed when the encapsulated potassium ferrocyanide was released

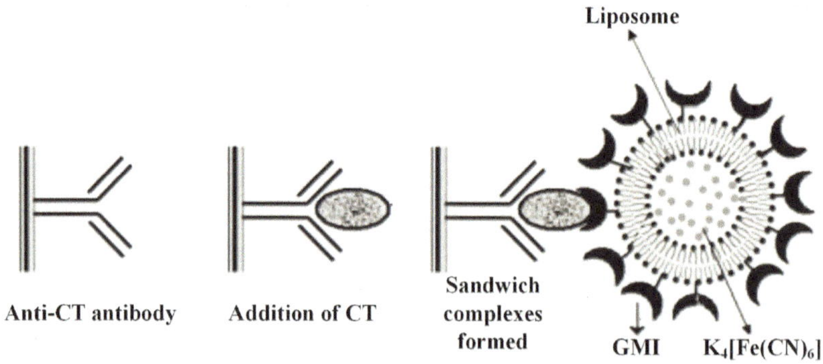

FIGURE 11.8 Schematic illustration of liposome-based immunosensor for cholera toxin monitoring [63].

from the captured liposomes after lyse treatment (Fig. 11.8). This immunosensor shows ultrasensitivity for cholera toxin detection with the detection limit of 10^{-15} g/mL. Furthermore, silica nanoparticles have high stability and versatility, and are easily modified with electroactive poly(guanine) to construct sensitive electrochemical immunosensors [65]. The developed sensor was applied to monitoring of tumor necrosis factor-alpha with detection limit of 2.0 pM, in which the poly(guanine)-modified silica acted as biological labels [66].

The modification of working electrodes using metal-based nanomaterials, such as silver [67], gold-manganese oxide [68], platinum [69], and gold-silver bimetallic nanoparticles [70], was also employed to obtain signal amplification of electrochemical immunoassays, attributing to their high electric conductivity, large surface area, strong absorption ability, and efficient catalytic activity. Moreover, two types of metallic nanomaterials simultaneously modified in one electrochemical immunosensor have been reported to maximize signal amplification. For instance, profiting from its large surface area and high electric conductivity, nanoporous silver was used to modify the working electrode, while nanoporous gold was adopted to label antibodies due to its intrinsic affinity to biomolecules [67]. Finally, this bimetallic nanoparticle-modified immunosensor was applied to electrochemical monitoring of CEA and α-fetoprotein (AFP) with the detection limits of 0.06 and 0.08 pg/mL, respectively.

AuNPs have gained a flourishing development in the past few decades due to their high biocompatibility, chemical stability, and tunable optical properties, and they have found wide applications in electrochemical immunoassays [57]. For example, a colloidal gold label–based electrochemical immunosensor was created to indirectly determine goat IgG with a detection limit of 3×10^{-12} M [71]. In this approach, the colloidal gold particles labeled with antibody were bonded to analytes through antigen–antibody reaction. Then, ionic gold (III) was released in acidic solution after oxidative treatment, which can be measured using a carbon electrode for reflecting the concentration of analytes. Munge et al. developed an extremely sensitive immunosensor based on AuNPs for electrochemical analysis of cancer biomarker interleukin 8 with the ultralow detection limit of 1 fg/mL [72]. As shown in Fig. 11.9, the AuNPs conjugated with glutathione were employed to modify electrodes to immobilize primary antibodies (Ab1) to capture targets. Then, superparamagnetic beads with multilabeled HRP conjugated with secondary antibodies (Ab2) were introduced to form the sandwiched immunosensors. In this method, the ultrahigh sensitivity and ultralow detection limit are attributed to the simultaneous use of AuNPs and superparamagnetic beads, because the AuNPs offer more surface area for Ab1 immobilization and the beads provide massive binding sites for HRP labels. For further signal amplification on electrochemical immunosensors, dual amplification strategies have been developed by introducing two kinds of nanomaterials. For instance, Wu et al. have reported a dual signal amplification strategy on a paper-based electrochemical

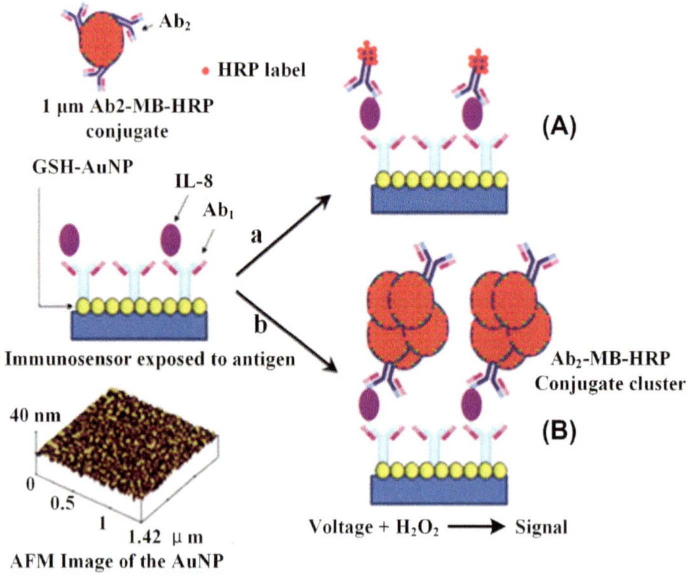

FIGURE 11.9 Detection principles of extremely sensitive immunoassay of interleukin 8 based on gold nanoparticles and superparamagnetic beads. (A) The traditional treatment with few enzyme labels. (B) The superparamagnetic beads treatment with massive enzyme labels [72].

immunosensor via using graphene and silica nanoparticles conjugated with horseradish peroxidase and antibody, which achieved multiplex and high-throughput detection of four cancer biomarkers in serum samples [73]. To amplify the signal, the graphene was used to accelerate the electron transfer on the electrode surface, while the silica nanoparticles were used to reduce the physical adsorption. In addition, another dual signal amplification strategy was realized through accelerating electron transfer by Au nanorods and quantifying target analytes with metal ion–coated Au/bovine serum albumin (Au/BSA) nanospheres [74].

5. FLUORESCENT IMMUNOASSAY

5.1 Heterogeneous Immunoassay

Fluorescent heterogeneous immunoassay also makes use of the high affinity between target analyte and two specific antibodies in aqueous phase. One of the specific antibodies is anchored on the substrate surface and the other is labeled with fluorescent nanoparticles to trace analyte. Generally, the fluorescence intensity of the fluorescent nanoparticles trapped on substrate is in direct proportion to the concentration of target analyte after several efficient washing steps, similar to the process of ELISA. Conventional fluorophores (i.e., organic dyes) with different emission wavelengths have been used in fluorescent immunoassays for simultaneous detection of multiple analytes for two decades [75]. However, the multicolor signal of organic dyes needs elaborate excitation and complicated data analysis [76], which limits its wide applications. To this end, QDs are recommended for fluorescent immunoassays because they are robust inorganic fluorophores that exhibit size-dependent fluorescence emission spectra that span the whole visible spectrum and possess high quantum yield and good photochemical stability [77]. For instance, simultaneous detection of four toxins was realized by using multicolor QDs with 510, 555, 590, 610 nm maximum emission in a single well of a microtiter plate [78], which could perform four or more multianalyte detection. Afterward, a greater detection capability of multicolor QD-based immunosorbent assay has been developed for detecting five illegally used drugs in China, Fig. 11.10A [79]. This technology can potentially provide significant help for defeating the substance abuse in China. However, it is still hard to use in source-limited areas, due to the dependence on expensive readout instruments. Therefore, a cellphone based hand held microplate reader was developed to mitigate this problem, Fig. 11.10B [80]. This reader is

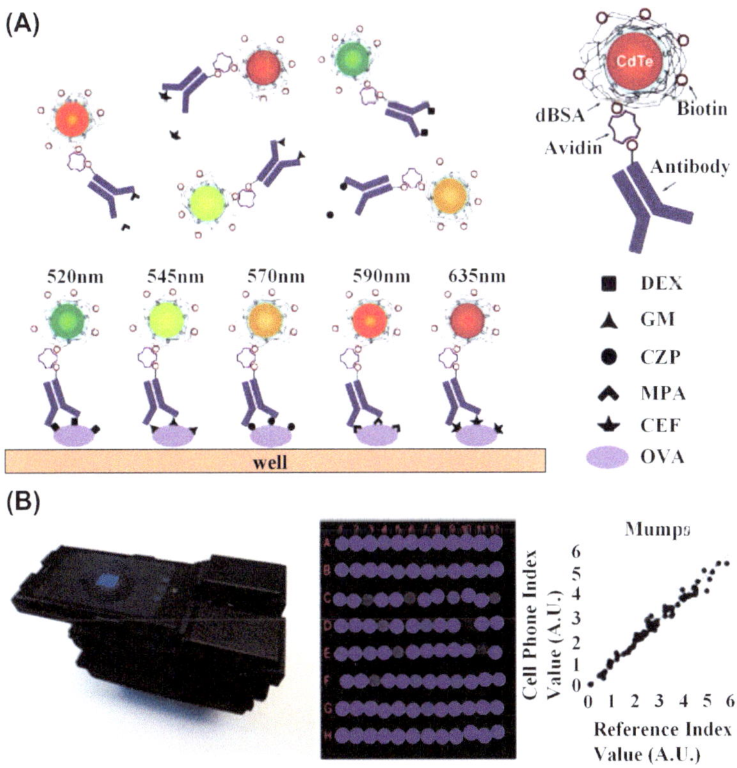

FIGURE 11.10 Fluorescent heterogeneous immunoassays. (A) Multicolor quantum dot-based immunoabsorbent assay for simultaneous five-drug detection. (B) Hand-held and cost-effective cellphone-based colorimetric microplate reader [79,80].

a 3D printed attachment with blue LED array, 96-well plate holder, and optical fiber bundle. The LED array could illuminate every well in the 96-well plate and then the images are captured through optical fiber bundle by cellphone camera. Additionally, a smart app with machine-learning algorithm was developed to analyze the results. This portable reader has been applied for reading FDA-approved mumps IgG, measles IgG, and herpes simplex virus IgG (HSV-1 and HSV-2) ELISA tests and achieved an accuracy of 99.6%, 98.6%, 99.4%, and 99.4%, respectively. However, these fluorescent ELISAs still need complicated operation progress and professional persons, and most reagents require cold storage and protection from light.

To address the problems, various POC diagnosis platforms are developed, which require several simplified operation steps; their reaction reagents are mostly positioned inside the device after freezing-drying. For example, a portable and fast fluorescence biosensor with high sensitivity has been established with QDs in a lateral flow test assay [81]. In this biosensor, the QD probes and dried powders of reaction reagents have been stored in the strip before use. When performing the detection, the user only needs to add samples into the sample pad. Then, the detection could be finished in 10 min and a detection limit of as low as 0.4 ng/mL was achieved for nitrated ceruloplasmin. Compared with AuNP-based LFA, the QDs not only achieve a higher sensitivity, but also are capable of quantifying the target concentration via measuring fluorescence intensity of QDs in test line, conquering a wide range from 1 ng/mL to 10 μg/mL. Recently, a fluorescence-enhanced QD-based nanosphere was used in LFA [82]. 332 ± 8 QDs are embedded in poly(styrene/acrylamide) copolymer nanospheres and then are conjugated with antibodies, resulting in a super-strong luminescence probe, Fig. 11.11A. With this probe, the LFA obtains a quantitative detection range from 0.178 to 11.4 nM with a detection limit of 27.8 pM, which is 257-fold more sensitive than traditional AuNP-based LFA for CRP detection.

Except for the immunoassay based on the proportional relationship between signal and analyte, competitive immunoassay also has widespread applications. In a competitive assay, the fluorescent probe generally is able to simultaneously bind with two substrates (i.e., analyte and alternate), of which analyte possesses higher affinity with the fluorescent probes. In absence of analytes, the probes would directly react with alternates preembedded at detection zone, resulting in accumulated fluorescence intensity. When analytes are present in sample, the fluorescent probes prefer to bind with analytes and the formed complex would pass through the detection zone, resulting in decreasing fluorescence intensity in detection zone as target analytes concentration increases. For example, a competitive LFA for multiple antibiotics detection is developed utilizing multicolored QDs [83]. This assay used three QDs with red, yellow, and green emission to respectively detect streptomycin, chloramphenicol, and ofloxacin in milk, achieving 80–200 times improvement of detection limit than traditional ELISA, Fig. 11.11B. Other POC diagnosis methods are also developed; for instance, microfluidic chip, which allows

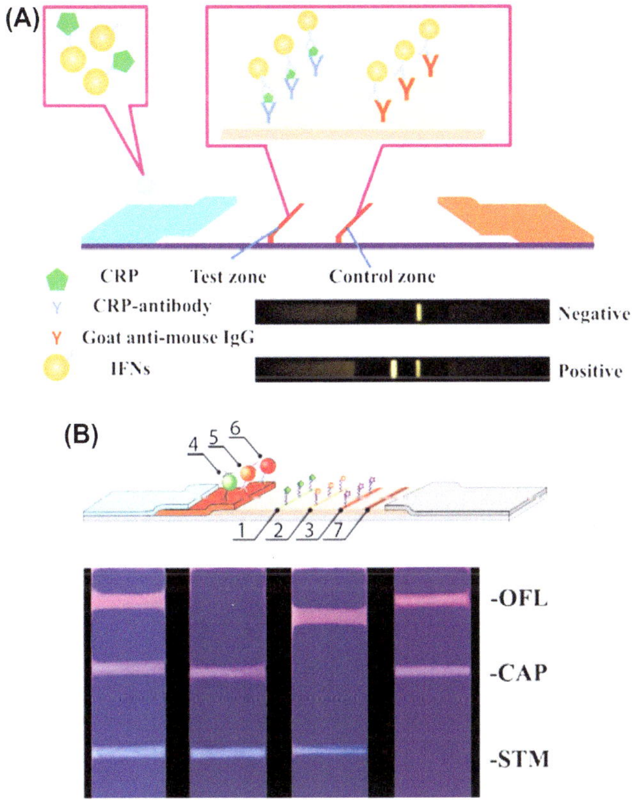

FIGURE 11.11 Fluorescent lateral flow immunoassays. (A) Quantum dot-embedded superstrong fluorescent nanosphere-based lateral flow assay (LFA) for sensitively detecting protein biomarkers. (B) Competitive fluorescent LFA for multiplex antibiotic detection [82,83].

ultrasensitive, multiplex, and high throughput detection [84]. In summary, whether in fluorescent immunoassay or competitive fluorescent immunoassay, the introduction of fluorescent NPs could bring two major advantages: (1) the great improvement of detection sensitivity, normally with two or three order of magnitudes; (2) the quantification capability for detecting analytes, benefited from its sensitive and sequential intensity change.

5.2 Fluorescence Resonance Energy Transfer Assay

Fluorescence resonance energy transfer (FRET) is the electronic energy transfer process between two fluorophores (i.e., a donor fluorophore and an acceptor fluorophore), leading to a change of the emission light, which is often used in sensitive fluorescent detections. A prerequisite to fulfill FRET is that the emission spectrum of donor must overlap with the absorption spectrum of acceptor. Therefore, the selected donor fluorophore confines the choice of acceptor fluorophore. Another prerequisite of FRET is that the distance between donor and acceptor should be shorter than 10 nm [85]. Thus, the response of the FRET system is very sensitive to the formation and disorganization of sandwich structure between donor and acceptor. Here, we classify the FRET system into two subclasses by different acceptors. The first subclass is the FRET system with fluorophores as acceptors, such as organic dyes, QDs, and so on, which display a matched absorption spectrum with donor's emission and transfer the absorbed energy to another wavelength emission. In the second subclass, the acceptor is fluorescence quencher, which can absorb and transfer the energy absorbed from donors to heat or other energy but without photons emitting. This kind of acceptor will reduce the emission intensity, even leading to a complete fluorescence quenching. In these cases, gold nanomaterials, carbon nanotubes, and graphene sheets always play the role of quenchers.

Benefitting from their broad and tunable absorption spectra and high quantum yields, QDs and organic dyes are often used as acceptors in FRET detection systems of the first subclass. For example, an efficient FRET system on the basis of QDs and dye-labeled sugar analogue was developed for maltose detection [86]. In this study, CdSe−ZnS core-shell nanocrystals conjugated with maltose binding protein (MBP) via metal-affinity coordination were used as donors. Two kinds of acceptors were used to form two different sensing FRET systems, respectively, denoted as β-cyclodextrin-QSY9 and β-cyclodextrin-Cy3.5. After β-cyclodextrin domain binding with MBP, the tailed QSY9 is able to quench the emission of QDs, Fig. 11.12A. The addition of maltose will liberate the β-cyclodextrin-QSY9 through competitive binding, leading to the fluorescence recovery. Similarly, in the case of β-cyclodextrin-Cy3.5, Cy3 serves as a bridge to transfer the energy from QDs to Cy3.5, resulting in red emission, Fig. 11.12B. This delicate design utilizes a two-step FRET mechanism to realize a long-distance FRET. Additionally, the analogical FRET system is further applied for trinitrotoluene detection with high specificity in an aqueous environment [87].

FRET is a powerful tool for sensing the tiny distance change between donor and acceptor fluorophores or the protein conformation folding in binding events. However, like other fluorescence technologies, traditional FRET is also suffering from photobleaching and autofluorescence. Therefore, a new resonance energy transfer strategy between chemiluminescent fluorophore and QDs is developed, called chemiluminescence resonance energy transfer (CRET). Compared with FRET, the occurrence of CRET is easily triggered through the oxidation of a luminescence substrate instead of needing an excitation laser. For example, the luminol/H_2O_2 CL reaction catalyzed by HRP could be selected to form a sensitive CRET system with QDs [88]. As illuminated in Fig. 11.12C, QDs modified with BSA can rapidly interact with HRP linked with BSA antibody (anti-BSA), where CRET happens. In this strategy, HRP as a catalyst was directly combined with the QDs, which could continuously catalyze the CL donor (i.e., luminol) emitting, finally achieving

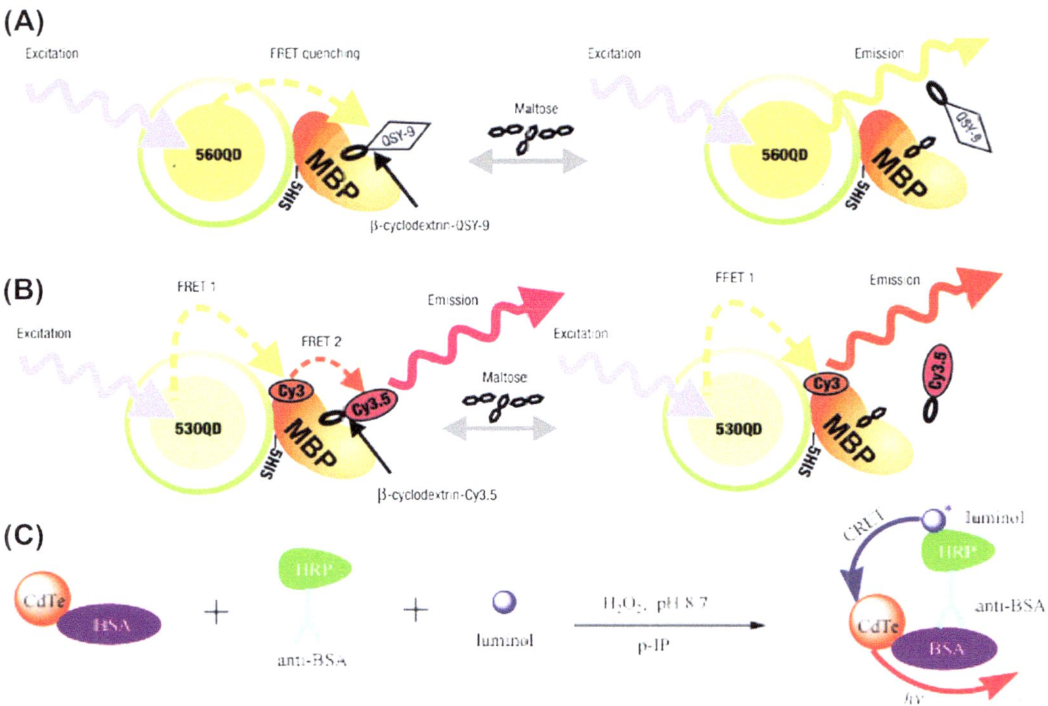

FIGURE 11.12 First subclass fluorescence resonance energy transfer (FRET) immunoassays. (A) 560QD-MBP and (B) 530QD-MBP-Cy3-β-CD-Cy3.5 FRET sensors. (C) Luminol donors and quantum dot accepter-based chemiluminescence resonance energy transfer [86,88].

an efficient CRET. However, CRET needs additional of H_2O_2, which complicates the reaction system, increases the operation processes, and is susceptible to environment conditions. Moreover, the emission of QDs could be quenched by high concentration of H_2O_2.

In the second FRET system subclass, Au nanomaterials are the most commonly used quenchers, such as Au nanoparticles and Au nanorods. Au nanomaterials are assumed to have high fluorescence quenching efficiency and a relative long effective distance [89]. First, the size difference between large Au nanomaterial donor and molecular acceptor improves ET efficiency from donor-acceptor pairs with distance. Second, their absorption cross sections and light scattering are both remarkably large, generally several orders larger than those of conventional dyes [90], therefore significantly reinforcing their absorption to transferred energy. Additionally, there is no requirement for the donor's orientation to proceed with ET when spherical Au nanoparticles are used as acceptors. As a consequence, on the basis of Au nanomaterials, many FRET detection systems are developed to conduct sensitive and accurate detections [89,91]. For instance, a sensitive and compatible AuNP-based fluorescence-activatable probe based on AuNPs has been proposed for ultrasensitive detection of PSA in patient sera (Fig. 11.13A). In this design, the AuNPs possess a high loading efficiency for organic dyes, which could be completely released by thiol-containing cysteamine. Therefore, the FRET detection system realized an ultrahigh detection sensitivity of 0.032 pg/mL for PSA, which overweighs 2 times lower than that of traditional fluorescence detection [92]. However, the assay conducted in microtiter plates requires the serum/plasma separation in whole blood detection and repeated washing processes to avoid fluorescent background and nonspecific absorption, which are labor-intensive and time-consuming.

A one-step in situ solid-substrate-based FRET system with upconversion and AuNPs is then developed for direct detection of whole blood [93]. As shown in Fig. 11.13B, the UCNPs are preembedded on a glass slide and AuNPs modified with IgG have been conjugated with UCNPs before detection. When mixed with analyte sample, the goat IgG presenting in samples would competitively bind with UCNPs and force the AuNPs off, leading the fluorescence recovery of UCNPs as an increase of analyte concentration. A similar UCNPs-AuNPs FRET system is also applied in ultrasensitive bacteria detection using aptamers in place of antibodies as recognition ligands on surface of nanoparticles [91].

Different from zero-dimensional AuNPs, graphene-like nanoscale two-dimensional nanomaterials (e.g., graphene, graphene oxide [GO], MoS_2, SnS_2, WS_2) possess a large cross section, two-dimensional distribution of disorders and random orientation of active sites, which have been used in FRET systems with tremendously increasing speed, Fig. 11.14. In a previous report, a nanoscale graphene and CdTe QDs based FRET immunoassay was

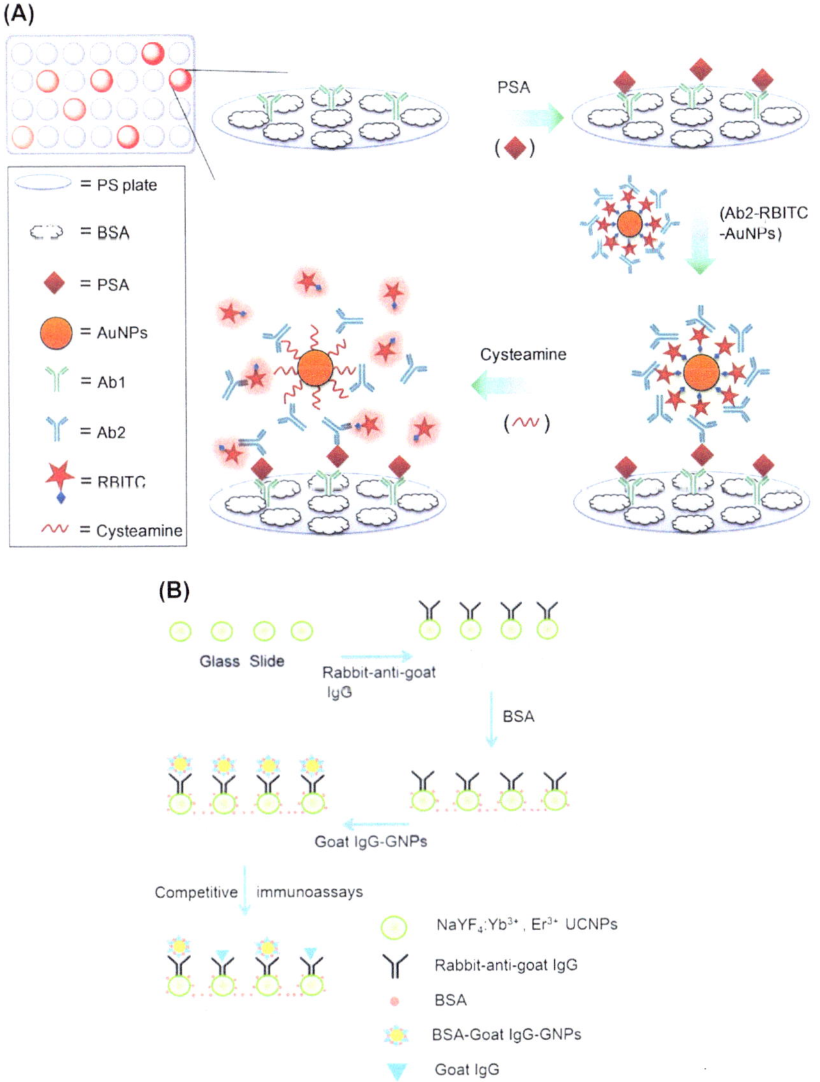

FIGURE 11.13 Fluorescence resonance energy transfer (FRET) immunoassays using gold nanoparticles as quencher. (A) Representation of the activatable dye-gold nanoparticle (AuNP) FRET immunoassay for prostate specific antigen. (B) One-step UCNPs-AuNPs FRET immunoassay for goat IgG [92,93].

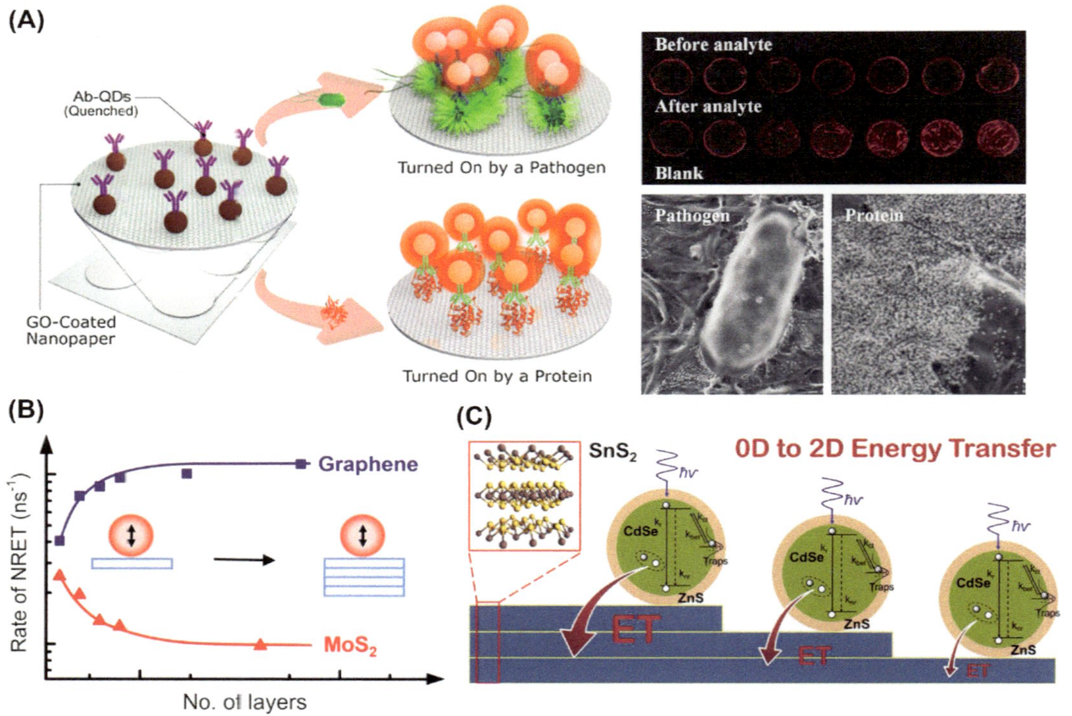

FIGURE 11.14 Fluorescence resonance energy transfer (FRET) immunoassays using 2D graphene as quencher. (A) Graphene oxide FRET immunosensor for pathogen and protein detection. Nonradiative energy transfer (NRET) from semiconducting quantum dots to adjacent two-dimensional sheets of (B) MoS_2 or (C) SnS_2 of single- and few-layer thickness [98–100].

developed to sensitively detect glycoprotein AFP for diagnosis of hepatocellular carcinoma [94]. In this system, the presence of target AFP could induce the construction of QDs/AFP/nanographene conjugates and decrease the fluorescence intensity of QDs, consequently achieving 66% ± 4% quenching efficiency and 0.15 ng/mL detection limit. More significantly, the authors found that the nanographene- and QD-based system broke the distance limit of traditional FRET systems. They derived an approximately 22.3 nm effective FRET distance, compared to the 10 nm maximum in traditional FRET systems; high flexibility and random arrangement of the two-dimensional graphene nanosheet facilitate their overlapping with QDs to form various effective QD-graphene configurations. The GO nanosheet was used together with green fluorescence protein modified SNAP-25 peptide to construct an ultrasensitive FRET biosensor [95]. In this biosensor, the recovery of fluorescence intensity displayed a linear relationship with the log concentration of target botulinum neurotoxins (BoNTs) in a wide detection range from 1 fg/mL to 1 pg/mL. Moreover, the novel FRET biosensor not only can detect the concentration of target analyte, but also can evaluate the protease activity of BoNT-LcA. Notably, it is the protease activity of target BoNT-LcA that brings an amplification effect to the detection and contributes to the ultrahigh sensitivity. Subsequently, graphene quantum dots (GQDs) with amine functionalization were utilized as donors in the FRET sensor with graphene nanosheet as acceptors for heart attack detection [96]. Cardiac Troponin I (cTnI) is the gold standard to diagnose myocardial infarction (MI), causing its high specificity for indicating MI. However, the reference limit of cTnI concentration in clinics is quite low at 0.04 ng/mL, which presents challenges to traditional detection methods. Here, the GQDs-graphene FRET system conquered this challenge, realizing a very low detection limit of 0.192 pg/mL and ultrawide detection range was achieved, across six orders of magnitude for target analyte concentration from 1 to 10^6 pg/mL.

6. CONCLUSION

Immunoassay has a long history of applications in clinical diagnosis. Stimulated by developments of new technologies and materials, various nano- and micromaterials, possessing various physical peculiarities, are continuously introduced in immunoassays to improve sensitivity, speed reaction rate, reduce cost, and simplify operation. They all exhibit outstanding performance by virtue of the unique properties of the nano- and micromaterials, some of which are even developed as commercial products for hospital or home use, such as Luminex xMAP system and pregnancy test paper. Also, a great deal of advanced immunoassays also achieved a breakthrough in one certain aspect, for instance plasmonic ELISA has reached an ultrahigh detection sensitivity of 10^{-3} fg/mL level. We believe these developing technologies will provide great possibilities for practical applications, especially in source-limited areas. Further development of nano- and micromaterials is able to accelerate the commercialization of more advanced immunoassays.

REFERENCES

[1] K.E. Rubenstein, et al., "Homogeneous" enzyme immunoassay. A new immunochemical technique, Biochem. Biophys. Res. Commun. 47 (1972) 846–851.

[2] G. Kohler, C. Milstein, Continuous cultures of fused cells secreting antibody of predefined specificity, Nature 256 (1975) 495–497.

[3] H.R. Schroeder, et al., Competitive protein binding assay for biotin monitored by chemiluminescence, Anal. Chem. 48 (1976) 1933–1937.

[4] A.H.B. Wu, A selected history and future of immunoassay development and applications in clinical chemistry, Clin. Chim. Acta 369 (2006) 119–124.

[5] G. Wang, et al., Highly efficient preparation of multiscaled quantum dot barcodes for multiplexed hepatitis B detection, ACS Nano 7 (2013) 471–481.

[6] Y. Zhao, et al., Quantum-dot-tagged bioresponsive hydrogel suspension array for multiplex label-free DNA detection, Adv. Funct. Mater. 20 (2010) 976–982.

[7] L. Jiseok, et al., Universal process-inert encoding architecture for polymer microparticles, Nat. Mater. 13 (2014) 524–529.

[8] D.C. Pregibon, et al., Multifunctional encoded particles for high-throughput biomolecule analysis, Science 315 (2007) 1393–1396.

[9] M.S. Akhras, et al., The sequencing bead array (SBA), a next-generation digital suspension array, PLoS One 8 (2013).

[10] H.D. Hill, C.A. Mirkin, The bio-barcode assay for the detection of protein and nucleic acid targets using DTT-induced ligand exchange, Nat. Protoc. 1 (2006) 324–336.

[11] A.R. Vaino, K.D. Janda, Euclidean shape-encoded combinatorial chemical libraries, Proc. Natl. Acad. Sci. U.S.A. 97 (2000) 7692–7696.

[12] G. Wang, et al., Highly efficient preparation of multiscaled quantum dot barcodes for multiplexed hepatitis B detection, Acs Nano 7 (2012) 471–481.

[13] M. Trau, B.J. Battersby, Novel colloidal materials for high-throughput screening applications in drug discovery and genomics, Adv. Mater. 13 (2001) 975–979.

[14] O. Akinfieva, et al., New directions in quantum dot-based cytometry detection of cancer serum markers and tumor cells, Crit. Rev. Oncol. Hematol. 86 (2012) 1–14.

[15] U. Resch-Genger, et al., Quantum dots versus organic dyes as fluorescent labels, Nat. Methods 5 (2008) 763–775.

[16] S.A. Dunbar, Applications of Luminex® xMAP™ technology for rapid, high-throughput multiplexed nucleic acid detection, Clin. Chim. Acta 363 (2006) 71–82.

[17] M. Han, et al., Quantum-dot-tagged microbeads for multiplexed optical coding of biomolecules, Nat. Biotech. 19 (2001) 631–635.

[18] Y. Li, et al., Synthesis and characterization of CdS quantum dots in polystyrene microbeads, J. Mater. Chem. 15 (2005) 1238–1243.

[19] Q. HaiSheng, et al., Multicolor polystyrene nanospheres tagged with up-conversion fluorescent nanocrystals, Nanotechnology 19 (2008) 255601.

[20] H.H. Gorris, O.S. Wolfbeis, Photon-upconverting nanoparticles for optical encoding and multiplexing of cells, biomolecules, and microspheres, Angew. Chem. Int. Ed. 52 (2013) 3584–3600.

[21] J.L. Ou, et al., Uniform polystyrene microspheres decorated with noble metal nanoparticles formed without using extra reducing agent, Colloids Surfaces A Physicochem. Eng. Asp. 305 (2007) 36–41.

[22] G. Chen, et al., Lanthanide-doped ultrasmall yttrium fluoride nanoparticles with enhanced multicolor upconversion photoluminescence, J. Mater. Chem. 22 (2012) 20190–20196.

[23] M. Lin, et al., Recent advances in synthesis and surface modification of lanthanide-doped upconversion nanoparticles for biomedical applications, Biotechnol. Adv. 30 (2012) 1551–1561.

[24] Y. Lu, et al., On-the-fly decoding luminescence lifetimes in the microsecond region for lanthanide-encoded suspension arrays, Nat. Commun. 5 (2011) 3741.

[25] B.H. Jun, et al., Surface-enhanced Raman scattering-active nanostructures and strategies for bioassays, Nanomedicine 6 (2011) 1463–1480.

[26] M.A.M. Gijs, et al., Microfluidic applications of magnetic particles for biological analysis and catalysis, Chem. Rev. 110 (2010) 1518–1563.

[27] Y.H. Tennico, et al., On-chip aptamer-based sandwich assay for thrombin detection employing magnetic beads and quantum dots, Anal. Chem. 82 (2010) 5591–5597.

[28] S. Centi, et al., Aptamer-based detection of plasma proteins by an electrochemical assay coupled to magnetic beads, Anal. Chem. 79 (2007) 1466–1473.

[29] R. De Palma, et al., Magnetic bead sensing platform for the detection of proteins, Anal. Chem. 79 (2007) 8669–8677.

[30] S. Wu, et al., Multiplexed fluorescence resonance energy transfer aptasensor between upconversion nanoparticles and graphene oxide for the simultaneous determination of mycotoxins, Anal. Chem. 84 (2012) 6263–6270.

[31] M. Herrmann, et al., Quantification of low-picomolar concentrations of TNF-α in serum using the dual-network microfluidic ELISA platform, Anal. Chem. 80 (2008) 5160–5167.

[32] H. Kuramitz, Magnetic microbead-based electrochemical immunoassays, Anal. Bioanal. Chem. 394 (2009) 61–69.

[33] R. De Palma, et al., The optimization of magnetosandwich assays for the sensitive and specific detection of proteins in serum, Anal. Chem. 79 (2007) 7540–7548.

[34] R. De Palma, et al., Magnetic particles as labels in bioassays: Interactions between a biotinylated gold substrate and streptavidin magnetic particles, J. Phys. Chem. C 111 (2007) 12227–12235.

[35] S.P. Mulvaney, et al., Rapid, femtomolar bioassays in complex matrices combining microfluidics and magnetoelectronics, Biosens. Bioelectron. 23 (2007) 191–200.

[36] J.R. Choi, et al., Lateral flow assay based on paper-hydrogel hybrid material for sensitive point-of-care detection of Dengue virus, Adv. Healthc. Mater. (2016). https://doi.org/10.1002/adhm.201600920.

[37] J. Hu, et al., Portable microfluidic and smartphone-based devices for monitoring of cardiovascular diseases at the point of care, Biotechnol. Adv. 34 (2016) 305–320.

[38] M. Chen, et al., Dual gold nanoparticle lateflow immunoassay for sensitive detection of *Escherichia coli* O157:H7, Anal. Chim. Acta 876 (2015) 71–76.

[39] M. Zhu, et al., Ultrasensitive detection of mercury with a novel one-step signal amplified lateral flow strip based on gold nanoparticle-labeled ssDNA recognition and enhancement probes, Biosens. Bioelectron. 61 (2014) 14–20.

[40] S. Girotti, et al., Development of a chemiluminescent ELISA and a colloidal gold-based LFIA for TNT detection, Anal. Bioanal. Chem. 396 (2010) 687–695.

[41] Q. Yu, et al., Gold nanoparticles-based lateral flow immunoassay with silver staining for simultaneous detection of fumonisin B1 and deoxynivalenol, Food Control 54 (2015) 347–352.

[42] D.H. Choi, et al., A dual gold nanoparticle conjugate-based lateral flow assay (LFA) method for the analysis of troponin I, Biosens. Bioelectron. 25 (2010) 1999–2002.

[43] J.R. Choi, et al., Polydimethylsiloxane-paper hybrid lateral flow assay for highly sensitive point-of-care nucleic acid testing, Anal. Chem. 88 (2016) 6254–6264.

[44] C. Parolo, et al., Enhanced lateral flow immunoassay using gold nanoparticles loaded with enzymes, Biosens. Bioelectron. 40 (2012) 412–416.

[45] R.-H. Shyu, et al., Colloidal gold-based immunochromatographic assay for detection of ricin, Toxicon 40 (2002) 255–258.

[46] J. Hu, et al., Oligonucleotide-linked gold nanoparticle aggregates for enhanced sensitivity in lateral flow assays, Lab Chip 13 (2013) 4352–4357.

[47] J. Hu, et al., Advances in paper-based point-of-care diagnostics, Biosens. Bioelectron. 54 (2014) 585–597.

[48] S.Y. Hong, et al., Quantitative lateral-flow immunoassay for the assessment of the cartilage oligomeric matrix protein as a marker of osteoarthritis, BioChip J. 6 (2012) 213–220.

[49] J. Hu, et al., Multiple test zones for improved detection performance in lateral flow assays, Sensors Actuators B Chem. 243 (2017) 484–488.

[50] R. de la Rica, M.M. Stevens, Plasmonic ELISA for the ultrasensitive detection of disease biomarkers with the naked eye, Nat. Nano 7 (2012) 821–824.

[51] L. Rodríguez-Lorenzo, et al., Plasmonic nanosensors with inverse sensitivity by means of enzyme-guided crystal growth, Nat. Mater. 11 (2012) 604–607.

[52] S.S. Aćimović, et al., LSPR chip for parallel, rapid, and sensitive detection of cancer markers in serum, Nano Lett. 14 (2014) 2636–2641.

[53] Y. Zhou, et al., Visual detection of copper(II) by azide- and alkyne-functionalized gold nanoparticles using click chemistry, Angew. Chem. 120 (2008) 7564–7566.

[54] D. Liu, et al., A highly sensitive gold-nanoparticle-based assay for acetylcholinesterase in cerebrospinal fluid of transgenic mice with Alzheimer's disease, Adv. Healthc. Mater. 1 (2012) 90–95.

[55] D. Liu, et al., Acetylcholinesterase-catalyzed hydrolysis allows ultrasensitive detection of pathogens with the naked eye, Angew. Chem. Int. Ed. 52 (2013) 14065–14069.

[56] G. Liu, Y. Lin, Nanomaterial labels in electrochemical immunosensors and immunoassays, Talanta 74 (2007) 308–317.

[57] A. Chen, S. Chatterjee, Nanomaterials based electrochemical sensors for biomedical applications, Chem. Soc. Rev. 42 (2013) 5425–5438.

[58] A. Hayat, et al., Current trends in nanomaterial-based amperometric biosensors, Sensors 14 (2014) 23439.

[59] F. Valentini, et al., Carbon nanotubes as electrode materials for the assembling of new electrochemical biosensors, Sensors Actuators B Chem. 100 (2004) 117–125.

[60] G. Liu, et al., A carbon nanotube-based high-sensitivity electrochemical immunosensor for rapid and portable detection of clenbuterol, Biosens. Bioelectron. 28 (2011) 308–313.

[61] Y. Wan, et al., Carbon nanotube-based ultrasensitive multiplexing electrochemical immunosensor for cancer biomarkers, Biosens. Bioelectron. 30 (2011) 93–99.

[62] D. Zang, et al., Electrochemical immunoassay on a 3D microfluidic paper-based device, Chem. Commun. 48 (2012) 4683–4685.

[63] S. Viswanathan, et al., Electrochemical immunosensor for cholera toxin using liposomes and poly(3,4-ethylenedioxythiophene)-coated carbon nanotubes, Anal. Chem. 78 (2006) 1115–1121.

[64] S. Viswanathan, et al., Disposable electrochemical immunosensor for carcinoembryonic antigen using ferrocene liposomes and MWCNT screen-printed electrode, Biosens. Bioelectron. 24 (2009) 1984–1989.

[65] J. Wang, et al., Electroactive silica nanoparticles for biological labeling, Small 2 (2006) 1134–1138.

[66] J. Wang, et al., Sensitive immunoassay of a biomarker tumor necrosis factor-α based on poly(guanine)-functionalized silica nanoparticle label, Anal. Chem. 78 (2006) 6974–6979.

[67] W. Li, et al., Development of a 3D origami multiplex electrochemical immunodevice using a nanoporous silver-paper electrode and metal ion functionalized nanoporous gold–chitosan, Chem. Commun. 49 (2013) 9540–9542.

[68] L. Li, et al., Growth of gold-manganese oxide nanostructures on a 3D origami device for glucose-oxidase label based electrochemical immunosensor, Biosens. Bioelectron. 61 (2014) 76–82.

[69] F. Liu, et al., Electrochemical device based on a Pt nanosphere-paper working electrode for in situ and real-time determination of the flux of H_2O_2 releasing from SK-BR-3 cancer cells, Chem. Commun. 50 (2014) 10315–10318.

[70] G. Sun, et al., Paper-based electrochemical immunosensor for carcinoembryonic antigen based on three dimensional flower-like gold electrode and gold-silver bimetallic nanoparticles, Electrochim. Acta (2014) 650–656.

[71] M. Dequaire, et al., An electrochemical metalloimmunoassay based on a colloidal gold label, Anal. Chem. 72 (2000) 5521–5528.

[72] B.S. Munge, et al., Nanostructured immunosensor for attomolar detection of cancer biomarker interleukin-8 using massively labeled superparamagnetic particles, Angew. Chem. Int. Ed. 50 (2011) 7915–7918.

[73] Y. Wu, et al., Paper-based microfluidic electrochemical immunodevice integrated with nanobioprobes onto graphene film for ultrasensitive multiplexed detection of cancer biomarkers, Anal. Chem. 85 (2013) 8661–8668.

[74] C. Ma, et al., 3D origami electrochemical immunodevice for sensitive point-of-care testing based on dual-signal amplification strategy, Biosens. Bioelectron. 63 (2015) 7–13.

[75] E.E. Swartzman, et al., A homogeneous and multiplexed immunoassay for high-throughput screening using fluorometric microvolume assay technology, Anal. Biochem. 271 (1999) 143–151.

[76] A. Buwe, et al., Multicolor spectral karyotyping of rat chromosomes, Science 273 (1996) 494–497.

[77] W.C.W. Chan, S. Nie, Quantum dot bioconjugates for ultrasensitive nonisotopic detection, Science 281 (1998) 2016–2018.

[78] E.R. Goldman, et al., Multiplexed toxin analysis using four colors of quantum dot fluororeagents, Anal. Chem. 76 (2004) 684–688.

[79] C. Peng, et al., Simultaneous and sensitive determination of multiplex chemical residues based on multicolor quantum dot probes, Biosens. Bioelectron. 24 (2009) 3657–3662.

[80] B. Berg, et al., Cellphone-based hand-held microplate reader for point-of-care testing of enzyme-linked immunosorbent assays, ACS Nano 9 (2015) 7857–7866.

[81] Z. Li, et al., Rapid and sensitive detection of protein biomarker using a portable fluorescence biosensor based on quantum dots and a lateral flow test strip, Anal. Chem. 82 (2010) 7008–7014.

[82] J. Hu, et al., Sensitive and quantitative detection of C-reaction protein based on immunofluorescent nanospheres coupled with lateral flow test strip, Anal. Chem. 88 (2016) 6577–6584.

[83] N.A. Taranova, et al., 'Traffic light' immunochromatographic test based on multicolor quantum dots for the simultaneous detection of several antibiotics in milk, Biosens. Bioelectron. 63 (2015) 255–261.

[84] M. Hu, et al., Ultrasensitive, multiplexed detection of cancer biomarkers directly in serum by using a quantum dot-based microfluidic protein chip, ACS Nano 4 (2010) 488–494.

[85] G. Bertolin, et al., A FRET biosensor reveals spatiotemporal activation and functions of aurora kinase A in living cells, Nat. Commun. 7 (2016) 12674.

[86] I.L. Medintz, et al., Self-assembled nanoscale biosensors based on quantum dot FRET donors, Nat. Mater. 2 (2003) 630–638.

[87] E.R. Goldman, et al., A hybrid quantum dot–antibody fragment fluorescence resonance energy transfer-based TNT sensor, J. Am. Chem. Soc. 127 (2005) 6744–6751.

[88] X. Huang, et al., A resonance energy transfer between chemiluminescent donors and luminescent quantum-dots as acceptors (CRET), Angew. Chem. 118 (2006) 5264–5267.

[89] S. Mayilo, et al., Long-range fluorescence quenching by gold nanoparticles in a sandwich immunoassay for cardiac troponin T, Nano Lett. 9 (2009) 4558–4563.

[90] W. Ni, et al., Tailoring longitudinal surface plasmon wavelengths, scattering and absorption cross sections of gold nanorods, ACS Nano 2 (2008) 677–686.

[91] B. Jin, et al., Upconversion nanoparticles based FRET aptasensor for rapid and ultrasenstive bacteria detection, Biosens. Bioelectron. 90 (2016) 525–533.

[92] D. Liu, et al., Gold nanoparticle-based activatable probe for sensing ultralow levels of prostate-specific antigen, ACS Nano 7 (2013) 5568–5576.

[93] C. Li, et al., One-step in situ solid-substrate-based whole blood immunoassay based on FRET between upconversion and gold nanoparticles, Biosens. Bioelectron. 92 (2016) 335–341.

[94] M. Liu, et al., Distance-independent quenching of quantum dots by nanoscale-graphene in self-assembled sandwich immunoassay, Chem. Commun. 46 (2010) 7909–7911.

[95] J. Shi, et al., A graphene oxide based fluorescence resonance energy transfer (FRET) biosensor for ultrasensitive detection of botulinum neurotoxin A (BoNT/A) enzymatic activity, Biosens. Bioelectron. 65 (2015) 238–244.

[96] D. Bhatnagar, et al., Graphene quantum dots FRET based sensor for early detection of heart attack in human, Biosens. Bioelectron. 79 (2016) 495–499.

[97] R. de la Rica, M.M. Stevens, Plasmonic ELISA for the detection of analytes at ultralow concentrations with the naked eye, Nat. Protoc. 8 (2013) 1759–1764.

[98] N. Cheeveewattanagul, et al., Straightforward immunosensing platform based on graphene oxide-decorated nanopaper: a highly sensitive and fast biosensing approach, Adv. Funct. Mater. (2017). https://doi.org/10.1002/adfm.201702741.

[99] A. Raja, et al., Energy transfer from quantum dots to graphene and MoS2: the role of absorption and screening in two-dimensional materials, Nano Lett. 16 (2016) 2328–2333.

[100] H. Zang, et al., Nonradiative energy transfer from Individual CdSe/ZnS quantum dots to single-layer and few-layer tin disulfide, ACS Nano 10 (2016) 4790–4796.

Chapter 12

Microcantilever-Based Sensors

Sandeep K. Vashist[1], John H.T. Luong[2]

[1]*IDS Immunodiagnostic Systems Deutschland GmbH, Frankfurt am Main, Germany;* [2]*University College Cork, Cork, Ireland*

1. INTRODUCTION

The recent advances in technologies have led to the miniaturization of biosensor devices, which is desirable toward the development of portable diagnostics for the point-of-care need to process minimal sample volumes. Critical drawbacks of the conventional technologies, including the requirement of large sample volume, expensive instruments, skilled analysts, and complex procedures, could be obviated by the deployment of microcantilever-based sensors. The use of microcantilevers for diverse bioanalytical applications became widely known by the innovative work of Binnig et al., the Nobel Prize in Physics, 1986, for the invention of scanning tunneling microscopy (STM) in 1981 and atomic force microscopy (AFM) in 1986 [1]. This was followed by the work of Eigler et al. [2,3] to create the world's smallest logo of IBM by positioning 35 xenon atoms using a STM. This discovery clearly demonstrated that individual atoms on a surface could be manipulated in a controlled manner.

Microcantilevers have several prospective advantages over the conventional technologies as they are highly sensitive, inexpensive, robust, high-throughput, require minimal sample and low energy, have a rapid response time and follow a nonhazardous label-free procedure [4–6]. The recently developed nanocantilevers present a major technology breakthrough that has given a rise to nanoelectromechanical systems and has pushed the sensitivity limits to the femto–attogram range. The technology is prospective for the development of miniaturized and highly sensitive, high-throughput, and rapid multianalyte sensors. The extensive research in the microcantilever sensors has demonstrated diverse bioanalytical applications in biomedical sciences, healthcare, environmental monitoring, and industrial and clinical analysis. The sensors have been employed for disease diagnosis, DNA hybridization, and detection of chemicals such as explosives, warfare agents, volatile organic compounds (VOCs), glucose, and ionic species. An array of several thousand cantilevers,

Handbook of Immunoassay Technologies. https://doi.org/10.1016/B978-0-12-811762-0.00012-8

enabling real-time, label-free, rapid, and parallel detection of many analytes has been made possible. The use of microcantilevers for next generation lab-on-a-chip technology has also been demonstrated. This chapter provides the developments in microcantilevers sensors, their working principle, readout systems, and the multifarious applications together with the future prospects.

2. MICROCANTILEVERS AND THEIR MODES OF OPERATION

A microcantilever is the miniaturized counterpart of a diving board that moves up and down at regular intervals. Thus, it is a cantilever beam that is fixed at one of its ends. The movement of the microcantilever changes after the adsorption of a specific mass of analyte on its surface. Most of the microcantilevers have dimensions in micrometers, which are some hundreds of micrometers in length, some tens of micrometers in width, and some hundreds of nanometers in thickness. They have various shapes (Fig. 12.1) and are typically composed of silicon, silicon derivatives, or polymers.

The adsorption of molecules on the microcantilever changes its vibrational frequency (dynamic mode) or surface stress, which leads to the nanomechanical bending of the microcantilever (static mode). The adsorption onto the sensing surface of microcantilever that is composed of two chemically different surfaces produces a differential stress between the two surfaces and induces bending [4]. The changes in the resonant frequency and the bending of the microcantilever can be monitored by optical, piezoresistive, capacitive, interferometric, or other detection techniques.

The sensitivity of a microcantilever is dependent on applying an appropriate coating on only one side of the microcantilever, which would make it highly selective for the detection of particular analyte molecules. Microcantilevers can detect very small surface stress of the order of 10^{-4} N/m but the exact nature of the surface stress response of microcantilever during molecular adsorption is

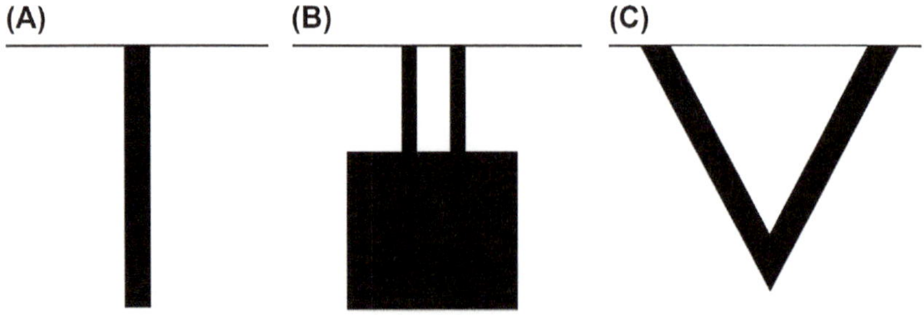

FIGURE 12.1 Various types of microcantilevers (top view): (A) Rectangular; (B) Double-legged; (C) Triangular. *Reprinted with permission from AZom. Com Pty Ltd, S.K. Vashist, A review of microcantilevers for sensing applications, AzoNano J. Nanotech. 3 (2007) 1–18.*

still unclear. Surface stress can be induced by conformational changes of the adsorbed molecules, electrostatic interactions between molecules, changes in the hydrophobicity of the surface, or other factors.

2.1 Operating Modes for Cantilever Mass Sensors

The mode of cantilever operation can be referred to as either static or dynamic depending on the measured parameter, i.e., resonant frequency or deflection, respectively (Fig. 12.2).

In the *static mode*, the adsorption of an analyte deflects the microcantilever due to the adsorption stress induced by molecular adsorption on just one side of the cantilever, which is converted to an electrical signal [7]. The deflection of the microcantilever is a result of the added mass and the resulting surface stress from the analyte [4]. However, the surface stress may not necessarily correlate with the amount of the bound analyte. The deflection can be upward or downward depending on the chemical bonding in the molecule. The static mode is simple as it does not require any actuation. It can be used in a liquid environment for the detection of single-base mismatch(es) in DNA sequence or biomolecular recognition events such as antibody–antigen and protein–protein. Most microcantilever-based sensors employ bimaterial (e.g., Au–Si) microcantilever beams as sensing elements, where the Au-coated side is usually coated with a specific receptor. The receptor surface is tensioned or relieved depending on the binding of the analyte with the receptor, which causes the nanomechanical deflection of microcantilever that can be measured by optical techniques.

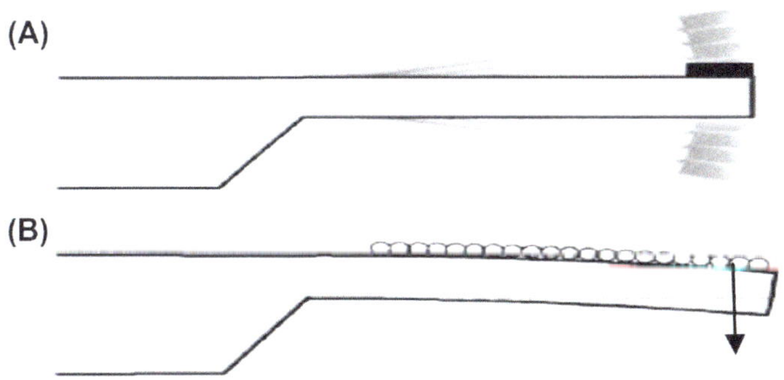

FIGURE 12.2 (A) Dynamic mode: absorption of analyte molecules in a sensor layer leads to shifting in resonance frequency. (B) Static mode: the cantilever bends owing to adsorption of analyte molecules and change of surface stress at the cantilever surface. *Reprinted with permission from Elsevier B.V, F. Battiston, J-P. Ramseyer, H. Lang et al., A chemical sensor based on a microfabricated cantilever array with simultaneous resonance-frequency and bending readout, Sens. Actuators B Chem. 77 (2001) 122–131.*

The *dynamic mode* (or resonant mode) involves the determination of mass changes during adsorption or desorption of the analyte by measuring the eigen frequency of the microcantilever [8]. However, it requires an excitation or actuation system to excite the cantilever at resonance. The microcantilever is used as a microbalance, which shifts to lower frequency on addition of mass on its surface. The adsorbed mass of the analyte on the cantilever can be calculated from the shift in its frequency [9] if the mechanical properties of the cantilever remain constant after the binding of the analyte. It was observed that a change of 1 Hz in the resonance frequency corresponds to a mass change of 1 pg on the microcantilever's surface [10]. The integrated electronics and simplicity of the setup make it prospective in comparison to the static method [11]. The dynamic mode is better than the static mode as it provides the absolute quantification of an analyte without any need for careful calibration and has better sensitivity. But it is complex and costly in comparison to the static mode and is not well suited for measurements in liquid media due to the liquid damping and consequent lowering of the Q factor [12].

The static and dynamic methods impose different constraints on the cantilever design for optimum sensitivity [11]. The static method requires long and soft cantilevers to achieve large deflections while the dynamic mode requires short and stiff cantilevers to achieve high operation frequencies.

3. DETECTION METHODS

Several detection methods have been developed for measuring the microcantilever deflection, as specified below.

3.1 Optical

Optical readout has been widely used to measure the microcantilever's deflection [13] (Fig. 12.3) and has been employed in AFM. It employs a low-power laser beam and a position-sensitive detector (PSD). The power of the laser beam is so low so that it does not affect the biomolecules bound to the surface of the microcantilever. The procedure involves the projection of the laser beam on the microcantilever and its reflection by the gold-coated microcantilever's surface so that the reflected beam falls on the PSD. The laser beam falls at a particular spot on the PSD when the microcantilever is undeflected, but it shifts position when the microcantilever is deflected due to the specific binding of analyte molecules.

The optical method is simple and sensitive and has the distinct advantage that it can detect the microcantilever's deflection in sub-nanometer range. However, it has several disadvantages based on which it cannot be used in all scenarios. The presence of a focused laser beam in a liquid cell environment can result in additional thermal management issues, which can lead to significant errors in analysis. Moreover, the laser alignment system is very costly

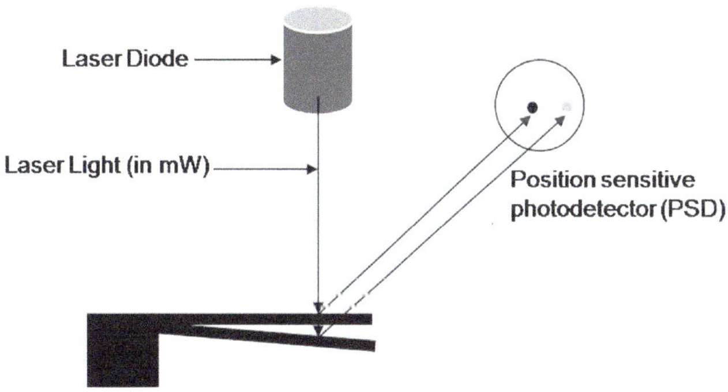

FIGURE 12.3 Optical detection system for detecting microcantilever's deflection. The reflected laser light from the deflected microcantilever's surface falls at a different position on the PSD. The deflection of the microcantilever is determined based on the distance between the two positions of the laser beam on the PSD. *Reprinted with permission from AZom. Com Pty Ltd, S.K. Vashist, A review of microcantilevers for sensing applications, AzoNano J. Nanotech. 3 (2007) 1–18.*

and requires high precision that can increase the cost of the measurement system, thereby making the analysis expensive. Furthermore, it reduces the portability and use of the system. Additionally, the optical detection techniques are sensitive to changes in the optical density of the sample and changes in the optical properties of the medium surrounding the microcantilever [14]. The optical readout of arrays is technologically challenging as it requires the same number of laser sources as the number of microcantilevers in an array. This technique is employed in commercial array platforms, where it requires sequential switching on and off of each laser source to avoid overlap of the reflected beams on the PSD. Furthermore, a scanning laser source could also be used, where a laser beam is scanned along the array to sequentially illuminate the free ends of each microcantilever [15].

3.2 Piezoresistive

The piezoresistive method [13,16,17] requires the embedding of a piezoresistive material near the top surface of the cantilever to determine the stress change occurring at the microcantilever's surface. The deflection of the microcantilever induces stress that puts a strain on the piezoresistor. It causes a change in resistance that can be measured electronically. The method has a unique advantage that the readout system can be integrated on the chip and it is compatible with microfabrication and miniaturization. Moreover, it works in nontransparent solutions and temperature controls can be easily implemented. But despite its widespread use for biosensor applications, it has several disadvantages. The intrinsic noise is the main concern as it directly affects the resolution and sensitivity of this method in comparison to the optical method [14]. But the reduction of the microcantilever's thickness might increase its sensitivity. The cross-sectional structure of a piezoresistive cantilever is complex as it requires a piezoresistor to be embedded in the cantilever, which makes the fabrication of such a cantilever with a composite structure more complicated. Therefore, the fabrication of thin, highly sensitive microcantilevers is extremely challenging. This is responsible for the lower sensitivity of piezoresistive readout based microcantilever sensors as its deflection resolution is only 1 nm, which is much lower in comparison to the deflection resolution of 1 A for the optical readout. Furthermore, the piezoelectric readout requires electrical connections to the microcantilever and their isolation from the solution.

The piezoresistor material in the microcantilever must be located closer to the surface of the microcantilever for maximum sensitivity. The piezoresistive coefficient of *n*-type silicon is higher than that for *p*-type as the type of doping used for the piezoresistive material affects its properties. The relative change in resistance of the piezoresistive material as a function of applied strain is represented by Eq. (12.1):

$$\frac{\Delta R}{R} = K_l \ \delta_l K_t \delta_t \tag{12.1}$$

where K denotes the gauge factor (GF) and subscripts l and t refer to the longitudinal and transverse parts, respectively, of the GF.

The sensitivity of a piezoresistor is related to the thickness t and the radius of curvature of microcantilever while the GF is proportional to Young's modulus E of the microcantilever. The GF can also be calculated directly by straining the cantilevers and measuring the change in resistance Eq. (12.2):

$$\text{GF} \cdot \delta = \frac{\Delta R}{R} \tag{12.2}$$

where δ is the strain in the material and R is the resistance. The GF should be of the order of 100 for a sensitive microcantilever.

The piezoresistive cantilever beam can be used as an arm of a Wheatstone bridge circuit (Fig. 12.4) where the resistance of the variable resistance arm $(R_0 + \Delta R)$ can be determined using the common voltage divider formula Eq. (12.3).

$$\begin{aligned} \Delta V &= V_0 \left\{ \frac{R_2}{R_1 + R_2} - \frac{R_3}{R_0 + \Delta R + R_3} \right\} \Rightarrow R_0 + \Delta R \\ &= R_3 \left\{ \frac{V_0(R_1 + R_2)}{R_2 V_0 - \Delta V(R_1 + R_2)} - 1 \right\} \end{aligned} \tag{12.3}$$

3.3 Capacitive

The capacitive method is based on the change in capacitance of a plane capacitor when the cantilever deflects due to adsorption of the analyte [18].

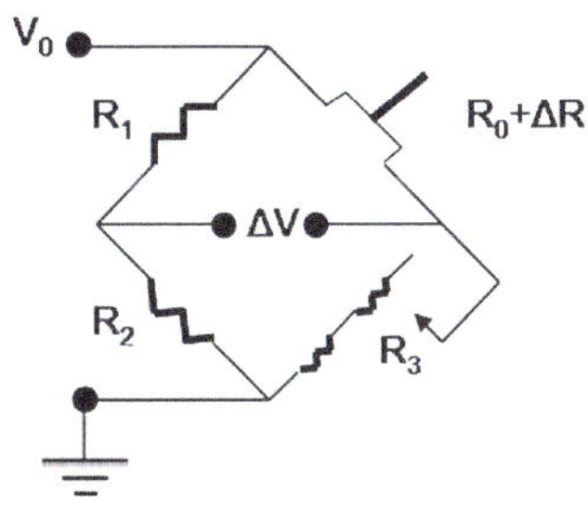

FIGURE 12.4 The Wheatstone Bridge Circuit used for the piezoresistive microcantilever. *Reprinted with permission from AZom. Com Pty Ltd, S.K. Vashist, A review of microcantilevers for sensing applications, AzoNano J. Nanotech. 3 (2007) 1–18.*

Here the microcantilever acts as one of two capacitor plates while a rigid beam with an electrode is the second capacitor plate [19,20]. The two electrodes are insulated from each other.

The capacitive readout is highly sensitive and provides absolute displacement. Capacitance sensing has low-temperature coefficients, low power dissipation, low noise, low cost of fabrication, and compatibility with very large-scale integration (VLSI) technology scaling, which is responsible for its widespread use. However, it is challenging to measure very small capacitance and the readout can only be performed in nonconductive environments. The method cannot be used to measure large displacements and does not work in electrolyte solutions due to faradic currents between the capacitive plates.

3.4 Piezoelectric

The piezoelectric method requires the deposition of a piezoelectric material, such as ZnO, on the cantilever. The deflection of the cantilever leads to the induction of transient charges in the piezoelectric layer due to the piezoelectric effect. However, large output signals can only be obtained if the piezoelectric film has a thickness well above the value corresponding to optimal mechanical characteristics.

3.5 Interferometry

The interferometry method is an optical detection technique [21,22] based highly sensitive method that provides a direct and absolute measurement of the displacement of the microcantilever. It is based on the interference of a reference laser beam with the laser beam reflected by the cantilever. The procedure involves bringing the cleaved end of an optical fiber in proximity to the cantilever's surface where a part of the light is reflected at the interface between the fiber and the surrounding medium while the other part is reflected at the cantilever back into the fiber. The two beams interfere with each other inside the fiber, resulting in an interference signal that is measured by a photodiode. The optical fiber located a few micrometers from the free end of the microcantilever can measure very small deflection in the range of 0.01 A but the positioning the fiber in such close proximity is very difficult. The method is not very prospective for sensing applications as it is less sensitive in liquids and works well only for small displacement.

3.6 Optical Diffraction Grating

The method is based on the formation of a diffraction pattern by the reflected laser light from the interdigitated cantilevers, where the intensity is proportional to the deflection of the microcantilever [23]. It could be used for infrared detection, chemical sensing, and AFM.

3.7 Charge-Coupled Device

The method is based on the use of a charge-coupled device (CCD) camera as the position-sensitive detector that measures the deflection of the microcantilever in response to the analyte detection [24]. However, it still requires extensive efforts to make it optimal for monitoring the small changes in microcantilever deflection. But as there have been striking improvements in the smartphone's camera and demonstration of its use for multifarious sensing applications [25–27], this method would be highly prospective for microcantilever sensors.

4. BENDING BEHAVIOR OF MICROCANTILEVERS

A uniform surface stress acting on an isotopic material will result in an increasing (in the case of compressive stress) and decreasing (in the case of tensile stress) surface area (Fig. 12.5), which results in the bending of the microcantilever. But there is a neutral plane in the microcantilever between the areas of compressive stress and tensile stress, which is not deformed. As a result of bending, a force F acting at a distance of x in the neutral plane results in a bending moment $M = Fx$. Therefore, the radius of curvature R is given by Eq. (12.4)

$$\frac{1}{R} = \frac{d^2z}{dx^2} = \frac{M}{EI} \tag{12.4}$$

where E and I are the apparent Young's modulus and the moment of inertia, respectively, for rectangular beams. I is calculated by Eq. (12.5)

$$I = \frac{bh^3}{12} \tag{12.5}$$

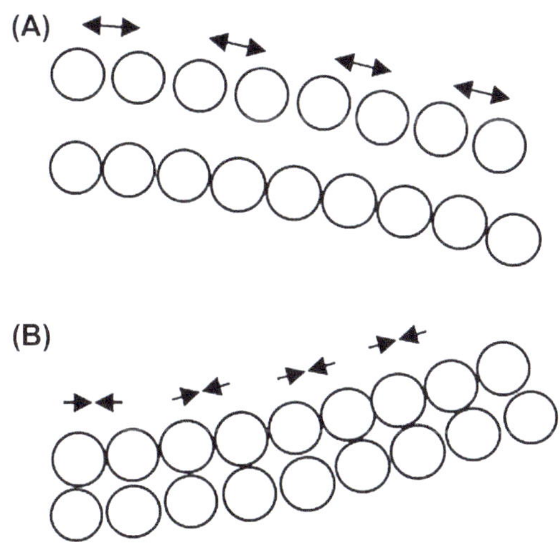

FIGURE 12.5 Bending of the cantilever beam in response to compressive and tensile stress: (A) compressive surface stress due to repulsion between the biomolecules leads to the downward (negative) deflection of the microcantilever; (B) tensile surface stress due to the attraction between molecules results in the upward (positive) deflection of the microcantilever. *Reprinted with permission from AZom. Com Pty Ltd, S.K. Vashist, A review of microcantilevers for sensing applications, AzoNano J. Nanotech. 3 (2007) 1–18.*

The change in the surface stress at one side of the microcantilever will cause its static bending, where the bending moment is given by Eq. (12.6):

$$M = \frac{\Delta\sigma bh}{2} \tag{12.6}$$

where $\Delta\sigma = \sigma_1 - \sigma_2$ is the differential surface stress, with σ_1 and σ_2 as the surface stresses at the upper and lower sides of the cantilever, respectively (Fig. 12.6). Inserting these values of I and M in Eq. (12.6) yields Stoney's formula [28] (Eq. 12.7).

$$\frac{1}{R} = \frac{6(1-\upsilon)\Delta\sigma}{Eh^2} \tag{12.7}$$

Taking into account the boundary conditions of a microcantilever ($R \gg L$), Eq. (12.7) can be solved and the resulting Eq. (12.8) can be used to calculate the displacement of the microcantilever.

$$s = \frac{3L^2(1-\upsilon)\Delta\sigma}{Eh^2} \tag{12.8}$$

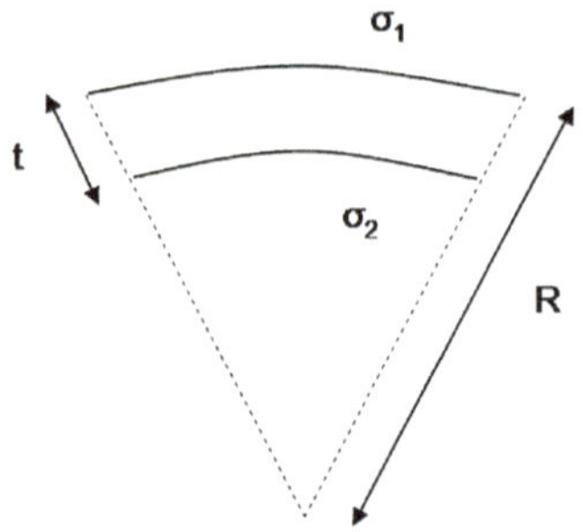

FIGURE 12.6 Lateral view of a thin cantilever beam of thickness t subjected to compressive stress. σ_1 is the stress at the upper surface and σ_2 is the stress at the lower surface of the cantilever. The cantilever beam bends with a constant radius of curvature R. *Reprinted with permission from AZom. Com Pty Ltd, S.K. Vashist, A review of microcantilevers for sensing applications, AzoNano J. Nanotech. 3 (2007) 1–18.*

The changes in surface stress can occur due to adsorption, electrostatic interactions between charged molecules on the surface, changes in the surface hydrophobicity, or conformational changes of the adsorbed molecules. The volume expansion of biomaterial cantilevers can also result in a static bending due to analyte adsorption due to the difference in the volume expansion coefficients of the two materials. Furthermore, adsorption-induced changes in surface charge density, the creation of ionized species resulting from receptor–analyte interaction, or swelling of the immobilized receptor layers, can also lead to microcantilever deflection [29]. But the deflection of the microcantilever, especially in case of biosensing applications, is a result of many effects that occur on its surface [14]. Physical variables such as changes in temperature, flow rate, pH, and ionic concentration can also result in the deflection of microcantilever [30].

Most of the sensing applications employ the reference cantilever so that the undesired artifacts and errors leading to the nonspecific bending of the microcantilever can be obviated [29].

5. FABRICATION OF MICROCANTILEVERS

The fabrication of microcantilevers requires the desired infrastructure and skills [4]. The microcantilever arrays have been developed commercially by various companies such as Concentris GmbH, Switzerland, and Micromotive Microtechnik, Germany.

Microcantilevers have been fabricated in various shapes, which usually depend on the type of detection method employed for the particular sensing application [4]. The square pads at the ends of cantilever beams are employed for capacitive detection-based sensors for increasing the detection sensitivity as the measured capacitance is proportional to the surface area of the parallel plates. In general, piezoresistive cantilevers are usually U-shaped, with components of the Wheatstone bridge circuit manufactured at their base points. The rectangular, paddle, and T-shaped cantilevers are employed for optical detection schemes. The shape and dimensions of the microcantilever determine the spring constant of the microcantilever while the resonance frequency is inversely proportional to its length. The sensitivity of the cantilever is directly proportional to its length-to-thickness ratio. The lower is the spring constant of the microcantilever, which is determined by the Young's modulus of the material, the higher is its detection sensitivity. Cantilevers are usually made from silicon or related materials having high Young's modulus but recently polymers like SU-8 having lower Young's modulus have also been used. Such polymer-based microcantilevers are more sensitive for the static deflection measurements in comparison with Si-based cantilevers.

6. MICROCANTILEVER-BASED SENSORS

Microcantilever-based sensors have very high sensitivity with detection limits lower than most of the conventional advanced techniques. They have been employed for the detection of a plethora of analytes (Table 12.1). Most commercial cantilevers, composed of silicon, silicon nitride, or silicon oxide, are available in various shapes and dimensions with varying force sensitivities. The use of recent integrated circuit and complementary metal oxide semiconductor technologies have further enabled the fabrication of extremely small microcantilever arrays. Therefore, microcantilever-based sensors have superior analytical performance in comparison with conventional analytical techniques, such as high-performance liquid chromatography and gas chromatography, which are complex, time-consuming, and expensive. They require bulky and expensive instrumentation, prolonged sample preparation, complex operational procedure, and highly skilled personnel. However, the microcantilever is compatible with microelectronics, which makes it feasible to cointegrate transducer and electronics onto a single chip [11,53]. Therefore, arrays of individual sensors can be assembled if numerous microcantilevers can be produced on a single chip. The placement of oscillator circuitry, amplifiers, and analog/digital converter near to the transducer significantly increases the bioanalytical performance via critically improved signal-to-noise ratio, which renders the chip less sensitive to external disturbances. Integrating the feedback eliminates the cross-talk (capacitive, inductive) between the

TABLE 12.1 Applications of Microcantilever-Based Sensors

Analytes	Detection Limit	Detection Method	References
Toluene	50 ppm	Capacitive	[31]
Octane	10 ppm	Capacitive	[31]
Carbonate	3–75 mg	Piezoelectric	[32]
Hydrofluoric acid	15 nmol	Optical	[33]
CrO_4^{2-}	10^{-9} M	Optical	[34]
Cs^{2+}	10^{-11} M	Optical	[35]
K^{+}	10^{-8} M	Optical	[35]
Hg^{2+}	10^{-11} M	Optical	[36]
Glucose	27–324 mg/dL	Optical	[37]
	0.2 mM	Optical	[38]
2-Mercaptoethanol	50 ppb	Optical	[39]
Hydrogen cyanide	ppm range	Piezoresistive	[40]
Freon-12	3.5 ng	Piezoresistive	[41]
Nitrotoluene	0.1 ng	Piezoelectric	[42]
2,4-Dinitrotoluene (DNT)	330 ppt	Optical	[43]
2,4,6-Trinitrotoluene (TNT)	70 pg	Piezoresistive	[44]
Pentaerythritol tetranitrate	ppt range	Piezoresistive	[45]
DDT	10 nM	Optical	[46]
Cd^{2+}	10^{-9} M	Optical	[47]
Hg^{+2}	0.1 mM	Optical	[48]
Zn^{+2}	0.1 mM	Optical	[48]
C-reactive proteins (CRP)	10 ng/mL	Piezoresistive	[49]
Myoglobin	85 ng/mL	Optical	[50]
PSA	0.2 ng/mL	Optical	[51]
Tributyrin	10 μM	Piezoelectric	[52]

connections supplying large heating currents for thermal actuation and the lines transmitting small signals of the piezoelectric Wheatstone bridge [11]. The integration also simplifies the packaging of the sensor device as it does not require shielding from external noise or electromagnetic interference. Such integrated microcantilever-based sensors have been designed for gaseous [11] and liquid-phase measurements [53].

6.1 Detection of Biomolecules

6.1.1 DNA

Microcantilever-based sensors were employed for the detection of specific DNA hybridization as the surface stress changes due to the specific binding of complementary oligonucleotides led to the nanomechanical deflection of microcantilevers [54]. The method was highly sensitive as it enabled the detection of a single base-pair mismatch between two 12-mer oligonucleotides. In another application, the hybridization was detected in a complex nonspecific background using complementary biomolecules in the concentration range of 75 nM to 2 μM [55]. Of interest is the microcantilever-based biosensing procedure using AuNPs-based DNA for the detection of specific DNA strands with a concentration of 23 pM and the detection of a single base-pair mismatch [56]. In another approach, the microcantilever-based sensor was employed for detection of thiol-functionalized single-stranded DNA that was adsorbed onto the surface [57].

The detection of a single 1587 base-pair DNA was demonstrated by employing microcantilevers that were bound to double-stranded DNA molecules via a thiol linker [58]. The piezoresistive microcantilevers [59,60] and polymer SU-8-based microcantilevers [61] have also been used for the sensing of DNA hybridization. The detection of 20-nM DNA hybridization was demonstrated using piezoresistive cantilevers [62]. In another application, microcantilever-based sensors were also used for the investigation of two DNA-binding proteins, i.e., transcription factors SP1 and NF-kappa B [63]. On the other hand, phase transition and stability of DNA have also been studied using microcantilever-based sensors [64]. An another potential application was the analysis of differential gene expression of gene 1−8U, a potential marker for cancer progression or viral infections, at the picomolar level, and detection of base-pair mismatches [65].

6.1.2 Prostate-Specific Antigen

A microcantilever-based biosensor was developed for the detection of prostate-specific antigen (PSA), which is present in patients having prostate cancer [66]. The procedure involves the binding of capture antibodies specific against PSA to the microcantilever's surface and the detection of PSA in the patient sample by the formation of immunocomplexes that lead to the nanomechanical bending of the microcantilever. The microcantilever deflection was detected with subnanometer precision using an optical method that employed a low-power laser beam and a photodetector. The developed method was cost-effective and detected PSA much lower than the clinically relevant range with a high precision equivalent to that of enzyme-linked immunosorbent assay (ELISA). In another study, the detection of PSA was demonstrated using piezoelectric nanomechanical microcantilevers based on the change in their resonant frequency [67].

6.1.3 Myoglobin

The detection of myoglobin in human serum was demonstrated using microcantilevers that were bound covalently to the anti–myoglobin monoclonal antibody on its upper surface via sulfosuccinimidyl 6-[3-(2-pyridyldithio)-propionamido] hexanoate (sulfo-LCSPDP) cross-linker [50]. The sensor detected 85 ng/mL myoglobin that falls within the physiological concentration range in healthy human serum.

6.1.4 Lipoproteins

The low-density lipoproteins (LDLs) and their oxidized form (oxLDLs) on heparin were differentiated by microcantilever sensing based on the measurement of surface stress on the microcantilevers [10]. The differentiation has potential utility in clinical diagnosis as oxLDLs leads to the accumulation of cholesterol in the aorta, which is associated with early stage coronary heart disease. The sensing procedure also detected the conformational changes in immunoglobulin G and blood serum albumin.

6.1.5 Glucose

A glucose oxidase bound microcantilever sensor was employed for the detection of glucose [38]. It was based on the deflection of microcantilever due to the change in surface stress induced by the reaction of glucose oxidase with glucose in solution. The microfluidic microcantilever-based sensor was unaffected by the common interferences that usually interfere with the blood glucose measurements. It detected glucose in the physiological range of 0–20 mM with an LOD of 0.2 mM, which enables its use for diabetic blood glucose monitoring. In another approach, a microcantilever-based sensor with prolonged stability was developed for the continuous monitoring of glucose in subcutaneous tissue [37]. It employed a biocompatible glucose-specific polymer, poly(acrylamide-ran-3-acrylamidophenylboronic acid) (PAA-ran-PAAPBA), which enabled the selective binding of glucose. The setup employs a magnetically driven vibrating perylene microcantilever in a microchamber filled with PAA-ran-PAAPBA solution, which was separated from the surroundings by a semipermeable membrane. The membrane allowed glucose to pass through and bind to the polymer. The determination of glucose concentration was done by the decrease in microcantilever's vibration in response to the reversible binding of glucose to the phenylboronic acid moiety of PAA-ran-PAAPBA. It detected glucose in the range of 27−324 mg/dL in just 3 min.

6.1.6 Tributyrin

A polysilicon microcantilever-based sensor was developed for hydrolyzed tributyrin, a triglyceride [52]. It detected up to 10 μM tributyrin with a good reproducibility and has better analytical performance in comparison with a potentiometric sensor that detected only 1 mM of tributyrin.

6.2 Detection of Gaseous Analytes

Palladium (Pd)-coated microcantilever has been used for the detection of hydrogen [68]. A prospective strategy involved the detection of hydrogen due to the differential stress induced by the adsorption of the Pd-based bimaterial cantilever [69], which resulted in the bending of the microcantilever. In another approach, an array of 10 Pd-coated microcantilevers [70] with capacitive readout was used for the detection of hydrogen. The capacitive readout circuit has wireless transmitters that can transmit data over a distance of 30 m. The coating of the microcantilever comprised of 90% Pd and 10% Ni, which could detect 0.4% hydrogen. Another approach also employed the Pd-coated microcantilevers for the detection of hydrogen and observed that the microcantilever's shape and surface roughness also impact the detection sensitivity [71]. Furthermore, the resonating microcantilever sensors were employed to determine the hydrogen storage capacity of carbon nanotubes (CNT) [72], which was found to be up to 6%, while the mass resolution was in the range of attograms.

The detection of CO was also demonstrated using a Ni-containing polyethylene oxide (PEO)−coated piezoresistive cantilevers [73]. Another prospective biosensing procedure involved in the detection of total carbonate in soil samples using piezoelectric microcantilever-based sensor based on the selective generation of CO_2 from soil samples [32]. The samples were put in the sample reaction chamber, while the microcantilever was placed in a flow cell. The developed sensor detected carbonate in the range of 3−75 mg with a precision of 1.7% and results agreeing well with those obtained by an established method. Furthermore, it was better than the quartz crystal microbalance (QCM) in sensitivity, resolution, and linearity. Of interest was the detection of Freon-12 using a piezoelectric microcantilever coated with a layer of microporous zeolite material [41,74]. The sensor was reversible, selective, highly sensitive, and reproducible, which detected Freon-12 down to 10 ppm [74] with the detection limit of 3.5 ng.

The detection of hydrogen peroxide at the nanomolar level was demonstrated using multilayer-coated microcantilevers that were functionalized through a layer-by-layer nanoassembly technique via intercalation of horseradish peroxidase. The deflection of the microcantilever was proportional to the concentration of hydrogen peroxide [75].

A polymer-coated cantilever array sensor was used to analyze mixtures of volatile organic compounds (VOCs) [76]. The sensor employed an electrostatically driven system to analyze a binary mixture of 1-butanol and *n*-octane. A magnetically actuated system was used for analyzing a ternary mixture of 1-butanol, toluene, and *n*-octane, and various alkanes and alcohols. In another approach, microcantilever sensor with capacitive readout was used to detect toluene and octane with LODs of 50 and 10 ppm, respectively [31].

6.3 Detection of Chemicals and Metal Ions

The microcantilever-based sensor has been developed for the detection of anhydrous HF using Si_3N_4 microcantilevers [77]. The microcantilever deflection was more sensitive at lower HF concentration, while the change in frequency could be employed for higher concentrations. The sensor detected HF with an LOD of 100 nmol, which was improved further to 15 nmol using SiO_x-coated microcantilevers. In another approach, a piezoresistive microcantilever-based sensor was demonstrated for the highly sensitive and specific detection of hydrogen cyanide gas (HCN) in the parts-per-million range [40]. The sensor had a response time of just 5 s and employed microcantilever coated with a keratin matrix functionalized with thiolated AuNPs. Microcantilevers have also been used to detect 2,4-dichlorophenoxyacetic acid (2,4-D) [78] by employing the specific biomolecular interactions between 2,4-D and monoclonal antibody against 2,4-D, which caused bending of the microcantilever. In another approach, microcantilever-based sensors detected dichloro diphenyl trichloroethane [46]. Of interest was the use of gold-coated silicon nitride microcantilevers for the detection of 2-mercaptoethanol vapors down to 50 ppb [39].

The metal ions have been detected using microcantilever-based sensors by employing a receptor molecule-coated surface that specifically recognizes the metal ions in solution. The binding of the metal ions to the receptors induces conformational changes and leads to the generation of interfacial stress, resulting in the deflection of the microcantilever. A prospective approach was the use of a microcantilever-based sensor for the detection of 10^{-9} M CrO_4^{2-} in a flow cell [34], which employed a self-assembled layer of triethyl-12-mercaptododecyl ammonium bromide on Au-coated microcantilever surface. In another approach, an antibody-coated microcantilever sensor selectively detected divalent Cd(II) ions [47]. The procedure involved the specific coating of antibody against Cd(II)-EDTA-BSA on the microcantilever surface that selectively detected Cd(II)-EDTA-BSA complex with an LOD of about 10^{-9} M. Of interest was the detection of the heavy metal ions using microcantilevers coated with a metal-binding protein, AgNt84-6, which can bind multiple atoms of Ni^{2+}, Zn^{2+}, Co^{2+}, Cu^{2+}, Cd^{2+}, and Hg^{2+} [48]. The metal-binding protein, bound to the Au-coated side of silicon nitride microcantilevers via linker groups, selectively discriminated between the metal ions that bind to AgNt84-6 and those that do not bind. A prospective strategy was the detection of mercury (Hg) vapor using Au-coated microcantilever-based sensor as Au chemisorbs specifically to Hg vapor with a high affinity [69,79]. In another approach, the piezoelectric cantilevers detected Hg down to a concentration of 50 ppb [79]. But the adsorption of Hg on Au is irreversible.

6.4 Detection of Humidity and pH

Microcantilevers have been used as humidity sensors by coating one side of the cantilever with a material that absorbs water. In one approach, one side of a silicon oxide cantilever was coated with Au, while the other side was untouched [80]. The water absorbs specifically to the untouched and exposed silicon oxide side of the cantilever that led to the microcantilever's deflection, while the Au surface was inert to water vapor absorption. In another approach, the humidity in the environment was measured by coating one side of the microcantilever with gelatin, which binds selectively to the water vapor and causes the microcantilever's deflection [81]. Furthermore, the microcantilevers were coated with hygroscopic materials like phosphoric acid that detected water vapor with picogram mass resolution [82]. The adsorption of water vapor on the coated surface of the microcantilever led to the change in its resonance frequency and caused the deflection of the microcantilever. The detection sensitivity can be increased by coating the microcantilevers with materials having a high affinity for the analyte.

Microcantilever coated with self-assembled monolayers of mercaptohexadecanoic acid (MHA, hydrophilic) and hexadecanethiol (HDT, hydrophobic) defects in the presence of hydrogen ions due to the induction of interfacial stress, which depends on pH and ionic strength [54,83]. The deflection of SiO_2

and silicon nitride microcantilevers coated with 4-aminobutyltriethoxysilane, 11-mercaptoundecanoic acid, and Au/Al was proportional over the pH range of 2–12. Aminosilane-modified SiO_2/Au microcantilevers detected pH in the range of 2–8 by demonstrating 49 nm deflection per pH unit. On the other hand, Si_3N_4/Au microcantilevers worked in the pH ranges of 2–6 and 8–12 by showing 30 nm deflection per pH unit [30,34]. In another approach, microcantilevers coated with poly(methacrylic acid) and poly(ethylene glycol) dimethacrylate were also sensitive to change pH [84]. Similarly, hydrogel coatings were also observed to be sensitive to pH [85].

6.5 Detection of Explosives and Monitoring of Ammunition

A lot of efforts have been demonstrated by researchers for the development of "nose on a chip" [86,87] using an array of microcantilevers, where each microcantilever is coated specifically to detect a particular organic compound. A prospective approach was the detection of trinitrotoluene (TNT) using microcantilever coated with platinum or a transition metal, which reacts with TNT when heated to 570°C and held at that temperature for 0.1 s. The reaction of TNT with Pt-coated microcantilever results in a miniexplosion, thereby enabling easy detection of explosives in airport luggage and landmines [88]. In another approach, Co-exchanged BEA zeolite-modified paddle-shaped cantilevers were used for the detection of nitrotoluene down to 1 ppm [42]. Of interest was the detection of organophosphorus compounds such as dimethyl methylphosphonate (DMMP) down to parts per billion using piezoresistive SiO_2 microcantilever coated with a self-assembled composite layer of Cu^{2+}/11-mercaptoundecanoic acid [89].

The detection of DMMP was also demonstrated using an array of lead zirconium titanate (PZT)/stainless-steel microcantilever sensors, which were coated with various metal oxide adsorbents at their tip [90]. Another prospective approach was the selective detection of peroxide vapors using microcantilever coated with a SAM, which undergoes chain polymerization in the presence of peroxide radicals that results in the deflection of microcantilever [91]. Of interest was the high-throughput chemical sensing and analysis using a multiplex cantilever array. The microcantilevers were coated with alkane thiols having different functional end groups, which were specific for the chemical sensing of toluene and water vapor [92]. Another potential application was the use of SXFA-[poly(1-(4-hydroxy-4-trifluoromethyl-5,5,5-trifluoro)pent-1-enyl)methylsiloxane]-polymer-coated microcantilevers to detect 2,4-dinitrotoluene (DNT) down to 300 ppt in a few seconds [43]. The sensor was reversible and could detect DNT continuously for a year as the sensor coating could withstand repeated exposure to varying levels of DNT concentrations.

Microcantilever-based sensors with remote wireless monitoring have been developed to monitor the missile storage and maintenance [36] to investigate the ammunition lifetime. This is based on the determination of humidity, temperature, pressure, shock, corrosion, and propellant degradation. Single chip detectors with electronics and telemetry with several hundred cantilevers in an array could simultaneously monitor, identify, and quantify many important analytical parameters to continuously monitor the ammunition.

7. ELECTRONIC NOSE

An electronic nose was designed using an array of silicon microcantilevers, where each cantilever was coated with various sensor layers such as metals, self-assembled monolayers, or polymers (Fig. 12.7A). The physisorption or chemisorption of analyte molecules on the microcantilevers lead to its deflection and/or a change in its resonance frequency. The microcantilever's deflection was read by an optical method (Fig. 12.7B). However, the readout of a large number of microcantilevers can only be done via multiplexing by switching on and off eight light sources at varied time intervals. The kinetics of the swelling process were related to the vapor pressure and the solubility characteristics of the analyte in the polymers, while the effects related to heat transfer or mass change could be used for analyte detection [86]. Thus, the exposure of "electronic nose" to a gaseous analyte mixture results in the individual electronic response from each microcantilever in the array. Of interest is the demonstration of an electronic nose for the characterization and identification of gaseous analytes [93], such as primary alcohols, natural flavors, and water vapor.

8. NANOCANTILEVERS

The trend toward decreasing the size of mechanical transducers into the submicrometer domain has initiated the quest for the fabricating ultrasensitive nanocantilevers. The smaller resonant cantilevers favor finer resolvable mass adsorption, higher resonance frequency, and lower air damping. However, at the nanoscale level, the mechanical behavior of nanocantilever starts resembling the vibrational modes of molecules and atoms. It is possible to attain extremely high fundamental frequencies approaching those of vibrational molecular modes in case of nanocantilevers [94]. The miniaturization of resonant cantilever improves its mass-detecting sensitivity, while specific nanomaterials such as carbon nanotubes could provide the desired superior mechanical characteristics to the silicon cantilever, i.e., low density, high elastic modulus, high Q factor, and prolonged chemical

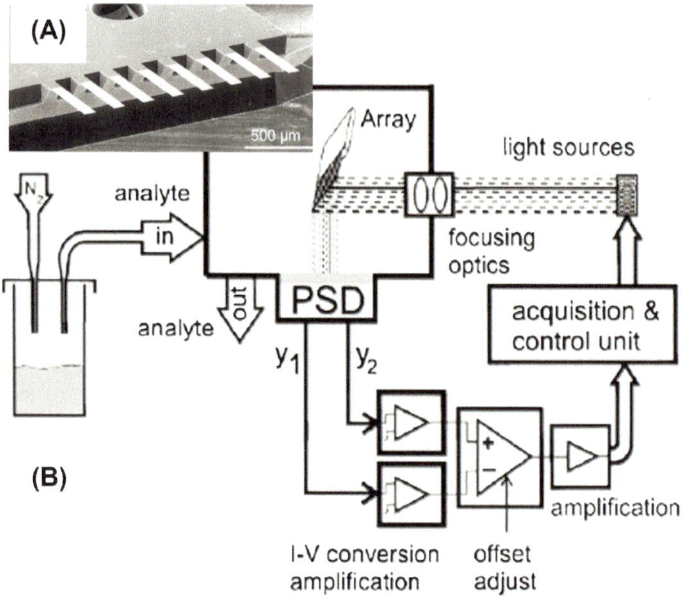

FIGURE 12.7 (A) Scanning electron microscopy image of a microcantilever array prior to deposition of sensor coatings. The array was produced from silicon by combined dry and wet etching techniques. (B) Schematic setup of the sample chamber with analyte inlet and outlet, cantilever deflection readout electronics, and control. The acquisition and control unit sequentially switches the light sources on and off via a time-multiplexing scheme (40 ms for each cantilever). The laser light ($\lambda = 760$ nm) is directed via focusing optics onto the apex of the cantilevers of the array, then reflected and collected by a linear position sensitive detector (PSD). The photocurrents from opposing electrodes are converted into voltages. The voltage difference is adjusted by an offset to allow optimal amplification and fed into the data acquisition and control unit. Data processing is performed offline on a personal computer. *Reprinted with permission from Elsevier B.V, M.K. Baller, H.P. Lang, J Fritz, et al., A cantilever array-based artificial nose, Ultramicroscopy 82 (2000) 1–9.*

stability in the atmosphere. Nanomechanical resonators with a mass of 2.34×10^{-18} g and a resonance frequency of 115 MHz have been fabricated [95]. Furthermore, mass sensitivity of only a few femtograms has been reported using nanocantilevers [96]. Of interest was the use of silicon nitride nanocantilevers 90 nm thick for detecting single 1578 base-pair DNA [58] with a mass of only 0.23 attograms. The procedure involved the placement of nanoscale Au dots at the ends of the nanocantilevers for capturing sulfide-modified double-stranded DNA, while scanning laser beams measured the vibrational frequency of the nanocantilevers. The nanocantilevers eliminated the need for polymerase chain reaction (PCR) amplification to detect DNA, which simplified the screening of specific gene sequences and mutations. Ultrathin resonance nanocantilevers of the order of 10 nm have also been developed in aluminum–molybdenum composites [97]. Furthermore, the detection of viruses has been demonstrated using an array of antibody-bound nanocantilevers of varying length and thickness of about 30 nm [98].

9. COMMERCIAL AVAILABILITY

Several companies provide the components of microcantilever-based sensors (Table 12.2), which include multifarious microcantilevers (various materials and dimensions), microcantilever-based sensors and analytical systems, and various other components. The functionalization of individual cantilevers in an array has been demonstrated using various instruments such as the CantiSpot platform by Cantion and the Cantisens FU-401 by Concentris.

TABLE 12.2 Companies Dealing in Various Components Pertaining to Microcantilever-Based Sensors

Company	Websites
Concentris GmbH	http://www.concentris.ch
Micromotive GmbH	http://www.micromotive.de/G/Octosensis_e.php
Nanoworld	http://www.nanoworld.com/probes_catalog.html
NanoAndMore GmbH	http://nanoandmore.com/index.php
NT-MDT	http://www.ntmdt-tips.com/
Atomic Force F&E GmbH	http://www.atomicforce.de/Olympus-Cantilevers.php
Novascan	http://www.novascan.com/products/afm_tips.php
Nanosensors	http://www.nanosensors.com/products_overview.html
Veeco Probes	https://www.veecoprobes.com/
Nanotools GmbH	http://nano-tools.com/
Asylum Research	http://www.asylumresearch.com/
Innovative Solutions Bulgaria Ltd.	http://www.budgetsensors.com/

10. CONCLUSIONS AND FUTURE TRENDS

Microcantilevers have been advocated for diversified biosensing, biomedical diagnostics, environmental monitoring, chemical sensing, etc. Their distinct advantages include high sensitivity, low cost, high throughput, minimal analyte volume requirement (in microliters), a rapid procedure with a fewer number of steps, and low power requirement. Furthermore, the fabrication of an array of microcantilevers enables simultaneous multiplex detection. Although the technology seems prospective, there are still several limitations that require further improvements such as the optimization of microcantilever designs, responsive coatings, and readout schemes. The differential pairs of microcantilevers can eliminate the interfering factors, while the use of microcantilevers with nanostructured coatings can enhance the signal. Furthermore, a side range of materials can be employed for the fabrication and modification of microcantilevers that can impart prolonged stability [99].

The rise of nanocantilevers has drastically increased the sensitivity down to attogram level, which has made it feasible to count the number of bound analyte molecules. A low-quality factor for resonating microcantilevers in liquid could be achieved by fabricating hollow cantilevers that can be filled with biochemical liquid. The confinement of fluid to the inside of a hollow cantilever would enable integration with the conventional microfluidic systems, which would significantly increase the sensitivity by eliminating high damping and viscous drag [100]. The use of nanowires [101], single-wall carbon nanotubes [102], graphene sheets, [103] or other nanocomposites will further increase the sensitivity of microcantilever-based sensors. Approaches to fabrication are also focused on nanostructures to increase the surface area. Furthermore, the use of nanostructured surfaces will substantially increase the surface area, thereby resulting in increased number of bound receptor molecules that can detect analyte with high sensitivity.

The future developments of microcantilever-based sensors should be directed toward the multiplex detection of biomarkers, which can effectively diagnose a particular disease via a specific clinical score. Taking into consideration the rapid developments in fabricating an array of microcantilevers and the successful integration of readout and microfluidic mechanisms for such arrays, it would be the most prominent application. However, extensive research efforts are required to develop a highly reproducible bioanalytical procedure with desired high precision, sensitivity, and specificity. Moreover, the development of a robust procedure would require extensive testing using clinical samples and methods compared with established clinically accredited technologies. The rapid advances in complementary technologies would further facilitate the development of critically improved microcantilever-based sensors.

REFERENCES

[1] G. Binnig, C.F. Quate, C. Gerber, Atomic force microscope, Phys. Rev. Lett. 56 (1986) 930–933.

[2] D.M. Eigler, E.K. Schweizer, Positioning single atoms with a scanning tunnelling microscope, Nature 344 (1990) 524–526.

[3] D.M. Eigler, P.S. Weiss, E.K. Schweizer, et al., Imaging Xe with a low-temperature scanning tunneling microscope, Phys. Rev. Lett. 66 (1991) 1189–1192.

[4] K.M. Goeders, J.S. Colton, L.A. Bottomley, Microcantilevers: sensing chemical interactions via mechanical motion, Chem. Rev. 108 (2008) 522–542.

[5] S.K. Vashist, A review of microcantilevers for sensing applications, AzoNano J. Nanotech. 3 (2007) 1–18.

[6] S.K. Vashist, H. Holthöfer, Microcantilevers for sensing applications, Meas. Control 43 (2010) 84–88.

[7] G. Wu, R.H. Datar, K.M. Hansen, et al., Bioassay of prostate-specific antigen (PSA) using microcantilevers, Nat. Biotech. 19 (2001) 856–860.

[8] P. Oden, G. Chen, R. Steele, et al., Viscous drag measurements utilizing microfabricated cantilevers, Appl. Phys. Lett. 68 (1996) 3814–3816.

[9] J. Barnes, R. Stephenson, C. Woodburn, et al., A femtojoule calorimeter using micromechanical sensors, Rev. Sci. Inst. 65 (1994) 3793–3798.

[10] F. Battiston, J.-P. Ramseyer, H. Lang, et al., A chemical sensor based on a microfabricated cantilever array with simultaneous resonance-frequency and bending readout, Sens. Actuators B Chem. 77 (2001) 122–131.

[11] D. Lange, C. Hagleitner, A. Hierlemann, et al., Complementary metal oxide semiconductor cantilever arrays on a single chip: mass-sensitive detection of volatile organic compounds, Anal. Chem. 74 (2002) 3084–3095.

[12] J. Tamayo, A.D. Humphris, A.M. Malloy, et al., Chemical sensors and biosensors in liquid environment based on microcantilevers with amplified quality factor, Ultramicroscopy 86 (2001) 167–173.

[13] G. Meyer, N.M. Amer, Novel optical approach to atomic force microscopy, Appl. Phys. Lett. 53 (1988) 1045–1047.

[14] L.G. Carrascosa, M. Moreno, M. Álvarez, et al., Nanomechanical biosensors: a new sensing tool, Trends Anal. Chem. 25 (2006) 196–206.

[15] L.M.L. Gomez, M.A. Sanchez, F.J.T. De Miguel, System and Method for Detecting the Displacement of a Plurality of Micro-and Nanomechanical Elements, Such as Microcantilevers, 2010. US Patent 7646494.

[16] J. Thaysen, A. Boisen, O. Hansen, et al., Atomic force microscopy probe with piezoresistive read-out and a highly symmetrical Wheatstone bridge arrangement, Sens. Actuators A Phys. 83 (2000) 47–53.

[17] M. Yang, X. Zhang, K. Vafai, et al., High sensitivity piezoresistive cantilever design and optimization for analyte-receptor binding, J. Micromech. Microeng. 13 (2003) 864.

[18] N. Blanc, J. Brugger, N. De Rooij, et al., Scanning force microscopy in the dynamic mode using microfabricated capacitive sensors, J. Vac. Sci. Technol. B 14 (1996) 901–905.

[19] J. Brugger, R. Buser, N. De Rooij, Micromachined atomic force microprobe with integrated capacitive read-out, J. Micromech. Microeng. 2 (1992) 218–220.

[20] T. Göddenhenrich, H. Lemke, U. Hartmann, et al., Force microscope with capacitive displacement detection, J. Vac. Sci. Technol. A 8 (1990) 383–387.

[21] R. Erlandsson, G. Mcclelland, C. Mate, et al., Atomic force microscopy using optical interferometry, J. Vac. Sci. Technol. A 6 (1988) 266–270.

[22] D. Rugar, H. Mamin, P. Guethner, Improved fiber-optic interferometer for atomic force microscopy, Appl. Phys. Lett. 55 (1989) 2588–2590.

[23] S. Manalis, S. Minne, A. Atalar, et al., Interdigital cantilevers for atomic force microscopy, Appl. Phys. Lett. 69 (1996) 3944–3946.

[24] B. Kim, O. Mader, U. Weimar, et al., Detection of antibody peptide interaction using microcantilevers as surface stress sensors, J. Vac. Sci. Technol. B 21 (2003) 1472–1475.

[25] S.K. Vashist, P.B. Luppa, L.Y. Yeo, et al., Emerging technologies for next-generation point-of-care testing, Trends Biotechnol. 33 (2015) 692–705.

[26] S.K. Vashist, O. Mudanyali, E.M. Schneider, et al., Cellphone-based devices for bioanalytical sciences, Anal. Bioanal. Chem. 406 (2014) 3263–3277.

[27] S.K. Vashist, E.M. Schneider, J.H. Luong, Commercial smartphone-based devices and smart applications for personalized healthcare monitoring and management, Diagnostics 4 (2014) 104–128.

[28] G.G. Stoney, The tension of metallic films deposited by electrolysis, Proc. R. Soc. Lond. A. Math. Phys. Sci. 82 (1909) 172–175.

[29] A. Boisen, T. Thundat, Design & fabrication of cantilever array biosensors, Mater. Today 12 (2009) 32–38.

[30] H.-F. Ji, K. Hansen, Z. Hu, et al., Detection of pH variation using modified microcantilever sensors, Sens. Actuators B Chem. 72 (2001) 233–238.

[31] J. Amírola, A. Rodríguez, L. Castaner, et al., Micromachined silicon microcantilevers for gas sensing applications with capacitive read-out, Sens. Actuators B Chem. 111 (2005) 247–253.

[32] M. Plata, J. Hernando, M. Zougagh, et al., Characterization and analytical validation of a microcantilever-based sensor for the determination of total carbonate in soil samples, Sens. Actuators B Chem. 134 (2008) 245–251.

[33] J. Mertens, E. Finot, M.-H. Nadal, et al., Detection of gas trace of hydrofluoric acid using microcantilever, Sens. Actuators B Chem. 99 (2004) 58–65.

[34] H.-F. Ji, T. Thundat, R. Dabestani, et al., Ultrasensitive detection of CrO_4^{2-} using a microcantilever sensor, Anal. Chem. 73 (2001) 1572–1576.

[35] H.-F. Ji, E. Finot, R. Dabestani, et al., A novel self-assembled monolayer (SAM) coated microcantilever for low level caesium detection, Chem. Commun. 6 (2000) 457–458.

[36] T. Thundat, P. Oden, R. Warmack, Microcantilever sensors, Microsc. Therm. Eng. 1 (1997) 185–199.

[37] X. Huang, S. Li, J.S. Schultz, et al., A MEMS affinity glucose sensor using a biocompatible glucose-responsive polymer, Sens. Actuators B Chem. 140 (2009) 603–609.

[38] J. Pei, F. Tian, T. Thundat, Glucose biosensor based on the microcantilever, Anal. Chem. 76 (2004) 292–297.

[39] P.G. Datskos, I. Sauers, Detection of 2-mercaptoethanol using gold-coated micromachined cantilevers, Sens. Actuators B Chem. 61 (1999) 75–82.

[40] T.L. Porter, T.L. Vail, M.P. Eastman, et al., A solid-state sensor platform for the detection of hydrogen cyanide gas, Sens. Actuators B Chem. 123 (2007) 313–317.

[41] J. Zhou, P. Li, S. Zhang, et al., Self-excited piezoelectric microcantilever for gas detection, Microelectron. Eng. 69 (2003) 37–46.

[42] M. Urbiztondo, I. Pellejero, M. Villarroya, et al., Zeolite-modified cantilevers for the sensing of nitrotoluene vapors, Sens. Actuators B Chem. 137 (2009) 608–616.

[43] L. Pinnaduwage, T. Thundat, J. Hawk, et al., Detection of 2, 4-dinitrotoluene using microcantilever sensors, Sens. Actuators B Chem. 99 (2004) 223–229.

[44] L. Pinnaduwage, A. Wig, D. Hedden, et al., Detection of trinitrotoluene via deflagration on a microcantilever, J. Appl. Phys. 95 (2004) 5871–5875.

[45] L. Pinnaduwage, V. Boiadjiev, J. Hawk, et al., Sensitive detection of plastic explosives with self-assembled monolayer-coated microcantilevers, Appl. Phys. Lett. 83 (2003) 1471–1473.

[46] M. Alvarez, A. Calle, J. Tamayo, et al., Development of nanomechanical biosensors for detection of the pesticide DDT, Biosens. Bioelectron. 18 (2003) 649–653.

[47] S. Velanki, S. Kelly, T. Thundat, et al., Detection of Cd (II) using antibody-modified microcantilever sensors, Ultramicroscopy 107 (2007) 1123–1128.

[48] S. Cherian, R.K. Gupta, B.C. Mullin, et al., Detection of heavy metal ions using protein-functionalized microcantilever sensors, Biosens. Bioelectron. 19 (2003) 411–416.

[49] K.W. Wee, G.Y. Kang, J. Park, et al., Novel electrical detection of label-free disease marker proteins using piezoresistive self-sensing micro-cantilevers, Biosens. Bioelectron. 20 (2005) 1932–1938.

[50] C. Grogan, R. Raiteri, G. O'Connor, et al., Characterisation of an antibody coated microcantilever as a potential immuno-based biosensor, Biosens. Bioelectron. 17 (2002) 201–207.

[51] T. Thundat, A. Majumdar, Microcantilevers for physical, chemical, and biological sensing, Sens. Sens. Biol. Eng. (2003) 337–355. Springer, Vienna.

[52] R.E. Fernandez, V. Hareesh, E. Bhattacharya, et al., Comparison of a potentiometric and a micromechanical triglyceride biosensor, Biosens. Bioelectron. 24 (2009) 1276–1280.

[53] C. Vancura, Y. Li, J. Lichtenberg, et al., Liquid-phase chemical and biochemical detection using fully integrated magnetically actuated complementary metal oxide semiconductor resonant cantilever sensor systems, Anal. Chem. 79 (2007) 1646–1654.

[54] J. Fritz, M. Baller, H. Lang, et al., Translating biomolecular recognition into nano-mechanics, Science 288 (2000) 316–318.

[55] R. Mckendry, J. Zhang, Y. Arntz, et al., Multiple label-free biodetection and quantitative DNA-binding assays on a nanomechanical cantilever array, Proc. Natl. Acad. Sci. USA 99 (2002) 9783–9788.

[56] M. Su, S. Li, V.P. Dravid, Microcantilever resonance-based DNA detection with nano-particle probes, Appl. Phys. Lett. 82 (2003) 3562–3564.

[57] M. Alvarez, L.G. Carrascosa, M. Moreno, et al., Nanomechanics of the formation of DNA self-assembled monolayers and hybridization on microcantilevers, Langmuir 20 (2004) 9663–9668.

[58] B. Ilic, Y. Yang, K. Aubin, et al., Enumeration of DNA molecules bound to a nano-mechanical oscillator, Nano Lett. 5 (2005) 925–929.

[59] R.L. Gunter, R. Zhine, W.G. Delinger, et al., Investigation of DNA sensing using piezoresistive microcantilever probes, IEEE Sens. J. 4 (2004) 430–433.

[60] R. Marie, H. Jensenius, J. Thaysen, et al., Adsorption kinetics and mechanical properties of thiol-modified DNA-oligos on gold investigated by microcantilever sensors, Ultramicroscopy 91 (2002) 29–36.

[61] M. Calleja, M. Nordstrom, M. Alvarez, et al., Highly sensitive polymer-based cantilever-sensors for DNA detection, Ultramicroscopy 105 (2005) 215–222.

[62] R. Mukhopadhyay, M. Lorentzen, J. Kjems, et al., Nanomechanical sensing of DNA sequences using piezoresistive cantilevers, Langmuir 21 (2005) 8400–8408.

[63] F. Huber, M. Hegner, C. Gerber, et al., Label free analysis of transcription factors using microcantilever arrays, Biosens. Bioelectron. 21 (2006) 1599–1605.

[64] S.L. Biswal, D. Raorane, A. Chaiken, et al., Using a microcantilever array for detecting phase transitions and stability of DNA. J. Lab. Autom. 11 (2006) 222–226.

[65] J. Zhang, H.P. Lang, F. Huber, et al., Rapid and label-free nanomechanical detection of biomarker transcripts in human RNA, Nat. Nanotechnol. 1 (2006) 214–220.

[66] G. Wu, H. Ji, K. Hansen, et al., Origin of nanomechanical cantilever motion generated from biomolecular interactions, Proc. Natl. Acad. Sci. USA 98 (2001) 1560–1564.

[67] J.H. Lee, K.S. Hwang, J. Park, et al., Immunoassay of prostate-specific antigen (PSA) using resonant frequency shift of piezoelectric nanomechanical microcantilever, Biosens. Bioelectron. 20 (2005) 2157–2162.

[68] H. Lang, A. Filippi, A. Tonin, et al., Towards a modular, versatile and portable sensor system for measurements in gaseous environments based on microcantilevers, Proced. Chem. 1 (2009) 208–211.

[69] Z. Hu, T. Thundat, R. Warmack, Investigation of adsorption and absorption-induced stresses using microcantilever sensors, J. Appl. Phys. 90 (2001) 427–431.

[70] D. Baselt, B. Fruhberger, E. Klaassen, et al., Design and performance of a microcantilever-based hydrogen sensor, Sens. Actuators B Chem. 88 (2003) 120–131.

[71] A. Fabre, E. Finot, J. Demoment, et al., Monitoring the chemical changes in Pd induced by hydrogen absorption using microcantilevers, Ultramicroscopy 97 (2003) 425–432.

[72] T. Ono, X. Li, H. Miyashita, et al., Mass sensing of adsorbed molecules in sub-picogram sample with ultrathin silicon resonator, Rev. Sci. Inst. 74 (2003) 1240–1243.

[73] A. Kooser, R. Gunter, W. Delinger, et al., Gas sensing using embedded piezoresistive microcantilever sensors, Sens. Actuators B Chem. 99 (2004) 474–479.

[74] J. Zhou, P. Li, S. Zhang, et al., Zeolite-modified microcantilever gas sensor for indoor air quality control, Sens. Actuators B Chem. 94 (2003) 337–342.

[75] X. Yan, X. Shi, K. Hill, et al., Microcantilevers modified by horseradish peroxidase intercalated nano-assembly for hydrogen peroxide detection, Anal. Sci. 22 (2006) 205–208.

[76] D. Then, A. Vidic, C. Ziegler, A highly sensitive self-oscillating cantilever array for the quantitative and qualitative analysis of organic vapor mixtures, Sens. Actuators B Chem. 117 (2006) 1–9.

[77] J. Mertens, E. Finot, T. Thundat, et al., Effects of temperature and pressure on microcantilever resonance response, Ultramicroscopy 97 (2003) 119–126.

[78] R. Raiteri, M. Grattarola, H.-J. Butt, et al., Micromechanical cantilever-based biosensors, Sens. Actuators B Chem. 79 (2001) 115–126.

[79] J. Adams, B. Rogers, L. Manning, et al., Piezoelectric self-sensing of adsorption-induced microcantilever bending, Sens. Actuators A. Phys. 121 (2005) 457–461.

[80] R. Berger, H. Lang, C. Gerber, et al., Micromechanical thermogravimetry, Chem. Phys. Lett. 294 (1998) 363–369.

[81] S. Singamaneni, M.C. Lemieux, H.P. Lang, et al., Bimaterial microcantilevers as a hybrid sensing platform, Adv. Mater. 20 (2008) 653–680.

[82] T. Thundat, G. Chen, R. Warmack, et al., Vapor detection using resonating microcantilevers, Anal. Chem. 67 (1995) 519–521.

[83] J. Fritz, M. Baller, H. Lang, et al., Stress at the solid-liquid interface of self-assembled monolayers on gold investigated with a nanomechanical sensor, Langmuir 16 (2000) 9694–9696.

[84] R. Bashir, J. Hilt, O. Elibol, et al., Micromechanical cantilever as an ultrasensitive pH microsensor, Appl. Phys. Lett. 81 (2002) 3091–3093.

[85] Y. Zhang, H.F. Ji, D. Snow, et al., A pH sensor based on a microcantilever coated with intelligent hydrogel, Instrum. Sci. Technol. 32 (2004) 361–369.

[86] M.K. Baller, H.P. Lang, J. Fritz, et al., A cantilever array-based artificial nose, Ultramicroscopy 82 (2000) 1–9.

[87] J. Yinon, Detection of explosives by electronic noses, Anal. Chem. 75 (2003) 98 A–105 A.

[88] G. Muralidharan, A. Wig, L.A. Pinnaduwage, et al., Adsorption-desorption characteristics of explosive vapors investigated with microcantilevers, Ultramicroscopy 97 (2003) 433–439.

[89] G. Zuo, X. Li, P. Li, et al., Detection of trace organophosphorus vapor with a self-assembled bilayer functionalized SiO2 microcantilever piezoresistive sensor, Anal. Chim. Acta 580 (2006) 123–127.

[90] Q. Zhao, Q. Zhu, W.Y. Shih, et al., Array adsorbent-coated lead zirconate titanate (PZT)/stainless steel cantilevers for dimethyl methylphosphonate (DMMP) detection, Sens. Actuators B Chem. 117 (2006) 74–79.

[91] J.P. Lock, E. Geraghty, L.C. Kagumba, et al., Trace detection of peroxides using a microcantilever detector, Thin Solid Films 517 (2009) 3584–3587.

[92] D. Raorane, S. Satyanarayana, A. Majumdar, Nano-chemo-mechanical sensor array platform for high-throughput chemical analysis, Sens. Actuators B Chem. 119 (2006) 466–474.

[93] H.P. Lang, Cantilever-Based Gas Sensing. Solid State Gas Sensing, Springer, 2009, pp. 1–24.

[94] N.V. Lavrik, M.J. Sepaniak, P.G. Datskos, Cantilever transducers as a platform for chemical and biological sensors, Rev. Sci. Inst. 75 (2004) 2229–2253.

[95] R.G. Knobel, A.N. Cleland, Nanometre-scale displacement sensing using a single electron transistor, Nature 424 (2003) 291–293.

[96] N.V. Lavrik, P.G. Datskos, Femtogram mass detection using photothermally actuated nanomechanical resonators, Appl. Phys. Lett. 82 (2003) 2697–2699.

[97] N. Nelson-Fitzpatrick, C. Ophus, E. Luber, et al., Synthesis and characterization of Au–Ta nanocomposites for nanomechanical cantilever devices, Nanotechnology 18 (2007) 355303.

[98] A.K. Gupta, P.R. Nair, D. Akin, et al., Anomalous resonance in a nanomechanical biosensor, Proc. Natl. Acad. Sci. USA 103 (2006) 13362–13367.

[99] M. Sepaniak, P. Datskos, N. Lavrik, et al., Microcantilever transducers: a new approach in sensor technology, Anal. Chem. 74 (2002) 568 A–575 A.

[100] T.P. Burg, S.R. Manalis, Suspended microchannel resonators for biomolecular detection, Appl. Phys. Lett. 83 (2003) 2698–2700.
[101] Y. Cui, Q. Wei, H. Park, et al., Nanowire nanosensors for highly sensitive and selective detection of biological and chemical species, Science 293 (2001) 1289–1292.
[102] G. Singh, P. Rice, R. Mahajan, Fabrication and mechanical characterization of a force sensor based on an individual carbon nanotube, Nanotechnology 18 (2007) 475501.
[103] A. Sakhaee-Pour, M. Ahmadian, A. Vafai, Applications of single-layered graphene sheets as mass sensors and atomistic dust detectors, Solid State Commun. 145 (2008) 168–172.

Chapter 13

Quartz Crystal Microbalance–Based Sensors

Sandeep K. Vashist[1], John H.T. Luong[2]
[1]*IDS Immunodiagnostic Systems Deutschland GmbH, Frankfurt am Main, Germany;* [2]*University College Cork, Cork, Ireland*

1. INTRODUCTION

Jacques and Pierre Curie observed in 1880 that a voltage proportional to the stress is produced by the application of a mechanical stress to certain materials, notably quartz [1]. Gunter Sauerbrey [2] coined the term "quartz crystal microbalance" (QCM) in the 1950s and pioneered the work that led to the use of quartz plate resonators as sensitive microbalances for thin films. The Sauerbrey's linear equation, $\Delta m = -C.\Delta f$, quantifies the mass change on the crystal's surface (Δm in ng/cm^2) to the frequency change (Δf) when the voltage is applied to oscillate the crystal at a specific frequency. The constant C is related to the resonant frequency, the active crystal area, the density of quartz, and the shear modulus of quartz.

Strictly, the above linear relationship is only valid for elastic subjects such as metallic coatings, metal oxides, and thin adsorbed layers, which do not affect energy dissipation during oscillation. Therefore, it does not apply to inelastic subjects including cells, polymers, and biomolecular systems, which lose energy due to viscous damping during oscillation. Soft or viscoelastic films also do not couple completely with the oscillating crystal, resulting in an underestimated mass change. Further, the Sauerbrey's equation is only applicable when the change in mass is less than 2% of the crystal mass. Consequently, QCM with dissipation monitoring (QCM-D) must be considered to measure the energy dissipation in biological samples.

As a shear mode device, QCM consists of a thin quartz disk with coated electrodes. The quartz crystal plate is cut to a specific orientation w.r.t. the crystal axes, i.e., AT- or BT-cut to facilitate the propagation of the acoustic wave perpendicular to the QCM's surface. The resonant frequency of the QCM depends on the angles w.r.t. the optical axis at which the wafer is cut from the crystal. The most commonly used angle is AT-cut with 35° 15′ from the Z axis of the crystal. The temperature dependence of the resonant frequency of the

Handbook of Immunoassay Technologies. https://doi.org/10.1016/B978-0-12-811762-0.00013-X

AT-cut crystal is zero at 25°C, reflecting its low-temperature coefficient at room temperature. Therefore, the AT-cut crystals are only subject to minimal changes in frequency due to variations in temperature.

In brief, the frequency of QCM decreases as the deposited mass on its surface increases as reflected by a negative sign in the Sauerbrey's equation. The mass sensitivity is dependent on the crystal thickness, which determines its resonant frequency. The thin QCMs with high resonant frequency and sensitivity are very fragile. Thus, most commercially available QCM crystals are 0.5 in. in diameter. The optically polished QCMs with reduced non-specificity are better for liquid-phase sensing. However, the changes in frequency and signal attenuation due to varying liquid viscosities require vigorous algorithms to extract the specific response signal from the noise. To date, QCMs have been widely used for the detection of biomarkers, clinically relevant analytes, cells, pathogens, environmental pollutants, and diversified bioanalytical applications (Table 13.1). The commercialization of this technique has received considerable attention owing to its portability and simplicity (Table 13.2).

Among immobilization strategies for binding an antibody to the QCM's surface as the detection molecule, passive adsorption has been the most widely used due to its simplicity [3]. However, the binding of antibodies often occurs in a random uncontrolled orientation, adversely affecting the detection sensitivity and specificity. The oriented immobilization of antibodies can be achieved by the use of crystallizable fragment (Fc)-binding proteins such as protein A, protein G, and protein A/G to bind a specific domain of the detection antibody. The antigen-binding sites (Fab region) of the detection antibody are completely free for binding antigens. Antibodies with amino or carboxyl groups can be covalently immobilized on the crystal's surface. In this context, the QCM's surface can be functionalized by a small organic molecule, e.g., 3-aminopropyltriethoxysilane (APTES), followed by glutaraldehyde activation to generate the aldehyde groups for binding the amino groups on the detection antibody. Other approaches involve polyelectrolytes, polymers, and Langmuir–Blodgett films. Antibodies can be thiolated, which then self-assemble on the crystal gold surface. Another popular procedure involves the binding of biotinylated antibodies to the streptavidin-coated QCM gold surface via the specific and strong biomolecular interaction between streptavidin and biotin.

Most of the conventional bioanalytical techniques such as Fourier transform infrared (FTIR) spectrometry, gas chromatography (GC), and mass spectrometry (MS) are time-consuming, need expensive instrumentation and specialized analysts, and can only perform the off-line analysis. Therefore, QCM is the most promising sensor technology that enables low-cost analysis, rapid response, portability, real-time label-free detection, and high sensitivity. It is a prospective technology for the highly sensitive detection.

TABLE 13.1 Commercial Quartz Crystal Microbalance Systems

Companies	Websites
Q-Sense	http://www.q-sense.com/
Elchema	http://www.elchema.com/EQCN.htm
Inficon	http://www.inficonthinfilmdeposition.com/en/index.html
Masscal	http://www.masscal.com
ElbaTech Srl	http://www.elbatech.com/
Resonant Probes	http://www.resonant-probes.de/index.htm
KSV	http://www.ksvltd.com/content/index/dissipativeqcm
Institute of Physical Chemistry, Polish Academy of Sciences	http://ichf.edu.pl/offers/instrum/quartz.htm
tectra GmbH	http://www.tectra-gmbh.com/qmb.htm
Stanford Research Systems, Inc.	http://www.thinksrs.com/products/QCM200.htm#
International Crystal Manufacturing Co, Inc	http://www.icmfg.com/qcm_crystals.html
QCM Research	http://www.qcmresearch.com/home.html
QCM Labs	http://www.qcmlab.com/index.htm
CH Instruments, Inc.	http://www.chinstruments.com/Products.html
Lap-Tech	http://www.laptech.com/
Sierra Sensors	http://www.sierrasensors.com/
Eco Chemie	http://www.ecochemie.nl/?pag=13/#EQCM

Reproduced with permission from Hindawi Publishing Corporation, S.K. Vashist, P. Vashist. Recent advances in quartz crystal microbalance-based sensors. J. Sens. (2011).

2. DETECTION OF BIOMOLECULES

An antibody-bound QCM is applicable for detecting prostate-specific antigen (PSA) and PSA-alpha 1-antichymotrypsin (ACT) complex in human serum (75%) [38]. The sensor employs the signal amplification provided by Au nanoparticles to detect PSA and PSA–ACT complex in 75% human serum with an LOD of 0.29 ng/mL with linearity up to 150 ng/mL. Of interest is the indirect competitive QCM immunosensor for the detection of C-reactive protein (CRP) [39]. The procedure involves the binding of 2 mg/mL of CRP

TABLE 13.2 Quartz Crystal Microbalance–Based Sensing Applications

Analytes	Detection Limits	References
Acetic acid, butyric acid, ammonia, dimethylamine, benzene, chlorobenzene	N.M.	[4]
Primary aliphatic alcohols such as ethanol, methanol, 1-propanol, 2-propanol	2–17 mg/L	[5]
Ethanol	1.73×10^{-2} μg	[6]
CH_3SH	100 ppb	[7]
DMMP, DMA, DCP, DCE	N.M.	[8]
DMMP	5–60 ppm	[9]
VOCs such as aromatics, chlororganics, ketones, and alcohols	N.M.	[10]
Toluene and p-xylene	N.M.	[11]
Toluene	N.M.	[12]
Xylene isomers	0–200 ppm	[13]
Alcohols, esters, acids, and aldehydes	N.M.	[14]
Bisphenol-A	0.01 ng/mL	[15]
TCDD	0.1–100 ng/mL	[16]
DDVP	6.5–32.5 ppm	[17]
1-NAP and 2-NAP	8.362×10^{-7} M for 1-NAP 2.146×10^{-7} M for 2-NAP	[18]
HCl	ppt level with a sensitivity of 0.1 ppm/Hz	[19]
	250 ppb/Hz sensitivity	[20]
Methane	0.05% (v/v)	[21]
	N.M.	[22]
NO	25 ppb/Hz sensitivity	[23]
NO_2	Sub-ppm level with a sensitivity of 1200 Hz/h for 50 ppm NO_2	[24]
Chloroform	N.M.	[25]
NH_3	N.M.	[26]

TABLE 13.2 Quartz Crystal Microbalance–Based Sensing Applications—cont'd

Analytes	Detection Limits	References
	N.M.	[27]
	0.3–15 ppm with LOD of 0.1 ppm	[27]
HCHO	3 ppm	[28]
L-mandelic acid	N.M.	[29]
Orange and melon flavors	N.M.	[30]
Ethanol, acetone, and trichloroethylene gas mixtures	N.M.	[31]
Humidity	Sensitivity of 12.32 Hz/RH%	[32]
Ag ions	N.M.	[33]
Olive oils	N.M.	[34]
Nanoparticles (Si and Ag nanopowder, rhodamine B, and ferrocene)	Linear up to 1300 μg/mL	[35]
Glucose	0.01–7.5 mM with LOD of 5 μM	[36]
Albumin	60–150 ppm	[37]
PSA and PSA-ACT	Linear dynamic range up to 150 ng/mL with LOD of 0.29 ng/mL	[38]
Total PSA	LOD 0.39 ng/mL	
CRP	N.M.	[39]
	0.003–200 ng/mL	[40]
EPGF	0.01–10 μg/mL	[41]
L-tryptophan	LOD 8.8 μM	[42]
Glycoproteins	50 μg/mL to 1 mg/mL	[43]
Folic acid	Linear range of 0–100 μM with LOD of 15.4 μM	[44]
Vibrio harveyi	10^3–10^7 CFU/mL	[45]
Campylobacter jejuni	N.M.	[46]
Desulfotomaculum	1.8 × 10^4–1.8 × 10^7 CFU/mL	[47]

Continued

TABLE 13.2 Quartz Crystal Microbalance–Based Sensing Applications—cont'd

Analytes	Detection Limits	References
Aflatoxin B1	0.1–4 ng/mL	[48]
Annexin A3	0.075–50 ng/mL	[49]
CA15.3	0.5–26 U/mL	
CD10	1×10^{-8} to 1×10^{-11} M	[50]
Human cardiac troponin T (cTnT)	0.5–4.5 ng/mL	[51]
Human IgG	0.005–20 pg/mL	[52]
Salmonella typhimurium	LOD 10–20 CFU/mL	[53]

N.M., not mentioned.
Reproduced with permission from Hindawi Publishing Corporation, S.K. Vashist, P. Vashist. Recent advances in quartz crystal microbalance-based sensors. J. Sens. (2011).

to the QCM's surface, followed by monitoring the frequency change when 200 μL of biotinylated antirat CRP antibody mixed with streptavidin-coated gold nanoparticle is introduced (Fig. 13.1). The competitive procedure shows 53% signal enhancement in comparison with the conventional

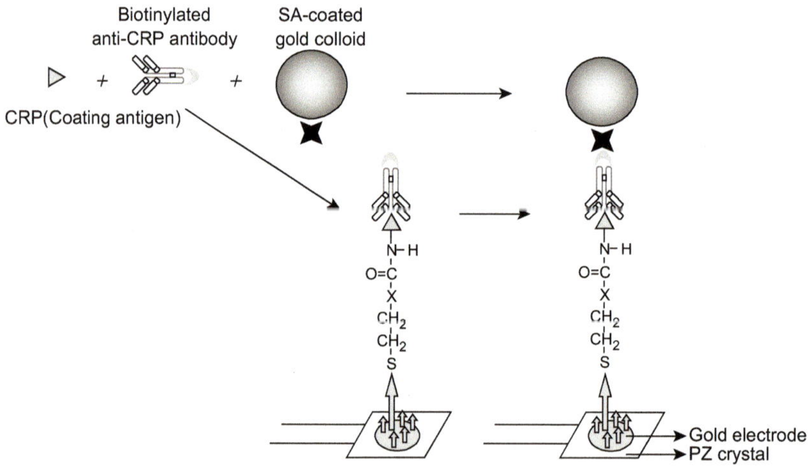

FIGURE 13.1 Schematic representation of the signal augmentation over QCM surface by the addition of gold colloid in the IC CRP measurement of this study. X is the residue remaining after antibody immobilization and represents the structure of $-CO-NH-(CH_2)_5-$ [39]. *Reproduced with permission from Hindawi Publishing Corporation, S.K. Vashist, P. Vashist, Recent advances in quartz crystal microbalance-based sensors, J. Sens. (2011).*

procedure based on the use of the unmodified antibody. CRP can also be detected using HRP-doped magnetic nanoparticles ($Fe_3O_4@SiO_2@Au$) as labels for signal enhancement [40]. The capture anti-CRP antibodies are bound to the QCM followed by the detection of CRP, which is detected by HRP, and anti-CRP detection antibodies bound $Fe_3O_4@SiO_2@Au$ nanocomposites. The sensor detects CRP in the range of 0.003–200 ng/mL with an LOD of 1 pg/mL. The results obtained by the sensor for the determination of CRP in diluted clinical human serum are in agreement with those obtained by ELISA.

A QCM-based biosensor for total PSA [54] involves the immobilization of anti-tPSA antibody to the QCM's surface, tPSA detection, and signal enhancement using anti-tPSA antibody-bound AuNPs. The sensor detects 0.29 ng/mL tPSA in 75% human serum when 40 nm AuNPs are employed. It precisely determines the tPSA concentration in patient samples. A dual-QCM immunosensor for human cardiac troponin T (cTnT) [51] is based on the functionalization of the QCM surface with SAM of cysteamine, followed by the immobilization of the monoclonal anti-cTnT antibody via glutaraldehyde cross-linking (Fig. 13.2). The sensor detects cTnT in human serum with an LOD of 0.008 ng/mL and high precision equivalent to that of electrochemiluminescent immunoassays. The Ab-bound QCM surface can be regenerated effectively using 1% (w/v) sodium dodecyl sulfate (SDS).

Of note is the detection of common acute lymphoblastic leukemia antigen (CD10) by CM-based immunobiosensor using anti-CD10 detection antibody-bound AuNPs for signal enhancement [50]. The sensor detects CD10 in the range of 1×10^{-8} to 1×10^{-11} M, using the bioanalytical procedure as specified in Fig. 13.3. QCM is coated with ZnO nanorods for the detection of breast cancer biomarker (CA15.3) [55]. Such nanorods are directly grown on the QCM's surface by wet chemistry and functionalized with APTES and glutaraldehyde for the binding of anti-CA15.3 antibodies (Fig. 13.4). It detects CA15.3 in the concentration range of 0.5–26 U/mL within 10 s. Of note is the detection of annexin A3 (ANXA3) by the QCM using polyclonal anti-ANXA3

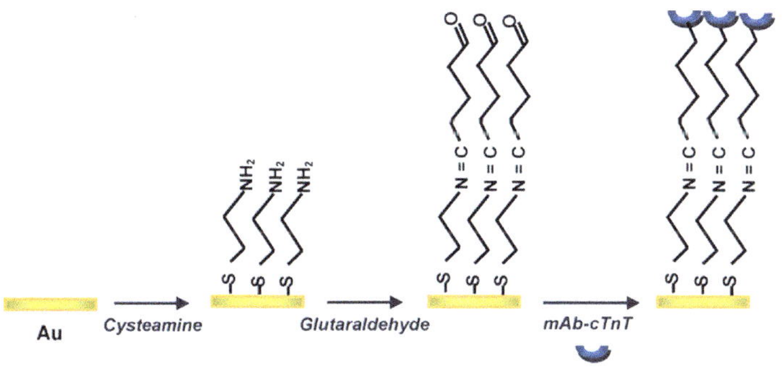

FIGURE 13.2 Immobilization procedure employed for the covalent binding of anti-cTnT antibodies to the QCM's surface [51]. *Reproduced with permission from Elsevier B.V.*

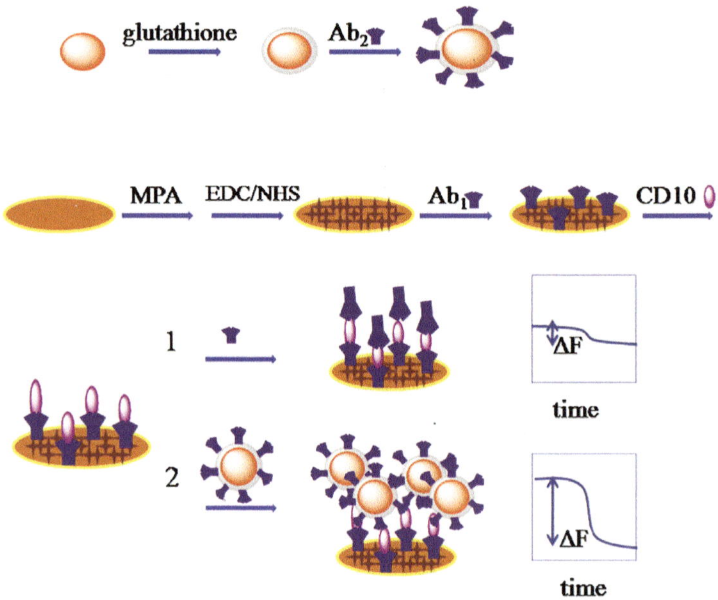

FIGURE 13.3 Bioanalytical procedure for the detection of CD10 by the quartz crystal microbalance–based sensing method [50]. *Reproduced with permission from Elsevier B.V.*

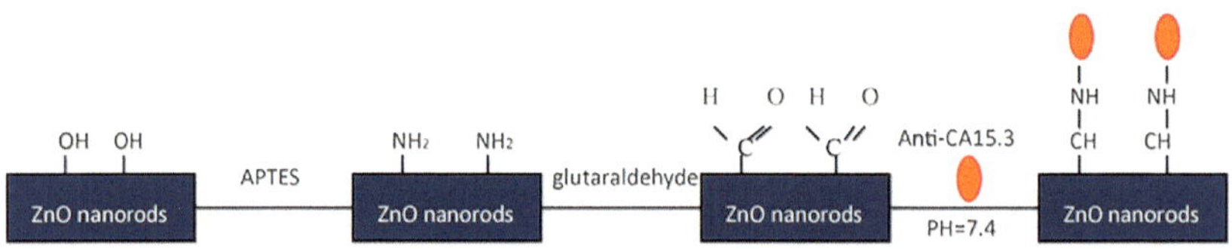

FIGURE 13.4 Development of anti-CA15.3 antibody-bound ZnO-functionalized quartz crystal microbalance's surface [55]. *Reproduced with permission from Elsevier B.V.*

antibody-bound CdS quantum dots (QDs), which is bound to the cystamine SAM bound QCM [49] (Fig. 13.5). The sensor detects ANXA3 in spiked human blood and urine samples in less than 15 min with an LOD of 0.075 ng/mL.

Concanavalin A (conA) molecules bound to the dextran-functionalized graphene-coated QCM can be used for the nonenzymatic glucose detection [36] (Fig. 13.6). Phenoxy-derived dextran (DexP) molecules are first bound to graphene-coated QCM surface by Π−Π stacking, followed by the specific binding of conA molecules to DexP. As glucose displaces con A, this sensing scheme detects glucose with a dynamic range and an LOD of 0.01−7.5 mM and 5 μM, respectively. A MIP (molecularly imprinted polymer)-coated QCM sensor is capable of detecting albumin from 60 to 150 ppm [37]. The albumin concentration in human serum detected by the sensor is also in agreement with

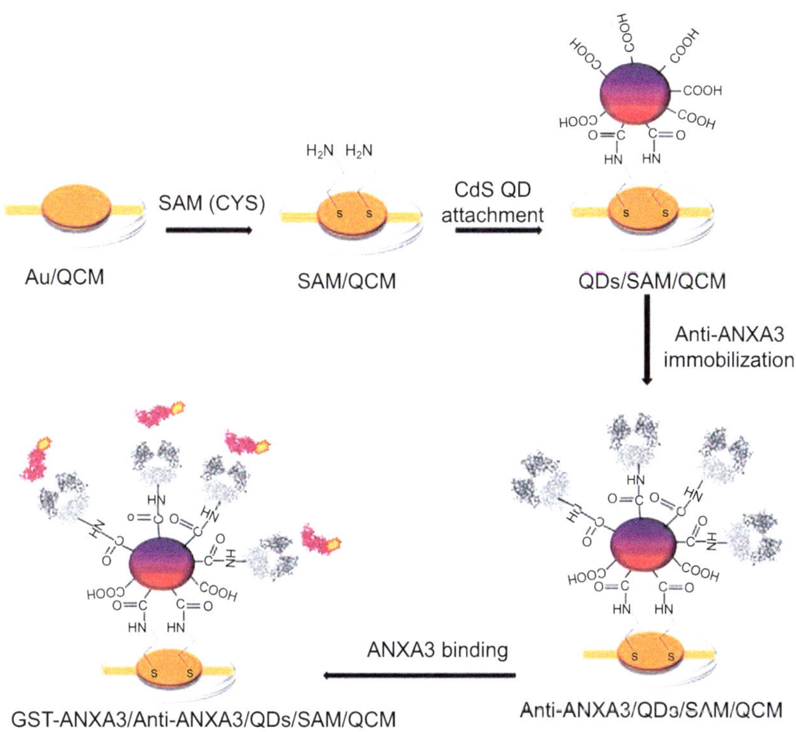

FIGURE 13.5 Quartz crystal microbalance–based sensing procedure for the detection of ANXA3 [49]. *Reproduced with permission from Elsevier B.V.*

that of the clinical assay. Albumin is imprinted with 3-dimethylaminopropyl methacrylamide-acrylate, followed by the coating of albumin MIP on QCM Au electrode. In the tetramethylene glycol dimethacrylate cross-linked polymer system, the adsorption capacities of different Au electrodes are in the decreasing order of Au–OH > Au–COOH > Au–NH_2 > Au, whereas the time taken to achieve a steady-state frequency is in the decreasing order of Au–NH_2<Au–OH < Au–COOH < Au. However, in the trimethylolpropane trimethacrylate cross-linking system, the adsorption capacities of different electrodes are in the decreasing order of Au > Au–NH_2>Au–OH > Au–COOH. The albumin MIP-QCM sensor exhibits a higher sensor response to albumin in comparison with that of a non-MIP-QCM sensor. The adsorption ratios of cytochrome *c*, lysozyme, albumin, and myoglobin for albumin MIP-QCM and non-MIP-QCM are 160:1:1942:30 and 13:1:249:86, respectively.

A QCM-based immunosensor for the epidermal growth factor receptor (EGFR) involves the activation of the anti-EGFR antibody with 1-[3-(dimethylamino)propyl]-3-ethylcarbodiimide hydrochloride (EDC) and *N*-hydroxysuccinimide (NHS), which is then immobilized on the gold surface of QCM

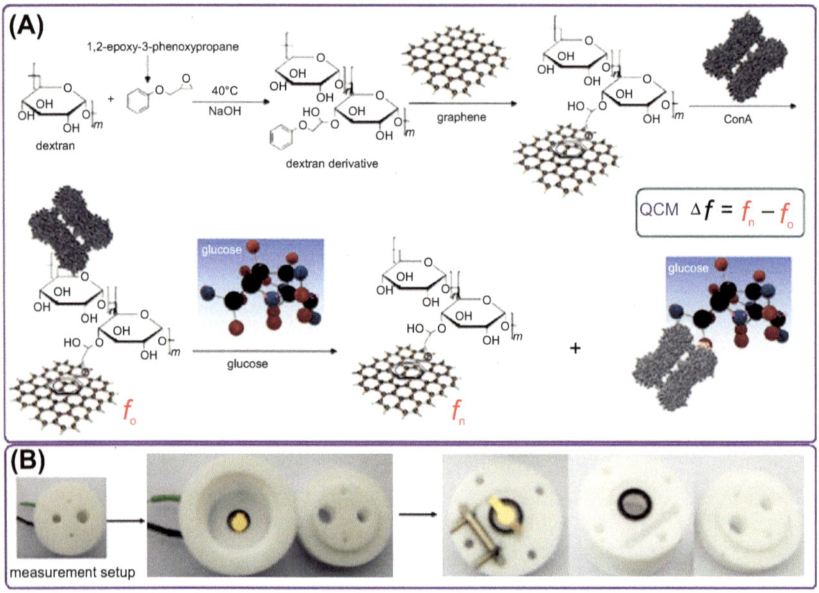

FIGURE 13.6 (A) Fabrication process and measurement principle of the displacement-based QCM glucose sensor and (B) the QCM measurement setup [36]. *Reproduced with permission from Hindawi Publishing Corporation, S.K. Vashist, P. Vashist, Recent advances in quartz crystal microbalance-based sensors, J. Sens. (2011).*

by the salicylic acid–based self-assembled monolayer (SAM) [41]. The protein G–based site-specific antibody immobilization strategy results in the oriented binding of the anti-EGFR antibody and offers a linear detection range of 0.01–10 μg/mL. The anti-EGFR antibody–coated QCM surface is effectively regenerated four times by treatment with 0.1 M NaOH and employed for many consecutive analyses. Of importance is the use of QCM-D for studying the biospecific glycoprotein–lectin interactions by employing four lectins covalently bound to the thiol-modified QCM's surface [43]. It detects glycoproteins in the linear range of 50 μg/mL to 1 mg/mL. The frequency and dissipation shifts of QCM provide pertinent information related to the analytical concentration and viscoelastic properties of the glycoprotein–lectin complex, respectively. Various glycoproteins can be determined via their unique lectin-binding patterns. The lectin-bound biosensor can be effectively regenerated using 10 mM glycine-HCl, pH 2.5 to perform repeated analyses for over 2 months.

Enantioselective MIP-coated QCM sensing is an elegant approach for detecting L-tryptophan with an LOD of 8.8 μM [42]. MIP is synthesized using acrylamide (AM) as a monomer and trimethylolpropane trimethacrylate (TRIM) as a cross-linking agent. The QCM is coated with thin permeable film

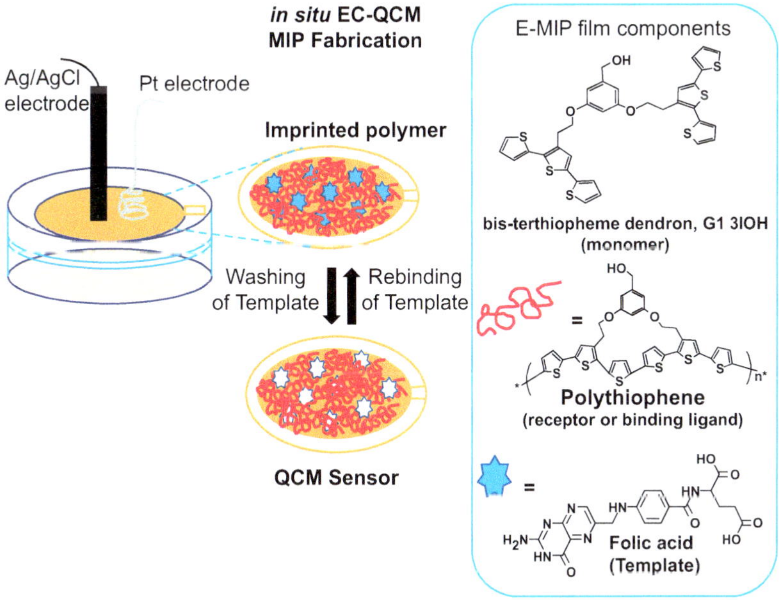

FIGURE 13.7 Schematic illustration of the fabrication of a polythiophene-based QCM sensor for folic acid [44]. *Reproduced with permission from Hindawi Publishing Corporation, S.K. Vashist, P. Vashist, Recent advances in quartz crystal microbalance-based sensors, J. Sens. (2011).*

coatings of MIP and employed for the selective detection of L-tryptophan w.r.t. D-tryptophan in citric acid buffer solutions. The sensor detects L-tryptophan with three- to fourfold higher signal than that from D-tryptophan and an enantiomeric selectivity coefficient of 6.4. MIP with a TRIM/AM molar ratio of 2.21 as the cross-linking monomer concentration shows the highest sensitivity and enantioselectivity for L-tryptophan among various MIPs produced with various molar ratios. Similarly, the QCM surface is coated with an electropolymerized MIP film of a bis-terthiophene dendron for the detection of folic acid with a linear range of 0–100 μM and an LOD of 15.4 μM (6.8 μg) [44] (Fig. 13.7). The relative cross-selectivity of the sensor for various structural analogues is in the decreasing order of pteroic acid (50%) > caffeine (10%) > and theophylline (6%)

3. DETECTION OF BACTERIA

Vibrio harveyi, a pathogen, is detected by a QCM-based immunosensor using the monoclonal antibody against *V. harveyi* bound covalently to the QCM's surface [45] (Fig. 13.8). The procedure employs a three-step antibody immobilization strategy: (1) the formation of SAM of carboxyl-terminated alkanethiol, i.e., 3-mercaptopropionic acid; (2) activation of carboxyl groups

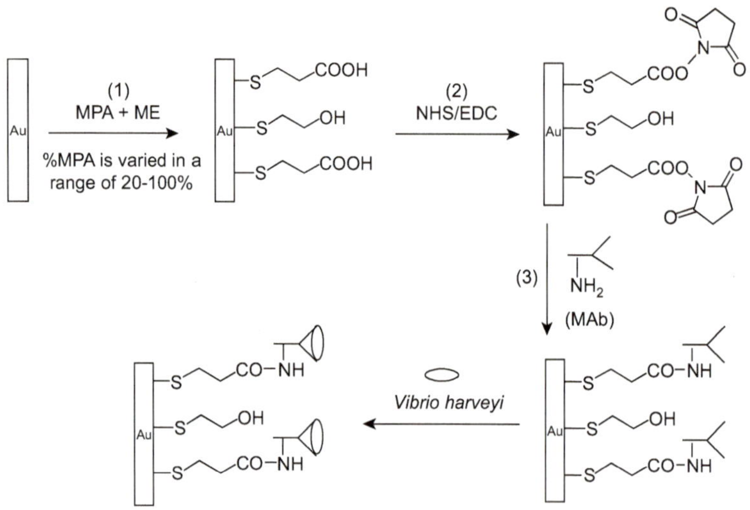

FIGURE 13.8 Schematic diagram showing the procedure for the preparation of monoclonal antibody-functionalized QCM-based immunosensor for *Vibrio harveyi* [45]. *Reproduced with permission from Hindawi Publishing Corporation, S.K. Vashist, P. Vashist, Recent advances in quartz crystal microbalance-based sensors, J. Sens. (2011).*

with N-NHS and 1-ethyl-3-(3-dimethylaminopropyl) carbodiimide (EDC); and (3) covalent binding of antibody by heterobifunctional cross-linking. The biosensor detects 10^3—10^7 CFU/mL of *V. harveyi* without any cross-reactivity to *Vibrio vulnificus* and *Vibrio parahaemolyticus*. The bacterial binding efficiency is further improved by blocking with 1% bovine serum albumin and the optimization of the monoclonal antibody immobilization density. Of interest is the QCM-based procedure for the recognition and discrimination of several *Campylobacter jejuni* strains using lectin-bound QCM [46]. The procedure involves the measurement of the dissipation shifts of the QCM in response to various bacterial strains and can be used for the discrimination of other pathogenic bacterial strains based on their selective binding to lectins. The sulfate-reducing bacterium, *Desulfotomaculum*, is detected by a QCM-based biosensor with a linear detection range of 1.8×10^4 to 1.8×10^7 CFU/mL [47]. The procedure involves the formation of conjugates of vancomycin-functionalized magnetic nanoparticles with bacteria after 30 min incubation under an external magnetic field formed. The sensing procedure is specific as there is no response from the vancomycin-resistant bacterium, *Vibrio anguillarum*. A QCM-based biosensor using nanoparticle-based signal enhancement is capable of detecting *Salmonella typhimurium* in raw and processed foods [53]. The procedure employs two QCMs: the first QCM is bound covalently to the specific monoclonal capture anti-*Salmonella* antibody using EDC-NHS heterobifunctional cross-linking while the second one is bound to mouse IgG antibody. The anti-*Salmonella* antibody–bound AuNPs detect *Salmonella* with an LOD of 10–20 CFU/mL and are capable of detecting this bacterium in food samples with high precision.

4. DETECTION OF VOLATILE ORGANIC COMPOUNDS

A QCM-based multianalyte detection system has been developed for the detection of acetic acid, butyric acid, ammonia, dimethylamine, benzene, chlorobenzene, and their mixtures [4]. It comprises a six-chip sensor module, where the chips are coated with synthetic polypeptides and conducting polymers. Principal component analysis (PCA) is applied for simple pattern recognition. The system provides sensitive and selective analyte detection and discriminates the characterized odor profiles of volatile organic compounds (VOCs) in various sensing applications. The detection of such VOCs, such as alcohols, ethers, esters, NO_2, ammonia, halocarbons, chemical warfare agents, and toxic gasses, is critically important due to the increasing terrorist threats and requirement of continuous online monitoring of military storage stations.

Primary aliphatic alcohols such as ethanol, methanol, 1-propanol, and 2-propanol vapors [5] are detected by a QCM-based method using a thin film of polyaniline (PANI) as the sensing material. PANI has high environmental stability, good performance at room temperature, high electrical conductivity, good reversibility, and good reproducibility. The adsorption of alcohol vapors onto the PANI-coated QCM results in increasing frequency shifts due to the hydrophilic nature of the film, electrostatic interactions, and the type of alcohol. The frequency shift of the QCM sensor exhibits a linear correlation with the concentration of primary aliphatic alcohol vapors in the range of 2–17 mg/L with good reproducibility and reversibility. In another approach, a QCM-based ethanol sensor with the limit of detection (1.73×10^{-2} μg) and the limit of quantitation (3.69×10^{-2} μg) is used for online monitoring of ethanol, 0.26%–0.70% (w/w) in whole meal bread and 0.68%–2.06% (w/w) in durum-wheat bread [6]. Therefore, this method might serve as a quality control for the online monitoring of ethanol in the bakery production.

A poly(vinylidene fluoride) (PVDF)-coated QCM sensor detects dimethyl methyl phosphonate (DMMP) vapors, a stimulant of nerve agents [9]. The frequency shifts show a perfect linear correlation (correlation coefficient of 0.997) with DMMP concentrations in the range of 5–60 ppm. Although the sensitivity to DMMP vapor is unaffected by humidity, it is higher in a low-temperature range. A QCM-based sensor for methyl mercaptan (CH_3SH) is based on the increase in the surface area of the sensing polymeric film [7]. Poly(ethylene imine) (PEI) serves as a polymeric layer and the Al_2O_3 porous film is employed to increase the surface area. The Al_2O_3 porous film is coated on the QCM substrate between the QCM electrode and polymeric film by the sol-gel method. The sensor detects 100 ppb of CH_3SH gas but shows interference with moisture, which can be counteracted using a humidity sensor as a feedback source. A QCM sensor array is fabricated for the detection of various gaseous chemical agents dimethyl methyl phosphonate (DMMP), N,N-dimethylacetamide (DMA), 1,5-dichloropentane (DCP), and dichloroethane (DCE) [8]. The four appropriate coating materials are screened out of 15 coating materials by applying PCA with hierarchical cluster methods to the sensor response data set collected from 15 QCM sensors for 12 analytes. The selected coating materials are coated on four QCMs in a sensor array to classify four simulant gasses. Similarly, miniaturized QCM arrays–based sensor, comprising 36 QCMs on a single AT-cut quartz blank, have been used for the detection of toluene in water by coating the QCMs with different layers such as polystyrene, amyl-calix[8]arene, and β-cyclodextrin [12]. The sensitivity of the polystyrene coating to the toluene in water is directly proportional to the resonant frequency of QCM. However, the higher is the sensitivity, the higher is the noise due to lower-quality factor. Therefore, the toluene detection must be performed at an optimum resonant frequency that is dependent on the stability of temperature and pressure control. Another study unravels the use of a QCM sensor array with four QCMs for the detection of xylene isomers [13]. It detects *m*-xylene and *p*-xylene in the range of 0–200 ppm with an accuracy of about 1%, high specificity, and no cross-sensitivity with residual water. Therefore, it enables precise xylene detection in variable ambient humidity. The low-cost sensor employs supramolecular host molecules–coated QCM resonators and multivariate data analysis. It is ideal for continuous online monitoring of gaseous analytes under harsh conditions owing to its high chemical and physical stability. Of interest is QCM-based lipid and lipopolymer odor sensing using chemisorbed PEGylated lipopolymer [14]. The sensor discriminates 10 odorants: alcohols (1-butanol, 1-hexanol, trans-2-hexenol), esters (butyl acetate, hexyl acetate, trans-2-hexenyl acetate), acids (isobutyric, hexanoic), and aldehydes (butanal, trans-2-hexenal).

Calixarene film–coated QCM sensors have been developed for the detection of aromatics, chlororganics, ketones, and alcohols [10]. They are based on the formation of host–guest complexes where the calixarene acts as host while metal ions and VOCs act as guests. Phosphorous-containing calixarenes show high sensitivity, i.e., they are better suited for sensing applications. The sensitivity and selectivity of calixarenes depend on the number of aryl fragments and the binding of different functional groups to upper and lower rims. A QCM-based sensor involves the use of MIPs for the detection of toluene and *p*-xylene [11]. The MIPs are prepared in the presence of toluene and *p*-xylene by conventional cross-linking polymerization. They are composed of methyl methacrylate (MMA) as a monomer and divinylbenzene (DVB) as a cross-linking agent. MIP powders are mixed with PMMA and coated on the QCM's surface. They bind reversibly to toluene and *p*-xylene, although the response is sluggish due to the presence of a matrix polymer around the MIP particles. Therefore, it requires significant improvements to improve the response time and the selectivity.

5. DETECTION OF CHEMICAL ANALYTES

A QCM-based bisphenol-A (BPA)/[4,4′-(1-methylethylidene) bisphenol] has been demonstrated by immobilizing anti-BPA monoclonal antibody [15], which binds specifically to BPA. The signal is enhanced by employing a secondary antibody tagged to a conjugated microsphere, which decreases the limit of detection from 0.1 ng/mL to 0.01 ng/mL. The anti-BPA antibody is conjugated to 2-methacryloyloxyethyl phosphorylcholine (MPC) polymeric nanoparticles and polystyrene nanoparticles of 202 nm in diameter. The MPC polymeric nanoparticles tagged anti-BPA show high signal enhancement with the stable signal. Moreover, the MPC polymeric nanoparticles exhibit excellent colloidal stability due to their strong solvation with water molecules, which resist their coagulation. Similarly, the detection of BPA is achieved by binding MIP to the SAM-coated QCM [57]. The SAM is composed of 2-aminoethanethiol and 11-mercaptoundecanoic acid, which show strong ion-pair interactions with the functional monomers in MIP for proper adhesions. The sensor selectively detects BPA in liquid form with a sensitivity of 100 ppb. A QCM-based biosensor for the detection of aflatoxin B1 (AFB1) in groundnut has been described [48]. The anti-AFB1 antibodies are bound to the QCM functionalized with SAM of 4-aminothiophenol, followed by blocking with BSA and detection of AFB1. With a linear detection range of 1–4 ng/mL, the results for the detection of AFB1 in extracts of contaminated groundnut are in agreement with those obtained by liquid chromatography-tandem MS.

A highly sensitive QCM sensor detects organophosphorus pesticide *o,o*-dimethyl-*o*-2,2-dichlorovinyl phosphate (DDVP) in the linear range of 6.5–32.5 ppm [17]. The procedure involves the deposition of poly(3,4-ethylenedioxythiophene) (PEDOT), a conducting polymer, on the electrode's surface. The change in the resonant frequency of QCM, a measure of the electrical loading, corresponds to the change in the electrostatic capacitance. The sensitivity of the developed sensors is directly proportional to the conductivity of polymer coating. The capacitance is noted in the range of 0.198–0.187 pF for PEDOT. The developed PEDOT film–coated sensor precisely detects DDVP in "real-world" samples. An interesting QCM sensing application is the simultaneous determination of 1-naphthol (1-NAP) and 2-naphthol (2-NAP) in human urine samples [18]. The procedure involves the immobilization of β-cyclodextrin (β-CD) onto the QCM's surface using a nanocrystalline TiO_2 film as the substrate and the excellent predictive ability of chemometrics. It detects 1-NAP with the LOD of 8.362×10^{-7} M and the correlation coefficient of 0.993. It also detects 2-NAP with the LOD and the correlation coefficient of 2.146×10^{-7} M and 0.990, respectively. For the detection of 2,3,7,8-tetrachlorodibenzo-p-dioxin (TCDD) [58], the most toxic molecule among dioxin-like compounds, the QCM surface is modified by treatment with cysteamine and glutaraldehyde. After the binding of the anti-TCDD monoclonal antibody, excess unreacted aldehyde groups are blocked with glycine. The MPC polymer is used as a stabilizer for reducing the nonspecific binding of the antigen solution and stabilizing the immunologic activity of antibody-functionalized QCM. The biosensor detects TCDD in fly ash samples [16,58] with linearity in the range of 0.1–100 ng/mL and a linear correlation coefficient of 0.99.

6. DETECTION OF GASEOUS ANALYTES

A QCM-based HCl gas sensor is based on morpholine-functional poly(styrene-co-chloromethyl styrene) copolymer coatings that bind irreversibly to HCl [19]. The chemical structure and composition of the copolymer determine the sensitivity of the sensor. It detects HCl selectively in ppt level in just 10 min with a sensitivity of 0.1 ppm/Hz. There is no interference from other gasses such as O_2, CO_2, and SO_2. The highest sensitivity is noticed when the sensor film coating copolymer has 40 mol% chloromethyl styrene cross-linked with 5 mol% divinylbenzene and the film thickness is above 2 μm. However, the sensor is subject to interference from changes in humidity and NO_2, which limits its use in potential industrial applications such as the analysis of combustion gasses. Another interesting approach for the detection of HCl gas in air involves three different kinds of poly(acrylamide) derivate coatings on QCM's surface [20]. The polymer poly(*N*,*N*-dimethylacrylamide) (PDMAA) shows the highest sensitivity for HCl sensing of about 250 ppb/Hz but incurs high irreversible response toward NO_2 gas and high interference to changes in test gas humidity. Therefore, it cannot be used for practical applications and require considerable improvements. A supramolecular cryptophane-A-based QCM sensor was developed for the selective and rapid detection of methane with a detection limit of 0.05% (v/v) and high recovery at room temperature [21]. The procedure involves the deposition of supramolecular cryptophane-A, synthesized from vanillyl alcohol by double trimerization method, on the QCM surface by electrospraying. A QCM-based sensor for methane is fabricated by depositing a PANI/PdO composite, synthesized by in situ chemical oxidative polymerization of aniline with PdO nanoparticles, on QCM's surface via layer-by-layer self-assembly [22]. The increase in relative humidity and temperature decreases the sensor response.

A QCM sensor for nitric oxide (NO) in the gaseous state involves the immobilization of cobalt phthalocyanine (CoPc) in a mesoporous silica matrix on the QCM's surface [23]. The CoPc/mesoporous silica hybrids are prepared by thermal calcination and thoroughly characterized by FTIR and scanning electron microscopy (SEM). The CoPc/mesoporous silica–coated QCM shows the highest sensitivity of up to 25 ppb/Hz with minimum cross-sensitivity to carbon monoxide (CO) in comparison with the nonmodified and CoPc-modified QCMs. Of interest is the detection of the sub-ppm level of nitrogen dioxide in the air by QCM coated with morpholine-functional cross-linked copolymer [24]. The cross-linked poly(styrene-co-chloromethyl styrene) copolymer reacts with morpholine to generate the morpholine-functional cross-linked copolymer. The highest sensitivity obtained by the developed sensor is 1200 Hz/h for 50 ppm of NO_2. The QCM-based detection of chloroform vapors is also demonstrated by coating the QCM's surface with the paracyclophanes B44TOS and CP44, which detects chloroform by the formation of host–guest complexes between chloroform and paracyclophanes [25]. Of interest is the detection of ammonia using an electrospun nanoporous membrane-coated QCM [26]. The electrospun nanofibers with diameters of 100–1400 nm were deposited on QCM by electrospinning. They are composed of cross-linkable poly(acrylic acid) (PAA) and poly(vinyl alcohol) (PVA). The increase in viscosity and conductivity of the homogeneous blend solution of PAA and PVA due to the increased PAA content results in an increase in the average diameter and rigidity of nanofibers. The sensitivity of nanofibrous membrane-coated QCM is much higher than that of continuous films-coated QCM but is affected by the content of PAA in nanofibrous membranes and relative humidity.

A ZnO nanorods–coated QCM sensor provides highly selective and reproducible detection of NH_3 at room temperature [27]. The procedure involves the synthesis of vertically aligned ZnO nanorods with a diameter of 100 nm and a height of 3 μm directly on the QCM Au electrode by wet chemistry at 90°C. The sensor shows good selectivity to NH_3 in comparison with liquefied petroleum gas, N_2O, CO, NO_2, and CO_2. A highly sensitive and rapid QCM-based detection of HCHO with an LOD of 3 ppm is achieved by the deposition of electrospun nanoporous polystyrene (PS) fibers with the high surface-area-to-volume ratio (~47.25 m^2/g) on a polyethyleneimine (PEI)-functionalized QCM [28]. The PS fibers, electrospun from a higher concentration of PS solution (13 wt%), have a larger pore size and greater surface area than PS fibers electrospun from a lesser concentration of PS. Therefore, the sensor response of PEI-PS (13 wt%) for the detection of 140 ppm HCHO is fourfold higher than that of PEI-PS (7 wt%). A QCM sensor for the detection of amine odors is based on an ultrathin film formed by the alternate adsorption of TiO_2 and polyacrylic acid (PAA) [59] (Fig. 13.9). The developed sensor exhibits a rapid response, high stability in a relative humidity range of 30%–70%, and linear detection in the range of 0.3–15 ppm with an LOD of 0.1 ppm for the detection of ammonia. It involves the binding of ammonia to the free carboxylic groups of PAA by acid–base interactions. The TiO_2/PAA_{400} ultrathin film condenses 15 ppm ambient ammonia concentration to ~20,000 ppm. The sensor detects n-butylamine and ammonia with a higher sensitivity in comparison with pyridine due to differences in their molecular weight and basicity.

7. SPECIAL ANALYTICAL APPLICATIONS

Odor sensing is challenging as odors are mixtures of hundreds of various kinds of molecules. However, fruit flavors can be sensed as they have characteristic

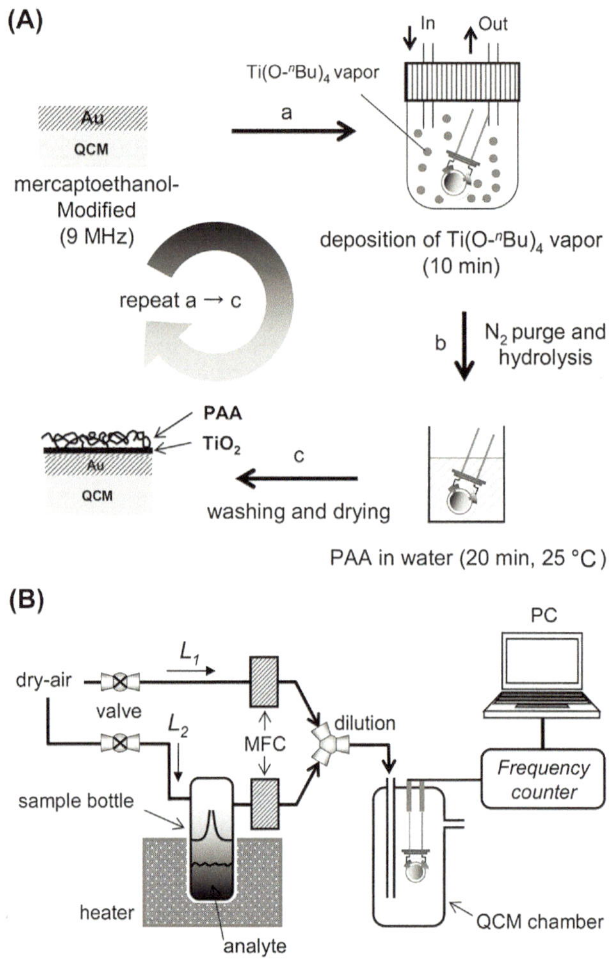

FIGURE 13.9 (A) Schematic illustration of the film preparation by the gas phase surface sol-gel process. (B) Schematic diagram of the QCM measurement setup [59]. *Reproduced with permission from Hindawi Publishing Corporation, S.K. Vashist, P. Vashist, Recent advances in quartz crystal microbalance-based sensors, J. Sens. (2011).*

impact odorants. An array of QCM odor sensors was coated with different materials such as lipids and stationary phase material of GC and was used for the odor sensing of orange and melon flavors [30]. The lipids used are divaleroyl phosphatidyl choline (DVPC), cardiolipin (CL), and cerebrosides (CS) while the stationary phase material of GC used is a piezo-L (Ap-L). The procedure involves the initial measurement of the steady-state sensor responses to single flavor components, followed by the estimation of important odor components by the computational two-level quantization method. The experimental approximation is made by the steady-state sensor responses and

PCA. The orange flavor is approximated using an array of sensors, where the odor is sensed by three odor components: linalool, decanal, and citral. However, the number of odor components increases in the case of melon flavor, so a DVPC-coated QCM is included as the additional component of the sensor array. A human sensory test is also performed to evaluate the similarities between the sensed odors and the original ones. These similarities are higher for the orange flavor. The specific approximation of melon flavor requires further improvement, which can be realized by increasing the number of odor components in approximate flavor.

An array of eight phthalocyanine-coated QCM sensors together with an artificial neural network is capable of detecting the composition of gas mixtures [31]. The preprocessing of digital data collected from the sensor responses is done by a sliding window algorithm followed by a three-layer artificial neural network to determine the composition of the gas mixtures. The various gas mixtures, such as ethanol-acetone, ethanol-trichloroethylene, and acetone-trichloroethylene, are analyzed selectively with a success rate of greater than 84% and an error of 10.6% sensor. However, the sensing procedure can only detect gaseous mixtures belonging to specific categories only. Another study demonstrates the highly selective chiral recognition of L-mandelic acid (L-MA) using an L-phenylalanine-coated QCM sensor and vapor-diffused molecular assembly (VDMA) reaction technique [29]. L-phenylalanine serves as a resolving agent for the selective detection of L-MA. The sensor has the chiral discrimination factor, i.e., L-MA/D-MA, of about 8 and is based on a simple analysis procedure. In contrast, the screening of resolving agents by the diastereomeric crystallization technique is very difficult. The results for the QCM-based chiral recognition of L-MA are in agreement with those obtained by diastereomeric crystallization. The selective detection of L-MA is further confirmed by contact angle measurements.

8. OTHER ANALYTICAL APPLICATIONS

A low-cost, stable, reproducible, and sensitive humidity sensor can be easily fabricated by coating mesoporous silicate SBA-15 with monodisperse hexagonal lamelliform on the QCM's surface [32] (Fig. 13.10). The sensitivity of the QCM sensor is proportional to the film's thickness. The 20-cm-thick SBA-15 film shows the best linearity with a sensitivity of 12.32 Hz/RH%. Silver ions in aqueous media [33] can be detected by a nanotubular PANI film–coated QCM. This sensing scheme can be used to monitor industrial wastewaters or search new resources of silver. Virgin and extravirgin olive oils from lampante oils are differentiated using eight QCM sensor arrays that are coated with five gas chromatographic stationary phases (OV-17, OV-275, PEG, Span 80, and Vaseline) as the sensing films and a PCA method [34]. The QCM-based method is cheap and easy to use, obviates the use of solvents, and does not require any sample pretreatment.

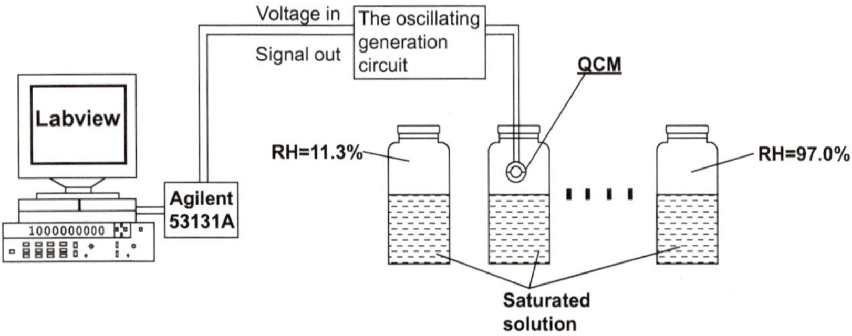

FIGURE 13.10 QCM setup for humidity testing [32]. *Reprinted with permission from Elsevier B.V, Hindawi Publishing Corporation, S.K. Vashist, P. Vashist, Recent advances in quartz crystal microbalance-based sensors, J. Sens. (2011).*

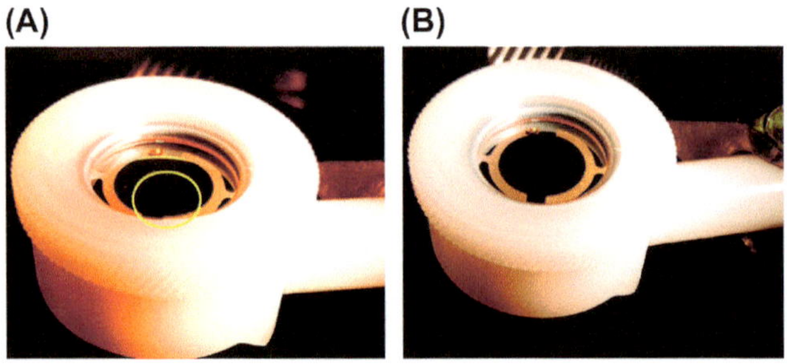

FIGURE 13.11 Quartz crystal images: (A) following the application of a nanocrystal colloidal suspension, a *circle* indicates residual solvent; (B) following complete solvent evaporation [35]. *Reproduced with permission from Hindawi Publishing Corporation, S.K. Vashist, P. Vashist, Recent advances in quartz crystal microbalance-based sensors, J. Sens. (2011).*

QCM can accurately determine the concentration of nanoparticles in a colloidal suspension. This simple procedure is based on the drop-casting of nanoparticle suspension in a volatile solvent, which leaves a dry nanoparticle residue after solvent evaporation on the QCM surface [35] (Fig. 13.11). The sensor response is linear for nanoparticle concentrations up to 1300 μg/mL in experiments performed with Si and Ag nanopowder in methanol, rhodamine B in methanol, and ferrocene in cyclohexane. The sensor determines the concentrations of octyl-terminated Si nanocrystal samples with median diameters in the range of 1.1–14.8 nm.

9. CONCLUSIONS AND FUTURE TRENDS

QCMs have been widely used for a broad range of analytical applications as the sensing procedure is simple, cost-effective, non-hazardous, real-time, and less time-consuming and labor-intensive. The ongoing research efforts have led to the emergence of innovative bioanalytical instruments that can perform both the surface-plasmon resonance (SPR) and QCM sensing [60]. In fact, QCM is an appropriate sensing platform for the online monitoring of analytes. Q-Sense has developed a new technology called QCM-D, which enables real-time, label-free measurements of molecular adsorption and/or interactions on various substrates. The dissipation parameter (D) provides novel insights into the viscoelastic properties of adsorbed layer. A device based on the simultaneous QCM and electrochemical measurements would be highly useful but it can only be used for studying electroplated, evaporated, or sputtered materials.

However, there are several challenges associated with QCM-based sensors as they still require stringent validation and clinical trials to analyze their bioanalytical performance before they can be approved for bioanalytical applications. It needs to be determined whether they have superior performance and advantages over the upcoming sensing technologies [61] such as real-time analytical devices, point-of-care systems [62], and smartphone-based devices [63,64], which have witnessed an exponential growth during the recent years. However, QCM might be appropriate for the development of potential bioanalytical systems that are based on the use of various sensing technologies into a single system. The integration of QCM with SPR and QCM with electrochemical detection has already been demonstrated by various companies and would provide highly useful for various applications [65]. The reuse of the crystal is another concern for low-cost applications. In this context, a simple approach to the development of a reusable piezoelectric crystal [66] should be considered. As an example, the crystal coated with protein A is allowed to react with antihuman albumin antibody to form a sensing layer for albumin. Instead of cleaning to remove bound albumin, the used crystals is simply saturated with albumin for the binding of a new anti-albumin antibody layer. In this manner, the albumin assay could be repeated up to five times using the same crystal.

REFERENCES

[1] J. Curie, P. Curie, An oscillating quartz crystal mass detector, Rendu 91 (1880) 294–297.

[2] G. Sauerbrey, Use of quartz vibration for weighing thin films on a microbalance, J. Phys. 155 (1959) 206–212.

[3] S.K. Vashist, E. Lam, S. Hrapovic, et al., Immobilization of antibodies and enzymes on 3-aminopropyltriethoxysilane-functionalized bioanalytical platforms for biosensors and diagnostics, Chem. Rev. 114 (2014) 11083–11130.

[4] H.-H. Lu, Y.K. Rao, T.-Z. Wu, et al., Direct characterization and quantification of volatile organic compounds by piezoelectric module chips sensor, Sens. Actuators B Chem. 137 (2009) 741–746.

[5] M.M. Ayad, N.L. Torad, Alcohol vapours sensor based on thin polyaniline salt film and quartz crystal microbalance, Talanta 78 (2009) 1280–1285.

[6] A. Bello, F. Bianchi, M. Careri, et al., Potentialities of a modified QCM sensor for the detection of analytes interacting via H-bonding and application to the determination of ethanol in bread, Sens. Actuators B Chem. 125 (2007) 321–325.
[7] M. Kikuchi, S. Shiratori, Quartz crystal microbalance (QCM) sensor for CH3SH gas by using polyelectrolyte-coated sol–gel film, Sens. Actuators B Chem. 108 (2005) 564–571.
[8] Z. Ying, Y. Jiang, X. Du, et al., Polymer coated sensor array based on quartz crystal microbalance for chemical agent analysis, Eur. Polym. J. 44 (2008) 1157–1164.
[9] Z. Ying, Y. Jiang, X. Du, et al., PVDF coated quartz crystal microbalance sensor for DMMP vapor detection, Sens. Actuators B Chem. 125 (2007) 167–172.
[10] I. Koshets, Z. Kazantseva, Y.M. Shirshov, et al., Calixarene films as sensitive coatings for QCM-based gas sensors, Sens. Actuators B Chem. 106 (2005) 177–181.
[11] M. Matsuguchi, T. Uno, Molecular imprinting strategy for solvent molecules and its application for QCM-based VOC vapor sensing, Sens. Actuators B Chem. 113 (2006) 94–99.
[12] J. Rabe, S. Buttgenbach, J. Schroder, et al., Monolithic miniaturized quartz microbalance array and its application to chemical sensor systems for liquids, IEEE Sens. J. 3 (2003) 361–368.
[13] F.L. Dickert, O. Hayden, M.E. Zenkel, Detection of volatile compounds with mass-sensitive sensor arrays in the presence of variable ambient humidity, Anal. Chem. 71 (1999) 1338–1341.
[14] B. Wyszynski, P. Somboon, T. Nakamoto, Chemisorbed PEGylated lipopolymers as sensing film supports for QCM odor sensors, Sens. Actuators B Chem. 130 (2008) 857–863.
[15] J.W. Park, S. Kurosawa, H. Aizawa, et al., Piezoelectric immunosensor for bisphenol A based on signal enhancing step with 2-methacrolyloxyethyl phosphorylcholine polymeric nanoparticle, Analyst 131 (2006) 155–162.
[16] J.-W. Park, S. Kurosawa, H. Aizawa, et al., Dioxin immunosensor using anti-2, 3, 7, 8–TCDD antibody which was produced with mono 6-(2, 3, 6, 7-tetrachloroxanthene-9-ylidene) hexyl succinate as a hapten, Biosens. Bioelect. 22 (2006) 409–414.
[17] H. Zeng, Y. Jiang, G. Xie, et al., Polymer coated QCM sensor with modified electrode for the detection of DDVP, Sens. Actuators B Chem. 122 (2007) 1–6.
[18] Y.-K. Yuan, X.-L. Xiao, Y.-S. Wang, et al., Quartz crystal microbalance with β-cyclodextrin/TiO_2 composite films coupled with chemometrics for the simultaneous determination of urinary 1-and 2-naphthol, Sens. Actuators B Chem. 145 (2010) 348–354.
[19] M. Matsuguchi, Y. Kadowaki, K. Noda, et al., HCl gas monitoring based on a QCM using morpholine-functional styrene-co-chloromethylstyrene copolymer coatings, Sens. Actuators B Chem. 120 (2007) 462–466.
[20] M. Matsuguchi, Y. Kadowaki, Poly (acrylamide) derivatives for QCM-based HCl gas sensor applications, Sens. Actuators B Chem. 130 (2008) 842–847.
[21] P. Sun, Y. Jiang, G. Xie, et al., A room temperature supramolecular-based quartz crystal microbalance (QCM) methane gas sensor, Sens. Actuators B Chem. 141 (2009) 104–108.
[22] G. Xie, P. Sun, X. Yan, et al., Fabrication of methane gas sensor by layer-by-layer self-assembly of polyaniline/PdO ultra thin films on quartz crystal microbalance, Sens. Actuators B Chem. 145 (2010) 373–377.
[23] A. Palaniappan, S. Moochhala, F.E. Tay, et al., Phthalocyanine/silica hybrid films on QCM for enhanced nitric oxide sensing, Sens. Actuators B Chem. 129 (2008) 184–187.
[24] M. Matsuguchi, Y. Kadowaki, M. Tanaka, A QCM-based NO_2 gas detector using morpholine-functional cross-linked copolymer coatings, Sens. Actuators B Chem. 108 (2005) 572–575.

[25] F. Dickert, A. Haunschild, V. Kuschow, et al., Mass-sensitive detection of solvent vapors. Mechanistic studies on host-guest sensor principles by FT-IR spectroscopy and BET adsorption analysis, Anal. Chem. 68 (1996) 1058–1061.

[26] B. Ding, J. Kim, Y. Miyazaki, et al., Electrospun nanofibrous membranes coated quartz crystal microbalance as gas sensor for NH 3 detection, Sens. Actuators B Chem. 101 (2004) 373–380.

[27] N. Van Quy, V.A. Minh, N. Van Luan, et al., Gas sensing properties at room temperature of a quartz crystal microbalance coated with ZnO nanorods, Sens. Actuators B Chem. 153 (2011) 188–193.

[28] C. Zhang, X. Wang, J. Lin, et al., Nanoporous polystyrene fibers functionalized by polyethyleneimine for enhanced formaldehyde sensing, Sens. Actuators B Chem. 152 (2011) 316–323.

[29] H.-S. Guo, J.-M. Kim, S.-M. Chang, et al., Chiral recognition of mandelic acid by L-phenylalanine-modified sensor using quartz crystal microbalance, Biosens. Bioelect. 24 (2009) 2931–2934.

[30] S. Muñoz-Aguirre, A. Yoshino, T. Nakamoto, et al., Odor approximation of fruit flavors using a QCM odor sensing system, Sens. Actuators B Chem. 123 (2007) 1101–1106.

[31] A. Ozmen, F. Tekce, M. Ebeoğlu, et al., Finding the composition of gas mixtures by a phthalocyanine-coated QCM sensor array and an artificial neural network, Sens. Actuators B Chem. 115 (2006) 450–454.

[32] Y. Zhu, H. Yuan, J. Xu, et al., Highly stable and sensitive humidity sensors based on quartz crystal microbalance coated with hexagonal lamelliform monodisperse mesoporous silica SBA-15 thin film, Sens. Actuators B Chem. 144 (2010) 164–169.

[33] M.M. Ayad, N. Prastomo, A. Matsuda, et al., Sensing of silver ions by nanotubular polyaniline film deposited on quartz-crystal in a microbalance, Synth. Met. 160 (2010) 42–46.

[34] M.E. Escuderos, S. Sánchez, A. Jiménez, Quartz Crystal Microbalance (QCM) sensor arrays selection for olive oil sensory evaluation, Food Chem. 124 (2011) 857–862.

[35] V. Reipa, G. Purdum, J. Choi, Measurement of nanoparticle concentration using quartz crystal microgravimetry, J. Phys. Chem. B 114 (2010) 16112–16117.

[36] D. Tang, Q. Li, J. Tang, et al., An enzyme-free quartz crystal microbalance biosensor for sensitive glucose detection in biological fluids based on glucose/dextran displacement approach, Anal. Chim. Acta 686 (2011) 144–149.

[37] T.Y. Lin, C.H. Hu, T.C. Chou, Determination of albumin concentration by MIP-QCM sensor, Biosens. Bioelect. 20 (2004) 75–81.

[38] Y. Uludağ, I.E. Tothill, Development of a sensitive detection method of cancer biomarkers in human serum (75%) using a quartz crystal microbalance sensor and nanoparticles amplification system, Talanta 82 (2010) 277–282.

[39] N. Kim, D.-K. Kim, Y.-J. Cho, Gold nanoparticle-based signal augmentation of quartz crystal microbalance immunosensor measuring C-reactive protein, Curr. Appl. Phys. 10 (2010) 1227–1230.

[40] N. Gan, P. Xiong, J. Wang, et al., A novel signal-amplified immunoassay for the detection of C-reactive protein using HRP-doped magnetic nanoparticles as labels with the electrochemical quartz crystal microbalance as a detector, J. Anal. Methods Chem. (2013).

[41] J.-C. Chen, S. Sadhasivam, F.-H. Lin, Label free gravimetric detection of epidermal growth factor receptor by antibody immobilization on quartz crystal microbalance, Process Biochem. 46 (2011) 543–550.

[42] F. Liu, X. Liu, S.-C. Ng, et al., Enantioselective molecular imprinting polymer coated QCM for the recognition of l-tryptophan, Sens. Actuators B Chem. 113 (2006) 234–240.

[43] M.E. Yakovleva, G.R. Safina, B. Danielsson, A study of glycoprotein–lectin interactions using quartz crystal microbalance, Anal. Chim. Acta 668 (2010) 80–85.
[44] D.C. Apodaca, R.B. Pernites, R.R. Ponnapati, et al., Electropolymerized molecularly imprinted polymer films of a bis-terthiophene dendron: folic acid quartz crystal microbalance sensing, ACS Appl. Mater. Inter. 3 (2011) 191–203.
[45] S. Buchatip, C. Ananthanawat, P. Sithigorngul, et al., Detection of the shrimp pathogenic bacteria, *Vibrio harveyi*, by a quartz crystal microbalance-specific antibody based sensor, Sens. Actuators B Chem. 145 (2010) 259–264.
[46] M.E. Yakovleva, A.P. Moran, G.R. Safina, et al., Lectin typing of *Campylobacter jejuni* using a novel quartz crystal microbalance technique, Anal. Chim. Acta 694 (2011) 1–5.
[47] Y. Wan, D. Zhang, B. Hou, Determination of sulphate-reducing bacteria based on vancomycin-functionalised magnetic nanoparticles using a modification-free quartz crystal microbalance, Biosens. Bioelect. 25 (2010) 1847–1850.
[48] R. Chauhan, P.R. Solanki, J. Singh, et al., A novel electrochemical piezoelectric label free immunosensor for aflatoxin B1 detection in groundnut, Food Control 52 (2015) 60–70.
[49] Y.J. Kim, M.M. Rahman, J.-J. Lee, Ultrasensitive and label-free detection of annexin A3 based on quartz crystal microbalance, Sens. Actuators B Chem. 177 (2013) 172–177.
[50] Z. Yan, M. Yang, Z. Wang, et al., A label-free immunosensor for detecting common acute lymphoblastic leukemia antigen (CD10) based on gold nanoparticles by quartz crystal microbalance, Sens. Actuators B Chem. 210 (2015) 248–253.
[51] A. Mattos, T. Freitas, V. Silva, et al., A dual quartz crystal microbalance for human cardiac troponin T in real time detection, Sens. Actuators B Chem. 161 (2012) 439–446.
[52] R. Akter, C.K. Rhee, M.A. Rahman, A highly sensitive quartz crystal microbalance immunosensor based on magnetic bead-supported bienzymes catalyzed mass enhancement strategy, Biosens. Bioelect. 66 (2015) 539–546.
[53] F. Salam, Y. Uludag, I.E. Tothill, Real-time and sensitive detection of *Salmonella typhimurium* using an automated quartz crystal microbalance (QCM) instrument with nanoparticles amplification, Talanta 115 (2013) 761–767.
[54] Y. Uludag, I.E. Tothill, Cancer biomarker detection in serum samples using surface plasmon resonance and quartz crystal microbalance sensors with nanoparticle signal amplification, Anal. Chem. 84 (2012) 5898–5904.
[55] X. Wang, H. Yu, D. Lu, et al., Label free detection of the breast cancer biomarker CA15. 3 using ZnO nanorods coated quartz crystal microbalance, Sens. Actuators B Chem. 195 (2014) 630–634.
[56] S.K. Vashist, P. Vashist, Recent advances in quartz crystal microbalance-based sensors, J. Sens. (2011).
[57] N. Tsuru, M. Kikuchi, H. Kawaguchi, et al., A quartz crystal microbalance sensor coated with MIP for "bisphenol A" and its properties, Thin Solid Films 499 (2006) 380–385.
[58] S. Kurosawa, H. Aizawa, J.W. Park, Quartz crystal microbalance immunosensor for highly sensitive 2,3,7,8-tetrachlorodibenzo-p-dioxin detection in fly ash from municipal solid waste incinerators, Analyst 130 (2005) 1495–1501.
[59] S.-W. Lee, N. Takahara, S. Korposh, et al., Nanoassembled thin film gas sensors. III. Sensitive detection of amine odors using TiO2/poly (acrylic acid) ultrathin film quartz crystal microbalance sensors, Anal. Chem. 82 (2010) 2228–2236.
[60] J. Kim, S. Kim, T. Ohashi, et al., Construction of simultaneous SPR and QCM sensing platform, Bioproc. Biosyst. Eng. 33 (2010) 39–45.
[61] S.K. Vashist, J.H.T. Luong, Trends in in vitro diagnostics and mobile healthcare, Biotechnol. Adv. 34 (2016) 137–138.

[62] S.K. Vashist, P.B. Luppa, L.Y. Yeo, et al., Emerging technologies for next-generation point-of-care testing, Trends Biotechnol. 33 (2015) 692–705.

[63] S.K. Vashist, O. Mudanyali, E.M. Schneider, et al., Cellphone-based devices for bioanalytical sciences, Anal. Bioanal. Chem. 406 (2014) 3263–3277.

[64] S.K. Vashist, E.M. Schneider, J.H.T. Luong, Commercial smartphone-based devices and smart applications for personalized healthcare monitoring and management, Diagnostics 4 (2014) 104–128.

[65] J.H.T. Luong, K.B. Male, J.D. Glennon, Biosensor technology: technology push versus market pull, Biotech. Adv. 26 (5) (2008) 492–500.

[66] E. Prusak-Sochaczewski, J.H.T. Luong, A new approach to the development of a reusable piezoelectric crystal biosensor, Anal. Lett. 23 (3) (1990) 401–409.

Chapter 14

Electrochemical Immunosensors: Fundamentals and Applications in Clinical Diagnostics

Kuldeep Mahato[1], Suveen Kumar[2], Ananya Srivastava[3], Pawan K. Maurya[4], Renu Singh[5], Pranjal Chandra[1]

[1]*Indian Institute of Technology Guwahati, Guwahati, India;* [2]*Ulsan National Institute of Science and Technology (UNIST), Ulsan, Republic of Korea;* [3]*Indian Institute of Technology Delhi, New Delhi, India;* [4]*Universidade Federal de Sao Paulo-UNIFESP, Sao Paulo, Brazil;* [5]*University of Minnesota, Saint Paul, MN, United States*

1. INTRODUCTION

Detection of pathogens, toxic elements, contaminants, and metabolic analytes plays a critical role in combating the disease preventions [1,2]. Thus, there is an urgent requirement of simple, effective, inexpensive, and quantitative detection tools to ensure the rapid appropriate diagnosis of counteracting pathogenic foreign agents and intrinsic metabolite imbalances [3,4]. Advancements in modern science paved the various techniques for detections since decades. In this context, soon after the novel discovery of antibody by Paul Ehrlich, its several applications have emerged. Immunoassay is one of them that has been appreciated globally for its potentials in clinical diagnostics. On the simplest note, an immunoassay is a technique that employs antibodies to detect a specific target producing either qualitative or quantitative or evaluation of both, usually in a sample mixture [5]. Recognition of molecular species in bioanalytical immunoassays is essentially based on the affinity-based antigen–antibody interactions. These assays, in general, use labeled antigens, haptens, or antibody for recognizing the target analytes [6]. Furthermore, different variants of immunoassays have been reported since the onset of such powerful efficient emerging technique, including radioimmunoassay (RIA), enzyme immunoassays (EIA), etc. The earliest developed immunoassay, RIA, uses radioactive isotope labeled antibody that facilitates the detection by forming radioactive antigen–antibody complexes, while in

Handbook of Immunoassay Technologies. https://doi.org/10.1016/B978-0-12-811762-0.00014-1

EIA a detecting signal has been generated by the enzymatic action of tagged enzyme that produces observable colors on exciting the chromogenic agent used. Moreover, instead of color producing agents, there are various other reporter molecules, which have been used in immunoassays such as DNA reporters, fluorogenic reporters, electrochemiluminescent tags, etc. Furthermore, a number of label-free immunoassays are also reported [7]. In addition, enzyme-linked immunosorbent assays (ELISA) a variant of EIA has become one of the effective immunosensing tools for a need of the hour since its postulation. So far, it has been exploited for several detection mechanisms, viz., human immunodeficiency virus (HIV) [8], west nile virus [9], and foodborne allergens [10]. Furthermore, the adoption of advanced bioanalytical technologies has molded these immunoassays to robust, quick, and efficient alternatives that imply for the better selectivity, sensitivity, and accuracy [11]. Electrochemical immunoassays, a hybrid technique composed by the amalgamation of EIA mechanism onto the functionalized electrode surfaces have widely been appreciated for its enhanced robustness, improved limit of detection (LODs), linear range (LR), limit of quantification (LOQ), and rapid nature. Nowadays this mechanism is extensively being explored to sense the various target analytes of clinical importance in miniaturized settings [12]. These miniaturized devices based on immunoassays are known as electrochemical immunosensors [13,14]. Essentially, these devices rely on the recognizing ability of biorecognition elements, i.e., antibodies, truncated antibodies, or the capturing molecules used against the specific target analytes. In the present chapter, we describe different type of immunosensing techniques, biorecognition elements, and various advanced approaches for electrochemical immunosensors, including electrical lateral flow immunoassays, receptor–ligand–receptor bifunctional-based assays, colorimetric immunosensing, and redox cycle–based approaches. Various applications, including electrochemical immunosensors based on protein, aptamer, cells, and enzyme-labeled, redox-labeled are also described with the adequate recent illustrations. The pictorial representation of the components associated with electrochemical immunosensor is depicted in Fig. 14.1.

2. ELECTROCHEMICAL TECHNIQUES IN IMMUNOSENSING

Various techniques have been employed to analyze the sample by using enzyme or enzyme-less colorimetric techniques. Electrochemical techniques (viz., voltammetry, amperometry, electrochemiluminescence (ECL), impedance, and photoelectrochemistry), and are extensively being used for their excellent capabilities to characterize the analytical performances of immunosensing probe. Electrochemical studies are, in general, carried out with trielectrode systems; where working, counter, and auxiliary electrodes (usually Ag/AgCl electrode or saturated calomel electrode) are used for building the

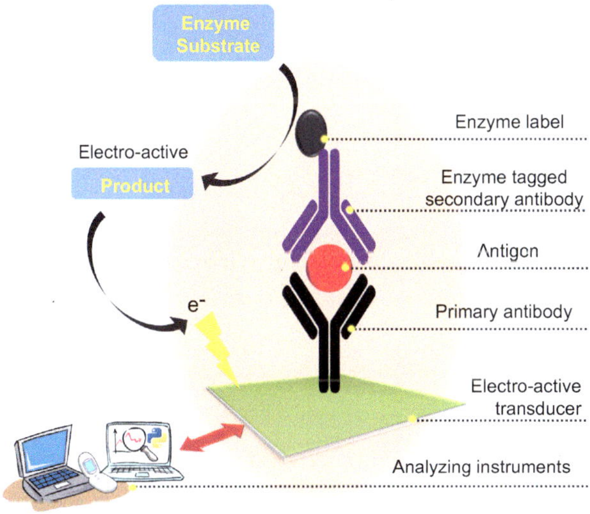

FIGURE 14.1 Schematic showing the basic components of electrochemical immunosensors.

analytical circuit [15]. In electrochemical immunosensors the working electrode is decorated with various biorecognitions elements, which are deposited over the working surface with the help of binder matrix. This section explains the techniques that have been employed for the evaluation of performances of such electrochemical immunosensors.

2.1 Voltammetry

Voltammetry, an electrochemical analytical method for assessing the current with a function of differential electrochemical reducing potentials is used for characterizing various immunosensors, possesses better sensitivity and applicability [16,17]. Voltammetric technique—based evaluation works on the fact that current flowing through the system is a function of potential applied across the electrodes. Briefly, in voltammetric methods, the range of potentials are scanned where the generated current is directly proportional to the concentration of electroactive species present in the sample. For the assessment of electrochemical immunosensors, various voltammetric methods, i.e., cyclic voltammetry (CV), differential pulse voltammetry (DPV), linear sweep voltammetry (LSV), square-wave voltammetry (SWV), stripping voltammetry (SV), have been adapted; however, CV, DPV, and LSV has frequently been used [18]. These voltammetric methods give both qualitative and quantitative information with excellent precisions ($<1\%$), sensitivity and wide linear dynamic range. There are several examples of immunosensors have been reported using voltammetric techniques, for instance, Nevas et al. have reported an immunosensor for detecting the gluten-induced autoimmune disorder termed as enteropathy, using voltammetry as a basic technique to evaluate the performances [19]. Characterization using voltammetric techniques determines various properties of the materials used in the biosensing probe such as equilibrium constant in coupled reactions, electrochemical reversibility, and electron transfer [18]. In analytical immunosensing studies, voltammetry is mainly used to find the LOD, LR, sensitivity, etc. [12,20]. However, the selections of the techniques are the most critical part, which is entirely dependent on the analyte samples and objectives. For example, to characterize the performances on sensitivity, SV with greater deposition time is preferred, which is achieved either by increasing the linear potential scan or adopting the differential pulse technique [21].

2.2 Amperometry

This technique is based on the evolution of current as a function of either time or potential applied across the electrodes, which provides the evaluation of various biosensing characteristics such as response time, dynamic range, and sensitivities similar to the other modes of electrochemical biosensing [22,23]. Usually, amperometric immunosensors consume the minute amount of the analytes, which strictly demands that analyte of either intrinsic or tagged redox moieties. Furthermore, the amperometric measurements probe fabrication greatly influences the process of immunosensing. Thus the construct used for probe decides shelf life, environmental stabilities pH, temperature, and various sensing parameters, viz., sensitivity, LOD, LR, and LOQ, etc. [24]. In addition, an amperometric technique that characterizes the reversible immunosensing probe has also been reported, which eventually checks the sensitivity dissipation of the probe in reusable immunosensors, where the pulsating potentials introduce the separation process [25,26]. For instance, using such type of probe a prototype has been reported by Gooding et al., which was fabricated with rabbit immunoglobulin as a biorecognition element on polypyrrole (PPy) layer over active electrode surface, shows the excellent reversibility when attributed by feeding of the pulsating potential of 200 ms during measurement. This treatment effectively stops the effects of stronger hydrogen bonding and hydrophobic interactions during the process providing enough interval for specific binding [25]. Alike the voltammetric techniques, amperometry are also being practiced under two specializations [1]: single potential amperometry and [2] pulsed amperometry. Single potential amperometry is carried out using a fixed potential across the electrodes. Usually, in this mode weak acid anions are detected, viz., cyanide, sulfide, iodide, sulfite, hydrazine, etc. whereas in the pulsed amperometric mode a pulse is introduced in the given setup to limit the electrode fouling as described in the above example [25,27].

2.3 Potentiometry

Potentiometry is another electrochemical method based on the fluctuation in concentration of ionic species and pH in the analyte sample [28]. Potentiometric evaluation of the biosensor is usually carried out using an ion-selective electrode, which has immobilized catalyst on the membrane. These immobilized catalysts on the membrane help to generate ionic species (generally hydrogen ions), which are then sensed by the potentiometric transducers. Potentiometric immunosensing is a most straightforward technique to examine the antigen–antibody interactions as no stimulus required during the measurement [29]. These techniques evaluate the performances of biosensors works by sensing of electrochemical, biological. or chemical borne ionic, pH or redox states at the probe electrode [30]. Classical potentiometric immunosensors work based on the change in electrical double layer due to the capping of biological elements on the probe surface, viz., antibody, aptamer complexes, etc. thus lowers the sensitivity [31,32]. Therefore, the biorecognition elements used in these immunosensors should essentially contain the antibody tagged with an enzymatic determinant to obtain quantifiable signals. After being recorded the signals are then interrelated with the analyte concentration present in the sample. For instance, Purvis et al. have reported PPy-based customizable potentiometric biosensors to detect the various target molecules, viz., troponin I, tumor necrosis factors, digoxin, and hepatitis B antigens using the enzyme-labeled detection mechanism on the PPy-coated screen-printed gold electrodes [30]. Another report on potentiometric immunosensors shown by Yuan and coworkers that have detected the hepatitis B antigen. In this mechanism, the biorecognition element, hepatitis B antibody conjugated with gold nanoparticle (AuNPs) and polyvinyl butyral is immobilized on the platinum (Pt) electrode systems [31].

2.4 Electrochemiluminescence

ECL is a phenomenon of light emission during the relaxation of electronically excited electrons of exergonic electrochemical reaction intermediates to the ground state. The wavelengths of the emitted light correspond to the energy gap of excited and relaxed states [33]. In electrochemical setup, ECL can be obtained by applying the voltage across the electrodes containing luminescent compounds in aprotic solution, viz., aromatic hydrocarbons, metal complexes, nanoparticles, and quantum dots, etc. [34,35]. ECL has been widely explored techniques in recent past due to its efficiency, ultrasensitivity, high specificity, and easy manipulation [36,37]. This technique has found a plethora of applications in various domains, viz., medical diagnostics [38], biohazards detections, [39] and warfare agents detection [40]. Conventionally, in ECL immunosensing techniques either polyclonal or monoclonal antibody labeled with luminescent molecule are used to fabricate the probe usually by covalent bonding; which can limit the bioactivity of enzyme labeled with the antibody by introducing the conformational changes [41]. To overcome such limitations metal nanoparticle-based nonenzymatic catalyzing systems have been incorporated to obtain the signals [42,43]. Based on ECL, various immunosensors have been reported [44,45]. Recently, Zhang et al. have reported an ultrasensitive ECL-based immunosensors to detect the tumor marker carbohydrate antigen 19-9. This proposed prototype has been fabricated using the sandwiched ECL technique where silver, bovine serum albumin (BSA) microspheres were functionalized with luminol. The tagging of luminol in the microsphere helps to decrease the distance between the luminophore and electrode surface that consequently increases the ECL intensity. The probe of the proposed prototype has been constructed by using the magnetic carbon nanotubes (CNT) decorated polyethyleneimine, which provides the excellent electrical and magnetic properties to this immunosensing platform. These properties of magnetic CNT, facilitate antibody immobilization and the deposition of magnetic entities over the electrode. This prototype has been reported with the excellent LR of 0.0005−150 U/mL with the LOD of 0.0002 U/mL in the presence of hydrogen peroxide [46]. In recent past, numerous ECL-based immunoassays have also been developed for the recognition of various molecules of clinical and environmental importance [41,47,48].

2.5 Electrochemical Impedance Spectroscopy

Electrochemical impedance spectroscopic (EIS) technique measures the impedance of electrochemical system, where it quantifies dielectric (resistive and capacitive) properties and impedance over a range of frequencies [49]. EIS measures the electron transfer and mass transfer depending on the alternating frequency used. The impedance of circuit in EIS is presented in either a Nyquist or a Bode plot. Nyquist plot generates a curve having two regimes: semicircular and linear; semicircular regime corresponds to the capacitance and resistance of the analyte complex, whereas the linear part is due to diffusive effect. In EIS, these impedances are analyzed for its linearity, causality, stability, and finiteness for obtaining good impedance values [50]. EIS technique is depending on the various sampling and intrinsic parameters such as the formation of a bilayer, resistivity, ion transportation, electron transfer through the electrode/electrolyte interface barrier, material, and geometry of electrode used, etc. [51]. Generally, the electrochemical systems show highly nonlinear behavior; linearization of such systems has been introduced by a small perturbation of alternating current amplitudes [50]. Usually, in EIS-based immunosensors the binding of antibody changes the impedance of system, which is directly dependent on the concentration of a target analyte present in the sample. Using this technique, extensive efforts have been given to obtain better sensitivity. For instance, Darkus et al. have characterized an immunosensing prototype by the EIS technique fabricated with the myelin basic protein antibodies as biorecognition elements and gelatin titanium dioxide–modified platinum electrode to detect multiple sclerosis autoantibodies [52]. This technique measures the analyte without any tags or labels, which makes it suitable for various applications, where the labeling of target analytes are tedious and time taking processes [53,54]. The EIS-based evaluation of immunosensors relies on physical blocking of the surface of a redox probe, which introduces impedance in the detection circuit, thereby resulting the decrease in signal strengths [55]. In addition, immunosensors are also reported based on the surface charge depletion from the probe, which is obtained either by attractive or repulsive force mediated ion migration that consequently introduces the impedances [56]. The major advantages associated with the EIS techniques are suitable for the biochip microfabrication as it shows very good compatibility to the analog and digital circuits allowing better signal processing, on-chip measurements, analysis, data storage, and further communications [51].

2.6 Photoelectrochemical Techniques

In photoelectrochemical (PEC) processes, photon to electrical signal conversion is triggered by excitation of electrons and charge transfer of a photoexcited material, which is produced by an applied light [57]. PEC-based techniques are another genre of the immunosensor evaluation that has studied for their greater sensitivity, which has been contributed by the ultra-reduced background noises. The amalgamation of two highly efficient analytical techniques, viz., PEC analysis and immunoassays, has led the domain flourished with various interesting sensory mechanisms [58]. So far, the PEC-based immunosensor has been extensively studied for the detection of various analytes, including dexamethasone (DXM), as proposed by Wang and coworkers based on PEC processes, where photocurrent responses on visible light conditions have been reported, where the competitive binding of DXM and BSA-DXM on the electrode surface decreases the specific attachment levels, resulting in the reduction of photocurrent intensity. This prototype has been reported with the significantly low LOD, apart from the high specificity, reproducibility, and acceptable stability [59]. However, various important factors in the mechanism of PEC biosensing such as photoexcited (charge) carrier generation, separation at nano–bio interfaces, etc. are not well explored. Furthermore, in PEC-based biosensing mechanism, sequence of charge transfer takes place between the analyte, photocatalytic material, and electrode before the generation of signal. Therefore the design of photocatalytic systems, electron capture, and transport manipulation in photoelectrodes play an important role in defining overall efficiency of PEC-based immunosensors [60]. PEC-based biosensing system assembled with a light harvesting semiconductor on a transparent working electrode, a metal electrocatalyst, a counterelectrode, and suitable electrolyte compatible for the target biomolecule, where additional photocurrent generated in the presence of target biomolecules giving the sensing of the analyte [60]. In general, PEC immunosensors face various detrimental effects due to the formation of an immunocomplex, which increases the steric hindrances toward the photogenerated holes on the electrode surface thus, photocurrent generation lowers down, which limits the PEC efficiency [61].

3. ADVANCED APPROACHES IN ELECTROCHEMICAL IMMUNOSENSORS

Although there exist several hundreds of reported unique prototypes, they essentially obey the rule for signal amplification as follows [1]: ultrafast electron transfer in between the signal-generating species and the electrode surface should occur [2], low or negligible background noises should be present [3], a sufficiently high concentration of the analyte at the active surface of electrode should be present [62]. With these empirical constraints, various strategies have been reported for the improved biosensing mechanisms, including electrochemical lateral flow assay, dual signal receptor-based bifunctional approaches, colorimetric, and redox cycle—based immunosensors.

3.1 Electrochemical Lateral Flow Assay

A conventional lateral flow immunoassay (LFIA) comprised of three types of antibodies commonly called "capture antibody," "detection antibody," and "secondary capture antibody," where the capture and detection antibodies bind to the antigen and secondary capture antibody binds to the capture antibody [63]. Detection and capture antibodies can bind to the analyte at different binding sites (epitopes). The control line of strip indicates the migration of sample across the membrane as intended, regardless of whether the analyte is present in the sample. The appearance of both the lines at test and control shows the positive result, whereas the single control line denotes the negative results. The integral components used in the LFIA are antibodies, membrane, tracer, pad, and plastic housing that give the support and integrity to the device. The performance is dependent on various parameters of the immunoassay: the biomolecule immobilization, its coverage on the tracer, and the environmental condition (buffer and pH). In general, LFIA works on two modes: first is based on noncompetitive and another is based on competitive immunoassay. In noncompetitive immunoassay format, the analyte is fixed in between the detection and capture antibodies, where the signal shows the direct dependency on concentration [63]. However, this is limited to the larger, multiantigenic (epitope) molecule. This format is not suitable for the small analytes having only one epitopic site. Also, the simultaneous binding of

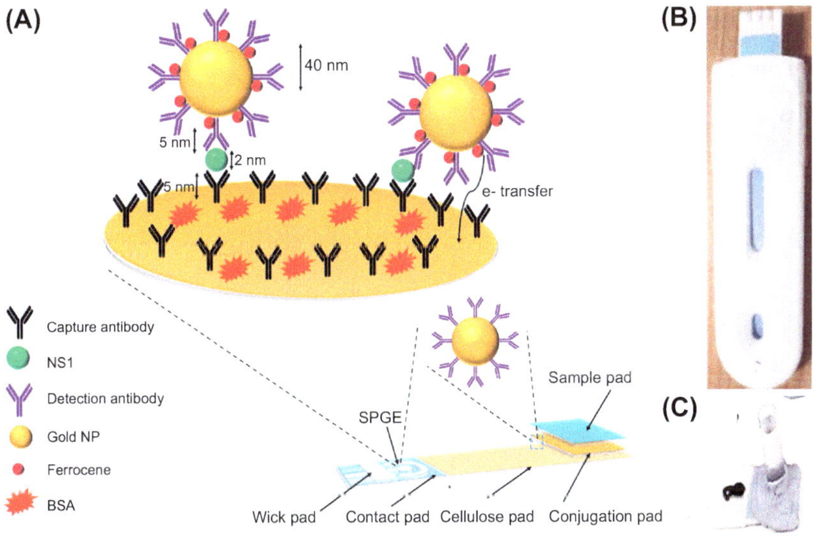

FIGURE 14.2 Electrochemical lateral flow immunosensing devices. (A) Showing the schematic reaction of the immunodetection. (B) Plastic casing of the immunosensor strip. (C) Sample loading holder. *Reprinted with permission from D. Sinawang Prima, V. Rai, R.E. Ionescu, R.S. Marks, Electrochemical lateral flow immunosensor for detection and quantification of dengue NS1 protein, Biosens. Bioelectron. 77 (2016) 400–408; copyright to Elsevier Inc.*

antibodies with complex (capture/tracer complex and detection antibody) to the small molecules may be prevented by the steric hindrance, which reduces the applicability of the performances of immunoassays. In competitive immunoassay-based LFIA, the analyte of unknown concentration competes to the analyte of known concentration for binding the single antibody. Unlike above, the signal in noncompetitive immunoassays format is contrariwise dependent on the analyte of unknown concentration that has been provided either by preimmobilized to the membrane or conjugate to the tracer. Thus, on sampling, the bright test line appears [18,64,65]. Furthermore, in contrast to the conventional LFIA, the coupling of electrochemical methods to the LFIA has shown a new path to introduce accurate, robust, onsite, handy, molecular diagnostics. It is termed as the electrochemical lateral flow immunoassay (shown in Fig. 14.2). Where the LFIA is paired with the screen-printed electrode [66]. It shows various advantages including quick, low-cost, single-step qualitative and quantitative results.

3.2 Approaches Based on Receptor–Ligand–Receptor Binding

Selectivity is the major concern in any mode of the biosensing prototypes [67]. To address this problem, several smart mechanisms have been adopted, including the dual receptor-based prototypes, where the analyte is targeted by the two receptor molecules following the sandwich pattern that introduces dual signals for highly selective confirmatory detection. For instance, a prototype reported by Chandra et al. to detect multidrug resistance cancer cells using a composite of conducting polymer with the AuNPs and functionalized multi-walled carbon nanotubes (MWCNT). This prototype has shown a wide dynamic range (50−100,000 cells/mL), which is comparable with the conventional ELISA-based techniques. Briefly, the immunosensing probe has been fabricated depositing the layer of AuNP polymeric conjugate on a glassy carbon electrode (GCE) followed by anti-P-glycoprotein immobilization on its surface. Then, it has been treated to the cell solution followed by the incubation in aminophenyl boronic acid, MWCNT, and hydrazine conjugate (Fig. 14.3A) [68]. Similarly, in other reports, cancer cells have been sensed by the same group using silver nanoparticles (AgNPs) and aptamer−AuNPs−hydrazine bioconjugate, where anti-HER2 (an antibody for human epidermal growth factor receptor 2) is covalently attached with 2,5-bis(2-thienyl)-1H-pyrrole-1-(p-benzoic acid) (DBP)-coated AuNPs (Fig. 14.3B). This immunosensor has not only reported the significantly low value of LOD (26 cells/mL) but also successfully delivered the visual and highly accurate electrochemical quantification [69]. Also, the binding of a ligand to multiple receptors introduces the better selectivity. In some cases, multiple interactions generate the signals that elicit different mode of characterization as proposed by Zhu et al. produces simultaneous electrochemical and optical signals, which gives the confirmatory detection of the analyte [69].

3.3 Colorimetric Immunosensing Approaches

Colorimetric immunosensing is another mode of detection that provides ease to the selective recognition and produces visual signals, which can be perceived by naked eyes, thus do not require any sophisticated instruments and are simple, economical, and portable for resource-limited settings [70]. In this system, tagged antibodies are used that add the specificity and selectivity. These tagged antibodies are immobilized with the help of the various nanomaterials such as the nanocomposite matrices, nanosheets, nanoribbons. Depending on the particles physical properties the matrix and immobilization techniques are adopted. For instance, Nafion, a proton-conducting polymer possessing an excellent film forming, high chemical stability, good adhesiveness, and biocompatibility is used for immobilizing the positively charged nanoconjugates in a number of electrochemical biosensing prototypes [71]. To date, several variants of colorimetric techniques have been proposed. However, these have the common approach in the fabrication process. The process, in general, involves the preparation of immobilization matrix, preparation of the nanoconjugate antibody tag, and immobilization of chromogenic

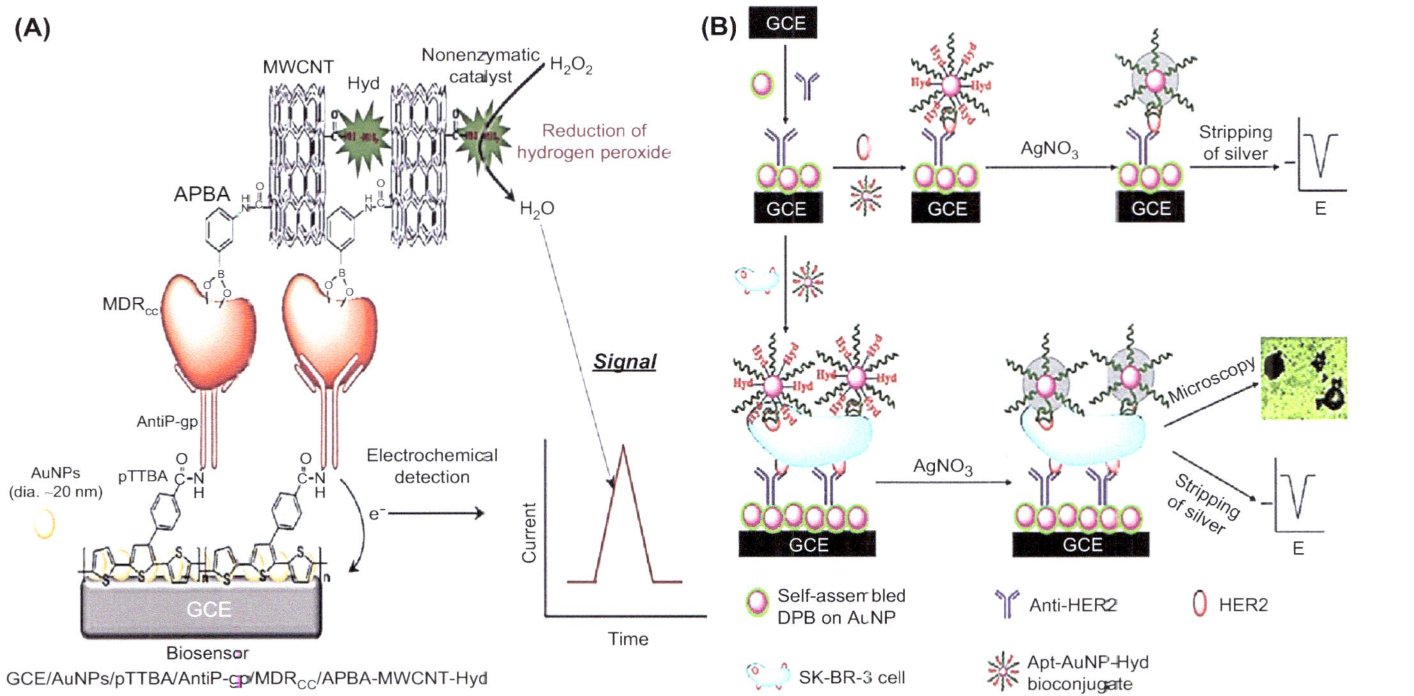

FIGURE 14.3 Schematic representation of dual receptor-based detection (A) shows the multiple drug resistant cancer cells immunosensing probe and target interaction assisted with the functionalized MWCNTs; (B) Immunosensor for HER2 protein and HER2-overexpressing SK-BR-3 breast cancer cells detections. *(A) Reprinted with permission from P. Chandra, H.-B. Noh, R. Pallela, Y.-B. Shim, Ultrasensitive detection of drug resistant cancer cells in biological matrixes using an amperometric nanobiosensor, Biosens. Bioelectron. 70 (2015):418–425. Copyright Elsevier Inc. (B) Reprinted with permission from Y. Zhu, P. Chandra, Y.-B. Shim, Ultrasensitive and selective electrochemical diagnosis of breast cancer based on a Hydrazine–Au nanoparticle–aptamer bioconjugate, Anal. Chem. 85 (2) (2013) 1058–1064. Copyright American Chemical Society.*

detection system on the detection site. These approaches are extensively used for their simplicity. However, such prototypes are suffering from the temperature sensitiveness mainly because of the presence of enzymatic moieties in the probes, which restricts the entire system to be operable in a wide range of temperatures. Thus, violating an essential requirement for the potential commercial biosensing volumes. Apart from this, the activation of chromogenic substances may be initiated by other electron-donating species, which are present in the sample solution competing with the analyte, therefore, the specificity of such prototypes is questionable. Recent trends of the colorimetric immunosensors have been exploited by silicon-based technologies, which are nowadays, frequently used in various engineering fields such as electronic devices and chip fabrications. Porous silica–based immunosensors are the recently added usage of the silicon-based materials.

The structurally originated optical properties of the porous nanopatterned material have exploited for the immunosensing, where microporous surfaces play an important role. A bare porous surface of silica is modified by functionalizing with the specific antibody, which essentially captures the analyte and thus introduces altered surface refractivity. This fluctuation in refractivity gives the selective detections. For instance, Massad-Ivanir et al. have reported the real-time *Escherichia coli* contaminant detection using the porous silicon–based platform [72]. Here anti–*E. coli* has been functionalized on the exposed surface of the porous silicon, where the bacterial contamination selectively adheres to the surface due to the antibody antigenic interaction on exposing with the sample. A detailed scheme has been shown in Fig. 14.4A [72]. Another example of colorimetric immunosensor has been reported by Fu et al. using iron oxide Prussian blue nanoparticles conversion strategy (scheme has been shown in Fig. 14.4B). In this colorimetric immunosensing prototype the properties of nanoparticle have been exploited where the antibody-tagged iron oxide nanoparticles are used to selectively detect the analyte [73].

3.4 Redox Cycling–Based Approaches

Redox cycling is inspired by the dual enzymatic activity of tyrosinase, i.e., oxygenation and oxidase activities. In redox cycling, enzyme-amplified electroactive species are rejuvenated after their electrochemical process (oxidation or reduction). Redox cycling–based immunosensing mechanisms work on the fact that the analytes get reduced after attaining oxidized state with the effect of additional catalyzing agent present along with the electrochemical cell other than the fundamental components. For instance, based on a redox cycle an immunosensing prototype reported by Akanda and Ju for antigenic proteins using tyrosinase-tagged primary antibody provided with phenol as substrate and NADH as reducing agent with wide dynamic window and high sensitivity; the schematic has been shown in Fig. 14.5 [62].

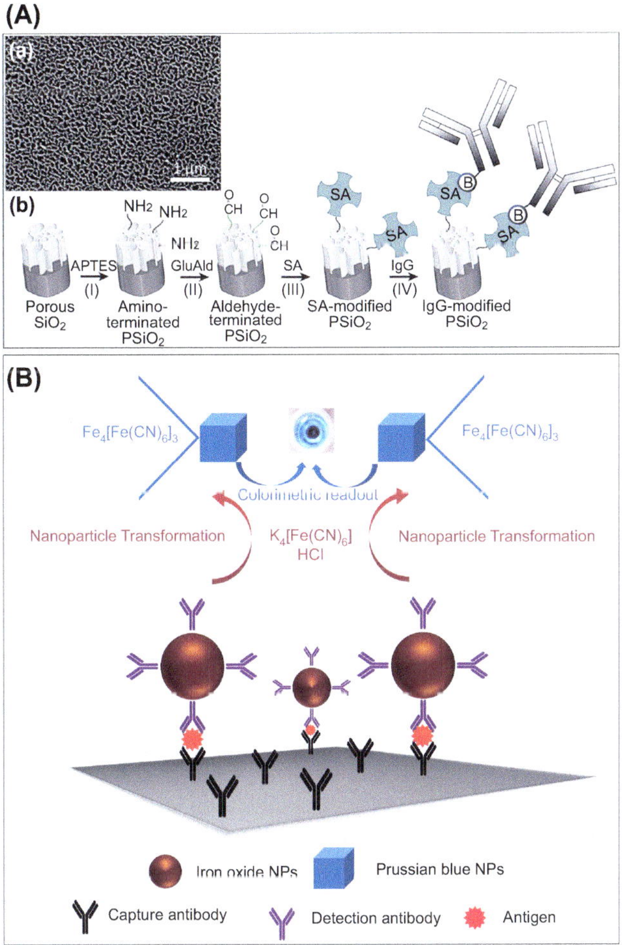

FIGURE 14.4 Different approach of colorimetric immunosensor (A) porous silica–based colorimetric immunosensing (B), complex compounds–assisted colorimetric immunosensing. *(A) Reprinted with permission from N. Massad-Ivanir, G. Shtenberg, N. Raz, C. Gazenbeek, D. Budding, M.P. Bos, et al., Porous silicon-based biosensors: towards real-time optical detection of target bacteria in the food industry, Sci. Rep. 6 (2016) 38099. Copyright Nature publishing group. (B) Reprinted with permission from G. Fu, S.T. Sanjay, X. Li, Cost-effective and sensitive colorimetric immunosensing using an iron oxide-to-Prussian blue nanoparticle conversion strategy, Analyst 141 (12) (2016) 3883–3889. Copyright Royal Society of Chemistry.*

Another example describes a high-throughput immunosensors based on multilabel strategy reported by Yang and coworkers, where they have developed a novel array electrode for simultaneous detection of multiple biomarkers in hepatocellular carcinoma [74]. In this published work, graphene nanosheet–PtPd bimetallic nanocomposites have been labeled with the horseradish peroxidase (HRP) and detector antibody to get amplified signal and antibody

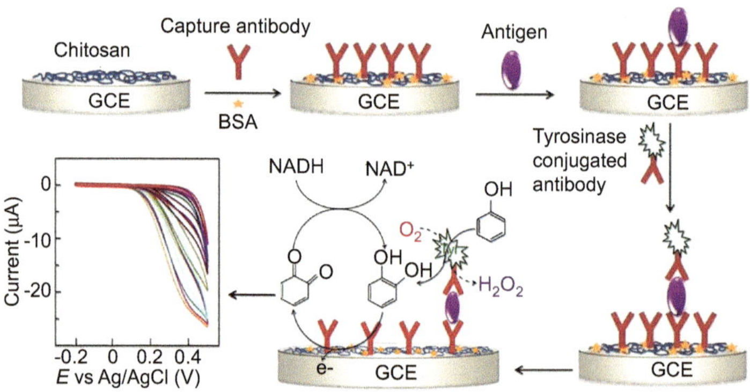

FIGURE 14.5 Redox cycling–based immunosensing prototypes. *Reprinted with permission from M.R. Akanda, H. Ju, A tyrosinase-responsive nonenzymatic redox cycling for amplified electrochemical immunosensing of protein, Anal. Chem. 88 (19) (2016) 9856–9861. Copyright American Chemical Society.*

loading. So far, various forms have been reported for redox cycling–based biosensors, including multiple electrode systems, multiple enzyme systems, electrochemical catalytic reaction systems, etc. [75]. Despite producing high sensitivity and wide dynamic ranges, technically these immunosensors suffer from extrinsic limitations associated with fabrication processes. For instance, in multiple electrode systems, improved signals can be achieved by introducing working coupled electrodes; however, in practice, achieving the high-yield redox cycle process requires the microfabricated microband electrodes or electrode arrays for amplified signals, which are, in general, very tedious to realize. In the multiple enzymatic systems, the rational design of the layer is necessary to achieve high redox efficiency [76]. Also, in the electrochemical systems, less stable reducing species exhibit low sensitivity. Moreover, the complexity associated with the fabrication processes limits the applicability and the sensitivity toward the particular analyte.

4. BIORECOGNITION ELEMENTS IN IMMUNOSENSORS

Biorecognition element is an integral component of immunosensors; in general, these elements sense the stimuli and feed the receptor signal to the transducer, which further converts a receptor signal to quantifiable electrical signals [77,78]. Antibodies or its fragment are used for this purpose in immunosensors. Nowadays several strategies of biorecognition are being employed where antibody determinant has been supplemented with aptamers to increase the sensitivity and temperature stability. This supplementary determinant is nothing but a synthetic oligonucleotide short chain of the DNA, which is obtained using the systemic evolution of ligands by exponential enrichment (SELEX) protocol [18]. This section describes the various bio-recognition elements, which are used for electrochemical immunosensing prototypes.

4.1 Antibodies: The Primary Recognition Elements of Immunoassays

Antibodies are the immune responsive macromolecules produced for the cellular defense to combat foreign pathogens and intrinsic threat to the cells. The agents that are being neutralized by antibodies are generally termed as antigens [79]. It is a complex biomolecule, comprising of different types of amino acids arranged in a highly ordered sequence. They are produced by plasma cells and are further used by the immune system for identification and neutralization of pathogens. It comprises plasma proteins called as immunoglobulins (Igs) [80]. The simplest antibodies can be defined as Y-shaped molecules with two identical binding sites for antigen known as fragment of antigen binding (Fab) sites, bearing amine ($-NH_3^+$) as a functional group and Fc fragment (fragment crystallizable, the lower side that crystallizes during cold storage) bearing carboxyl ($-COOH$) group that helps in antibody immobilization onto the specific surface [80]. Fig. 14.6 shows the typical structure of antibody molecule with antigen binding site and $-COOH$ groups. During the recognition process, antibody antigen forms a complex via Fab variable region. Another part is

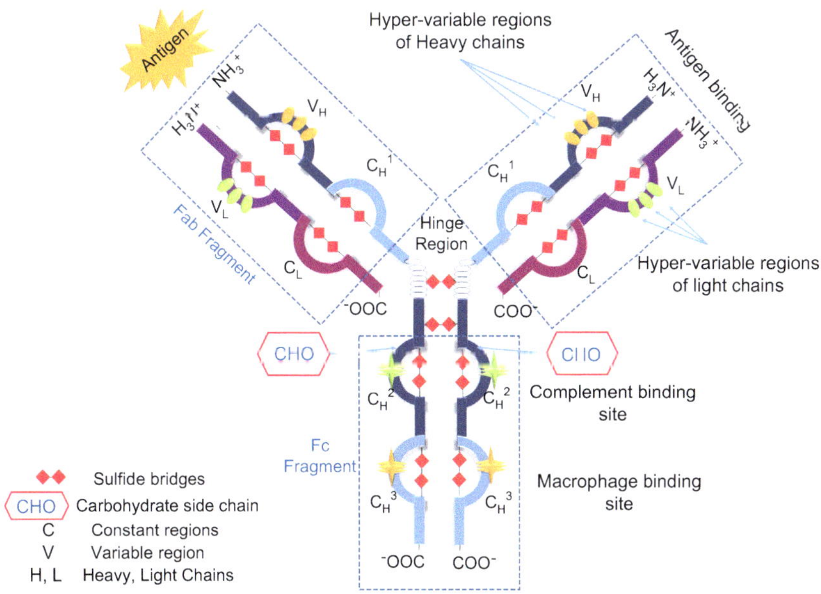

FIGURE 14.6 Typical structure of antibody molecule with antigen binding site.

complemented by the Fc region, which facilitates the antibody to communicate with the other components [81]. Antibodies are composed of four domains with two light and two heavy peptide chains bridged by disulfide linkages. Modern advances in immunology have paved the utilization of the antibodies in the various applied applications, including therapeutics and diagnostics. In biosensors, antibodies provide the enhanced specificity, selectivity, and sensitivity also called immunosensors [14].

The antibody in immunosensors serves as biorecognition elements, which are coupled to the transducer (a working electrode in the case of electrochemical biosensors). Where the coupling is achieved by various techniques of immobilization, including biochemical affinity, covalent bonding, and molecular imprinting [82]. Since the antibodies are highly specific to the epitope (target site of antigens) in immunosensors the binding of the probe to the target is highly selective that enhances the selectivity by screening out the other interfering molecules present in the sample. However, in the case of antigens having multiepitopic determinants, antibody production for highly specific immunosensors becomes impractical since the process of isolating a particular type of antibody is quite expensive. To address such shortcomings, recombinant antibodies are synthesized with the help of computational data-mining for epitope determination and consequently using the recombinant technology, where the target of the interest has been selected from libraries or docked by the computational calculation [83,84]. In short, recombinant antibodies introduce better selectivity and specificity by targeting the complementarity-determining regions using site-directed mutagenesis or chain shuffling [85,86]. Further advantages associated with the recombinant antibodies that these can be tagged with the molecular tags (GFP, His, and C-myc), which are useful in easy isolation in further downstream and various biosensor fabrication processes such as immobilization, labeling, or tagging, etc. [83,87].

4.2 Truncated (Fragmented) Antibodies

Truncated antibody essentially means that fragmented part of antibody obtained by using the proteolytic enzymes (pepsin or papain) to generate fragmented Fab and Fc regions for chopping out the redundant fragments of the antibody as in some cases a larger size of immunoglobulin molecules tends to compromise the specificity in immunosensing [88]. Therefore, to achieve better sensing, it is required to have smaller determinants of the same specification. The truncations are, in general, introduced by destabilizing peptide backbone as these bonds in antibodies are highly stable. Thus, fragmentation is carried out using various techniques depending on the desired sizes such as site-targeted fragmentation of peptides marking particular amino acid residues (asparagine, serine, threonine, cysteine, etc.), disulfide bridges dissolution between heavy—heavy and heavy—light chains, etc. [89]. Also, it is possible to make single-chain variable fragments after digesting the Fab and fusion of variable region of both heavy and light chains. In this configuration, the fused chains have a smaller size and yet exhibit the similar specificity [90]. In addition to this, immunoglobulin macromolecules are more likely to bind asymmetrically on the operating platform, which gives rise to the inappropriate orientation that usually suppresses the specificity and reproducibility of the immunosensors. Thus, the truncation of such bulky moieties is required to make suitable for appropriately oriented binding. These fragmented antibodies are, in general, produced by the recombinant antibody technologies, where the monoclonal antibodies are being utilized to produce smaller fragments of antibodies. Although these fragmented antibodies are very promising, fragmentation may cause affinity loss, aggregation, and sensitivity toward the temperature. Thus, preferred the recombinant antibody technology, which is based on the combinatorial approaches where a selection of chains and fragment achieved using expression libraries. These antibody fragment libraries are derived from the single scaffold such as Fab, single-chain Fv, and VH determinants [91].

4.3 Aptamer Antibodies Hybrid

Aptamers are double-stranded DNA or single-stranded RNA or DNA molecules that bind specifically to molecular targets. These "designer" molecules were first developed in 1990 and were followed by the development SELEX, which is the technology for isolation of highly specific aptamers [23,92,93]. However, a study in 2000s revealed the presence of "riboswitches"—a type of naturally occurring RNA aptamers that function as gene regulators in the cell nucleus. Furthermore, these possess properties comparable of capturing the molecule to the monoclonal antibodies. Hence, these are used alternative to the specific biorecognition elements for various diagnostics [94,95] and therapeutics [96,97]. Structurally, aptamers have spatial pockets and cavities of nucleic acid folds, which help to obtain high-affinity binding with molecular targets, viz., metal ions, protein complex, whole cell, viruses, etc. [98]. Owing to its advantages over close competent monoclonal antibodies and wider flexibilities in applications; nowadays, aptamers are commonly used in biosensing [99,100]. Recent years, bifunctionalization approaches for biorecognition elements have been emphasized for signal amplification, which is an urgent requirement of the robust ultrasensitive bioassays [101]. Adoption of bifunctionalized elements obtained using aptamer with antibody in immunosensors provides better selectivity, sensitivity, temperature resistance, and thus improved diagnosis. Ocana et al. have reported an electrochemical impedance biosensor for cytochrome *c* using aptamer–antibody sandwich biorecognition element on MWCNT screen-printed electrodes that have shown an ultralow LOD (12 pM) under optimized conditions [102]. The major advantages of such recognition elements are they provide interference-free detection. In the above example, the interfering molecules that are copresent in the sample have been found that the potential interfering protein immunoglobulin, which has prominent interferences on the control assays, shows no effect on the sensing with adopted sandwich variants [102]. In addition, such recognition element has been reported by Ilkhani et al., where the antibody−aptamer hybrid was used for epidermal growth receptor detection in biological fluids using electrochemical immunosensing methods and reported the significantly low LOD (50 pg/mL) [103]. It is anticipated that these hybrid elements are more efficient to eliminate the close interfering competitor. Also, when it comes to the commercial feasibilities, the diagnostics should have robustness and wider applicability. An aptamer in such hybrids not only compliments the antibodies in context to selectivity but also provides low cost and environmental stability to such diagnostics.

5. ELECTROCHEMICAL IMMUNOSENSORS

Electrochemical immunosensors are propitious substitutes to the present immunological methods for the advancement of handheld devices, which can be used for the point-of-care tests. Immunosensors use specific antigen−antibody complex formation for analytical purposes as antibody contains paratope at the Fab region that binds to an epitope of an antigen, which makes antibody highly specific in nature. An antigen usually contains different epitopes along its surface arranged randomly and dominant epitopes known as determinants. The antibody binds to antigen through the spatial complementary mechanism, and it is reversible [80,104]. During the interaction between antigen and antibody, several weak forces, i.e., *Van der Waals*, hydrogen bonding, electrostatic, hydrophobic interaction, etc., are involved [80,104]. Based on binding sites of Fab region, the antibody has been categorized into the monoclonal and polyclonal types. Monoclonal antibodies are made in a laboratory rather than by a person's immune system unlike polyclonal antibodies and possess the affinity to attach with the same epitope of an antigen. However, polyclonal antibodies bind to different epitopes of same or different antigens. Besides this, monoclonal antibodies are highly specific and less cross-reactive with other proteins [80,105]. These exclusive properties of monoclonal antibodies make them a potential candidate for detection of specific antigen in disease/analyte detection. Thc main source of the antibody production is specialized white blood cells, which are also called B lymphocytes. When an antigen is encountered by these B-cells, the antigen binding occurs on its surface, which triggers them to divide and mature into a group of identical cells called a clone. Antibodies are now commercially available globally. However, their production takes months making it expensive and with different quality [80]. In the current scenario, the production has been streamlined and transposed to bacteria, yeast, etc. to some extent [106]. Antibodies are being used as a bioreceptor in the fabrication of immunosensor and here in this chapter efforts have been made to discuss various types of immunosensors based on protein, bacterial/cancer, enzyme-labeled, redox-labeled, and aptamer.

5.1 Label-Free Protein-Based Immunosensors

This section covers the fabrication of label-free immunosensor for protein biomolecules detection. In label-free immunosensor, electrochemical response either increases or decreases on the concentration of the biomolecules without tagging the primary or secondary antibodies or antigens [107–110]. Owing to the lots of advantages such as cost-effectiveness, ease of fabrication, etc., efforts have been given for fabricating numerous label-free electrochemical immunosensors for several biomarkers such as alpha-synuclein autoantibody, cardiac troponin-I, troponin T, angiogenin, CA-125 (cancer antigen-125), carbohydrate antigen 15-3, CEA (carcino embryonic antigen), CYFRA-21-1, C-reactive protein, EGFR (epidermal growth factor receptor), hepatitis B surface antigen, interferon-6, 8, 10, P53 protein, PSA (prostate-specific antigen) for detection of various diseases, including Parkinson's, cardiovascular, autoinflammatory, cancer, Alzheimer disease, foodborne and airborne diseases [106–119]. For instance, Wang and coworkers developed a simple label-free immunosensor to detect CEA by modulating the surface-bound redox species thionine. For the fabrication of this immunosensor, GCE was used as a substrate, and a mixture of thionine and chitosan was immobilized over it by drop cast method. This film was further covered with a Nafion membrane containing SiO_2 nanoparticles. Polyethyleneimine was used as linker molecules to immobilize anti-CEA biomolecules. This label-free immunosensor showed LR in between 1 fg/mL and 100 pg/mL with LOD of 1 fg/mL [120]. Another work by Feng and his coworkers also described efficient detection of CEA using electrodeposited Prussian blue (PB) nanoparticles by incorporating graphene and MWCNTs onto the GCE surface. They improved the stability of biosensor by reducing the leakage of PB from working surface. MWCNTs–graphene composite was used to enhance the working surface area and electron transfer rate of PB; AuNPs were used to immobilize CEA antibody. The fabricated immunoelectrode displayed two LR (0.2–1.0, and 1.0–40.0 ng/mL) with LOD of 60 pg/mL [121]. In another work by Gao et al., Nile Blue A (NB) was used as a replacement of PB for the development of label-free immunosensor for CEA detection [116]. In this proposed immunosensor, AuNPs and NB were electrochemically hybridized onto the electrochemically reduced graphene oxide (NB-ERGO). Pep stacking interaction process was used in the synthesis of NB–graphene oxide (NB-GO) composite, and AuNPs/NB–ERGO nanocomposite was synthesized by reducing chloroauric acid ($HAuCl_4$) and nanocomposites of NB-GO using chronoamperometry (CA). Under optimal conditions, the proposed immunoelectrode exhibited linearity in the range 0.001–40 ng/mL with LOD of

0.00045 ng/mL. Kumar et al. developed several conducting papers–based immunosensors for the detection of CEA [109,122–124]. In this process, poly (3,4-ethylenedioxythiophene): poly(styrene sulfonate) (PEDOT:PSS) was used to make paper conducting. After physically absorbing anti-CEA onto conducting paper, they found that the fabricated immunosensor could detect CEA within LR 6–20 ng/mL but with low sensitivity of 3.6 Ω mL/ng with LOD of 2.68 ng/mL [122]. In another report, PEDOT: PSS/polyvinylalcohol nanofiber ornamented conducting paper-based immunosensor shows LR of 0.2–25 ng/mL, the sensitivity of 14.2 μA/ng mL cm^2 with shelf life of 22 days [123]. To improve sensitivity of the immunosensor, conducting paper was modified with rGO and the conductivity of rGO-modified paper was increased by treating it with ethylene glycol ($\sim 1.16 \times 10^{-4}$ S/cm up to $\sim 3.57 \times 10^{-2}$ S/cm ~300 times) due to conformational reordering in the polymer because of robust noncovalent cooperative communication between PEDOT and cellulose. rGO-modified conducting paper–based immunoelectrode displayed a higher sensitivity (25.8 μA/ng mL cm^2) with LR 1–10 ng/mL [109]. To improve the LR, CNTs were used instead of rGO, and synthesized PEDOT: PSS-CNT nanocomposite eventually leading to an improved LR of 2–15 ng/mL with a sensitivity of 7.8 μA/ng mL cm^2 [125]. Furthermore, for improvement in biosensing characteristics, the same group used gold to make the paper more conducting and deposited polyaniline onto conducting paper for making a crack-free surface. This also helped in immobilization of anti-CEA biomolecules on the conducting paper. The fabricated immunoelectrode showed a dynamic detection range in between 2 and 20 ng/mL and sensitivity of 13.9 μA/ng mL cm^2 along with a shelf life of up to 22 days (Fig. 14.7) [124].

Paper-based diagnostics are further being integrated with microfluidic technology to enhance the sensitivity for point-of-care applications. Recently, Wang et al. fabricated a paper-based microfluidic immunosensor. In this approach, amino functionalized graphene/thionine/gold nanoparticles screen-printed working electrode (SPWE) was used and reported with LR and LOD of 50–500 ng/mL and 10 pg/mL, respectively, for CEA [126]. Moreover, for the improvement in dynamic range of immunoelectrode for CEA detection, nitrogen-doped graphene quantum dots (N-GQDs) supported PtPd bimetallic nanoparticles (PtPd/N-GQDs) were synthesized through the hydrothermal process, and self-assembly approach was utilized to functionalize Au nanoparticles with the synthesized materials. Owing to the synergistic effect of Pt, Pd, N-GQD, and Au, this novel nanocomposite showed excellent electrocatalytic activity and biocompatibility. The excellent conductivity and large surface area of PtPd/N-GQDs@Au attracted attention to using this nanomaterial as a transducing material that can also efficiently conjugate capture antibodies. This fabricated immunoelectrode showed a wide dynamic range of 5 fg/mL–50 ng/mL with LOD of 2 fg/mL for CEA detection. This

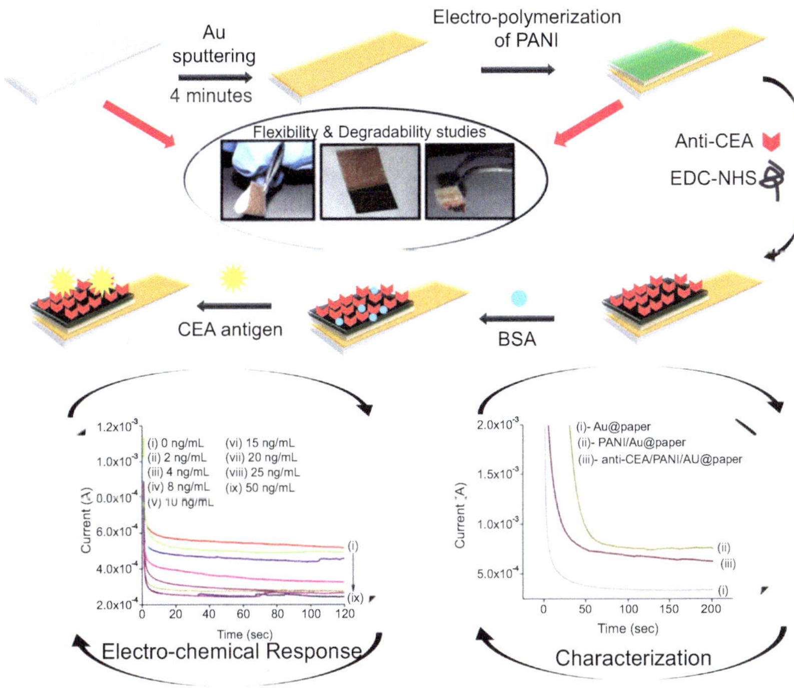

FIGURE 14.7 Schematic relating to fabrication of the Au/PANI-based immunosensor for carcino embryonic antigen (CEA) detection [124]. *BSA*, bovine serum albumin.

immunoelectrode also exhibited high sensitivity, special selectivity, and long-standing stability, which shows a potential application in bioassay analysis for CEA biomolecules [127].

Several other biomarkers, such as α-fetoprotein (AFP), EGFR, murine double minute 2 (MDM2), CYFRA-21-1, have been used for the detection of various types of cancer [108,113,117,128,129]. A label-free EIS immunosensor for the detection of AFP has been fabricated by using wheat-germ agglutinin (WGA) lectin as molecular recognition component by adsorbing carboxyl-functionalized single-walled CNTs onto a screen-printed carbon electrode and covalently binding WGA onto the SWNTs-modified electrode. Immunosensor showed the LOD of 0.1 ng/L with the detection range from 1 to 100 ng/L [128]. Sun et al. coimmobilized hydroxyapatite nanoparticles and graphene oxide onto the electrode surface and found an extensive LR of 0.01–10 ng/mL with LOD of 5 pg/mL for AFP detection and further evaluated in human serum samples [129]. Vashudev and coworkers have detected EGFR by using immobilized anti-EGFRs on a dithiobissuccinimidyl propionate self-assembled monolayer (SAM) on an Au electrode. The anti-EGFR and EGFR complex formed on the immunoelectrode demonstrated a high association constant of 5×10^{12} L/mol that resulted in high affinity of anti-EGFR to EGFR, which resulted in improved LR of 1 pg/mL−100 ng/mL and remarkable LOD of 1 pg/mL with high sensitivity of 2.02 μA/M [117].

MDM2 is a brain cancer biomarker secreted in the serum of affected patients [112]. It is widely being used in detection and prognosis of brain cancer patients. Recently, a label-free impedimetric immunosensor has been developed that is based on an cysteamine-modified Au electrode. Antibodies against MDM2 were covalently immobilized onto this modified Au electrode using homobifunctional 1, 4-phenylene diisothiocyanate linker. Changes in impedance of the sensor surface on exposure to varied concentrations of MDM2 protein were recorded, thus yielding a dynamic detection range of 1 pg/mL−1 mg/mL. The LOD was obtained to be 0.29 pg/mL in phosphate buffer saline and 1.3 pg/mL in mouse brain tissue homogenate [112].

In serum sample of breast cancer patients, numerous biomarkers such as carbohydrate antigen 15-3 (CA15-3), human epidermal growth factor receptor 2 (HER-2), etc. are released that helps in detection and monitoring of breast cancer [114,115]. Recently, a N-doped highly conductive graphene sheets−based immunosensor was reported for the detection of CA15-3 biomarker and exhibited lower LOD of 0.012 U/mL and LR of 0.1−20 U/mL [114]. In addition to this, Ravalli et al. reported a sensitive impedimetric biosensor for HER-2 detection in the serum sample of breast cancer patients. In this report, gold nanostructured screen-printed graphite was used as a sensor platform for the detection of HER-2. Cysteine was used for surface modification of AuNPs; anti-HER-2 was immobilized onto these cysteine-modified nanoparticles. This immunosensor showed good analytical performance for detection of HER-2 in the concentration range from 0 to 40 μg/L and LOD of 6.0 μg/L [115].

The focus has increased on noninvasive detection of secreted biomarkers to facilitate painless detection of various diseases [107,108]. Recently, the United States Food and Drug Administration has approved the nuclear matrix protein 22 (NMP22) biomarker assay for the evaluation of bladder cancer (BC) patients. Li et al. have fabricated a sensitive label-free electrochemical immunosensor for NMP22 detection based on Au@Pd/Ag yolk-bimetallic shell nanoparticles (Au@Pd/AgNPs) deposited onto GCE by amino-functionalization of graphene to immobilize anti-NMP22 antibodies. The fabricated immunosensor showed high selectivity in LR because of high surface area and outstanding electrical conductivity [110]. A similar yet simplified biosensor design was demonstrated by Jia et al. wherein they developed an Au−Pt bimetallic nanostructures-based amperometric immunosensor for sensitive detection of NMP22 with LOD of 3.33 pg/mL and an LR from 0.01 to 10 ng/mL [107]. Efforts are also being made toward label-free detection of salivary biomarkers of oral cancer [108,130]. CYFRA-21-1 has been recognized as a novel promising biomarker that is secreted in high concentration in saliva of oral cancer patients. In the last couple of years, efforts have been made to detect this biomarker electrochemically, which can be promising for noninvasive oral cancer detection [108]. Kumar et al. have developed nanostructured zirconia ($nZrO_2$)-based immunosensors for salivary CYFRA-21-1 biomarker detection. This nanomaterials-based immunosensor displayed LR from 2 to 16 ng/mL with a sensitivity of 2.2 μA mL/ng and stability of 6 weeks [108]. Furthermore, they used nanostructured hafnia instead of $nZrO_2$ and found improved LR of 2−18 ng/mL. However, the detection of a whole physiological range of CYFRA-21-1 ($0-17.46 \pm 1.46$ ng/mL) remained a challenge [131]. To overcome this problem, $nZrO_2$ was uniformly decorated onto rGO surface resulting in $nZrO_2$-rGO nanocomposites. This $nZrO_2$-rGO-based immunoelectrode showed a wide LR of 2−22 ng/mL with LOD of 0.122 ng/mL having a sensitivity of 0.756 μA mL/ng [130]. However, lower LOD below 2 ng/mL could not be achieved with these fabricated immunoelectrodes. Thus in the next work, they improved the loading of anti-CYFRA-21-1 antibodies onto $nZrO_2$ by replacing 3-aminopropyl triethoxysilane (APTES) by serine amino acid as a zero-linker and achieved the LR from 0.01 to 29 ng/mL [132]. Moving ahead, efforts have also been made to detect several other diseases or biomolecules by using label-free immunosensors, such as hepatitis B, cardiovascular disease, etc. [133−136] (Table 14.1).

5.2 Bacterial/Cancer Cells−Based Immunosensors

In broad applications of antibodies, whole cell detection via electrochemical approach is one of the major applications [137]. Several bacterial, viral, fungal cells have been recognized as being pathogenic, i.e., such microbes can potentially harm the human biological function units on exposure leading to an abnormal, diseased state. Epidemics of food- or waterborne diseases claim millions of lives annually, which remains a major challenge to public health providers worldwide. Thus, efforts are underway to develop robust, sensitive techniques and assays that can detect pathogens in remote settings (i.e., outside of a centralized laboratory environment). Special attention is being given to pathogens, i.e., *E. coli* O157:H7, *Salmonella*, *Vibrio cholera*, and *Campylobacter jejuni*, that are more common to contaminate fresh produce directly or indirectly by various mode of contaminations such as animals, insects, soil, water, dirty equipment, and human handling [138,139]. Both labeled and label-free electrochemical biosensors have been fabricated for whole cell microbial detection. A gram-negative bacterium *E. coli* causes dysentery and hemolytic uremic syndrome, can colonize the gastrointestinal tracts of animals, and can be responsible for food contamination at multiple steps along the food chain, which essentially includes the production, processing, distribution, retail marketing, and handling or preparation process. To prevent the food contamination by entering the pathogen in food, an ideal detection

TABLE 14.1 Label-Free Protein-Based Immunosensors

Type of Immunosensor	Protein Detected	Method of Immobilization	Measurement Conditions	Limit of Detection/Sensitivity	Linear Range	Transduction Technique	[References] (Year)
Au–Pt nanostructures-based immunosensor	NMP22	Direct immobilization	PBS, pH 7.5	3.33 pg/mL	0.01–10 ng/mL	Amperometry	[107] (2015)
$nZrO_2$-based immunosensors	CYFRA-21-1	Covalent binding	PBS, pH 7.0	0.08 ng/mL	2–16 ng/mL	CV	[108] (2015)
rGO-modified conducting paper–based immunosensor	CEA	Physical absorption	50 mM PBS, pH 7.4 containing 5 mM $[Fe(CN)_6]^{3-/4-}$	25.8 μA/ng mL cm^2	1–10 ng/mL	CA	[110] (2015)
Au@Pd/Ag yolk-bimetallic shell nanoparticles and aminated graphene-based immunosensor	NMP22	Covalent binding	PBS, pH 7.4	3.3 pg/mL	0.01–18 ng/mL	DPV	[111] (2014)
Au electrode using homobifunctional 1, 4-phenylene diisothiocyanate linker–based immunosensor	MDM2	Covalent binding	10 mM PBS, pH 7.4 containing 10 mM $[Fe(CN)_6]^{3-/4-}$	0.29 pg/mL in PBS and 1.3 pg/mL in mouse brain tissue homogenate	1 pg/mL–1 mg/mL	EIS	[112] (2013)
N-doped graphene sheets–based immunosensor	Carbohydrate antigen 15-3 (CA 15-3)	Covalent binding	PBS, pH 7.4 containing 5 mmol/L potassium ferricyanide	0.012 U/mL	0.1–20 U/mL	DPV	[114] (2013)

Gold nanostructured screen-printed graphite-based immunosensor	HER 2	Physical adsorption	0.1 M PBS buffer, pH 7.4	6.0 μg/L	0–40 μg/L	EIS	[115] (2015)
AuNPs and Nile blue A hybridized electrochemically onto the rGO-based immunosensor	CEA	Electrostatic interaction	0.1 M phosphate buffer containing 0.1 M KCl, pH 6.5	0.00045 ng/mL	0.001–40 ng/mL	DPV	[116] (2016)
Dithiobissuccinimidyl propionate SAM-based immunosensor	EGFR	Covalent binding	PBS solution (10 mM, pH 7.4)	1 pg/mL	1 pg/mL–100 ng/mL	CV	[117] (2013)
GCE surface coated with a mixture of thionine and chitosan, and covered with Nafion membrane–containing SiO_2 nanoparticles–based immunosensor	CEA	Electrostatic absorption via polyethyleneimine and glutaraldehyde cross-linking	0.01 M Phosphate buffer, pH 7.4	1 fg/mL	1 fg/mL–100 pg/mL	DPV	[120] (2013)
GCE modified with electrodeposited Prussian Blue, a graphene and MWCNT assembly–based immunosensor	CEA	Electrostatic interaction	PBS containing 0.1 mol/L KCl, pH 6.0	60 pg/mL	(0.2–1.0, and 1.0–40.0 ng/mL)	DPV	[121] (2013)
Poly (3,4-ethylenedioxythiophene): poly(styrenesulfonate) (PEDOT: PSS)-based conducting paper–based immunosensor	CEA	Physical absorption	50 mM PBS, pH 7.4 containing 5 mM $[Fe(CN)_6]^{3-/4-}$ at a 0.01 V potential over the frequency range from 100 KHz to 10 Hz	2.68 ng/mL	6–20 ng/mL	EIS	[122] (2016)

Continued

TABLE 14.1 Label-Free Protein-Based Immunosensors—cont'd

Type of Immunosensor	Protein Detected	Method of Immobilization	Measurement Conditions	Limit of Detection/ Sensitivity	Linear Range	Transduction Technique	[References] (Year)
PEDOT:PSS/ polyvinylalcohol nanofiber decorated conducting paper–based immunosensor	CEA	Physical absorption	50×10^{-3} M PBS, pH 7.0 containing 5×10^{-3} M $[Fe(CN)_6]^{3-/4-}$	14.2 μA/ ng mL cm^2	0.2–25 ng/mL	CA	[123] (2016)
Au–Polyaniline–modified conducting paper–based immunosensor	CEA	Covalent immobilization via an EDC-NHS coupling reaction	50 mM PBS, pH 7.4 containing 5 mM $[Fe(CN)_6]^{3-/4-}$	13.9 μA/ ng mL cm^2	2–20 ng/mL	CA	[124] (2016)
CNT-modified conducting paper–based immunosensor	CEA	Covalent immobilization via an EDC-NHS coupling reaction	50 mM PBS, pH 7.4 containing 5 mM $[Fe(CN)_6]^{3-/4-}$	7.8 μA/ ng mL cm^2	2–15 ng/mL	CA	[125] (2015)
Graphene/thionine/gold nanoparticles nanocomposites deposited onto SPWE-based microfluidic immunosensor	CEA	Physical absorption	0.1 M PBS solutions (pH 7.4)	10 pg/mL	50–500 ng/mL	DPV	[126] (2016)
N-GQDs supported PtPd bimetallic nanoparticles–based immunosensor	CEA	Covalent binding	PBS (0.1 M, pH 6.47)	2 fg/mL	5 fg/mL–50 ng/mL	CA	[127] (2017)

Carboxyl-functionalized single-wall CNTs onto a screen-printed carbon electrode–based immunosensor	Alpha-fetoprotein (AFP)	Covalent binding	10 mM PBS, pH 7.4, 10 mM Na_2HPO_4/ NaH_2PO_4 and 100 mM NaCl	0.1 ng/mL	1–100 ng/mL	EIS	[128] (2013)
Hydroxyapatite nanoparticles and GO-based immunosensor	AFP	Physical absorption	Phosphate buffer	5 pg/mL	0.01–10 ng/mL	SWV	[129] (2016)
APTES-nZrO_2-rGO nanocomposite–based immunosensor	CYFRA-21-1	Covalent binding	PBS	0.122 ng/ mL	2–22 ng/mL	DPV	[130] (2016)
Nanostructured hafnia-based immunosensor	CYFRA-21-1	Covalent binding	PBS	9.28 μA/ mL ng cm^2	2–18 ng/mL	DPV	[131] (2016)
nZrO_2-rGO nanocomposite with serine amino acid as a zero-linker-based immunosensor	CYFRA-21-1	Covalent binding	PBS	0.01 ng/mL	0.01–29 ng/mL	DPV	[132] (2016)

CEA, carcino embryonic antigen; *CV*, cyclic voltammetry; *DPV*, differential pulse voltammetry; *EIS*, electrochemical impedance spectroscopic; *LSV*, linear sweep voltammetry; *PBS*, phosphate buffered saline; *SV*, stripping voltammetry; *SWV*, square-wave voltammetry.

technique should be capable of processing many samples in a rapid way. Keeping this in view, several electrochemical biosensors have been fabricated for *E. coli* detection. A disposable electrochemical immunosensor for the simultaneous measurements of *E. coli* O157:H7, *Campylobacter,* and *Salmonella* have been developed [140–142]. This immunosensor was based on a MWCNTs–polyallylamine-modified screen-printed electrode onto which anti–*E. coli*, anti-*Campylobacter,* and anti-*Salmonella* antibodies were immobilized. The obtained calibration curves for all three bacteria were found in the range of 1×10^3–5×10^5 cells/mL with the LOD being 400 cells/mL for *Salmonella* and *Campylobacter* and 800 cells/mL for *E. coli* [142]. Furthermore, in another study, MWCNTs/sodium alginate/carboxymethyl chitosan composite onto carbon screen-printed disposable electrochemical immunosensor was fabricated for *E. coli* O157:H7 and *Enterobacter sakazakii* detection. The fabricated immunoelectrode showed the LR between 10^4–10^{10} CFU/mL (for both microbes) with an LOD of 4.57×10^3 CFU/mL for *E. sakazakii* and 3.27×10^3 CFU/mL for *E. coli*, respectively [140]. For *E. coli* O157:H7 detection, biofunctional nanoporous alumina membrane–based immunosensor has been developed. Antibody-modified magnetic beads (MBs) were utilized to concentrate *E. coli* O157:H7 from samples and MB-conjugated *E. coli* O157:H7 cells were arrested on the nanoporous alumina membrane with the LOD of 10 CFU/mL [143]. To improve the dynamic range of *E. coli* O157:H7, hyaluronic acid was used with alumina nanoporous membrane to successfully reduce the nonspecific binding of biomolecules and allowed effective antibodies immobilization. Furthermore, EIS was used to detect *E. coli* in an LR of 10–10^5 CFU/mL with LOD of 10 CFU/mL. *E. coli* O157:H7 was detected in whole milk samples with the LOD of 83.7 CFU/mL [144]. In an attempt to develop a biosensor with improved electrochemical properties for detection of *E. coli* O157:H7, a novel 3D Ag nanoflower-based EIS immunosensor has been developed that displayed LR of 3.0×10^2–3.0×10^8 CFU/mL [145]. Gayathri et al. used HRP-tagged secondary antibodies for uropathogenic *E. coli* detection [146]. In this approach, electrochemical immunosensor was fabricated by using thionine dye immobilized onto a functionalized MWCNT–chitosan composite on GCE. This labeled fabricated amperometric immunoelectrode showed excellent linearity of 10^2–10^9 CFU/mL with a sensitivity of 7.162 μA $[\log(\text{CFU/mL})]^{-1}$. For multipathogenic microorganism detection, a labeled impedimetric immunosensor has been developed. In this approach, biotinylated antibodies-functionalized MBs were used to capture target bacteria, and a screen-printed interdigitated microelectrode (SP-IDME) was used for detection of *E. coli* O157:H7 and *Salmonella typhimurium.* Furthermore, MBs-Ab-cell complexes have been labeled by glucose oxidase (GOx)-Ab conjugates resulting to MBs-Ab-cell-Ab-GOx biomass, which was mixed with the glucose solution to trigger an enzymatic reaction that produced gluconic acid and decreased the impedance of the solution as measured on the SP-IDME. This proposed immunoelectrode have shown the specific detection of the *E. coli* O157:H7 and *S. typhimurium* with LOD and LR of 2.05×10^3 CFU/g (*E. coli* O157:H7), 1.04×10^3 CFU/mL (*S. typhimurium*), and 10^2-10^6 CFU/mL (range for both the strains), respectively (Fig. 14.8) [147].

The graphene-wrapped copper oxide–cysteine (rGO-CysCu) hierarchical structure has been used to fabricate impedimetric immunosensor. In this study, the self-assembly of rGO-CysCu molecules onto gold electrode provided a high surface area and high electron transfer rate constant that helped in the development of ultrasensitive label-free electrochemical immunosensor for quantitative determination of *E. coli* O157:H7. This proposed immunosensor showed a wide detection range of $10-10^8$ CFU/mL with an LOD of 3.8 CFU/mL [148]. In addition to *E. coli*, several other microorganisms such as *Listeria monocytogenes* (LM), *Pantoea stewartii*, *Streptococcus aureus* cause several troublesome diseases [149–151]. Efforts have been made to fabricate electrochemical immunosensors for rapid monitoring of these pathogens. Cheng et al. developed labeled SAM modified Au electrodes for the detection of foodborne gram-positive *L. monocytogenes* pathogen. The fabricated immunoelectrode could detect LM in detection range from 10^2-10^6 CFU/mL [150]. In the detection of *P. stewartii*, a dual amplified electrochemical sandwich immunosensor has been developed. In this report, the large surface area of AuNPs and catalytic ability of HRP were mentioned as helping in deducing an LOD of 7.8×10^3 CFU/mL, 20 times lower than conventional ELISA technique for *P. stewartii* detection [151]. Efforts have also been made to develop immunosensor for *Staphylococcus aureus* detection. In one study, SAM of 3-mercaptopropionic acid was covalently bound to Au electrode surface, and

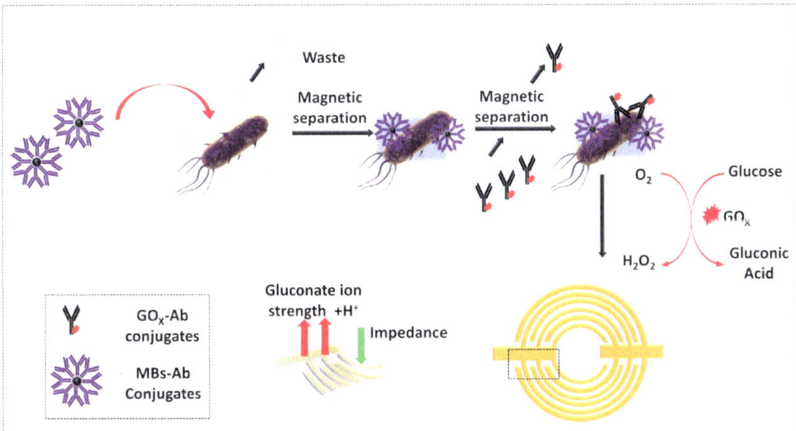

FIGURE 14.8 Illustration of the mechanism and construction of the immunosensor for rapid detection of *Escherichia coli O157:H7* and *Salmonella typhimurium* in foods [147].

EIS technique was used to investigate the biosensing parameters and observed LR of $10-10^6$ CFU/mL with LOD of 10 CFU/mL [149].

Cancer is a foremost root of death globally, and the incidence of melanoma is continuing to increase worldwide and in the United States. Biosensors have the potential to detect a wide range of cancer biomarkers, which are overexpressed on the cancer cell surface, sensitively and specifically and simply. In this regard, electrochemical immunosensing transducer systems using the characteristics of various materials for cancer diagnosis have taken their road into a wide diversity of approaches capable of enlightening the detecting efficiency of melanoma cells. Herein, the various reduced graphene oxide–tetrasodium 1, 3, 6, 8-pyrenetetrasulfonic acid/metal hexacyanoferrates (rGO-TPA/MHCFnano) nanocomposites based a high sensitive sandwich-type electrochemical immunosensor was fabricated for the determination of SKBR-3 breast cancer cell. The authors found the LOD of 21 cells/mL in the concentration range of 500–30,000 cells/mL [152]. In another report, silica nanoparticles–PPy nanocomposite was used to diagnose melanoma cells and found LOD of 20 cells/mL in the detection range of 50–7500 cells 2.5/mL [153]. Rare circulating tumor cells (CTCs) in blood was detected using a semiintegrated immunosensor by sample enrichment with a combination of immunomagnetic isolation and size filtration and found recovery rate above 70%. Impedance was measured by spiking three levels of CTC number of 30 ± 2, 124 ± 29, 273 ± 23 in 10 μL of a filtered blood sample [154]. A tridimensional biosensor platform constructed on a series of covalent linkages between lectin molecules and a cysteine layer immobilized over gold-coated TiO_2 butterfly-like tridimensional nanomembranes was proposed for the first time to be used as an electrochemical point-of-care device. Through the use of vegetal lectins, they managed to take advantage of the markedly a typical glycomic profile of the cancerous mammalian cell membrane and successfully made a distinction between highly invasive (T47D) and less invasive (MCF7) cancer cell lines. The selectivity of the biosensor was tested by using normal human skin-fibroblast. The proposed cytosensor demonstrated LOD as low as 10 cells/mL for every cell line and an LR from $10-1.0 \times 10^6$ cells/mL [155]. Multivalent identification and signal amplification approaches were used to detect human prostate metastatic cancer cells (Du-145) using a graphene/gold nanoparticle/horseradish peroxidase (HRP)-conjugated trastuzumab antibody (G/GNP/Ab-HRP) hybrid nanostructure and showed low LOD of 20 cells/mL with the LR from 10^2-10^6 cells/mL [156] (Table 14.2).

5.3 Enzyme-Labeled Immunosensors

Enzymatic-labeled electrochemical immunosensors are based on "sandwich" mode of operation. These types of immunosensors help in the enhancement of electrochemical biosensing parameters. There are several enzymes, i.e., alkaline phosphatase (AP), horseradish peroxidase (HRP), tyrosinase, etc., that

TABLE 14.2 Bacterial/Cancer Cells-Based Immunosensors

Type of Immunosensor	Cancer/Bacterial Cell Detected	Method of Immobilization	Measurement Conditions	Limit of Detection	Linear Range	Transduction Technique	[References] (Year)
MWCNTs–polyallylamine-modified screen-printed electrode–based immunosensor	*Escherichia coli* O157:H7, *Campylobacter* and *Salmonella*	Covalent immobilization	20 mM PBS, pH 8.6	400 cells/mL for *Salmonella* and *Campylobacter* and 800 cells/mL for *E. coli*	1×10^3–5×10^5 cells/mL	SV	[142] (2012)
MWCNTs/sodium alginate/carboxymethyl chitosan composite–based immunosensor	*E. coli* O157:H7 and *Enterobacter sakazakii* (*E. sakazakii*)	Covalent immobilization	0.01 M PBS, pH 7.4	4.5710^3 CFU/mL for *E. sakazakii* and 3.27×10^3 CFU/mL for *E. coli*,	10–10^{10} CFU/mL	CV	[140] (2013)
Biofunctional nanoporous alumina membrane–based immunosensor	*E. coli* O157:H7	Physical adsorption	PBS	10 CFU/mL	100 to 10^4 CFU/mL	EIS	[143] (2013)
Alumina nanoporous membrane with hyaluronic acid–based immunosensor	*E. coli* O157:H7	Covalent immobilization	PBS, pH 7.4	10 CFU/mL 83.7 CFU/mL & in whole milk	10–10^5 CFU/mL	EIS	[144] (2013)
3D Ag nanoflower–based immunosensor	*E. coli* O157:H7	Covalent conjugation	10 mM PBS, pH 7.4 containing 5 mM $[Fe(CN)_6]^{3-/4-}$ 0.1 M KCl	100 CFU/mL	3.0×10^2 to 3.0×10^8 CFU/mL	EIS	[145] (2016)

Continued

TABLE 14.2 Bacterial/Cancer Cells-Based Immunosensors—cont'd

Type of Immunosensor	Cancer/ Bacterial Cell Detected	Method of Immobilization	Measurement Conditions	Limit of Detection	Linear Range	Transduction Technique	[References] (Year)
Thionine dye immobilized onto a functionalized MWCNT–chitosan composite on GCE-based immunosensor	Uropathogenic *E. coli*	Covalent immobilization	PBS, pH 7.0	10^2 CFU/mL	10^2–10^9 CFU/mL	EIS	[146] (2016)
Streptavidin-coated MBs were functionalized with corresponding biotinylated antibodies	*E. coli* O157:H7 and *S. typhimurium*	Avidin–biotin coupling	0.1 mol/L PBS, pH 7.4	be 2.05×10^3 CFU/g and 1.04×10^3 CFU/mL	10^2–10^6 CFU/m	Impedance	[147] (2016)
Graphene-wrapped copper oxide–cysteine onto gold electrode–based immunosensor	*Escherichia coli* O157: H7	Covalent immobilization	100 mM PBS, pH 7.4 containing 5 mM $[Fe(CN)_6]^{3-/4-}$ 0.9% NaCl)	3.8 CFU/mL	10–10^8 CFU/mL	EIS	[148] (2017)
SAM of 3-mercaptopropionic acid was covalently bound to Au electrode–based immunosensor	*Staphylococcus aureus*	Covalent immobilization	8 mM $[Fe(CN)_6]^{3-/4-}$ PBS	10 CFU/mL	10–10^6 CFU/mL	EIS	[149] (2012)

Labeled SAM–modified Au electrodes–based immunosensor	*Listeria monocytogenes*	Electrostatic interaction	0.01 PBS, pH 7.0	1.0×10^{1} CFU/mL	10^{2}–10^{6} CFU/mL	EIS	[150] (2014)
AuNPs and catalytic ability of HRP-based sandwich immunosensor	*Pantoea stewartii*	Covalent immobilization	PBS, pH 7.0 containing 1.5 mM HQ and 2.0 mM H_2O_2	7.8×10^{3} CFU/mL	2.0×10^{7}–4.0×10^{4} CFU/mL	LSV	[151] (2014)
rGO-tetrasodium 1, 3, 6, 8-pyrenetetrasulfonic acid/metal hexacyanoferrates nanocomposites included rGO-TPA/FeHCF, rGO-TPA/CoHCF, rGO-TPA/NiHCF, and rGO-TPA/CuHCF nanocomposites–based immunosensor	Breast cancer	Covalent immobilization	5.0 mM $[Fe(CN)_6]^{3-/4-}$	21 cells/mL	500–30,000 cells/mL	DPV	[152] (2017)
Silica nanoparticles–PPy nanocomposite–based immunosensor	Melanoma cancer cells	Amino-functionalization	1 × PBS, pH 7.4, 1 mM ferricyanide; $[Fe(CN)_6]^{3-}$, 0.1 M KCl, 100 mM EDTA	20 cells/mL	50–7500 cells/mL	CV	[153] (2015)

Continued

TABLE 14.2 Bacterial/Cancer Cells-Based Immunosensors—cont'd

Type of Immunosensor	Cancer/ Bacterial Cell Detected	Method of Immobilization	Measurement Conditions	Limit of Detection	Linear Range	Transduction Technique	[References] (Year)
Semiintegrated electrochemical sensor	Circulating tumor cells	–	–	10 spiked MCF7 cells in 0.5 mL of blood	–	EIS	[154] (2011)
Gold-coated TiO_2 butterfly-like tridimensional nanomembranes–based immunosensor	Breast cancer	Covalent immobilization	PBS, pH 7.4, $Fe(CN)_6]^{3-/4-}$,	10 cells/mL	10–1.0 × 10^6 cells/mL	EIS	[155] (2016)
Graphene–gold nanoparticle–based immunosensor	Human prostate metastatic cancer cells	Covalent immobilization		20 cells	10^2–10^6 cells/mL	DPV	[156] (2017)

CV, cyclic voltammetry; *DPV*, differential pulse voltammetry; *EIS*, electrochemical impedance spectroscopic; *LSV*, linear sweep voltammetry; *PBS*, phosphate buffered saline; *SV*, stripping voltammetry.

have been used in the labeling of antigen or antibodies molecules [62,157,158]. In this type of immunosensor, change in the biorecognition event into an observable analytical response can be completed by one of two main ways: (1) redox species, which works as an electron-transfer indicator for enzymatic reaction (HRP enzyme as a label in the presence of H_2O_2 catalyzes the oxidation of a mediator), can be electrochemically monitored and (2) enzyme labels (alkaline phosphatase [ALP] converts aromatic monoester phosphates to phenolic derivatives) can convert an electrochemically inactive substrate into a measurable electroactive product [159]. Tissue transgultaminase (tTG) is an antigen that is secreted in very low concentration in serum of patients diagnosed with celiac disease. Efforts are being made to develop a sensitive, clinically relevant device to detect tTG antigen [160,161]. Neves et al. developed CNTs- and AuNPs-based disposable screen-printed immunosensors for detection of tTG antigen in celiac disease by immobilizing tTG and electrochemically detected using ALP-labeled antihuman IgG antibodies. Under optimized conditions, this fabricated immunoelectrode was shown to achieve saturation after 40 U/mL [19]. To improve the biosensing characteristics, the same group designed a double working screen-printed electrode and used a mixture of 3-indoxyl phosphate with silver ions immobilized with ALP-labeled antihuman antibodies for gliadin (serum antigen for celiac disease) and tTG. The proposed immunoelectrode displayed cumulative LR for both antigens in between 1 and 100 U/mL. The LOD and LOQ were found to be 2.45 U/mL and 20.42 U/mL, respectively, for tTG and 3.16 U/mL and 46.77 U/mL, respectively, for AGA. ALP has also been used in the fabrication of several other immunosensors for ovarian cancer biomarker detection, as well as for HIV-1 gp 41 or HIV-2 gp 36 protein detection [162].

HRP is one of the most prominently used enzymes has been used in tagging of secondary antibodies for fabrication of selective and sensitive immunosensing platform [163]. Wu et al. have fabricated an Au−chitosan composite−based HRP-labeled screen-printed carbon electrode for detection of CEA molecules. The LR and LOD were found to be 0.50−25 and 0.22 ng/mL, respectively, for CEA biomolecules [164]. To improve the sensing properties, Tang et al. used thionine (TH)-labeled magnetic AuNPs for fabricating electrode surface and HRP as a signal enhancer for CEA molecules. They conducted experiments with and without tagging of secondary antibodies, they obtained the LR of 1.2−125 ng/mL for unlabeled electrode and 0.01−160 ng/mL for HRP-labeled immunosensor. Also, the lower LOD was found to be 0.01 ng/mL in HRP-labeled immunosensor [165]. Further to detect trace levels of CEA biomolecules in clinical samples, Zhong et al. synthesized nanogold-enwrapped graphene nanocomposite (NGGNs) and used this composite in the fabrication of immunoelectrode surface. The dynamic working range and lower LOD were found to be 0.005−350 and 0.01 ng/mL, respectively, for CEA molecules (Fig. 14.9) [166].

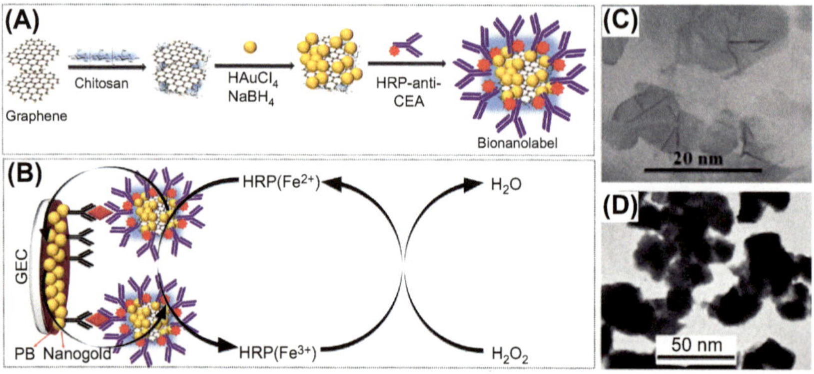

FIGURE 14.9 Schematic for HRP-anti-CEA-NGGN (A) bionanolabel fabrication process, (B) measurement protocol, and TEM images of (C) chitosan-protected graphene and (D) the prepared NGGNs [166].

In recent times, HIV p24 antigen has been used in the detection of HIV [163,167]. Several conventional techniques such as RIA, fluoroimmunoassay, ELISA, and electrochemical methods have been used, but these techniques possess several limitations, i.e., lengthy analysis times, costly instruments, and complex operating methods. To overcome these limitations, HRP-labeled immunosensing platforms have been developed for detection of HIV p24 antigen. Gan et al. fabricated a portable HRP-labeled electrochemical immunosensor for detection of HIV p24 antigen. This proposed immunosensor was based on gold colloid, and silicon dioxide-coated magnetic Fe_3O_4 nanoparticles (MNPs) tagged with the primary anti-p24 antibodies helped in enhancing the sensitivity. Under optimized conditions, this immunoelectrode showed 1000 times higher sensitivity on ELISA technique with LR of 0.001−10.00 ng/mL and LOD of 0.5 pg/mL [167]. Further to improve its electrochemical biosensing parameters, Fang et al. developed an immunoelectrode wherein graphene oxide (GO) worked as a nanocarrier and CNTs−silica were used as a matrix. The double layer of HRP enzyme helped in enhancing the biosensing parameters of the proposed labeled immunoelectrode for HIV-p24 biomolecules. This immunoelectrode showed the range of LR in between 0.5 pg/mL and 8.5 ng/mL with LOD of 0.15 pg/mL [163]. HRP has also been used as a label in the detection of other biomarkers such as AFP protein [158,168], transforming growth factor β1 (TGF-β1) cytokine [169], carbohydrate antigen 19-9 (CA19-9) [170], ofloxacin [171], interleukin-6 (IL-6) [172], microcystin−LR (MC-LR) [173] (Table 14.3).

TABLE 14.3 Enzyme-Labeled Immunosensors

Type of Immunosensor	Enzyme Labeled	Analyte Detected	Method of Immobilization	Measurement Conditions	Limit of Detection	Linear *n* Range	Transduction Technique	[References] (Year)
CNTs- and AuNPs-based disposable screen-printed electrochemical immunosensor	Alkaline phosphatase	Celiac disease	Physical adsorption	–	40 U/mL	0–100 U/mL	CV	[19] (2012)
Double working screen-printed electrode and mixture of 3-indoxyl phosphate with silver ions	Alkaline phosphatase	Celiac disease	Physical adsorption	–	2.45 U/mL	1–100 U/mL	CV	[162] (2013)
Graphene oxide–enzyme encapsulated in CNT–silica-based immunosensor	HRP	Human immunodeficiency virus p24	Covalent immobilization	7.0 PBS containing 3mM H_2O_2,	0.15 pg/mL	0.5 pg/mL–8 5 ng/mL	DPV	[163] (2015)
Au–chitosan composite-based immunosensor	HRP	CEA	Physical adsorption	0. 2M PBS, pH 7.0	0.22 ng/mL	0.50–25 ng/mL	DPV	[164] (2006)
Thionine-labeled magnetic AuNPs–based immunosensor	HRP	CEA	–	PBS, pH 7.4	0.01 ng/mL	0.01–160 ng/mL	Amperometry	[165] (2008)

Continued

TABLE 14.3 Enzyme-Labeled Immunosensors—cont'd

Type of Immunosensor	Enzyme Labeled	Analyte Detected	Method of Immobilization	Measurement Conditions	Limit of Detection	Linear *n* Range	Transduction Technique	[References] (Year)
Nanogold-enwrapped graphene nanocomposite–based immunosensor	HRP	CEA	–	PBS, pH 7.0 containing 0.8 mM H_2O_2	0.01 ng/mL	0.005–350 ng/mL	CV	[166] (2010)
Gold colloid, and silicon dioxide–coated magnetic Fe_3O_4 nanoparticles–based immunosensor	HRP	Human immunodeficiency virus p24	–	0.01 mol/L PBS, 0.1 mol/L KCl, pH 7.4	0.5 pg/mL	0.001–10.00 ng/mL	DPV	[167] (2013)

CEA, carcino embryonic antigen; *CV*, cyclic voltammetry; *DPV*, differential pulse voltammetry; *EIS*, electrochemical impedance spectroscopic; *HRP*, horseradish peroxidase *PBS*, phosphate buffered saline.

5.4 Redox-Labeled Immunosensors

The previous section described the significance of enzyme labels in immunosensor fabrication in the context of signal amplification and device sensitivity. However, the sensitivity achieved by using enzyme labels is still not sufficient to enable the use of immunosensor in clinical practice [174]. Thus, redox labeling has been introduced as a supplementary mechanism to enzymatic reactions for signal amplification in biosensing devices. There are several key features that have required in a potential candidate of a redox species for the development of sensitive immunosensing platform [175]: (1) redox species should be electroactive in between 0 and 200 mV of potential; (2) it does not show any cross reactions with the substrate; (3) it should be stable in buffer used for electrochemical studies; (4) it should not cause electrode fouling; (5) it should have chemical groups for coupling.

Several redox species such as ferrocene and its derivatives (ferrocene monocarboxylic acid, 1,1′-ferrocene dicarboxylic acid), heme-modified antibodies, platinum nanoparticles infinite coordination polymer (PtNP@ICP), biotin, streptavidin, etc. have been used for development of immunosensing platform [176–179]. Wang et al. developed three redox probes based on ferrocene derivatives for use as secondary antibody-conjugated labels in a novel gold nanoparticle/graphene (AuNP/Gr)-based immunosensor. The three ferrocene derivatives used in this study were ferrocene monocarboxylic acid (Fc [1]), β-ferrocenyl-propenoic acid (Fc [2]), and 1,10-ferrocene dicarboxylic acid (Fc [3]), and IgG was used. The immunosensor displayed LR of 1–300 ng/mL with the best sensitivity found with Fc [3] label. This study is also significant in the sense that it depicts the correlation between the structure of label, its electrochemical characteristics, and the performance of a biosensor (Fig. 14.10) [176].

Zhang et al. [29] utilized the structural tailorability of infinite coordination polymers (ICP) and catalytic ability of Pt nanoparticles to synthesize an "all-in-one" ferrocene dicarboxylic acid (FcDA)–mediated PtNP@ICP redox and catalysis polymer for use as a cross-linking antibody label in a sandwich-type immunosensor. This immunosensor, when using PtNP@ICP-cross-linked antibodies (Ab_2) as labels, displayed LR of 0.001–60 ng/mL with an LOD of 0.3 pg/mL against prostate-specific antigen (PSA) biomolecules, while the use of only ICP Ab_2 as secondary antibodies yielded a detection range of 0.01–50 ng/mL. This made clear that the synthesized PtNP@ICP labels have improved electrocatalytic and functional properties [177]. Tang and Ma developed a novel ratiometric electrochemical immunosensor, wherein, in place of the conventional sandwich assay, carboxymethyl cellulose–Au–Pb^{2+} (CMC–Au–Pb^{2+}) and carbon–Au–Cu^{2+} (C–Au–Cu^{2+}) were used as redox labeling and immunoprobe, respectively. The rationale behind this design strategy was that both the redox and immunoprobe give strong current signals and that the corresponding current values would display a specific ratio

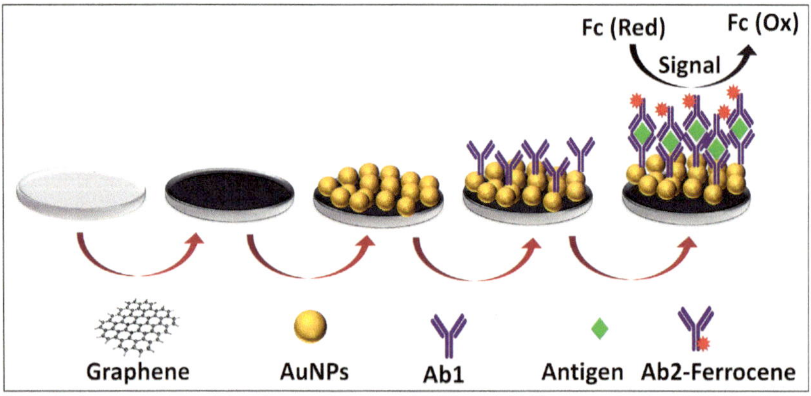

FIGURE 14.10 Schematic illustration of fabrication of the electrochemical immunosensor with graphene/gold nanoparticles platform and ferrocene derivatives label immunosensor. *Adopted from G. Wang, X. Gang, X. Zhou, G. Zhang, H. Huang, X. Zhang, et al., Electrochemical immunosensor with graphene/gold nanoparticles platform and ferrocene derivatives label, Talanta 103 (2013) 75–80.*

($I_{probe}/I_{substrate}$) that would change on exposure to different amounts of the target analyte. Using IgG as an analyte, this novel sandwich immunosensor displayed a wide LR of 1 fg/mL–100 ng/mL, which was found to be two orders of magnitude higher than other previous works. The LOD was found to be 0.26 fg/mL and a good correlation with ELISA was shown regarding analyzing real samples [178]. Tirado et al. reported an amperometric sandwich assay-based immunosensor for the detection of TGF-β1, which has been proposed as a biomarker of renal disease. This immunosensor was based on carboxylic acid–functionalized magnetic nanoparticles supported onto screen-printed carbon electrode, covalently immobilized specific antibodies for TGF-β1 and biotin-labeled anti-TGF secondary antibodies. These secondary antibodies were further conjugated to a novel HRP-labeled streptavidin polymer (poly-HRP-strept) that consist of multi-HRP enzymes for signal amplification via glutathione as a substrate. The rationale behind the use of poly-HRP enzyme in place of conventional HRP labels was that presence of multiple HRP molecules would lead to higher biocatalysis of enzyme substrate, and hence, the sensitivity of the biosensor would increase drastically. To test this hypothesis, authors compared the current response produced by the immunosensor in the presence of conventional HRP-strept and that of poly-HRP-strept. The current response of the biosensor was found to be significantly higher for poly-HRP-strept as compared with HRP-strept. In addition, the best specific-to-unspecific current ratio for poly-HRP-strept was found to be at lower dilution factor (1/500) as compared with the conventional label (1/2000). Furthermore, this immunosensor displayed linearity in a wide range of 15–3000 pg/mL with an LOD of 10 pg/mL, which is an improvement over those of conventional ELISA. The usefulness of the biosensor was further evaluated by analyzing urine samples spiked with different concentrations of TGF-β1, and matrix effects were analyzed [179]. These recent studies demonstrated that electrochemical redox labeled–based immunosensors indeed have the potential of outperforming conventional ELISA assays regarding incubation time, sensitivity, and reproducibility. However, use of multienzymes and many chemical reagents as redox probes and substrates have somewhat limited the practical use of redox-labeled biosensors. Thus, more studies are required that would simplify the design of redox labels and thus would offer more robust biosensing devices (Table 14.4).

5.5 Aptamer-Based Immunosensors

As described earlier in this chapter the aptamers are the oligonucleotides either of dsDNA or ssDNA obtained by the selective enrichment methods. Owing to its small size, high affinity for targets, and high pH and temperature stability, aptamers have been under intense investigation for a number of applications, including targeted therapeutics and diagnostics [180]. One of the promising applications for aptamers is as detection probes in immunosensors. The small size of aptamers, the presence of predictable secondary structures and the easy tunability in their spatial structures (due to conformational changes) allow for various novel biosensing strategies that have not been realized with enzymes or antibodies as an exclusive biorecognition molecules [181–183]. The hybrid system of antibodies–aptamers has been found with improved biosensing parameters. This improvement in immunosensor may be due to a number of advantages of aptamers such as high selectivity even for those targets (e.g., ions, small molecules, etc.), which are not easily recognized by another bioreceptors, ease of chemical modification, flexibility in structure and conformation, high pH stability and reversible thermal denaturation, cheaper production costs, etc. [183,184]. Moreover, aptamer antibodies–based immunosensing assays are also more reproducible, robust, and economical [185].

Taleat et al. reported on the development of an electrochemical biosensor based on poly o-aminobenzoic acid–modified graphite electrodes and utilized the antibody target aptamer sandwich strategy for the detection of MUC1 cancer biomarker. This study utilized methylene blue as redox indicator that directly interacts with the G residues present in the MUC1-specific aptamers, thus bypassing the need for additional chemical labeling steps and simplifying the fabrication process. This aptamer-mediated immunosensor was shown to detect MUC1 in the range of 1–12 ppb with an LOD of 0.62 ppb [186]. Similarly, Ilkhani et al. reported an electrochemical aptamer/antibody-based sandwich assay for the detection of a cancer biomarker EGFR. This work

TABLE 14.4 Redox-Labeled Immunosensors

Type of Immunosensor	Analyte Detected	Method of Immobilization	Redox-Labeled Used	Measurement Conditions	Limit of Detection	Linear Range	Transduction Technique	[References] (Year)
Gold nanoparticle/graphene–based immunosensor	Human IgG	Covalent immobilization	Ferrocene derivatives-ferrocene monocarboxylic acid (Fc [$_1$]), β-ferrocenyl-propenoic acid (Fc [$_2$]) and 1,10-ferrocene dicarboxylic acid (Fc [$_3$])	PBS, pH 7.4	0.4 ng/mL	1–300 ng/mL	DPV	[176] (2013)
Coordination polymers and Pt nanoparticles–based immunosensor	Prostate-specific antigen (PSA),	Physical adsorption	Redox and catalysis "all-in-one" infinite coordination polymer	PBS	0.3 pg/mL	0.01–50 ng/mL	DPV	[177] (2015)
Carboxymethyl cellulose–Au–Pb^{2+}-based immunosensor	Human IgG	Physical adsorption	CMC-Au with Pb^{2+} and CNP–Au with Cu^{2+},	PBS	0.26 fg/mL	1 fg/mL–100 ng/mL	SWV	[178] (2016)
Carboxylic acid—functionalized magnetic nanoparticles–based immunosensor	Transforming growth factor- β1	Covalent immobilization	HRP-labeled streptavidin polymer (poly-HRP-Strept)	0.1 M PBS, pH 7.2	10 pg/mL	15–3000 pg/mL	Amperometry	[179] (2017)

DPV, differential pulse voltammetry; *PBS*, phosphate buffered saline; *SWV*, square-wave voltammetry.

utilized EGFR-specific aptamers immobilized onto streptavidin-coated MBs as capture probes and gold nanoparticles (AuNP)-labeled anti-EGFR antibodies as the detector/signal probe. This biosensor was shown to display a wide range of detection of 1−40 ng/mL with good LOD of 50 pg/mL. The performance of the biosensor was also analyzed against serum samples from breast cancer patients [103]. Another work by Kavosi et al. utilized aptamers as detection probes and developed a triple signal amplification strategy for ultrasensitive detection of PSA in real samples. In this approach, anti-PSA antibodies were covalently attached to graphene oxide/chitosan-modified GCE, and thionine was used as redox label. A sandwich assay was achieved by the addition of HRP-labeled PSA-aptamers that were covalently linked to polyamidoamine (PAMA)-encapsulated gold nanoparticles (Au-PAMA). The use of labeled aptamers enhanced the specificity and reduced interference by other moieties present in the matrix, and hence, the biosensor displayed an excellent LOD of 10 fg/mL in a wide LR [187]. Furthermore, Xiang et al. developed a sandwich antibody-target-aptamer assay for the detection of microcystin−leucine−arginine (MC-LR), which is a hepatotoxin and a potential bioweapon. A limited number of epitopic determinants and steric hindrance among the MC-LR molecules have hindered the development of efficient immunoassays against MC-LR. Thus, the authors replaced MC-LR specific aptamers immobilized onto the glass surface of a portable analyzer as biorecognition molecules. Anti-MC-LR antibodies labeled with HRP were used as detector probes. The biosensor showed fairly good sensitivity of 0.3 μg/mL, but the main highlight of this device was its excellent specificity toward MC-LR when tested against other MCs (e.g., MC-LA, MC-YR, nodularin-R), which are structurally nearly similar to MC-LR and are cross-reactive against anti-MC-LR monoclonal antibodies [188]. This suggested that aptamers played a significant role in the specificity of the biosensor. Other studies have also pointed toward the advantages of using aptamer/antibody sandwich assays in place of traditional ELISA and antibody/antibody sensors. To date, very few reports are available in the development of aptamer antibodies−based immunosensors. Efforts should be made to explore this eminence field for various immunosensors development (Table 14.5).

6. CONCLUSIONS AND OUTLOOK

Immunosensors are the most successful format of biosensing devices due to their high specificity and quick recognitions of the target analyte. These are widely being investigated using various types of transducers such as optical, electrochemical, etc. Owing to the better, quick, and authenticated signal generation capabilities, the electrochemical immunosensors are considered as a potential biosensor for next-generation diagnostics. The performance of electrochemical immunosensors is usually assessed by various techniques, including the voltammetry, amperometry, ECL, electrochemical impedance

TABLE 14.5 Aptamer-Based Immunosensors

Type of Immunosensor	Analyte Detected	Method of Immobilization	Measurement Conditions	Limit of Detection	Linear Range	Transduction Technique	[References] (Year)
Functionalized conductive polymer–based immunosensor	Cancer biomarker human MUC1 protein	Covalent immobilization	PBS, pH 7.4	0.62 ppb	1–12 ppb	DPV	[186] (2014)
Gold nanoparticles–based immunosensor	Epidermal growth factor receptor	Biotin–streptavidin coupling	–	50 pg/mL	1 to 40 ng/mL	DPV	[103] (2015)
Polyamidoamine encapsulated gold nanoparticles–based immunosensor	Prostate-specific antigen tumor marker	Covalent immobilization	PBS, pH 7, 0.1 M KCl, 2.5 mM Fe(CN) 63-/4-	10 fg/mL	0.1 pg/mL–90 ng/mL	DPV, EIS	[187] (2015)
Glass surface aptamer–based immunoassay	Microcystin–leucine–arginine	–	–	0.3 μg/L	0.5–4.0 μg/L	Chemiluminescence	[188] (2014)

DPV, differential pulse voltammetry; *EIS*, electrochemical impedance spectroscopic; *PBS*, phosphate buffered saline.

spectrometry, etc. In this chapter, such techniques have been described by examining the different aspects of the characterization of biosensing prototypes regarding its analytical performances, viz., LODs, dynamic ranges, sensitivity, etc. Furthermore, this chapter has also described the various classes of biorecognition elements, which are an integral part of any immunosensor that supports to achieve the selectivity for analytes, including antibody, truncated antibody, and aptamer–antibody hybrid biorecognition elements. This chapter has highlighted the various strategies adopted for advanced electrochemical immunosensors for achieving the better performance such as lateral flow assays, colorimetric, receptor–ligand–receptor, and redox cycling–based strategies. Furthermore, several advancements of the electrochemical immunosensors have been comprehended with the various recent examples. Conclusively, this chapter gives the glimpses of the wide range of electrochemical biosensors, technological and engineering interventions in the immunosensing mechanisms, and various excellent strategies for the fabrications of such immunosensors. At the same time, this chapter opens the window for advanced and directional improvements of immunosensors that essentially comprised of the fundamental knowledge of different fields, including material science, bioscience, electronic physics, etc. Electrochemical immunosensors being a multidisciplinary arena of the recent advancements and engineering excellences are also not remained untouched by the limitations. The shortcomings of these prototypes include the sensitivities toward high temperature and challenges in fabrications. Although these are widely investigated, only a few have been successfully translated into commercial products. Thus, this area is still open for several improvements to make the mechanistic simple, highly efficient, affordable, user-friendly, robust, equipment free, and deliverable volumes for all end users.

ACKNOWLEDGMENTS

This work is supported by DST Ramanujan Fellowship (SB/S2/RJN-042/2015) and DBT project no. BT/PR16634/NER/95/225/2015, Government of India awarded to Dr. Pranjal Chandra. One of the authors, Renu Singh acknowledges the Department of Bioproducts and Biosystems Engineering, the University of Minnesota for support.

REFERENCES

[1] F. Ricci, G. Adornetto, G. Palleschi, A review of experimental aspects of electrochemical immunosensors, Electrochim. Acta 84 (2012) 74–83.

[2] P. Chandra, J. Singh, A. Singh, A. Srivastava, R.N. Goyal, Y.B. Shim, Gold nanoparticles and nanocomposites in clinical diagnostics using electrochemical methods, J. Nanopart. (2013) 1–12, 535901.

[3] R.M. Lequin, Enzyme immunoassay (EIA)/Enzyme-Linked immunosorbent assay (ELISA), Clin. Chem. 51 (12) (2005) 2415.

[4] R. Singh, M.D. Mukherjee, G. Sumana, R.K. Gupta, S. Sood, B.D. Malhotra, Biosensors for pathogen detection: a smart approach towards clinical diagnosis, Sens. Actuators B Chem. 197 (2014) 385–404.

[5] S. Pelsue, Fundamentals of Immunoassays Portland Maine, Maine Biotechnology Services, 2016. Available from: http://mainebiotechnology.com/.

[6] M.J. O'Sullivan, J.W. Bridges, V. Marks, Enzyme immunoassay: a review, Ann. Clin. Biochem. 16 (1–6) (1979) 221–239.

[7] W. Contributors, Immunoassay: Wikipedia, the Free Encyclopedia, 2016. Available from: http://en.wikipedia.org.

[8] E. Mylonakis, M. Paliou, M. Lally, T.P. Flanigan, J.D. Rich, Laboratory testing for infection with the human immunodeficiency virus: established and novel approaches, Am. J. Med. 109 (7) (2000) 568–576.

[9] P.-Y. Shi, S.J. Wong, Serologic diagnosis of West Nile virus infection, Expert Rev. Mol. Diagn. 3 (6) (2003) 733–741.

[10] A.J. van Hengel, Food allergen detection methods and the challenge to protect food-allergic consumers, Anal. Bioanal. Chem. 389 (1) (2007) 111–118.

[11] K. Mahato, A. Prasad, P. Maurya, P. Chandra, Nanobiosensors: next generation point-of-care biomedical devices for personalized diagnosis, J. Anal. Bioanal. Tech. 7 (2016) e125.

[12] S.K. Yadav, B. Agrawal, P. Chandra, R.N. Goyal, In vitro chloramphenicol detection in a *Haemophilus influenza* model using an aptamer-polymer based electrochemical biosensor, Biosens. Bioelectron. 55 (2014) 337–342.

[13] C. Moina, G. Ybarra, Fundamentals and applications of immunosensors, Adv. Immunoass. Technol. (2012) 65–80.

[14] P. Chandra, W.C.A. Koh, H.-B. Noh, Y.-B. Shim, In vitro monitoring of i-NOS concentrations with an immunosensor: the inhibitory effect of endocrine disruptors on i-NOS release, Biosens. Bioelectron. 32 (1) (2012) 278–282.

[15] S.K. Yadav, P. Chandra, R.N. Goyal, Y.-B. Shim, A review on determination of steroids in biological samples exploiting nanobio-electroanalytical methods, Anal. Chim. Acta 762 (2013) 14–24.

[16] R. Pallela, P. Chandra, H.-B. Noh, Y.-B. Shim, An amperometric nanobiosensor using a biocompatible conjugate for early detection of metastatic cancer cells in biological fluid, Biosens. Bioelectron. 85 (2016) 883–890.

[17] K.K. Mistry, K. Layek, A. Mahapatra, C. RoyChaudhuri, H. Saha, A review on amperometric-type immunosensors based on screen-printed electrodes, Analyst 139 (10) (2014) 2289–2311.

[18] P. Chandra, Nanobiosensors for personalized and onsite biomedical diagnosis, IET (2016).

[19] M.M.P.S. Neves, M.B. González-García, H.P.A. Nouws, A. Costa-García, Celiac disease detection using a transglutaminase electrochemical immunosensor fabricated on nano-hybrid screen-printed carbon electrodes, Biosen. Bioelectron. 31 (1) (2012) 95–100.

[20] M. Choudhary, P. Yadav, A. Singh, S. Kaur, J. Ramirez-Vick, P. Chandra, et al., CD 59 targeted ultrasensitive electrochemical immunosensor for fast and noninvasive diagnosis of oral cancer, Electroanalysis 28 (10) (2016) 2565–2574.

[21] D. Harvey, Voltammetric Methods, 2016. Available from: http://chem.libretexts.org/ http://chem.libretexts.org/.

[22] S.-Y. Won, P. Chandra, T.S. Hee, Y.-B. Shim, Simultaneous detection of antibacterial sulfonamides in a microfluidic device with amperometry, Biosens. Bioelectron. 39 (1) (2013) 204–209.

[23] P. Chandra, H.-B. Noh, M.-S. Won, Y.-B. Shim, Detection of daunomycin using phosphatidylserine and aptamer co-immobilized on Au nanoparticles deposited conducting polymer, Biosens. Bioelectron. 26 (11) (2011) 4442–4449.

[24] T.R.J. Holford, F. Davis, S.P.J. Higson, Recent trends in antibody based sensors, Biosens. Bioelectron. 34 (1) (2012) 12–24.

[25] J. Piehler, A. Brecht, T. Giersch, B. Hock, G. Gauglitz, Assessment of affinity constants by rapid solid phase detection of equilibrium binding in a flow system, J. Immunol. Methods 201 (2) (1997) 189–206.

[26] H.-B. Noh, P. Chandra, Y.-J. Kim, Y.-B. Shim, A simple separation method with a microfluidic channel based on alternating current potential modulation, Anal. Chem. 84 (22) (2012) 9738–9744.

[27] F. Settle, Handbook of Instrumental Techniques for Analytical Chemistry Settle F, Prentice Hall, 1997.

[28] B.J. Sanghavi, A.K. Srivastava, Simultaneous voltammetric determination of acetaminophen, aspirin and caffeine using an in situ surfactant-modified multiwalled carbon nanotube paste electrode, Electrochim. Acta 55 (28) (2010) 8638–8648.

[29] Q. Zhang, A. Prabhu, A. San, J.F. Al-Sharab, K. Levon, A polyaniline based ultrasensitive potentiometric immunosensor for cardiac troponin complex detection, Biosens. Bioelectron. 72 (2015) 100–106.

[30] D. Purvis, O. Leonardova, D. Farmakovsky, V. Cherkasov, An ultrasensitive and stable potentiometric immunosensor, Biosens. Bioelectron. 18 (11) (2003) 1385–1390.

[31] R. Yuan, D. Tang, Y. Chai, X. Zhong, Y. Liu, J. Dai, Ultrasensitive potentiometric immunosensor based on SA and OCA techniques for immobilization of HBsAb with colloidal Au and polyvinyl butyral as matrixes, Langmuir 20 (17) (2004) 7240–7245.

[32] D.P. Tang, R. Yuan, Y.Q. Chai, X. Zhong, Y. Liu, J.Y. Dai, et al., Novel potentiometric immunosensor for hepatitis B surface antigen using a gold nanoparticle-based biomolecular immobilization method, Anal. Biochem. 333 (2) (2004) 345–350.

[33] R.J. Forster, P. Bertoncello, T.E. Keyes, Electrogenerated chemiluminescence, Annu. Rev. Anal. Chem. 2 (1) (2009) 359–385.

[34] E. Rampazzo, S. Bonacchi, D. Genovese, R. Juris, M. Marcaccio, M. Montalti, et al., Nanoparticles in metal complexes-based electrogenerated chemiluminescence for highly sensitive applications, Coord. Chem. Rev. 256 (15–16) (2012) 1664–1681.

[35] G. Valenti, C. Bruno, S. Rapino, A. Fiorani, E.A. Jackson, L.T. Scott, et al., Intense and tunable electrochemiluminescence of corannulene, J. Phys. Chem. C 114 (45) (2010) 19467–19472.

[36] K. Muzyka, Current trends in the development of the electrochemiluminescent immunosensors, Biosens. Bioelectron. 54 (2014) 393–407.

[37] J. Li, H. Ma, D. Wu, X. Li, Y. Zhao, Y. Zhang, et al., A label-free electrochemiluminescence the immunosensor based on KNbO3 Au nanoparticles@Bi2S3 for detection of prostate specific antigen, Biosens. Bioelectron. 74 (2015) 104–112.

[38] D. Zhu, Y. Tang, D. Xing, W.R. Chen, Pcr-free quantitative detection of genetically modified organism from raw materials. an electrochemiluminescence-based bio bar code method, Anal. Chem. 80 (10) (2008) 3566–3571.

[39] V.R. Rivera, F.J. Gamez, W.K. Keener, J.A. White, M.A. Poli, Rapid detection of *Clostridium botulinum* toxins A, B, E, and F in clinical samples, selected food matrices, and buffer using paramagnetic bead-based electrochemiluminescence detection, Anal. Biochem. 353 (2) (2006) 248–256.

[40] A. Wolter, R. Niessner, M. Seidel, Detection of *Escherichia coli* O157:H7, *Salmonella typhimurium*, and *Legionella pneumophila* in water using a flow-through chemiluminescence microarray readout system, Anal. Chem. 80 (15) (2008) 5854–5863.

[41] X. Mu, Z. Tong, Q. Huang, B. Liu, Z. Liu, L. Hao, et al., An electrochemiluminescence immunosensor based on gold-magnetic nanoparticles and phage displayed antibodies, Sensors (Basel, Switz.) 16 (3) (2016) 308.

[42] J. Tang, D. Tang, Q. Li, B. Su, B. Qiu, G. Chen, Sensitive electrochemical immunoassay of carcinoembryonic antigen with signal dual-amplification using glucose oxidase and an artificial catalase, Anal. Chim. Acta 697 (1) (2011) 16–22.

[43] C. Ding, W. Zhang, W. Wang, Y. Chen, X. Li, Amplification strategies using electrochemiluminescence biosensors for the detection of DNA, bioactive molecules and cancer biomarkers, Trends Anal. Chem. Trac 65 (2015) 137–150.

[44] Y. Cai, N. Li, D.M. Kong, H.X. Shen, Fluorogenic substrate screening for G-quadruplex DNAzyme-based sensors, Biosens. Bioelectron. 49 (2013) 312–317.

[45] C.A. Marquette, A. Degiuli, L. Blum, Electrochemiluminescent biosensors array for the concomitant detection of choline, glucose, glutamate, lactate, lysine and urate, Biosens. Bioelectron. 19 (5) (2003) 433–439.

[46] A. Zhang, H. Xiang, X. Zhang, W. Guo, E. Yuan, C. Huang, et al., A novel sandwich electrochemiluminescence immunosensor for ultrasensitive detection of carbohydrate antigen 19-9 based on immobilizing luminol on Ag@BSA core/shell microspheres, Biosens. Bioelectron. 75 (2016) 206–212.

[47] Q. Xu, J. Li, S. Li, H. Pan, A highly sensitive electrochemiluminescence immunosensor based on magnetic nanoparticles and its application in CA125 determination, J. Solid State Electr. 16 (9) (2012) 2891–2898.

[48] D. Wu, Y. Liu, Y. Wang, L. Hu, H. Ma, G. Wang, et al., Label-free electrochemiluminescent immunosensor for detection of prostate specific antigen based on aminated graphene quantum dots and carboxyl graphene quantum dots, Sci. Rep. 6 (2016) 20511.

[49] J.-G. Guan, Y.-Q. Miao, Q.-J. Zhang, Impedimetric biosensors, J. Biosci. Bioeng. 97 (4) (2004) 219–226.

[50] C. Gabrielli, M. Keddam, H. Takenouti, New trends in the investigation of electrochemical systems by impedance techniques: multi-transfer function analysis, Electrochim. Acta 35 (10) (1990) 1553–1557.

[51] Y. Ram, T. Yoetz-Kopelman, Y. Dror, A. Freeman, Y. Shacham-Diamand, Impact of molecular surface charge on biosensing by electrochemical impedance spectroscopy, Electrochim. Acta 200 (2016) 161–167.

[52] B. Derkus, E. Emregul, C. Yucesan, K. Cebesoy Emregul, Myelin basic protein immunosensor for multiple sclerosis detection based upon label-free electrochemical impedance spectroscopy, Biosens. Bioelectron. 46 (2013) 53–60.

[53] J.S. Daniels, N. Pourmand, Label-free impedance biosensors: opportunities and challenges, Electroanalysis 19 (12) (2007) 1239–1257.

[54] S.J. Lee, V. Anandan, G. Zhang, Electrochemical fabrication and evaluation of highly sensitive nanorod-modified electrodes for a biotin/avidin system, Biosens. Bioelectron. 23 (7) (2008) 1117–1124.

[55] Y. Huang, M.C. Bell, Suni II, Impedance biosensor for peanut protein Ara h 1, Anal. Chem. 80 (23) (2008) 9157–9161.

[56] T. Komura, T. Yamaguchi, H. Shimatani, R. Okushio, Interfacial charge-transfer resistance at ionizable thiol monolayer-modified gold electrodes as studied by impedance spectroscopy, Electrochim. Acta 49 (4) (2004) 597–606.

[57] W.-W. Zhao, J.-J. Xu, H.-Y. Chen, Photoelectrochemical bioanalysis: the state of the art, Chem. Soc. Rev. 44 (3) (2015) 729–741.

[58] W. Wen, X. Yan, C. Zhu, D. Du, Y. Lin, Recent advances in electrochemical immunosensors, Anal. Chem. (2016).

[59] X. Wang, T. Yan, Y. Li, Y. Liu, B. Du, H. Ma, et al., A competitive photoelectrochemical immunosensor based on a CdS induced signal amplification strategy for the ultrasensitive detection of dexamethasone, Sci. Rep. 5 (2015) 17945.

[60] A. Devadoss, P. Sudhagar, C. Terashima, K. Nakata, A. Fujishima, Photoelectrochemical biosensors: new insights into promising photoelectrodes and signal amplification strategies, J. Photochem. Photobiol. C. 24 (2015) 43–63.

[61] Q. Kang, L. Yang, Y. Chen, S. Luo, L. Wen, Q. Cai, et al., Photoelectrochemical detection of pentachlorophenol with a multiple hybrid CdSexTe1-x/TiO2 nanotube structure-based label-free immunosensor, Anal. Chem. 82 (23) (2010) 9749–9754.

[62] M.R. Akanda, H. Ju, A tyrosinase-responsive nonenzymatic redox cycling for amplified electrochemical immunosensing of protein, Anal. Chem. 88 (19) (2016) 9856–9861.

[63] G. Brett, S. Chambers, L. Huang, M. Morgan, Design and development of immunoassays for detection of proteins, Food Control 10 (6) (1999) 401–406.

[64] M.B. Esch, A.J. Baeumner, R.A. Durst, Detection of *Cryptosporidium parvum* using oligonucleotide-tagged liposomes in a competitive assay format, Anal. Chem. 73 (13) (2001) 3162–3167.

[65] J.A.A. Ho, R.D. Wauchope, A strip liposome immunoassay for aflatoxin B1, Anal. Chem. 74 (7) (2002) 1493–1496.

[66] D. Sinawang Prima, V. Rai, R.E. Ionescu, R.S. Marks, Electrochemical lateral flow immunosensor for detection and quantification of dengue NS1 protein, Biosens. Bioelectron. 77 (2016) 400–408.

[67] P. Chandra, Electrochemical nanobiosensors for cancer diagnosis, J. Anal. Bioanal. Tech. 6 (1) (2015) e119.

[68] P. Chandra, H.-B. Noh, R. Pallela, Y.-B. Shim, Ultrasensitive detection of drug resistant cancer cells in biological matrixes using an amperometric nanobiosensor, Biosens. Bioelectron. 70 (2015) 418–425.

[69] Y. Zhu, P. Chandra, B. Shim Y-, Ultrasensitive and selective electrochemical diagnosis of breast cancer based on a Hydrazine–Au nanoparticle–aptamer bioconjugate, Anal. Chem. 85 (2) (2013) 1058–1064.

[70] W. Liu, H. Yang, Y. Ding, S. Ge, J. Yu, M. Yan, et al., Paper-based colorimetric immunosensor for visual detection of carcinoembryonic antigen based on the high peroxidase-like catalytic performance of ZnFe2O4-multiwalled carbon nanotubes, Analyst 139 (1) (2014) 251–258.

[71] E. Canbay, B. Sahin, M. Kıran, E. Akyilmaz, MWCNT–cysteamine–Nafion modified gold electrode based on myoglobin for determination of hydrogen peroxide and nitrite, Bioelectrochemistry 101 (2015) 126–131.

[72] N. Massad-Ivanir, G. Shtenberg, N. Raz, C. Gazenbeek, D. Budding, M.P. Bos, et al., Porous silicon-based biosensors: towards real-time optical detection of target bacteria in the food industry, Sci. Rep. 6 (2016) 38099.

[73] G. Fu, S.T. Sanjay, X. Li, Cost-effective and sensitive colorimetric immunosensing using an iron oxide-to-Prussian blue nanoparticle conversion strategy, Analyst 141 (12) (2016) 3883−3889.

[74] Z.-H. Yang, Y. Zhuo, Y.-Q. Chai, R. Yuan, High throughput immunosenor based on multi-label strategy and a novel array electrode, Sci. Rep. 4 (2014) 4747.

[75] J. Das, K. Jo, J.W. Lee, H. Yang, Electrochemical immunosensor using p-aminophenol redox cycling by hydrazine combined with a low background current, Anal. Chem. 79 (7) (2007) 2790−2796.

[76] B. Limoges, D. Marchal, F. Mavré, J.-M. Savéant, High amplification rates from the association of two enzymes confined within a nanometric layer immobilized on an electrode: modeling and illustrating example, J. Am. Chem. Soc. 128 (18) (2006) 6014−6015.

[77] R. Singh, A. Sharma, S. Hong, J. Jang, Electrical immunosensor based on dielectrophoretically-deposited carbon nanotubes for detection of influenza virus H1N1, Analyst 139 (21) (2014) 5415−5421.

[78] A. Sharma, S. Hong, R. Singh, J. Jang, Single-walled carbon nanotube based transparent immunosensor for detection of a prostate cancer biomarker osteopontin, Anal. Chim. Acta 869 (2015) 68−73.

[79] B.J.A. Alberts, J. Lewis, The Adaptive Immune System, fourth ed., Garland Science, New York, 2002. Available from: https://www.ncbi.nlm.nih.gov/books/NBK21070/.

[80] T.J. Kindt, R.A. Goldsby, B.A. Osborne, J. Kuby, Kuby Immunology, Macmillan, 2007.

[81] B. Al-Lazikani, A.M. Lesk, C. Chothia, Standard conformations for the canonical structures of immunoglobulins, J. Mol. Biol. 273 (4) (1997) 927−948.

[82] Y. Liu, J. Yu, Oriented immobilization of proteins on solid supports for use in biosensors and biochips: a review, Microchim. Acta 183 (1) (2016) 1−19.

[83] B. Byrne, E. Stack, N. Gilmartin, R. O'Kennedy, Antibody-based sensors: principles, problems and potential for detection of pathogens and associated toxins, Sensors 9 (6) (2009) 4407−4445.

[84] H.R. Hoogenboom, Selecting and screening recombinant antibody libraries, Nat. Biotechnol. 23 (9) (2005) 1105−1116.

[85] J. Maynard, G. Georgiou, Antibody engineering, Annu. Rev. Biomed. Eng. 2 (1) (2000) 339−376.

[86] P.J. Hudson, C. Souriau, Engineered antibodies, Nat. Med. 9 (1) (2003) 129−134.

[87] W. Finlay, N. DeVore, E. Dobrovolskaia, A. Gam, C.S. Goodyear, J. Slater, Exploiting the avian immunoglobulin system to simplify the generation of recombinant antibodies to allergenic proteins, Clin. Exp. Allergy 35 (8) (2005) 1040−1048.

[88] S. Sharma, H. Byrne, J. O'Kennedy Richard, Antibodies and antibody-derived analytical biosensors, Essays Biochem. 60 (1) (2016) 9−18.

[89] J. Vlasak, R. Ionescu, Fragmentation of monoclonal antibodies, mAbs 3 (3) (2011) 253−263.

[90] L. Torrance, A. Ziegler, H. Pittman, M. Paterson, R. Toth, I. Eggleston, Oriented immobilisation of engineered single-chain antibodies to develop biosensors for virus detection, J. Virol. Methods 134 (1−2) (2006) 164−170.

[91] D. Saerens, L. Huang, K. Bonroy, S. Muyldermans, Antibody fragments as probe in biosensor development, Sensors (Basel, Switz.) 8 (8) (2008) 4669−4686.

[92] A.D. Ellington, J.W. Szostak, In vitro selection of RNA molecules that bind specific ligands, Nature 346 (6287) (1990) 818−822.

[93] Y. Zhu, P. Chandra, K.-M. Song, C. Ban, Y.-B. Shim, Label-free detection of kanamycin based on the aptamer-functionalized conducting polymer/gold nanocomposite, Biosens. Bioelectron. 36 (1) (2012) 29−34.

[94] J.C. Cox, A.D. Ellington, Automated selection of anti-Protein aptamers, Bioorg. Med. Chem. Lett. 9 (10) (2001) 2525–2531.

[95] H. Cai, T.M.-H. Lee, I.M. Hsing, Label-free protein recognition using an aptamer-based impedance measurement assay, Sens. Actuators B Chem. 114 (1) (2006) 433–437.

[96] G. Biesecker, L. Dihel, K. Enney, R.A. Bendele, Derivation of RNA aptamer inhibitors of human complement C5, Immunopharmacology 42 (1–3) (1999) 219–230.

[97] B.J. Hicke, C. Marion, Y.-F. Chang, T. Gould, C.K. Lynott, D. Parma, et al., Tenascin-C aptamers are generated using tumor cells and purified protein, J. Biol. Chem. 276 (52) (2001) 48644–48654.

[98] S.D. Jayasena, Aptamers: an emerging class of molecules that rival antibodies in diagnostics, Clin. Chem. 45 (9) (1999) 1628–1650.

[99] B. Shen, J. Li, W. Cheng, Y. Yan, R. Tang, Y. Li, et al., Electrochemical aptasensor for highly sensitive determination of cocaine using a supramolecular aptamer and rolling circle amplification, Microchim. Acta 182 (1) (2015) 361–367.

[100] L. Zhang, P. Cui, B. Zhang, F. Gao, Aptamer-based turn-on detection of thrombin in biological fluids based on efficient phosphorescence energy transfer from Mn-Doped ZnS quantum dots to carbon nanodots, Chem. Eur. J. 19 (28) (2013) 9242–9250.

[101] Y. Zhu, P. Chandra, C. Ban, Y.-B. Shim, Electrochemical evaluation of binding affinity for aptamer selection using the microarray chip, Electroanalysis 24 (5) (2012) 1057–1064.

[102] C. Ocaña, S. Lukic, M. del Valle, Aptamer-antibody sandwich assay for cytochrome c employing an MWCNT platform and electrochemical impedance, Microchim. Acta 182 (11) (2015) 2045–2053.

[103] H. Ilkhani, M. Sarparast, A. Noori, S. Zahra Bathaie, M.F. Mousavi, Electrochemical aptamer/antibody based sandwich immunosensor for the detection of EGFR, a cancer biomarker, using gold nanoparticles as a signaling probe, Biosens. Bioelectron. 74 (2015) 491–497.

[104] B.L. Feringa, W.R. Browne, Molecular Switches: Wiley Online Library, 2001.

[105] N.S. Lipman, L.R. Jackson, L.J. Trudel, F. Weis-Garcia, Monoclonal versus polyclonal antibodies: distinguishing characteristics, applications, and information resources, ILAR J. 46 (3) (2005) 258–268.

[106] X. Luo, J.J. Davis, Electrical biosensors and the label free detection of protein disease biomarkers, Chem. Soc. Rev. 42 (13) (2013) 5944–5962.

[107] H. Jia, P. Gao, H. Ma, D. Wu, B. Du, Q. Wei, Preparation of Au–Pt nanostructures by combining top-down with bottom-up strategies and application in label-free electrochemical immunosensor for detection of NMP22, Bioelectrochemistry 101 (2015) 22–27.

[108] S. Kumar, S. Kumar, S. Tiwari, S. Srivastava, M. Srivastava, B.K. Yadav, et al., Biofunctionalized nanostructured zirconia for biomedical application: a smart approach for oral cancer detection, Adv. Sci. 2 (8) (2015).

[109] S. Kumar, S. Kumar, S. Srivastava, B.K. Yadav, S.H. Lee, J.G. Sharma, et al., Reduced graphene oxide modified smart conducting paper for cancer biosensor, Biosens. Bioelectron. 73 (2015) 114–122.

[110] N. Li, Y. Wang, Y. Li, W. Cao, H. Ma, D. Wu, et al., A label-free electrochemical immunosensor based on Au@ Pd/Ag yolk-bimetallic shell nanoparticles and amination graphene for detection of nuclear matrix protein 22, Sens. Actuators B Chem. 202 (2014) 67–73.

[111] H. Wang, H. Li, Y. Zhang, Q. Wei, H. Ma, D. Wu, et al., Label-free immunosensor based on Pd nanoplates for amperometric immunoassay of alpha-fetoprotein, Biosens. Bioelectron. 53 (2014) 305–309.

[112] R. Elshafey, C. Tlili, A. Abulrob, A.C. Tavares, M. Zourob, Label-free impedimetric immunosensor for ultrasensitive detection of cancer marker Murine double minute 2 in brain tissue, Biosens. Bioelectron. 39 (1) (2013) 220–225.

[113] H. Fan, Y. Zhang, D. Wu, H. Ma, X. Li, Y. Li, et al., Construction of label-free electrochemical immunosensor on mesoporous carbon nanospheres for breast cancer susceptibility gene, Anal. Chim. Acta 770 (2013) 62–67.

[114] H. Li, J. He, S. Li, A.P. Turner, Electrochemical immunosensor with N-doped graphene-modified electrode for label-free detection of the breast cancer biomarker CA 15-3, Biosens. Bioelectron. 43 (2013) 25–29.

[115] A. Ravalli, C.G. da Rocha, H. Yamanaka, G. Marrazza, A label-free electrochemical affisensor for cancer marker detection: the case of HER2, Bioelectrochemistry 106 (2015) 268–275.

[116] Y.-S. Gao, X.-F. Zhu, J.-K. Xu, L.-M. Lu, W.-M. Wang, T.-T. Yang, et al., Label-free electrochemical immunosensor based on Nile blue A-reduced graphene oxide nanocomposites for carcinoembryonic antigen detection, Anal. Biochem. 500 (2016) 80–87.

[117] A. Vasudev, A. Kaushik, S. Bhansali, Electrochemical immunosensor for label free epidermal growth factor receptor (EGFR) detection, Biosens. Bioelectron. 39 (1) (2013) 300–305.

[118] A. Ravalli, G.P. Dos Santos, M. Ferroni, G. Faglia, H. Yamanaka, G. Marrazza, New label free CA125 detection based on gold nanostructured screen-printed electrode, Sens. Actuators B Chem. 179 (2013) 194–200.

[119] R. Singh, S. Kashyap, S. Kumar, S. Abraham, T.K. Gupta, A.M. Kayastha, et al., Excellent storage stability and sensitive detection of neurotoxin quinolinic acid, Biosens. Bioelectron. 90 (2017) 224–229.

[120] R. Wang, X. Chen, J. Ma, Z. Ma, Ultrasensitive detection of carcinoembryonic antigen by a simple label-free immunosensor, Sens. Actuators B Chem. 176 (2013) 1044–1050.

[121] D. Feng, X. Lu, X. Dong, Y. Ling, Y. Zhang, Label-free electrochemical immunosensor for the carcinoembryonic antigen using a glassy carbon electrode modified with electrodeposited Prussian blue, a graphene and carbon nanotube assembly and an antibody immobilized on gold nanoparticles, Microchim. Acta 180 (9–10) (2013) 767–774.

[122] S. Kumar, S. Kumar, C.M. Pandey, B.D. Malhotra (Eds.), Conducting Paper Based Sensor for Cancer Biomarker Detection. Journal of Physics: Conference Series, IOP Publishing, 2016.

[123] S. Kumar, P. Rai, J.G. Sharma, A. Sharma, B.D. Malhotra, PEDOT: PSS/PVA-Nanofibers-decorated conducting paper for cancer diagnostics, Adv. Mater. Technol. (2016).

[124] S. Kumar, A. Sen, S. Kumar, S. Augustine, B.K. Yadav, S. Mishra, et al., Polyaniline modified flexible conducting paper for cancer detection, Appl. Phys. Lett. 108 (20) (2016) 203702.

[125] S. Kumar, M. Willander, J.G. Sharma, B.D. Malhotra, A solution processed carbon nanotube modified conducting paper sensor for cancer detection, J. Mater. Chem. B 3 (48) (2015) 9305–9314.

[126] Y. Wang, H. Xu, J. Luo, J. Liu, L. Wang, Y. Fan, et al., A novel label-free microfluidic paper-based immunosensor for highly sensitive electrochemical detection of carcinoembryonic antigen, Biosens. Bioelectron. 83 (2016) 319–326.

[127] Y. Yang, Q. Liu, Y. Liu, J. Cui, H. Liu, P. Wang, et al., A novel label-free electrochemical immunosensor based on functionalized nitrogen-doped graphene quantum dots for carcinoembryonic antigen detection, Biosens. Bioelectron. 90 (2017) 31–38.

[128] H. Yang, Z. Li, X. Wei, R. Huang, H. Qi, Q. Gao, et al., Detection and discrimination of alpha-fetoprotein with a label-free electrochemical impedance spectroscopy biosensor array based on lectin functionalized carbon nanotubes, Talanta 111 (2013) 62–68.

[129] J. Sun, D. Tian, Q. Guo, L. Zhang, W. Jiang, M. Yang, A label-free electrochemical immunosensor for the detection of cancer biomarker [small alpha]-fetoprotein (AFP) based on hydroxyapatite induced redox current, Anal. Methods 8 (40) (2016) 7319–7323.

[130] S. Kumar, J.G. Sharma, S. Maji, B.D. Malhotra, Nanostructured zirconia decorated reduced graphene oxide based efficient biosensing platform for non-invasive oral cancer detection, Biosens. Bioelectron. 78 (2016) 497–504.

[131] S. Kumar, S. Kumar, S. Tiwari, S. Augustine, S. Srivastava, B.K. Yadav, et al., Highly sensitive protein functionalized nanostructured hafnium oxide based biosensing platform for non-invasive oral cancer detection, Sens. Actuators B Chem. 235 (2016) 1–10.

[132] S. Kumar, J.G. Sharma, S. Maji, B.D. Malhotra, A biocompatible serine functionalized nanostructured zirconia based biosensing platform for non-invasive oral cancer detection, RSC Adv. 6 (80) (2016) 77037–77046.

[133] L. Wang, J. Shan, F. Feng, Z. Ma, Novel redox species polyaniline derivative-Au/Pt as sensing platform for label-free electrochemical immunoassay of carbohydrate antigen 199, Anal. Chim. Acta 911 (2016) 108–113.

[134] H. Deng, J. Li, Y. Zhang, H. Pan, G. Xu, A new strategy for label-free electrochemical immunoassay based on "gate-effect" of β-cyclodextrin modified electrode, Anal. Chim. Acta 926 (2016) 48–54.

[135] H.D. Jang, S.K. Kim, H. Chang, J.-W. Choi, 3D label-free prostate specific antigen (PSA) immunosensor based on graphene–gold composites, Biosens. Bioelectron. 63 (2015) 546–551.

[136] F. Li, Y. Yu, H. Cui, D. Yang, Z. Bian, Label-free electrochemiluminescence immunosensor for cardiac troponin I using luminol functionalized gold nanoparticles as a sensing platform, Analyst 138 (6) (2013) 1844–1850.

[137] J. Pancrazio, J. Whelan, D. Borkholder, W. Ma, D. Stenger, Development and application of cell-based biosensors, Ann. Biomed. Eng. 27 (6) (1999) 697–711.

[138] M.R. Akanda, V. Tamilavan, S. Park, K. Jo, M.H. Hyun, H. Yang, Hydroquinone diphosphate as a phosphatase substrate in enzymatic amplification combined with electrochemical–chemical–chemical redox cycling for the detection of *E. coli* O157: H7, Anal. Chem. 85 (3) (2013) 1631–1636.

[139] J. Dong, H. Zhao, M. Xu, Q. Ma, S. Ai, A label-free electrochemical impedance immunosensor based on AuNPs/PAMAM-MWCNT-Chi nanocomposite modified glassy carbon electrode for detection of *Salmonella typhimurium* in milk, Food Chem. 141 (3) (2013) 1980–1986.

[140] W. Dou, W. Tang, G. Zhao, A disposable electrochemical immunosensor arrays using 4-channel screen-printed carbon electrode for simultaneous detection of *Escherichia coli* O157: H7 and *Enterobacter sakazakii*, Electrochim. Acta 97 (2013) 79–85.

[141] S. Liébana, D. Brandão, S. Alegret, M.I. Pividori, Electrochemical immunosensors, genosensors and phagosensors for Salmonella detection, Anal. Methods 6 (22) (2014) 8858–8873.

[142] S. Viswanathan, C. Rani, J.-A. Ho, Electrochemical immunosensor for multiplexed detection of food-borne pathogens using nanocrystal bioconjugates and MWCNT screen-printed electrode, Talanta 94 (2012) 315–319.

[143] K.Y. Chan, W.W. Ye, Y. Zhang, L.D. Xiao, P.H. Leung, Y. Li, et al., Ultrasensitive detection of *E. coli* O157: H7 with biofunctional magnetic bead concentration via

nanoporous membrane based electrochemical immunosensor, Biosens. Bioelectron. 41 (2013) 532–537.

[144] C.-K. Joung, H.-N. Kim, M.-C. Lim, T.-J. Jeon, H.-Y. Kim, Y.-R. Kim, A nanoporous membrane-based impedimetric immunosensor for label-free detection of pathogenic bacteria in whole milk, Biosens. Bioelectron. 44 (2013) 210–215.

[145] H. Huang, M. Liu, X. Wang, W. Zhang, D.-P. Yang, L. Cui, et al., Label-free 3D Ag nanoflower-based electrochemical immunosensor for the detection of *Escherichia coli* O157: H7 pathogens, Nanoscale Res. Lett. 11 (1) (2016) 507.

[146] C.H. Gayathri, P. Mayuri, K. Sankaran, A.S. Kumar, An electrochemical immunosensor for efficient detection of uropathogenic *E. coli* based on thionine dye immobilized chitosan/functionalized-MWCNT modified electrode, Biosens. Bioelectron. 82 (2016) 71–77.

[147] M. Xu, R. Wang, Y. Li, Rapid detection of *Escherichia coli* O157:H7 and Salmonella Typhimurium in foods using an electrochemical immunosensor based on screen-printed interdigitated microelectrode and immunomagnetic separation, Talanta 148 (2016) 200–208.

[148] C.M. Pandey, I. Tiwari, V.N. Singh, K. Sood, G. Sumana, B.D. Malhotra, Highly sensitive electrochemical immunosensor based on graphene-wrapped copper oxide-cysteine hierarchical structure for detection of pathogenic bacteria, Sens. Actuators B Chem. 238 (2017) 1060–1069.

[149] M. Braiek, K.B. Rokbani, A. Chrouda, B. Mrabet, A. Bakhrouf, A. Maaref, et al., An electrochemical immunosensor for detection of *Staphylococcus aureus* bacteria based on immobilization of antibodies on self-assembled monolayers-functionalized gold electrode, Biosensors 2 (4) (2012) 417–426.

[150] C. Cheng, Y. Peng, J. Bai, X. Zhang, Y. Liu, X. Fan, et al., Rapid detection of *Listeria monocytogenes* in milk by self-assembled electrochemical immunosensor, Sens. Actuators B Chem. 190 (2014) 900–906.

[151] Y. Zhao, L. Liu, D. Kong, H. Kuang, L. Wang, C. Xu, Dual amplified electrochemical immunosensor for highly sensitive detection of *Pantoea stewartii* sbusp. *stewartii*, ACS Appl. Mater. Inter. 6 (23) (2014) 21178–21183.

[152] M. Amouzadeh Tabrizi, M. Shamsipur, R. Saber, S. Sarkar, N. Zolfaghari, An ultrasensitive sandwich-type electrochemical immunosensor for the determination of SKBR-3 breast cancer cell using rGO-TPA/FeHCFnano labeled Anti-HCT as a signal tag, Sens. Actuators B Chem. 243 (2017) 823–830.

[153] R. Seenivasan, N. Maddodi, V. Setaluri, S. Gunasekaran, An electrochemical immunosensing method for detecting melanoma cells, Biosens. Bioelectron. 68 (2015) 508–515.

[154] Y.-K. Chung, J. Reboud, K.C. Lee, H.M. Lim, P.Y. Lim, K.Y. Wang, et al., An electrical biosensor for the detection of circulating tumor cells, Biosens. Bioelectron. 26 (5) (2011) 2520–2526.

[155] F. Zanghelini, I.A.M. Frías, M.J.B.M. Rêgo, M.G.R. Pitta, M. Sacilloti, M.D.L. Oliveira, et al., Biosensing breast cancer cells based on a three-dimensional TIO2 nanomembrane transducer, Biosens. Bioelectron. 92 (2017) 313–320.

[156] A. Yadegari, M. Omidi, F. Yazdian, H. Zali, L. Tayebi, An electrochemical cytosensor for ultrasensitive detection of cancer cells using modified graphene-gold nanostructures, RSC Adv. 7 (4) (2017) 2365–2372.

[157] M.M.P. da Silva Neves, M.B.G. García, D.H. Santos, P. Fanjul-Bolado, Hydroquinone diphosphate/Ag+ as an enzymatic substrate for alkaline phosphatase catalyzed silver deposition, Electrochem. Commun. 60 (2015) 1–4.

[158] J. Tang, B. Su, D. Tang, G. Chen, Conductive carbon nanoparticles-based electrochemical immunosensor with enhanced sensitivity for α-fetoprotein using irregular-shaped gold nanoparticles-labeled enzyme-linked antibodies as signal improvement, Biosens. Bioelectron. 25 (12) (2010) 2657–2662.

[159] C. Kokkinos, A. Economou, M.I. Prodromidis, Electrochemical immunosensors: critical survey of different architectures and transduction strategies, Trends Anal. Chem. Trac. 79 (2016) 88–105.

[160] M. Pividori, A. Lermo, A. Bonanni, S. Alegret, M. Del Valle, Electrochemical immunosensor for the diagnosis of celiac disease, Anal. Biochem. 388 (2) (2009) 229–234.

[161] T. Balkenhohl, F. Lisdat, Screen-printed electrodes as impedimetric immunosensors for the detection of anti-transglutaminase antibodies in human sera, Anal. Chim. Acta 597 (1) (2007) 50–57.

[162] M.M.P.S. Neves, M.B. González-García, C. Delerue-Matos, A. Costa-García, Multiplexed electrochemical immunosensor for detection of celiac disease serological markers, Sens. Actuators B Chem. 187 (2013) 33–39.

[163] Y.-S. Fang, X.-J. Huang, L.-S. Wang, J.-F. Wang, An enhanced sensitive electrochemical immunosensor based on efficient encapsulation of enzyme in silica matrix for the detection of human immunodeficiency virus p24, Biosens. Bioelectron. 64 (2015) 324–332.

[164] J. Wu, J. Tang, Z. Dai, F. Yan, H. Ju, N. El Murr, A disposable electrochemical immunosensor for flow injection immunoassay of carcinoembryonic antigen, Biosens. Bioelectron. 22 (1) (2006) 102–108.

[165] D. Tang, R. Yuan, Y. Chai, Ultrasensitive electrochemical immunosensor for clinical immunoassay using thionine-doped magnetic gold nanospheres as labels and horseradish peroxidase as enhancer, Anal. Chem. 80 (5) (2008) 1582–1588.

[166] Z. Zhong, W. Wu, D. Wang, D. Wang, J. Shan, Y. Qing, et al., Nanogold-enwrapped graphene nanocomposites as trace labels for sensitivity enhancement of electrochemical immunosensors in clinical immunoassays: carcinoembryonic antigen as a model, Biosens. Bioelectron. 25 (10) (2010) 2379–2383.

[167] N. Gan, X. Du, Y. Cao, F. Hu, T. Li, Q. Jiang, An ultrasensitive electrochemical immunosensor for HIV p24 based on Fe3O4@ SiO2 nanomagnetic probes and nanogold colloid-labeled enzyme–antibody copolymer as signal tag, Materials 6 (4) (2013) 1255–1269.

[168] H. Yu, F. Yan, Z. Dai, H. Ju, A disposable amperometric immunosensor for α-1-fetoprotein based on enzyme-labeled antibody/chitosan-membrane-modified screen-printed carbon electrode, Anal. Biochem. 331 (1) (2004) 98–105.

[169] E. Sánchez-Tirado, A. González-Cortés, P. Yáñez-Sedeño, J.M. Pingarrón, Carbon nanotubes functionalized by click chemistry as scaffolds for the preparation of electrochemical immunosensors. Application to the determination of TGF-beta 1 cytokine, Analyst 141 (20) (2016) 5730–5737.

[170] D. Du, F. Yan, S. Liu, H. Ju, Immunological assay for carbohydrate antigen 19-9 using an electrochemical immunosensor and antigen immobilization in titania sol–gel matrix, J. Immunol. Methods 283 (1) (2003) 67–75.

[171] Z. He, S. Zang, Y. Liu, Y. He, H. Lei, A multi-walled carbon nanotubes-poly (l-lysine) modified enantioselective immunosensor for ofloxacin by using multi-enzyme-labeled gold nanoflower as signal enhancer, Biosens. Bioelectron. 73 (2015) 85–92.

[172] R. Malhotra, V. Patel, J.P. Vaqué, J.S. Gutkind, J.F. Rusling, Ultrasensitive electrochemical immunosensor for oral cancer biomarker IL-6 using carbon nanotube forest electrodes and multilabel amplification, Anal. Chem. 82 (8) (2010) 3118–3123.

[173] J. Zhang, J. Lei, C. Xu, L. Ding, H. Ju, Carbon nanohorn sensitized electrochemical immunosensor for rapid detection of microcystin-LR, Anal. Chem. 82 (3) (2010) 1117–1122.

[174] H. Yang, Enzyme-based ultrasensitive electrochemical biosensors, Curr. Opin. Chem. Biol. 16 (3–4) (2012) 422–428.

[175] A. Warsinke, A. Benkert, F.W. Scheller, Electrochemical immunoassays, Fresenius J. Anal. Chem. 366 (6–7) (2000) 622–634.

[176] G. Wang, X. Gang, X. Zhou, G. Zhang, H. Huang, X. Zhang, et al., Electrochemical immunosensor with graphene/gold nanoparticles platform and ferrocene derivatives label, Talanta 103 (2013) 75–80.

[177] B. Zhang, B. Liu, G. Chen, D. Tang, Redox and catalysis "all-in-one" infinite coordination polymer for electrochemical immunosensor of tumor markers, Biosens. Bioelectron. 64 (2015) 6–12.

[178] Z. Tang, Z. Ma, Ratiometric ultrasensitive electrochemical immunosensor based on redox substrate and immunoprobe, Sci. Rep. 6 (2016).

[179] E. Sanchez-Tirado, G. Martinez-Garcia, A. Gonzalez-Cortes, P. Yanez-Sedeno, J.M. Pingarron, Electrochemical immunosensor for sensitive determination of transforming growth factor (TGF) – beta1 in urine, Biosens. Bioelectron. 88 (2017) 9–14.

[180] Z. Xiao, O.C. Farokhzad, Aptamer-functionalized nanoparticles for medical applications: challenges and opportunities, ACS Nano. 6 (5) (2012) 3670–3676.

[181] S. MacKay, D. Wishart, J.Z. Xing, J. Chen, Developing trends in aptamer-based biosensor devices and their applications, IEEE Trans. Biomed. Circuits Syst. 8 (1) (2014) 4–14.

[182] W. Zhou, P.-J.J. Huang, J. Ding, J. Liu, Aptamer-based biosensors for biomedical diagnostics, Analyst 139 (11) (2014) 2627–2640.

[183] Y. Du, B. Li, E. Wang, "Fitting" makes "sensing" simple: label-free detection strategies based on nucleic acid aptamers, Acc. Chem. Res. 46 (2) (2012) 203–213.

[184] K. Han, Z. Liang, N. Zhou, Design strategies for aptamer-based biosensors, Sensors 10 (5) (2010) 4541–4557.

[185] P. Jolly, P. Damborsky, N. Madaboosi, R.R.G. Soares, V. Chu, J.P. Conde, et al., DNA aptamer-based sandwich microfluidic assays for dual quantification and multi-glycan profiling of cancer biomarkers, Biosens. Bioelectron. 79 (2016) 313–319.

[186] Z. Taleat, C. Cristea, G. Marrazza, M. Mazloum-Ardakani, R. Săndulescu, Electrochemical immunoassay based on aptamer–protein interaction and functionalized polymer for cancer biomarker detection, J. Electroanal. Chem. 717–718 (2014) 119–124.

[187] B. Kavosi, A. Salimi, R. Hallaj, F. Moradi, Ultrasensitive electrochemical immunosensor for PSA biomarker detection in prostate cancer cells using gold nanoparticles/PAMAM dendrimer loaded with enzyme linked aptamer as integrated triple signal amplification strategy, Biosens. Bioelectron. 74 (2015) 915–923.

[188] A. Xiang, X. Lei, F. Ren, L. Zang, Q. Wang, J. Zhang, et al., An aptamer-based immunoassay in microchannels of a portable analyzer for detection of microcystin-leucine-arginine, Talanta 130 (2014) 363–369.

Chapter 15

Lab-on-a-Chip (LOC) Immunoassays

Sandeep K. Vashist[1], John H.T. Luong[2]
[1]*IDS Immunodiagnostic Systems Deutschland GmbH, Frankfurt am Main, Germany;* [2]*University College Cork, Cork, Ireland*

1. INTRODUCTION

The outreach of healthcare has been obstructed by the lack of resources, insufficient healthcare professionals, and unaffordable costs of clinical diagnostics. Therefore, there are increasing efforts toward the development of low-cost and easy-to-use in vitro diagnostic (IVD) platforms for the diagnosis of diseases in the developing countries and remote settings. To date, there are numerous developments in the field of lab-on-a-chip (LOC) technologies in the last two decades. Cost-effective LOC platforms are diversified based on immunobiosensing concepts, microfluidic (MF) protocols, materials, reagent storage and release mechanisms, and low-cost readout devices. The current generation of LOC platforms and the rapid improvements in immunoassays (IAs) and complementary technologies are paving the way to highly advanced LOC IAs to target point-of-care testing (POCT). The evolving trend is strongly inclined to the development of LOC IAs for personalized healthcare monitoring and management. The innovative advances made in cellphone (CP) detection–based bioanalytical technologies have further started a new era of mobile healthcare. Considering over 7 billion CP users with 70% in the developing countries, the interface of LOC IA with CPs is indeed revolutionary, enabling personalized POCT and better healthcare management [1,2]. CPs would provide spatiotemporal tagged data to offer real-time healthcare monitoring and management. The mobile health solutions market is growing at a rapid pace and is expected to reach USD 90.5 billion by 2022 [3]. During the recent years, several LOC IAs and devices have been successfully implemented in central laboratories and physician office laboratories for POCT. Apparently, LOC IAs play a critical role in centralized and decentralized, remote, and personalized settings. Their deployment would be immensely useful in epidemics and emergencies that require immediate detection of analytes at the point of need.

Handbook of Immunoassay Technologies. https://doi.org/10.1016/B978-0-12-811762-0.00015-3

LOC-based POCT would enable rapid and cost-effective clinical diagnosis using minimal sample volumes [4] at the bedside of a patient, in a physician's office, at home, or in paramedical support vehicles, as well as secondary and tertiary care settings. It would provide the results just within a few minutes using a highly simplified protocol with one or two process steps to process whole blood, serum, plasma, urine, or other samples. The point-of-care (POC) diagnostics' global market will reach USD 37 billion by 2021 [5]. India and China would account for the most predominant market share due to enormous populations with increased prevalence of chronic and infectious diseases. Consequently, LOC IAs and POCT devices must be affordable for those at risk of infection or chronic diseases. They must also be robust and unrefrigerated in addition to providing the desired sensitivity and specificity, minimal steps, and short turnaround time.

This chapter provides an overview of various LOC IAs along with the global trend toward next-generation POCT, the challenges, and outlook. Table 15.1 summarizes some key features of various LOC IA formats.

2. LAB-ON-A-CHIP IMMUNOASSAYS

2.1 Microfluidic

Besides blood glucose sensing strips as the most predominant MFs-based LOC device in the market [6], numerous LOC formats have emerged during the last two decades [7]. Piccolo Xpress whole blood chemistry analyzer from Abaxis Inc., USA, is one of such LOC-based POC devices [8] (Fig. 15.1A). The system is based on centrifugal MFs-based LabDisk (CML) platform to carry out up to 14 tests on a single reagent LabDisk (8 cm diameter and bar-coded) (Fig. 15.1A). The company has developed numerous Clinical Laboratory Improvement Amendments (CLIA) waived tests. Similar CML platforms, i.e., Gyrolab Bioaffy CDs [9] and analyzer instruments have been developed by Gyros Protein Technologies, Sweden. Gyrolab xPlore performs a single CD-based IA, whereas Gyrolab xP workstation can run up to five CDs simultaneously [10].

CML-based LOC IAs are available for the detection of biomarkers, toxins, nucleic acid, pathogens, and other analytes [11]. Of notice is the rapid CML-based sandwich enzyme immunoassay with a sample-to-answer time of just 25 min [12]. It involves the detection of C-reactive protein (CRP) by using capture antihuman CRP Ab-bound superparamagnetic Dynabeads (M-280 tosyl activated) and horseradish peroxidase (HRP)-labeled antihuman CRP Ab in an IA chamber of CML. The sandwich immune complexes formed in the presence of CRP are transferred to the wash chambers and the detection chamber. The enzyme—substrate reaction triggers a chemiluminescent signal, which is measured by the miniaturized detector integrated into the LabDisk reader.

TABLE 15.1 The Conceptual Potential of Emerging Technologies for Next-Generation Lab-on-a-Chip (LOC) Immunoassays (IAs) for Point-of-Care Testing (POCT)

Parameters	Cellphone (1)	Paper-based assay (2)	LOC (3)	Next-generation LOC IAs for POCT[a] (1 + 2 + 3)
Performance				
Suitability for POCT	●	●	●	●
Technology penetration	●	●	●	N.A.[b]
Utility in epidemics and emergencies	●	●	●	●
Prerequisite of prolonged storage of reagents	✓	✓	✓	✓
Prerequisite of rapid assay	✓	✓	✓	✓
Portability	●	●	●	●
Cost-effectiveness of consumables	●	●	●	●
Overall cost-effectiveness	●	●	●	●
Quantitative	✓	✗	✓	✓
Sensitivity	●	●	●	●
Specificity	●	●	●	●
Throughput	●	●	●	●
Precision	●	●	●	●
Reproducibility	●	●	●	●
Capable of mass production	✓	✓	✓	✓
Compliance with regulatory guidelines	✓	✗	✓	✓
Ease of Operation				
Ease of operation	●	●	●	●
Labor intensiveness	●	●	●	●

Continued

TABLE 15.1 The Conceptual Potential of Emerging Technologies for Next-Generation Lab-on-a-Chip (LOC) Immunoassays (IAs) for Point-of-Care Testing (POCT)—cont'd

Parameters	Cellphone (1)	Paper-based assay (2)	LOC (3)	Next-generation LOC IAs for POCT[a] (1 + 2 + 3)
Need for power supply	✗	✓	✓	✗
Need for readout instruments	✗[c]	✗	✓	✗[c]
Stand-alone analysis	✓	✓	✓	✓
Personalized	✓	✗	✗	✓
Accessibility of POCT results anytime anywhere	✓	✗	✗	✓
Basic skill set required for operation	✓	✓	✗	✓
Connectivity				
Connectivity to Cloud	✓	✗	✗	✓
Smart applications and portal services	✓	✗	✗	✓
Test history and data patterns	✓	✗	✓ (Limited)	✓
Spatiotemporal mapping	✓	✗	✗	✓
Demographic data and statistics	✓	✗	✗	✓
Telemedicine support	✓	✗	✗	✓
Text alerts	✓	✗	✗	✓
Preventive healthcare tools	✓	✗	✗	✓

High ●; Medium ●; Low ●.

[a]*This column has been computed conceptually taking into account the characteristics of the component technologies.*

[b]*N.A., not applicable.*

[c]*A CP attachment or interfaced instrument would be required for nonoptical signal detection such as in the case of electrochemical readout.*

Reproduced With Permission From Elsevier B.V, S.K. Vashist, et al., Emerging technologies for next-generation point-of-care testing, Trends Biotechnol. 33 (11) (2015) 692–705.

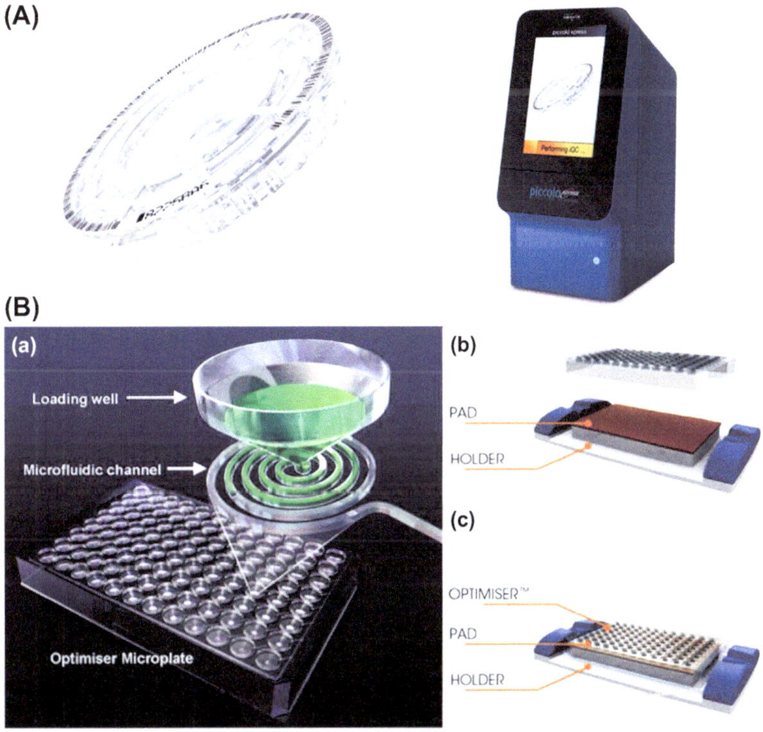

FIGURE 15.1 (A) (left) Centrifugal microfluidics-based LabDisk (CML) platform for fully automated IA; (right) Piccolo Xpress chemistry analyzer. (B) Optimiser microplates developed by Siloam Biosciences Inc., USA. *(A) Reproduced with permission from Elsevier B.V, S.K. Vashist, et al., Emerging technologies for next-generation point-of-care testing, Trends Biotechnol. 33 (11) (2015) 692–705. (B) Reproduced with permission from the Royal Society of Chemistry, J. Kai, et al., A novel microfluidic microplate as the next generation assay platform for enzyme linked immunoassays (ELISA), Lab Chip 12 (21) (2012) 4257–4262.*

The Optimiser ELISA by Siloam Biosciences Inc. [13], USA, is a prospective MFs-based LOC format using a novel MF microtiter plate (MTP) (Fig. 15.1B). It requires only 4.5 μL of the sample and critically reduced number of steps, resulting in a sample-to-answer time of just a few minutes. The system is based on a standard 96-well polystyrene MTP with each MTP well connected to a spiral MF channel via a through-hole at the bottom. The IA is performed using various flow steps, where the liquid dispensed into the MTP is drawn into the microchannel by capillary forces exerted by the absorbent pad at the other end of the microchannel. The Optimiser ELISA involves the sequential dispensing of capture Ab, blocking buffer, target analyte, fluorescently labeled detection Ab, and wash buffer, followed by the fluorescence readout of Optimiser MTP in a plate reader.

2.2 Paper

Paper-based IAs (PBIAs) with the readout by CP are the most cost-effective and easy-to-use assay formats for the primary healthcare professionals in the developing countries. The widespread prevalent PBIAs are lateral flow IAs (LFIAs), dipstick IAs, and MF paper-based analytical devices (μPADs) [15,16]. Besides the most used substrates for PBIA filter paper, other attempted substrates are chromatography paper, nitrocellulose membrane, and paper/polymer or paper/nanomaterial composites [17]. The porosity, surface chemistry, and optical properties of the substrates play a predominant role in the immunodiagnostic applications of PBIAs [18]. The nitrocellulose membranes are available in various pore sizes (0.05−12 μm), whereas polyethylene has controlled pore sizes [19].

LFIA uses nitrocellulose test strips along which the sample dispensed on to a sample pad flows laterally over a conjugate pad, where the analyte molecules bind to the conjugate particles (e.g., gold nanoparticles [AuNPs]) due to the capillary action provided by the absorbent pad (Fig. 15.2A). The home pregnancy test based on the detection of human chorionic gonadotropin (hCG) in urine is the most widely used paper-based LFIA, despite LFIAs being developed for primary hepatic carcinoma biomarkers [20], HIV diagnosis [21], and other target analytes. Two- (2D) or three-dimensional (3D) μPADs have been used for PBIAs [22,23] (Fig. 15.2B). The 3D μPADs are obtained by stacking layers of a 2D-patterned paper [24] or using other strategies such as spray adhesives [25], thermal adhesives (toner) [26], cutting and lamination [27], and origami [28]. Various applications are feasible by using PBIAs on μPADs, e.g., sandwich ELISAs [29,30] and MF paper-based electrochemical devices (μPEDs) prepared by printing electrodes on paper [31,32].

The lack of detection sensitivity for POCT applications by conventional PBIAs can be improved by signal enhancement strategies based on the use of enzymes [33] and nanomaterials, e.g., AuNPs [34]. Other strategies involve the plasmonic nanoparticle (NP) conjugation or aggregation [35], thermal contrasting methods [36], and optimization of test strip dimensions [37].

The colorimetric PBIAs can be quantitatively visualized by comparing the colorimetric signal against a predetermined score chart [38]. If required, the precise quantification of the colorimetric signal is done by using cameras, scanners, commercial test strip readers, or hand-held colorimeters. The most recent is the cost-effective highly precise quantification of the colorimetric signal by CP-based image analysis [1,39]. The ongoing efforts focus on the development of PBIAs for multiplex analyte detection [40,41] and the fabrication of fully integrated LOC platforms with sample pretreatment and separation [42]. The PBIAs are discussed comprehensively in a separate chapter of this book.

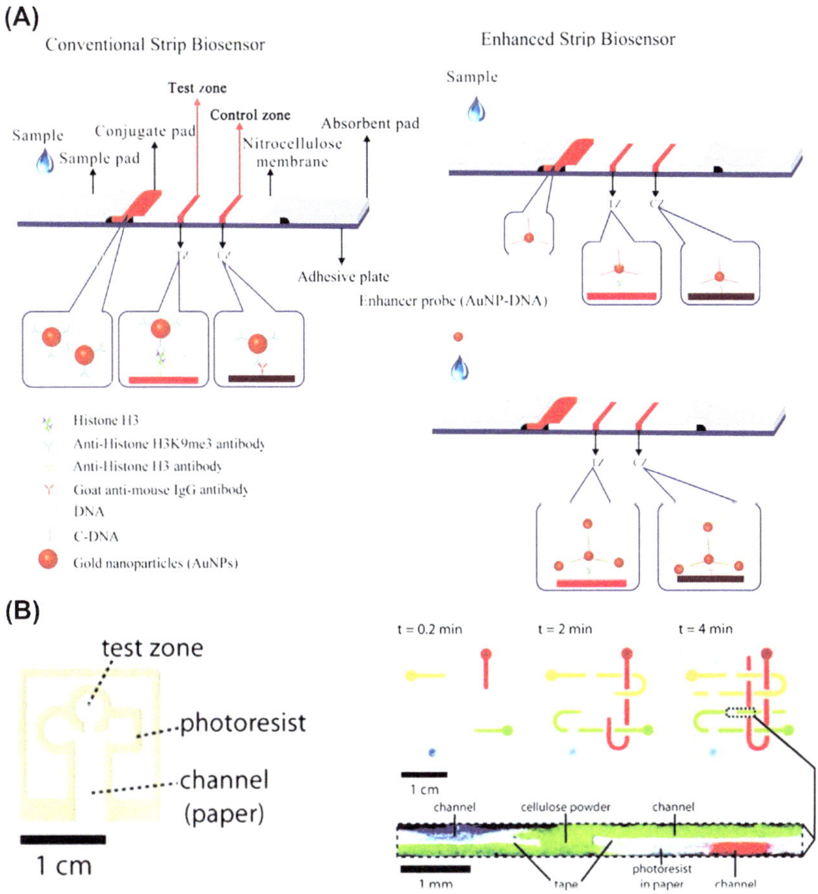

FIGURE 15.2 (A) Conventional and enhanced lateral flow immunoassay for the visual detection of trimethylated lysine 9 of histone H3 (H3K9me3) in 20 ng of histone extract from HeLa cells within 15 min [43]. (B) A typical 2D [23] (left) and 3D [24] (right) microfluidic paper-based analytical device (μPAD). *Reproduced with permission from the American Chemical Society.*

2.3 Cellphone

CP-based devices have been developed recently for the readout of IAs [44]. In brief, these devices use an optically opaque attachment equipped with light-emitting diodes (LEDs) and diffusers, a sample chamber [45–47], and apertures together with a plano-convex lens in front of the camera to adjust spatial resolution and the field of view [46]. The setup can be customized for fluorescent and colorimetric measurements [48]. A CP-based colorimetric reader for the colorimetric readout of a 96-well MTP uses the device's screen for

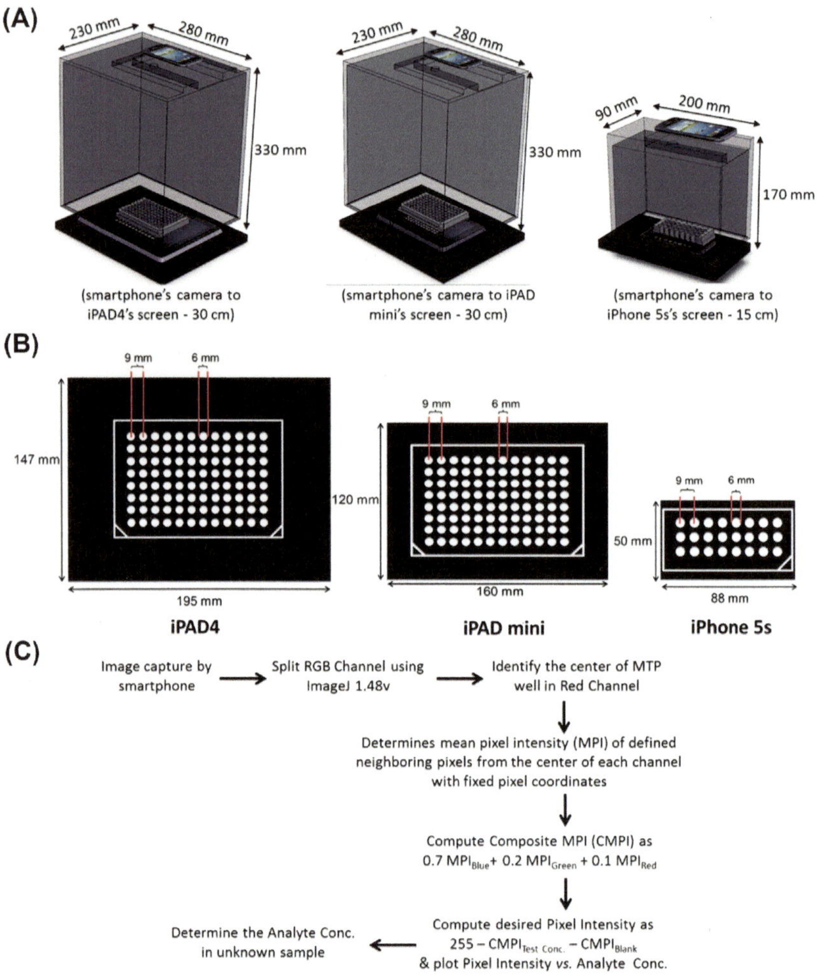

FIGURE 15.3 (A) Cellphone-based colorimetric readers developing using the gadgets' (iPad4, iPad mini, or iPhone 5s) screen-based bottom illumination, Samsung Galaxy SIII mini's back camera (5 MP)–based imaging, and a custom-made polyamide dark hood and polyamide base holder assembly. (B) Screensavers used for the screen-based bottom illumination of the 96-well microtiter plate in gadgets. (C) Image processing algorithm used for the analysis of smartphone-captured colorimetric images. *Reproduced with permission from Springer Nature, A.G. Venkatesh, et al., A smartphone-based colorimetric reader for human C-reactive protein immunoassay, Methods Mol. Biol. 1571 (2017) 343–356.*

bottom illumination of the MTP inside a dark hood and CP-based imaging [44,49] (Fig. 15.3). As a few examples, the system is tested for the highly sensitive detection of CRP [44]. A CP-based colorimetric reader using an optomechanical CP attachment can detect 1–25 part per million of peanut concentrations in commercial cookies with a limit of detection (LOD) of

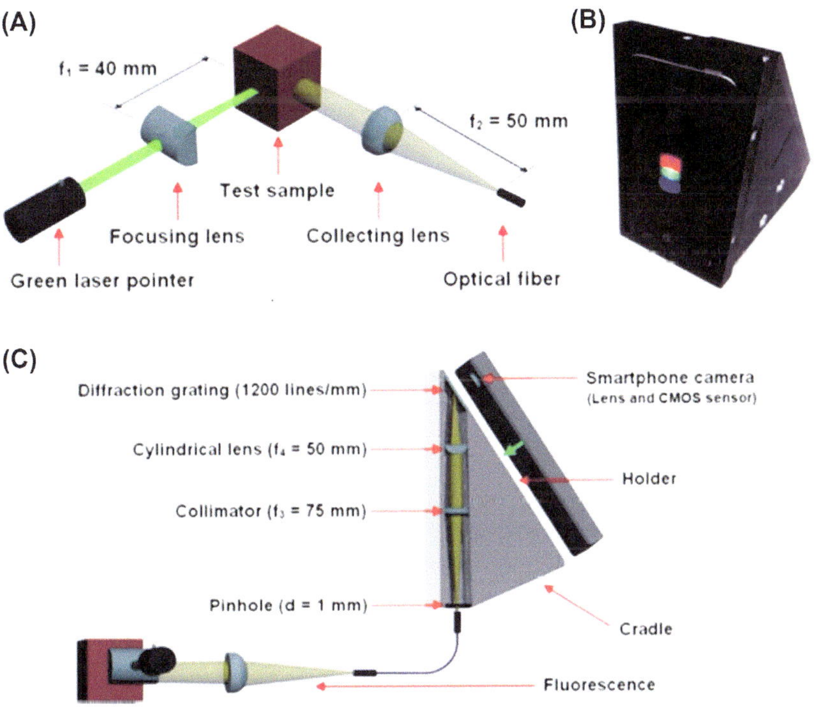

FIGURE 15.4 An illustration of a cellphone (CP)-based fluorimeter system. (A) Schematic illustrating the optical setup; (B) optical cradle installed on the CP; and (C) fluorescent detection scheme. *Reproduced with permission from the American Chemical Society, H. Yu, Y. Tan, B.T. Cunningham, Smartphone fluorescence spectroscopy, Anal. Chem. 86 (17) (2014) 8805–8813.*

~1 ppm in ~20 min [50]. Furthermore, a smartphone (SP)-based microchip ELISA detected HE4 cancer biomarker in urine [51]. Of interest is the development of a CP-based fluorimeter [52] (Fig. 15.4), which measures the full emission spectrum of any light emitter using a cradle and the CP's camera as a spectrometer for the readout of ELISAs [53].

A colorimetric detection–based cellphone-based IA (CPIA) is reported for 25-hydroxyvitamin D using an AuNP-based IA [54] and the total blood cholesterol [55]. Similarly, colorimetric detection–based CPIA using a quantum dot (QD)–based assay is used for determining the activity of trypsin, chymotrypsin, and enterokinase [48]. Biochemiluminescent detection–based CPIA using an LFIA minicartridge format is tested for total bile acid [56], total cholesterol [56], and salivary cortisol [57].

Commercial CPIAs are also available for disease biomarkers and clinically relevant analytes [1,2]. Cellmic, USA, offers a compact optomechanical attachment–based rapid diagnostic test reader for colorimetric and fluorometric LFIAs [39,58]. Its other product is a low-cost SP-based Holomic Rapid

Diagnostic Test Reader (HRDR-200) [59] for the readout of LFIAs. This system has an SP, an integrated reader housing, smart application, and access to secure Cloud Services and Test Developer. HRDR-300 is the recently developed fluorescent reader for the multicolor imaging–based readout of lateral flow assay (LFA) with high accuracy using an advanced image processing algorithm. The forthcoming HRDR-400 reads both chromatographic and fluorescent LFAs. Another application is based on a QD-based sandwich IA for the detection of *Escherichia coli* O157: H7 using anti–*E. coli* O157: H7 Ab-functionalized capillary array and CP-based detection [60] with a detection limit of ~5–10 CFU/mL in water and milk samples. Furthermore, a CP-based electrochemical sensing platform [61] is integrated with a disposable capillary flow MF chip. An angle-resolved surface plasmon resonance (SPR) system uses CP's screen-based illumination and detection of SPR signal by CP's camera [62]. It detects β_2 microglobulin, a biomarker for cancer, kidney disease, and inflammatory disorders, in humans with an LOD of ~0.1 μg/mL.

A CP-based LFIA (Fig. 15.5) relies on the illumination provided by SP's flash via an optical fiber and a collimating lens. The signal is detected at an optimized angle that maximizes the Mie scattering from a nitrocellulose membrane [17], resulting in maximal Rayleigh scattering from AuNPs in the LFIA bands. It detects thyroid-stimulating hormone (TSH) in 15 min with an LOD of 0.31 mIU/L, i.e., below the minimum accepted TSH concentration of 0.4 mIU/L for hyperthyroidism. GENTAG, Inc., USA, [63] together with MacroArray Technologies, LLC, USA, introduces a CP-based urine analysis platform and a disposable IA to detect prostate cancer diagnostic marker 1.

A novel compact CP-based electrochemical detection system [61] involves a small embedded circuit for signal processing and data analysis, a disposable SIM card-size MF chip for fluidic handling and sensing, a sample loading capillary, and processing and pumping (Fig. 15.6). For the detection of

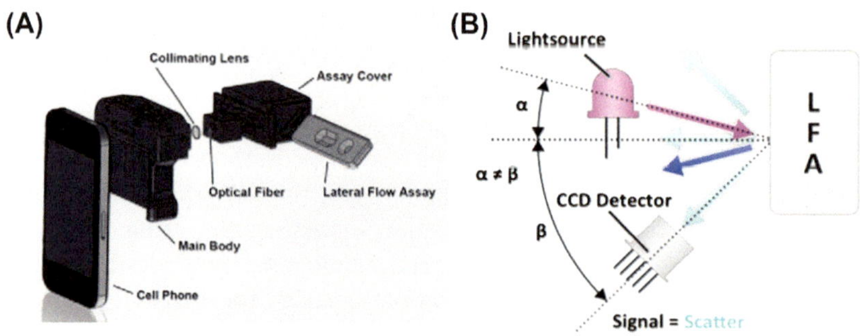

FIGURE 15.5 (A) Cellphone-based lateral flow immunoassay reader showing the placement of collimating lens and an optical fiber set at specific angles in reference to the LFA cassette [64]. (B) The detector is oriented at an angle, β, optimized for decreased light scattering from a nitrocellulose membrane, which differs from the angle of the light source to surface, α, providing an improved signal-to-noise ratio [64]. *Reproduced with permission from Elsevier B.V.*

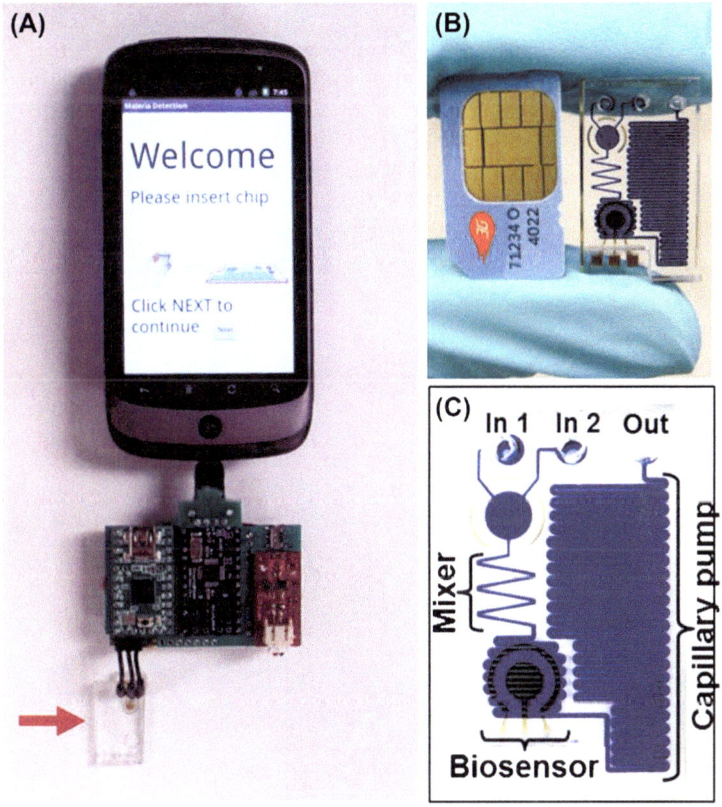

FIGURE 15.6 (A) A cellphone-based electrochemical sensor with the microfluidic (MF) chip [61]. (B) Comparison of the size of a MF chip with the CP SIM card. (C) Magnified view of a dye-filled MF chip with labeled components. *Reproduced with permission from the Royal Society of Chemistry.*

Plasmodium falciparum histidine-rich protein 2 (*Pf*HRP2), a smart application is included to provide step-by-step instructions on the screen.

The comprehensive description of various CPIAs is not provided as they are discussed in depth in another chapter of this book.

3. CRITIQUES AND OUTLOOK

Highly simplified and rapid IA procedures are the prerequisites for the development of rapid LOC-based IAs. Some such prospective IA formats include Perkin Elmer's AlphaLISA, Meso Scale Diagnostics LLC's electrochemiluminescent ELISA, and CP-based EasyELISA. The Perkin Elmer's AlphaLISA [65] is based on the interaction of streptavidin-coated Alpha donor beads with antianalyte Ab-conjugated AlphaLISA acceptor beads to form a sandwich immune complex in the presence of the analyte and the generation of chemiluminescence signal. Similarly, Meso Scale Diagnostics LLC's electrochemiluminescent ELISA [66] facilitates multiplex analyte detection in a microwell with high throughput. It enables highly sensitive wash-free sandwich IAs on carbon electrode surface in microwell plates using specific SULFO-TAG label–tagged detection Ab, which results in the emission of light on electrochemical stimulation. Of notice is one-step kinetics-based IA formats [67,68], which enable rapid and highly sensitive analyte detection in just 30 min. CP EasyELISA platforms have also been reported as rapid and highly simplified IAs with readout by CP's camera using a dark hood and a simple mechanical attachment [69].

The immobilization chemistries, surface modification and bioconjugation strategies, signal transduction or biosensing schemes, and signal enhancement strategies must be refined to develop analytically superior LOC IAs [70]. The ongoing efforts focus on robust new biorecognition elements, e.g., aptamers, antibody fragments, and molecular imprinted polymers. Such new biorecognition elements would improve the stability, sensitivity, and specificity of the LOC IAs and exhibit a higher affinity for the target analyte than the conventional Ab.

The prolonged storage of IA components is a key requirement for most LOC IAs. Various strategies involve stabilizers, freeze drying, sugars, sugar alcohols, reagent pouches, and encapsulating in natural polysaccharides [71]. Pullulan, a nonionic natural polysaccharide, forms an oxygen-tight solid tablet on drying, which enables the prolonged storage of biomolecules in their native 3D form at room temperature. Pullulan pellets are also water-soluble, and pullulan does not interfere with the specific IA after solubilization. Thus, the stored IA reagents are immediately released into aqueous solution [71]. The HRP-labeled Ab and its substrate (diaminobenzidine) can be stored for a long time by drying them onto a glass fiber pad [72]. Other strategies for LOC IAs encompass lyophilization and the use of trehalose, sugars with transition metal ions, and sugar alcohols with cationic polymers or zinc ions.

Despite the above achievements, the development of clinically viable LOC IAs still encounters technical challenges. Their analytical detection precision and reproducibility must be validated for a large number of patient samples without significant interference from interfering and cross-reacting substances [73,74]. The LOC IA must be corroborated with clinically accredited IA and other pending issues involve batch-to-batch variability and quality control for PBIA [75,76].

A fully integrated LOC platform for IAs requires the realization of MF operational steps for all process steps in an IA. Emerging nanomaterials and nanocomposites can greatly improve detection sensitivity and analytical performance of an IA. Hence, the physicochemical properties and behavior of these materials in different solutions and sample matrices must be sufficiently characterized.

About the use of CP-based devices and IAs for personalized healthcare monitoring and management [1,14,77], several challenges need to be tackled before they can be used routinely for clinical diagnosis. The foremost being the hygiene aspects, disposal, and contamination issues of CP-based devices. Among numerous applications based on CPs, only a selected few have been successfully commercialized, especially the personalized healthcare devices for general healthcare, diabetic, and weight management. Integrated personalized smart applications developed by iHealth Labs, Runtastic, Apple, Garmin, Samsung, etc. provide an overview of the general health status of the users [77]. Apparently, similar smart applications can be developed for CP-based LOC IAs and interfaced to personalized healthcare devices. However, the frequently changing technical specifications of CPs, particularly the specifications of the CP's camera, are limiting their ubiquitous applications and regulatory compliance. The emerging trend of creating modular CP designs by Google Inc. and developing stand-alone devices for reproducible imaging using fixed camera, which can be interfaced or attached to the CP, will mitigate such shortcomings [4].

The exponential growth of cloud computing has raised concerns about data security and ownership. Advanced encryption algorithms must secure the patient's personal health records. The need to establish an international cloud computing standard to harmonize the regulatory procedure has been addressed by the creation of EuroCloud and Google's Data Liberation Front [77].

The economic feasibility of POCT is another determining factor for the commercialization and wide-scale adoption of POCT devices. Testing requirements posed by healthcare organizations can also play a significant role in these decisions. Complex instrumentation involving multiple cartridges and multiplex operations require extensive and costly training as well as competency assessment, which neither practitioners nor healthcare organizations may be able to afford or are willing to undertake.

The critical role played by the health insurance industry in the adoption of POC IAs has been overlooked. It is important to involve the health insurance at a very early stage of LOC IA development to assess whether the IA could be reimbursed. This decision is made by the insurances based on multiple factors, especially the cost and utility of the LOC IA, ease of use, and improvement in healthcare monitoring and/or management. Economically viable LOC IA must be mass produced and readily adopted by the market. An ideal LOC IA must be simplified and avoid costly readers, complex instruments, and laborious IA procedures.

4. CONCLUSIONS

Numerous LOC platforms have emerged together with innovative complementary technologies for immunodiagnostic applications. The most prominent LOC platforms involve MF, rapid IA, automated operational procedure, integrated low-cost reader, interface with CP and gadgets, and prolonged storage of IA reagents. Although LOC IAs were initially intended for the developing countries with limited resources and lack of healthcare professionals, the emerging trend toward personalized mobile healthcare shows a universal need of such IAs. LOC IAs revolutionize the healthcare monitoring and management to provide an ideal solution for personalized and preventive healthcare using the emerging mobile health technologies and smart applications. The next-generation LOC IAs will make their way into healthcare to drastically reduce the healthcare costs. The continuous improvements in complementary technologies and new breakthroughs in IA concepts would lead to critically improved LOC IAs. The evolution of LOC IAs during the recent years has posed several new challenges in their implementation and use and requires pertinent research endeavors.

REFERENCES

[1] S.K. Vashist, et al., Cellphone-based devices for bioanalytical sciences, Anal. Bioanal. Chem. 406 (14) (2014) 3263–3277.

[2] A. Ozcan, Mobile phones democratize and cultivate next-generation imaging, diagnostics and measurement tools, Lab Chip 14 (17) (2014) 3187–3194.

[3] mHealth Solutions Market, 2017. http://www.marketsandmarkets.com/PressReleases/mhealth-apps-and-solutions.asp.

[4] P.B. Luppa, et al., Point-of-care testing (POCT): current techniques and future perspectives, Trends Anal. Chem. 30 (6) (2011) 887–898.

[5] Point-of-care Diagnostics Market, 2016. http://www.marketsandmarkets.com/PressReleases/point-of-care-diagnostic.asp.

[6] A.P. Turner, Biosensors: sense and sensibility, Chem. Soc. Rev. 42 (8) (2013) 3184–3196.

[7] C.D. Chin, V. Linder, S.K. Sia, Commercialization of microfluidic point-of-care diagnostic devices, Lab Chip 12 (12) (2012) 2118–2134.

[8] Piccolo Xpress, 2017. http://www.abaxis.com/medical/piccolo-xpress.

[9] Gyrolab CDs, 2017. http://www.gyros.com/products/gyrolab-cds/.

[10] Gyrolab XPlore, 2017. http://www.gyros.com/products/systems/gyrolab-xplore/.

[11] R. Gorkin, et al., Centrifugal microfluidics for biomedical applications, Lab Chip 10 (14) (2010) 1758–1773.

[12] G. Czilwik, et al., Magnetic chemiluminescent immunoassay for human C-reactive protein on the centrifugal microfluidics platform, RSC Adv. 5 (76) (2015) 61906–61912.

[13] J. Kai, et al., A novel microfluidic microplate as the next generation assay platform for enzyme linked immunoassays (ELISA), Lab Chip 12 (21) (2012) 4257–4262.

[14] S.K. Vashist, et al., Emerging technologies for next-generation point-of-care testing, Trends Biotechnol. 33 (11) (2015) 692–705.

[15] X. Mao, T.J. Huang, Microfluidic diagnostics for the developing world, Lab Chip 12 (8) (2012) 1412–1416.

[16] X. Li, D.R. Ballerini, W. Shen, A perspective on paper-based microfluidics: current status and future trends, Biomicrofluidics 6 (1) (2012) 11301–1130113.

[17] J. Hu, et al., Advances in paper-based point-of-care diagnostics, Biosens. Bioelectron. 54 (2014) 585–597.

[18] R. Pelton, Bioactive paper provides a low-cost platform for diagnostics, Trends Anal. Chem. 28 (8) (2009) 925–942.
[19] C. Fernandez-Sanchez, et al., One-step immunostrip test for the simultaneous detection of free and total prostate specific antigen in serum, J. Immunol. Methods 307 (1–2) (2005) 1–12.
[20] Q. Yang, et al., Quantum dot-based immunochromatography test strip for rapid, quantitative and sensitive detection of alpha fetoprotein, Biosens. Bioelectron. 30 (1) (2011) 145–150.
[21] G.F. van den Berk, et al., Evaluation of the rapid immunoassay determine HIV 1/2 for detection of antibodies to human immunodeficiency virus types 1 and 2, J. Clin. Microbiol. 41 (8) (2003) 3868–3869.
[22] A. Nilghaz, et al., Flexible microfluidic cloth-based analytical devices using a low-cost wax patterning technique, Lab Chip 12 (1) (2012) 209–218.
[23] A.W. Martinez, et al., Diagnostics for the developing world: microfluidic paper-based analytical devices, Anal. Chem. 82 (1) (2009) 3–10.
[24] A.W. Martinez, S.T. Phillips, G.M. Whitesides, Three-dimensional microfluidic devices fabricated in layered paper and tape, Proc. Natl. Acad. Sci. USA 105 (50) (2008) 19606–19611.
[25] G.G. Lewis, et al., High throughput method for prototyping three-dimensional, paper-based microfluidic devices, Lab Chip 12 (15) (2012) 2630–2633.
[26] K.M. Schilling, D. Jauregui, A.W. Martinez, Paper and toner three-dimensional fluidic devices: programming fluid flow to improve point-of-care diagnostics, Lab Chip 13 (4) (2013) 628–631.
[27] C.L. Cassano, Z.H. Fan, Laminated paper-based analytical devices (LPAD): fabrication, characterization, and assays, Microfluid. Nanofluid. 15 (2) (2013) 173–181.
[28] H. Liu, R.M. Crooks, Three-dimensional paper microfluidic devices assembled using the principles of origami, J. Am. Chem. Soc. 133 (44) (2011) 17564–17566.
[29] C.M. Cheng, et al., Paper-based ELISA, Angew. Chem. Int. Ed. Engl. 49 (28) (2010) 4771–4774.
[30] A. Apilux, et al., Development of automated paper-based devices for sequential multistep sandwich enzyme-linked immunosorbent assays using inkjet printing, Lab Chip 13 (1) (2013) 126–135.
[31] Z. Nie, et al., Integration of paper-based microfluidic devices with commercial electrochemical readers, Lab Chip 10 (22) (2010) 3163–3169.
[32] J. Lu, et al., Electrochemical DNA sensor based on three-dimensional folding paper device for specific and sensitive point-of-care testing, Electrochim. Acta 80 (2012) 334–341.
[33] C. Parolo, A. de la Escosura-Muniz, A. Merkoci, Enhanced lateral flow immunoassay using gold nanoparticles loaded with enzymes, Biosens. Bioelectron. 40 (1) (2013) 412–416.
[34] J. Hu, et al., Oligonucleotide-linked gold nanoparticle aggregates for enhanced sensitivity in lateral flow assays, Lab Chip 13 (22) (2013) 4352–4357.
[35] D.H. Choi, et al., A dual gold nanoparticle conjugate-based lateral flow assay (LFA) method for the analysis of troponin I, Biosens. Bioelectron. 25 (8) (2010) 1999–2002.
[36] Z. Qin, et al., Significantly improved analytical sensitivity of lateral flow immunoassays by using thermal contrast, Angew. Chem. 124 (18) (2012) 4434–4437.
[37] C. Parolo, et al., Simple paper architecture modifications lead to enhanced sensitivity in nanoparticle based lateral flow immunoassays, Lab Chip 13 (33) (2013) 386–390.
[38] M.A. Dineva, et al., Simultaneous visual detection of multiple viral amplicons by dipstick assay, J. Clin. Microbiol. 43 (8) (2005) 4015–4021.

[39] O. Mudanyali, et al., Integrated rapid-diagnostic-test reader platform on a cellphone, Lab Chip 12 (15) (2012) 2678–2686.

[40] S.J. Vella, et al., Measuring markers of liver function using a micropatterned paper device designed for blood from a fingerstick, Anal. Chem. 84 (6) (2012) 2883–2891.

[41] N.R. Pollock, et al., A paper-based multiplexed transaminase test for low-cost, point-of-care liver function testing, Sci. Transl. Med. 4 (152) (2012) 152ra129.

[42] X. Yang, et al., Integrated separation of blood plasma from whole blood for microfluidic paper-based analytical devices, Lab Chip 12 (2) (2012) 274–280.

[43] C. Ge, et al., An enhanced strip biosensor for rapid and sensitive detection of histone methylation, Anal. Chem. 85 (19) (2013) 9343–9349.

[44] S.K. Vashist, et al., A smartphone-based colorimetric reader for bioanalytical applications using the screen-based bottom illumination provided by gadgets, Biosens. Bioelectron. 67 (2015) 248–255.

[45] A.F. Coskun, et al., Albumin testing in urine using a smart-phone, Lab Chip 13 (21) (2013) 4231–4238.

[46] Q. Wei, et al., Detection and spatial mapping of mercury contamination in water samples using a smart-phone, ACS Nano 8 (2) (2014) 1121–1129.

[47] K. Su, et al., High-sensitive and high-efficient biochemical analysis method using a bionic electronic eye in combination with a smartphone-based colorimetric reader system, Sens. Actuators B Chem. 216 (2015) 134–140.

[48] E. Petryayeva, W.R. Algar, Multiplexed homogeneous assays of proteolytic activity using a smartphone and quantum dots, Anal. Chem. 86 (6) (2014) 3195–3202.

[49] A.G. Venkatesh, et al., A smartphone-based colorimetric reader for human C-reactive protein immunoassay, Methods Mol. Biol. 1571 (2017) 343–356.

[50] A.F. Coskun, et al., A personalized food allergen testing platform on a cellphone, Lab Chip 13 (4) (2013) 636–640.

[51] S. Wang, et al., Integration of cell phone imaging with microchip ELISA to detect ovarian cancer HE4 biomarker in urine at the point-of-care, Lab Chip 11 (20) (2011) 3411–3418.

[52] H. Yu, Y. Tan, B.T. Cunningham, Smartphone fluorescence spectroscopy, Anal. Chem. 86 (17) (2014) 8805–8813.

[53] K.D. Long, H. Yu, B.T. Cunningham, Smartphone instrument for portable enzyme-linked immunosorbent assays, Biomed. Opt. Express 5 (11) (2014) 3792–3806.

[54] S. Lee, et al., A smartphone platform for the quantification of vitamin D levels, Lab Chip 14 (8) (2014) 1437–1442.

[55] V. Oncescu, M. Mancuso, D. Erickson, Cholesterol testing on a smartphone, Lab Chip 14 (4) (2014) 759–763.

[56] A. Roda, et al., Integrating biochemiluminescence detection on smartphones: mobile chemistry platform for point-of-need analysis, Anal. Chem. 86 (15) (2014) 7299–7304.

[57] M. Zangheri, et al., A simple and compact smartphone accessory for quantitative chemiluminescence-based lateral flow immunoassay for salivary cortisol detection, Biosens. Bioelectron. 64 (2015) 63–68.

[58] Rapid Assay Reader, 2017. http://www.cellmic.com/content/rapid-test-readers/.

[59] W. Liu, et al., Nanoparticle coated paper-based chemiluminescence device for the determination of L-cysteine, Talanta 120 (2014) 336–341.

[60] H. Zhu, U. Sikora, A. Ozcan, Quantum dot enabled detection of *Escherichia coli* using a cell-phone, Analyst 137 (11) (2012) 2541–2544.

[61] P.B. Lillehoj, et al., Rapid electrochemical detection on a mobile phone, Lab Chip 13 (15) (2013) 2950–2955.

[62] P. Preechaburana, et al., Surface plasmon resonance chemical sensing on cell phones, Angew. Chem. Int. Ed. Engl. 51 (46) (2012) 11585−11588.

[63] D.J. Litman, et al., An internally referenced test strip immunoassay for morphine, Clin. Chem. 29 (9) (1983) 1598−1603.

[64] D.J. You, T.S. Park, J.Y. Yoon, Cell-phone-based measurement of TSH using Mie scatter optimized lateral flow assays, Biosens. Bioelectron. 40 (1) (2013) 180−185.

[65] L. Beaudet, et al., AlphaLISA immunoassays: the no-wash alternative to ELISAs for research and drug discovery, Nat. Methods 5 (12) (2008).

[66] MSD Technology Platform, 2017. https://www.mesoscale.com/~/media/files/brochures/techbrochure.pdf.

[67] S.K. Vashist, et al., One-step kinetics-based immunoassay for the highly sensitive detection of C-reactive protein in less than 30min, Anal. Biochem. 456 (2014) 32−37.

[68] S.K. Vashist, et al., Graphene-based rapid and highly-sensitive immunoassay for C-reactive protein using a smartphone-based colorimetric reader, Biosens. Bioelectron. 66 (0) (2015) 169−176.

[69] S.K. Vashist, G. Czilwik, A.G. Alagarswamy, Elisa System and Related Methods, 2014. WIPO Patent Pub No WO/2014/198836.

[70] S.K. Vashist, et al., Immobilization of antibodies and enzymes on 3-aminopropyltriethoxysilane-functionalized bioanalytical platforms for biosensors and diagnostics, Chem. Rev. 114 (21) (2014) 11083−11130.

[71] S. Jahanshahi-Anbuhi, et al., Pullulan encapsulation of labile biomolecules to give stable bioassay tablets, Angew. Chem. Int. Ed. Engl. 53 (24) (2014) 6155−6158.

[72] S. Ramachandran, et al., Long-term dry storage of an enzyme-based reagent system for ELISA in point-of-care devices, Analyst 139 (6) (2014) 1456−1462.

[73] Guidance for Industry − Bioanalytical Method Validation, 2001. https://www.fda.gov/downloads/Drugs/Guidances/ucm070107.pdf.

[74] Guideline on Bioanalytical Method Validation, 2011. http://www.ema.europa.eu/docs/en_GB/document_library/Scientific_guideline/2011/08/WC500109686.pdf.

[75] K. Abe, et al., Inkjet-printed paperfluidic immuno-chemical sensing device, Anal. Bioanal. Chem. 398 (2) (2010) 885−893.

[76] C.Z. Li, et al., Paper based point-of-care testing disc for multiplex whole cell bacteria analysis, Biosens. Bioelectron. 26 (11) (2011) 4342−4348.

[77] S.K. Vashist, E.M. Schneider, J.H.T. Luong, Commercial smartphone-based devices and smart applications for personalized healthcare monitoring and management, Diagnostics 4 (3) (2014) 104−128.

Chapter 16

Smartphone-Based Immunoassays

Sandeep K. Vashist[1], John H.T. Luong[2]
[1]*IDS Immunodiagnostic Systems Deutschland GmbH, Frankfurt am Main, Germany;* [2]*University College Cork, Cork, Ireland*

1. INTRODUCTION

Smartphones (SPs) have witnessed continuous improvements during 2000−17. They have become omnipresent as evinced by about 7 billion cell phone subscribers, i.e., 94% of the world population [1] with over 70% from developing countries. The current generation of SPs is equipped with sophisticated features, advanced processors, increased memory, enhanced resolution camera, high-end security, personalization, numerous smart applications, and a variety of built-in sensors. As a result of all these improvements, they have become very user-friendly. SPs have been transformed into a personalized in vitro diagnostic (IVD) system that has the capability of delivering mobile healthcare (mH) to remote, resource-deficient, private, and public settings [2−5]. Individuals can perform several diagnostic tests at their personalized settings and receive the feedback from healthcare professionals in real time. As the diagnostic results gathered by the SP device are tagged with spatiotemporal information, the monitoring and management of epidemics and emergency cases are greatly facilitated [6]. Significant advances have been made to improve signal response, robustness, and cost-effectiveness of SP-based immunoassays (SPIAs) with increasing efforts being devoted to applications in healthcare and bioanalytical sciences [7].

There are substantial efforts in the development of SPIA based on lateral flow assays (LFAs) [8−12], microscopy [13−17], electrochemical sensing [18−21], immunoassays (IAs) [22−25], surface plasmon resonance (SPR)-based biosensing [26,27], flow cytometry [15,28], and optical detection [29−32]. Several companies have emerged and most of the healthcare and diagnostic giants and SP manufacturers have also participated in this competitive and emerging field. Consequently, there are numerous commercially available SP-based devices for general healthcare and fitness, which are being used by millions of customers worldwide [5]. These devices are manufactured by iHealth, AliveCor,

Handbook of Immunoassay Technologies. https://doi.org/10.1016/B978-0-12-811762-0.00016-5

GENTAG, Mobile Assay, CellScope, Holomic, Runtastic, Nonin, etc. Therefore, the coming SP-based mH era will revolutionize personalized healthcare monitoring and management via highly efficient and cost-effective SPIA. This chapter provides a critical overview of SPIA together with technical challenges, trends, and future possibilities.

2. SIGNAL DETECTION IN SMARTPHONE-BASED IMMUNOASSAYS

The SP's high-resolution rear camera, ranging from a few megapixels (MP) to 41 MP, can be transformed to a spectrophotometer or microscope using an interface with necessary optical components and a sample compartment. The reusable or disposable interface links the bioassay to the SP and holds the camera in alignment with the optical components. The light source is usually an internal light emitting diode (LED), the phone flash [33], or the SP's illuminated screen [30,31].

2.1 Colorimetric, Fluorescent, and Luminescent Detection

Colorimetric detection—based SPIAs use an optically opaque attachment that consists of LEDs and diffusers to provide uniform illumination, apertures to provide desired imaging field of view (FOV), and a sample chamber [23,34] (Fig. 16.1A and B). The FOV is enhanced by placing a plano-convex lens in front of the camera [34]. This setup equipped with a monochromatic filter is extended for imaging the fluorescent signal of a labeled dye [35]. The ultraviolet (UV) LEDs or UV transilluminators are also needed for the desired excitation wavelength, e.g., for quantum dots (QDs) [24] or EvaGreen, a dye used in the polymerase chain reaction (PCR) [36]. For luminescent detection, the measurement is simply performed in the dark [37] using a plano-convex lens in front of the camera to discriminate similar signals [38] (Fig. 16.1A and C). In photoluminescence-based SPIAs using QDs and Förster resonance energy transfer, the setup requires a UV lamp (365 nm) for excitation, a long-pass filter (400 nm cutoff), and a short-pass filter (650 nm cutoff), inserted in front of the SP camera [39]. For multisample imaging, a microtiter plate (MTP) with 96 wells is placed on the gadget's screen saver to provide uniform bottom illumination inside a dark hood. The SP is simply placed on top of the hood in a containment with an aperture in alignment with the SP's back camera [31] (Fig. 16.1D).

An SP colorimetric reader—based food allergen testing platform "iTube" [5] was developed. It contains a cost-effective, compact, and lightweight (~40 g) optomechanical attachment that comprises an inexpensive plano-convex lens, two diffused LEDs for vertical illumination, sample tubes, and circular apertures for spatial control of the imaging FOV. Allergen test and control assays absorb the incident LED light and alter the intensity of the transmitted

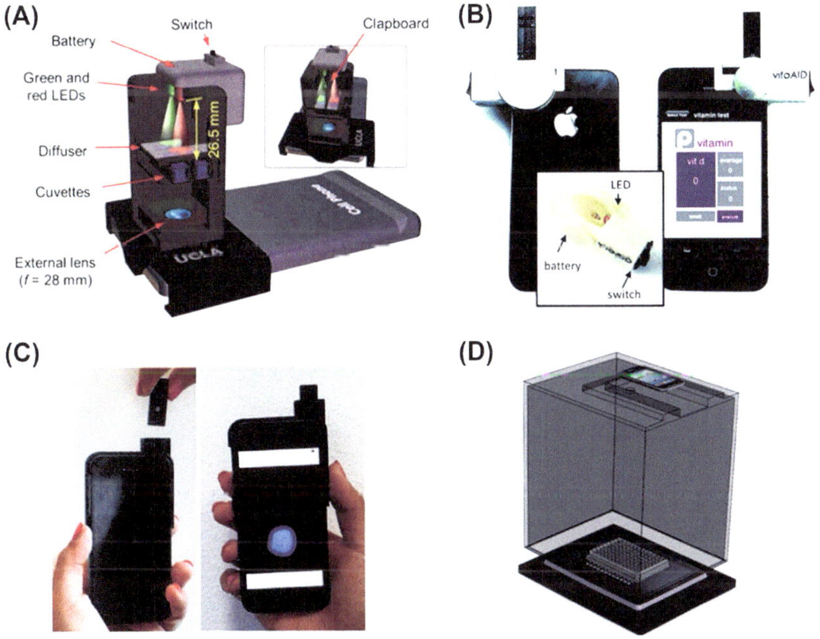

FIGURE 16.1 Smartphone (SP)-based devices for colorimetric SP-based immunoassay. (A) 3D schematic for colorimetric detection of Hg^{2+} [34]. (B) vitaAID accessory on an iPhone for colorimetric detection of vitamin D [40]. (C) SP accessory and minicartridge for the biochemiluminescent detection of cholesterol [38]. (D) An SP-based colorimetric reader based on the screen-based bottom illumination provided by gadgets [31]. *Reproduced with permission from Elsevier B.V.*

light, which is recorded by the SP camera. A smart application instantaneously processes the raw camera images and provides the digitally quantified allergen concentration in food samples.

An SP fluorometer for the measurement of a full emission spectrum of any light emitter [41] uses a cradle that incorporates a diffraction grating, cylindrical lens, and a collimator (Fig. 16.2). The transmission diffraction grating is placed in front of the camera with optics integrated to collimate the light emitted from the sample onto the grating. The emission spectrum is measured by a complementary metal oxide semiconductor (CMOS) image sensor.

2.2 Electrochemical Detection

A compact SP-based electrochemical (EC) detection system [18] is based on a small embedded circuit for signal processing and data analysis using a disposable SIM card–size microfluidic chip for fluidic handling and sensing; a sample loading capillary; and processing and pumping (Fig. 16.3). The sample is loaded

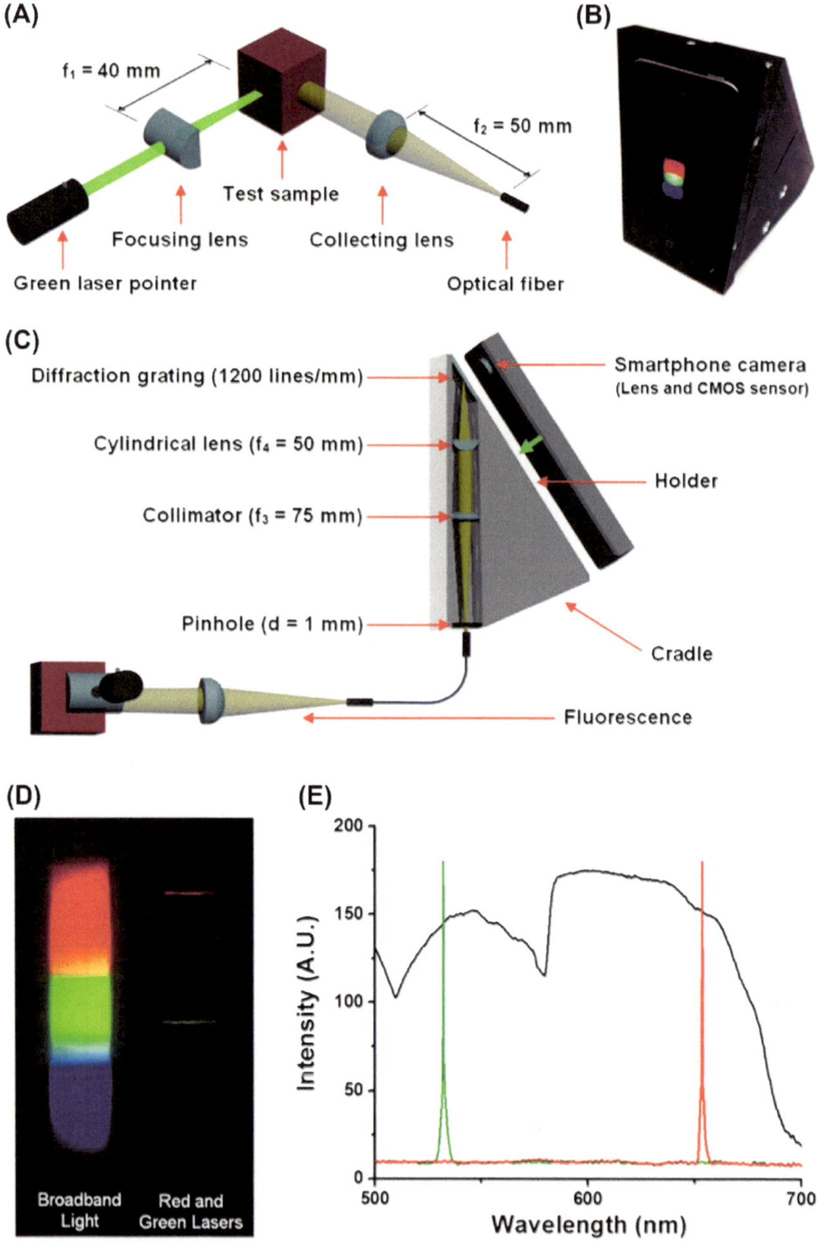

FIGURE 16.2 A smartphone (SP) fluorometer [41]. (A) The excitation laser (532 nm) illuminates a cuvette containing the probe–target duplexes. Fluorophores linked to one end of the molecular beacon probes are excited to emit fluorescence in all directions. A portion of the emission is gathered by using a collecting lens oriented perpendicular to the laser. The emitted light enters the cradle, where it is further collimated before incidence on a diffraction grating placed in front of the camera. (B) An optical cradle installed on the SP. The screen of the SP shows a spectrum of the transmitted light dispersed by the grating. (C) Illustration of the SP fluorometer system. (D) Images of broadband light, red and green laser beams. (E) Light intensity distributions of two lasers (*solid green line* with $\lambda_{green} = 532$ nm and *solid red line* with $\lambda_{red} = 653$ nm) and a broadband light source (*solid black line*) measured by the SP fluorometer. *Reproduced with permission from the American Chemical Society.*

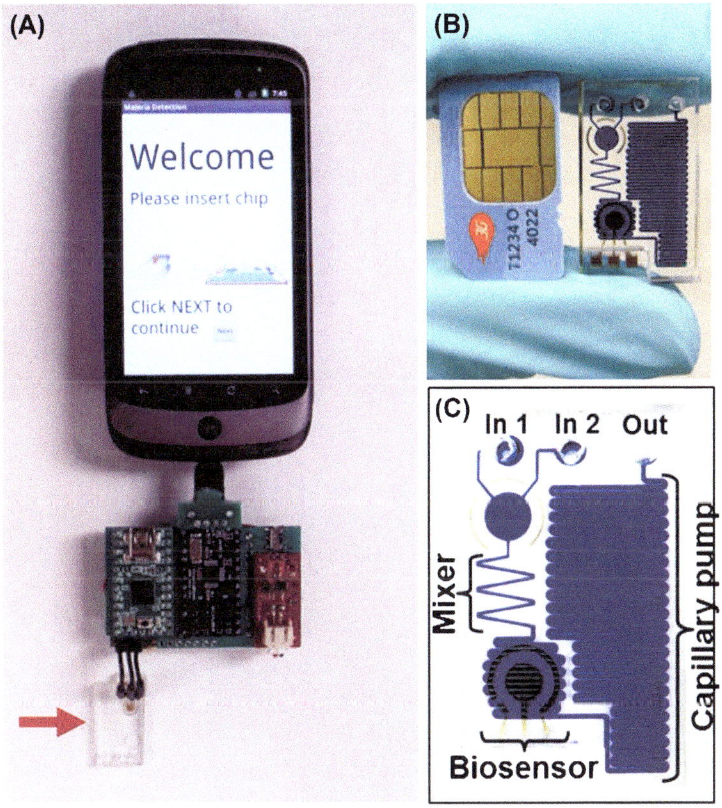

FIGURE 16.3 (A) Smartphone-based electrochemical sensor with the *arrow* indicating the microfluidic chip [18]. (B) Comparison of the size of microfluidic chip with the smartphone SIM card. (C) Magnified view of a dye filled microfluidic chip with labeled components. *Reproduced with permission from the Royal Society of Chemistry.*

by a pipette but subsequently driven by the flow capillary. The microfluidic chip (measuring 25 mm × 15 mm) has a modified polydimethylsiloxane (PDMS) surface coating that enables an autonomous capillary-driven flow with a constant flow rate. It has two inlets and an outlet (1 mm in diameter) and a 2-mm chamber downstream to merge the appropriate volumes of the sample (0.5 μL) and reporter (2 μL) solutions before their entrance to a zigzag mixer. Thereafter, the serpentine channels with a spacing of 40 μm enable a uniform flow of a sample–reporter mixture over the EC sensor. A secondary serpentine channel (25 cm in length, 400 μm in width, and 100 μm in height) acts as a capillary pump and a waste reservoir. It accommodates the tetramethylbenzidine/hydrogen peroxide (TMB/H_2O_2) substrate for washing and the enzymatic reaction for subsequent amperometric measurement. *Plasmodium falciparum* histidine–rich protein 2 (*Pf*HRP2) capture antibodies are uniformly bound to the polypyrrole-coated electrode surface. In the absence of *Pf*HRP2, the antibodies cannot bind to the electrode and are washed away when the TMB/H_2O_2 substrate is loaded. In the presence of *Pf*HRP2, the detection antibodies bind to the electrode and result in EC current, which is proportional to the *Pf*HRP2 concentration when the electrode is poised at a fixed potential. The user performs the measurement through a smart application that provides step-by-step instructions on the screen. After each measurement, the results are displayed on the SP's screen and stored in its memory for subsequent data transmission and analysis.

SP-based blood glucose meters deserve a special mention here. iHealth has developed Wireless Smart Gluco-Monitoring System and iHealth Align (Fig. 16.4), which are Food and Drug Administration (FDA) approved, Conformité Européenne (CE) compliant, and meet the ISO 15197:2003 in vitro blood glucose monitoring requirements. These blood glucose meters are operated by iHealth Gluco, a smart application available for free at iOS and Android stores, which requires users to log in using their specific details. The iHealth Align, the smallest glucose meter available commercially, plugs into the headphone jack of the SP and is connected to the SP by Bluetooth technology. The Wireless Smart Gluco-Monitoring System is a miniaturized stand-alone blood glucose meter that can be purchased online for US$ 29.95. It is interfaced wirelessly with the SP via the Bluetooth technology and the smart app. It comes with an LED display and is powered by a built-in, rechargeable battery.

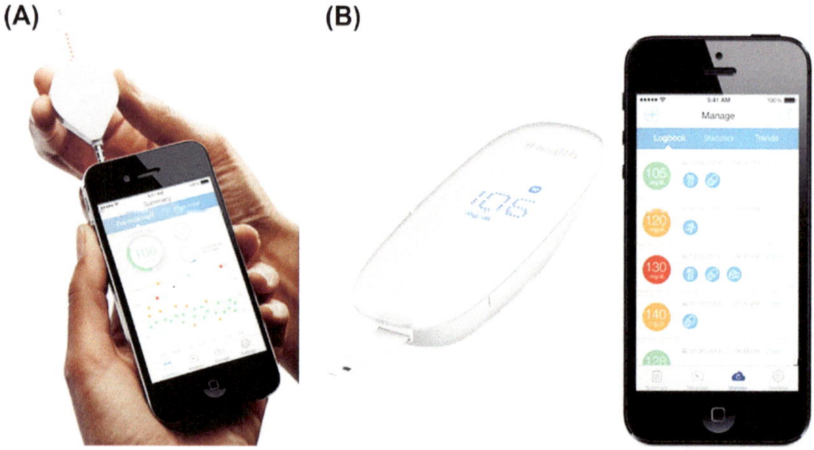

FIGURE 16.4 Smartphone-based blood glucose meters developed by iHealth Inc. (A) iHealth Align. (B) Wireless Smart Gluco-Monitoring System. *Reproduced with permission from iHealth Labs, Inc.*

2.3 Surface Plasmon Resonance Detection

In an angle-resolved SPR-based SPIA [26], a disposable SPR coupler is attached to the SP screen. The SPR coupler is a single disposable device, which employs the conditioned illumination provided by the cell phone's screen and the cell phone's front camera–based optical detection of the SPR signal. The procedure employs the gentle adherence of the SPR coupler to the cell phone's screen to utilize the controlled illumination provided by the cell phone display, which provides the wide-angle illumination for angle-resolved SPR. The angle-resolved SPR signals deflected by total internal deflection are recorded by the front camera (video graphics array [VGA] resolution of ~0.31 or 2 MP with intensity measurements limited to 256 levels/channel). An image of a red rectangle displayed on the cell phone's screen provides the illumination while the placement of SPR coupler was outlined by an image of a white frame displayed on the cell phone. The SPR coupler, made up of PDMS prism and epoxy, has a refractive index of 1.5 matching that of glass. The epoxy layer achieves the desired illumination angles on the gold (Au)/water interface under red illumination. The angle-resolved SPR signal in the light reflected from the Au interface was conditioned by a PDMS plano-cylindric element and then deflected by total internal reflection within the PDMS prism. The sensing operation is controlled by a custom-developed software, which sets the exposure of the camera to assure that the SPR signals are measured in identical/repeatable conditions. The custom-made SPR chip has two calibration channels for high and low concentrations and a test channel. An iOS-based smart application is employed for image acquisition to control the exposure time and ISO number of camera's exposure so that the signals are measured under identical conditions.

2.4 Lateral Flow Assays

An SP-based lateral flow IA (LFIA) format (Fig. 16.5A) is based on the signal detection at an optimized angle to maximize Mie scattering from a nitrocellulose membrane [18], resulting in maximal Rayleigh scattering from gold nanoparticles (AuNPs) in the LFIA bands (Fig. 16.5B). The LFIA is illuminated by the SP embedded flash via an optical fiber and a collimating lens.

Cellmic, formerly Holomic LLC, developed the SP-based Holomic Rapid Diagnostic Test Reader (HRDR-200) [42] (Fig. 16.6A) for the readout of LFIA with high sensitivity and precision. HRDR-200 is equipped with an SP, an integrated reader housing, smart application, and access to secure Cloud Services and Test Developer. It is lightweight, handheld, economical, compliant with ISO 13485, and registered with FDA as a Class 1 medical device. The reader enables real-time diagnostic data integration via a secure Cloud Service and integration with electronic health records (EHRs),

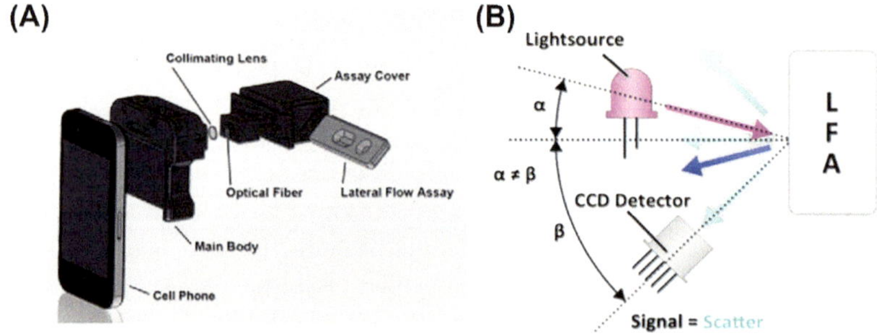

FIGURE 16.5 (A) An exploded view of smartphone-based lateral flow assay (LFA) reader showing the placement of the collimating lens and an optical fiber set at specific angles in reference to the LFA cassette [8]. (B) The detector is oriented at an angle, β, that is optimized for decreased light scattering from a nitrocellulose membrane and differs from the angle of the light source to surface, α, which provides an improved signal-to-noise ratio [8]. *(B) Reproduced with permission from Elsevier B.V.*

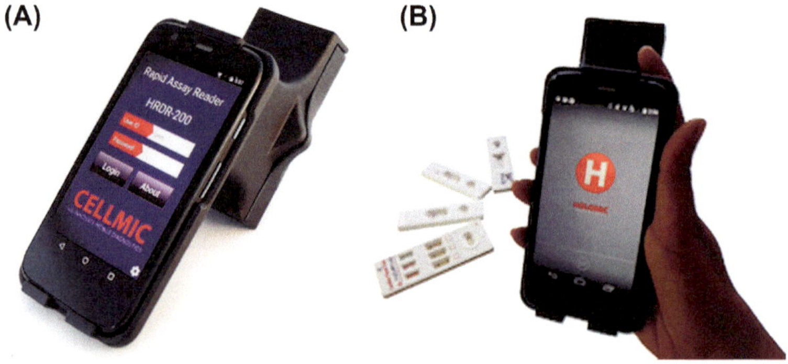

FIGURE 16.6 Smartphone-based diagnostic readers developed by Cellmic. (A) Holomic Rapid Diagnostic Test Reader (HRDR-200). (B) HRDR-300. *Reproduced with permission from Cellmic.*

laboratory information system, and hospital information system. HRDR-200 is a universal SP reader that can read chromatographic and fluorescent assays in lateral flow, flow-through, and dipstick test formats with high precision. It can also read tests of various sizes or formats, including strips, cassettes, and multiplexed tests. It is used either stand-alone or connected to a network. The device can identify the test, lot number, and patient data as it is equipped with QR code scanner for automatic entry of test and patient data and analyzes the lot expiry. The data analysis is performed on a PC with Test Explorer Software while the test results can be printed, e-mailed, or texted. It is connected to laptops, PCs, printers, routers, and EHRs via wireless, Bluetooth, and USB. The device prototype and its working have been demonstrated elsewhere [9].

HRDR-300 (Fig. 16.6B) is the recently developed fluorescent reader for the multicolor imaging−based readout of the control and test lines of LFA with high accuracy using an advanced image processing algorithm. The users can select the excitation and emission in the regions between 350−700 nm and 300−800 nm, respectively. The device can also be customized according to the needs of the multiplexed fluorescent assay, such as the requirement of multiple excitation sources. All other features are the same as those of chromatographic HRDR-200. The device is also ISO 13485 compliant and registered with the FDA as a Class I medical device. The company is working on the development of HRDR-400, a dual reader for chromatographic and fluorescent rapid tests in a single device. Holomic Substance Test Assistant (HSTA) is a handheld, economical, and high-performance tool, which acts as a single chromatographic and fluorescent viewer for alcohol and substance abuse tests of different sizes or formats, including saliva tests. All the device features are similar to those of HRDR-200.

The rapid diagnostic test (RDT) reader prototype [9] used for HRDR-200 comprises a plano-convex lens, a microcontroller, and three LED arrays (two located underneath the RDT tray for reflection imaging and one at the top for transmission imaging). It analyzes various types of LFIAs and RDTs in reflection or transmission imaging modes under diffused LED illumination and provides a quantitative determination of analytes. The device can run on external batteries or the SP battery via a USB connection. The highly sensitive RDT reader can detect even the minor color signal variation that cannot be observed by visualization. A smart application processes the raw images taken by the SP's camera within <0.2 s and shares the results with a central server also (in addition to storing the information on the SP), which can be assessed by a remote computer using web browsers. The software application also provides a dynamic spatiotemporal map and real-time statistics for various diseases that can be diagnosed by RDTs, thereby enabling healthcare professionals and policy makers to monitor, track, and analyze emerging diseases and outbreaks.

Mobile Assay Inc. provides a cost-effective cell phone−based rapid lateral flow test reader together with Instantaneous Analysis software. It employs Mobile Image Ratiometry (MIR) to quantify the rapid lateral flow test strips in 5−10 min. The quantification is done by image analysis of dye signal on the test strips, where the MIR subtracts background noise, selects the signal bands, and provides their pixel density ratio. The test strip immunoassay involves the dispensing of sample at the specified area on the test strip. The capillary action then takes the liquid sample together with the dye-conjugated antibodies, which are specific for the analyte to be detected, across the test and control areas. In the absence of analyte in the sample, two bands appear as the dye will bind at both the test and the control areas. However, in the presence of analyte, the relative intensity of test band increases directly with the concentration of the analyte. The intensity of the test band determined by MIR is proportional to the analyte concentration. The Data Collection Module enables real-time data transfer to the Mobile Assay Cloud that can be accessed locally through Tracker or remotely through Mobile Assay Cloud. Text alerts can then be pushed to the clients using the Tracker Manager.

2.5 Microscopy

Fig. 16.7A and B shows a ~28 g SP attachment for the wide-field fluorescent and dark field imaging [43]. The device employs battery-powered LEDs, a

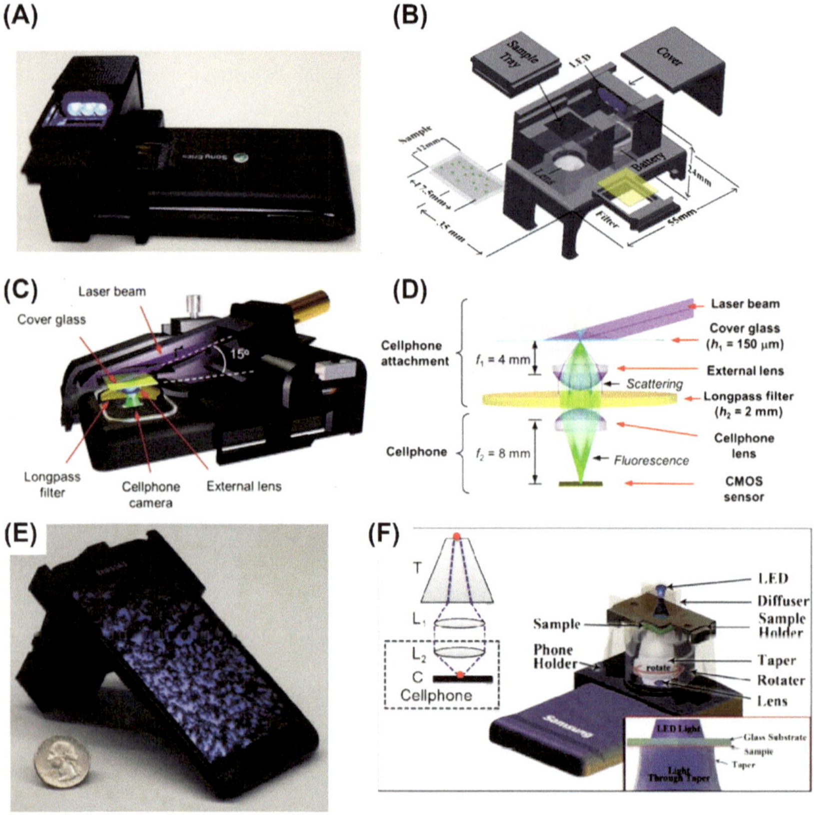

FIGURE 16.7 (A) Compact and lightweight (~28 g) smartphone attachment for wide-field fluorescent and dark field imaging, and (B) its schematic [43]. (C) Schematic of the smartphone-based fluorescence microscope and (D) ray-tracing diagram: excitation and scattered beams (*solid blue rays*) and the fluorescent emission (*solid green rays*) [44]. (E) smartphone-based contact microscopy platform and (F) schematic illustration of the smartphone attachment [45]. *(B) Reproduced with permission from The Royal Society of Chemistry. (D) Reproduced with permission from the American Chemical Society. (F) Reproduced with permission from The Royal Society of Chemistry.*

plastic color filter to create a dark background, and an additional lens in front of the camera to image the fluorescent emission. The dark field imaging is acquired using the same setup but without the filter. The development of a portable (~186 g) SP attachment for fluorescence microscopy, which enables imaging of 100 nm nanoparticles and individual human cytomegaloviruses, is also noticed [44] (Fig. 16.7C and D). It consists of a compact laser-diode (450 nm) and an optical setup with a long-pass interference filter, a low numerical aperture lens, and a coarse mechanical translation stage (Fig. 16.7D). Highly dense samples in the transmission mode are imaged [45] by a contact microscopic platform (76 g) using a glass-based tapered fiber array illuminated by an LED (Fig. 16.7E and F). The sample transmission image is delivered from the taper's top-to-bottom with 3× magnification in each direction. The image is projected on the CMOS sensor of the SP via two lenses. A "shift and add" algorithm is used for the digital fusion of multiframe images to improve resolution and obviate spatial artifacts.

2.6 Cytometry

An SP-based optofluidic fluorescent imaging flow cytometer, in the form of a compact, lightweight (~18 g), and cost-effective SP attachment, has been developed for analyzing large sample volumes of >0.1 mL [14]. It employs an inexpensive lens, a plastic color filter (cost <0.1 USD), two LEDs (cost ~ 0.6 USD), and coin cell batteries (cost ~0.5 USD). Using a syringe pump, the sample is continuously delivered to the imaging volume through a disposable microfluidic channel, which acts as a multilayered optofluidic waveguide (Fig. 16.8). The dark-field background is created by using an inexpensive plastic absorption filter as the guided excitation light propagates perpendicular to the detection path. The cell movement through the microfluidic channel is recorded and processed to determine the cell count and density. With a fluorescence resolution of ~2 μm, this platform can be used for whole blood analysis and screening of waterborne parasites in drinking waters. Subsequently, the same group developed a miniaturized and low-cost SP imaging–based cytometry platform along with a smart application to measure the density of RBCs and WBCs that are imaged in bright-field and fluorescent modes. This platform can also simultaneously determine the hemoglobin (Hb) concentration in human blood samples [13] (Fig. 16.9). The system for rapid blood analysis by using only ~10 μL of the sample provides a satisfactory analytical performance, as validated using a standard benchtop Sysmex KN21 hematology analyzer. Its universal base is attached to the cell phone camera unit and three "add-on" components for WBC/RBC counting and Hb density measurement. The desired "add-on" components can be clicked into the same port on the base attachment to perform the specific blood test via the smart application, running on the same cell phone for image analysis and cell counting.

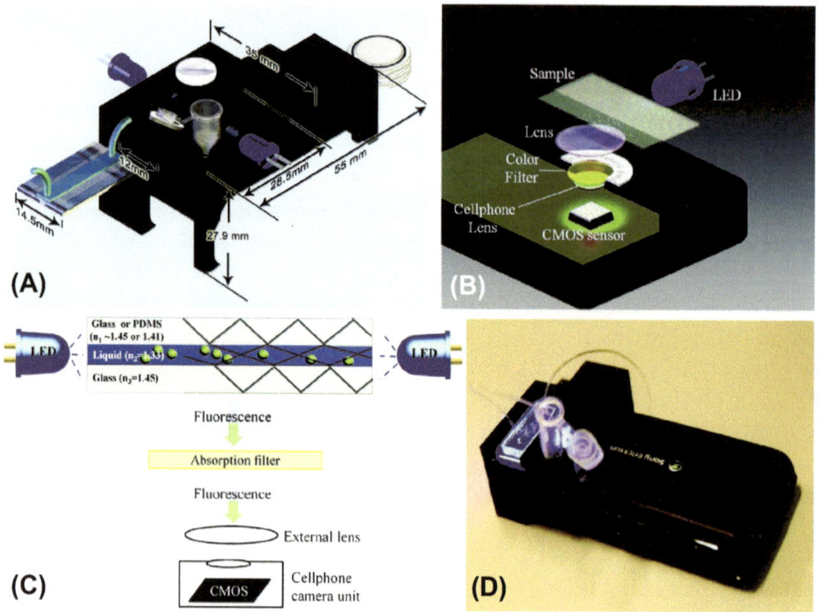

FIGURE 16.8 Smartphone-based flow cytometer [28]. (A–C) Schematics of the lightweight (~18 g) and compact (~35 × 55 × 27.9 mm^3) optical attachment for optofluidic fluorescent imaging cytometry on a cell phone, which can be repeatedly attached or detached to the cellphone body without the need for fine alignment. (D) Actual photograph of optofluidic fluorescent imaging cytometer on a cell phone. *Reproduced with permission from the American Chemical Society.*

3. APPLICATIONS OF SMARTPHONE-BASED IMMUNOASSAYS

3.1 Detection of Biomolecules and Microorganisms

A colorimetric detection-based SPIA has been reported for 25 hydroxyvitamin D using an AuNP-based IA [40] (Fig. 16.1B) and the total blood cholesterol using the CardioCheck test strips [32]. An SP-based colorimetric reader (SPCR) based on the bottom illumination of a 96-well MTP by the gadget's screen detects C-reactive protein (CRP) [31] (Fig. 16.1D). The SP-based imaging of colorimetric CRP IA has also been performed by placing the MTP on a white paper. However, its detection range and limit of detection (LOD) are inferior to those of SPCR (Table 16.1). Of interest is the SPIA for the activity of trypsin, chymotrypsin, and enterokinase, which is based on a QD-based assay [39] with colorimetric detection. A compact optomechanical SP attachment using magnetic polystyrene microspheres and QD-labeled Ab detects recombinant bovine somatotropin [51]. A different format is the paper microfluidic-based fluorescent IA of β-D-galactosidase [52].

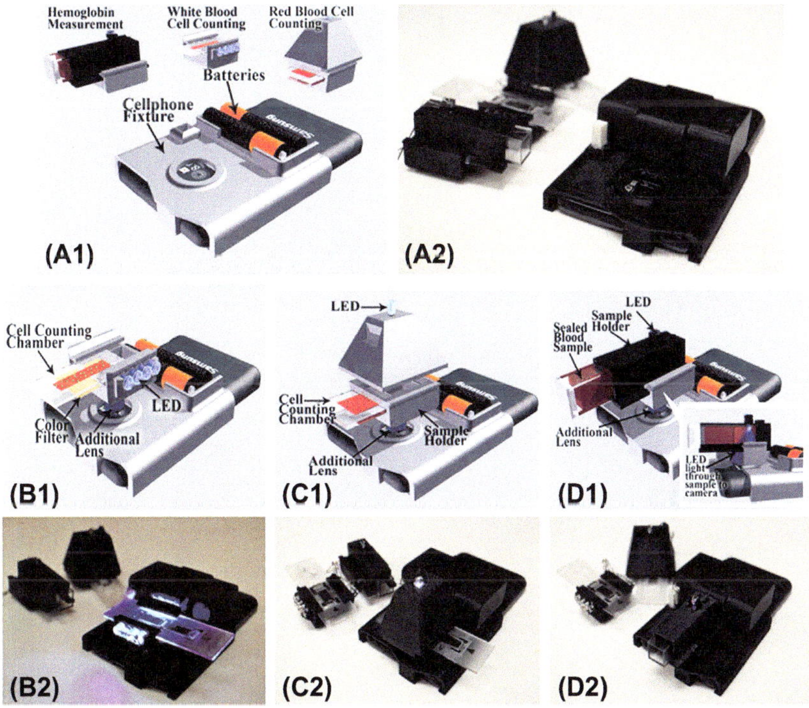

FIGURE 16.9 Smartphone-based cytometry platform [15]. (A1) and (A2) Illustration and picture of smartphone-based blood analysis platform comprising a base attachment with two AA batteries and a universal port for adapting three different add-on components for WBC counting, RBC counting, and hemoglobin measurements, respectively. (B1) and (B2) Illustration and picture of the WBC counting device. (C1) and (C2) Illustration and picture of the RBC counting device. (D1) and (D2) Illustration and picture of Hb measurement device. *Reproduced with permission from the Royal Society of Chemistry.*

Biochemiluminescent detection—based SPIA using a LFA minicartridge format has been developed for total bile acid [38] (Fig. 16.1C), total cholesterol [38] (Fig. 16.1C), and salivary cortisol [37]. A rapid diagnostic test reader (65 g) provides the readout of LFIAs for malaria, tuberculosis (TB), and human immunodeficiency virus (HIV) [9].

An SP-based Mie scattering-optimized LFIA (Fig. 16.5A) detected thyroid-stimulating hormone (TSH) in less than 15 min [18] with an LOD of 0.31 mIU/L, which is below the minimum accepted TSH concentration of 0.4 mIU/L for hyperthyroidism. The results obtained for the detection of TSH in human serum samples agreed well with those obtained by the clinical IA.

An angle-resolved SPR detection system [26] detected β_2 microglobulin (β_2M), a biomarker for cancer, kidney disease, and inflammatory disorders. The developed SPR system was tested for the detection of β_2M using the commercial Biacore sensor chip CM5 and a custom-made chip, which were

TABLE 16.1 Smartphone-Based Immunoassays (SPIAs)

Types of SPIAs	Analytes Detected	Bioanalytical Performance	References
Colorimetric detection	Peanut	Detects 1–25 ppm of peanut within 20 min.	[23]
	Multianalyte sensing arrays	An image processing algorithm for ascorbic acid, glucose, protein, ketones, urobilinogen, bilirubin, and RBC in paper-based urine strips.	[46]
	25-Hydroxyvitamin D	AuNP-based IA, detection range: 15–110 nM.	[40]
	Herpes virus nucleic acid	DNA sequences (1 mM) from KSHV using an NP assay in microfluidic chips.	[47]
	CRP	CRP (0.035–0.182 μg/mL) with an LOD of 0.026 μg/mL.	[25]
	CRP	SBCR-based IA for CRP (0.3–81 ng/mL) with an LOD of 0.4 ng/mL.	[31]
	Cholesterol	The total cholesterol level in blood (140–400 mg/dL) within 60 s.	[32]
	Proteases	A QD-based multiplex assay for the activity of trypsin, chymotrypsin, and enterokinase (pM-nM)	[39]
	Cocaine	AuNP and aptamer-based assay for cocaine (0.2–0.8 mg/mL).	[48]
	HE4	Microchip ELISA for HE4 in urine with the linearity of 19.5–1250 ng/mL and an LOD of 19.5 ng/mL.	[49]
	Hg	Plasmonic AuNP and aptamer-based assay for Hg^{2+} with an LOD of ~3.5 ppb.	[34]
Fluorescent detection	*Mycobacterium tuberculosis*	*M. tuberculosis* bacilli in Auramine O-stained sputum smears.	[17]
	Salmonella typhimurium DNA	A fluorescent microscope and a paper microfluidics-based assay for detecting *S. typhimurium* down to 10^4 CFU/mL.	[50]
	rbST Abs	rbST Abs in milk extracts by an IA using magnetic polystyrene microspheres and QD-labeled detection Ab.	[51]

	Salmonella and *Escherichia coli O157*	LFIA strips for *Salmonella* and *E. coli O157* with an LOD of 10^5 CFU/mL.	[35]
	β-D-galactosidase	A paper microfluidics-based IA for β-D-galactosidase, with linearity (0.7–12 nM) and LOD (0.7 nM).	[52]
	E. coli	To detect *E. coli O157* by QD-based sandwich IA in glass capillaries with an LOD of ~5–10 CFU/mL.	[24]
Luminescent detection	Bile acid	An LFA minicartridge format for total bile acid in serum and oral fluid with linearity and an LOD of 0.5–100 and 0.5 μM.	[38]
	Cholesterol	Using an LFA minicartridge format for total cholesterol in serum with linearity and an LOD of 20–386 and 20 mg/dL.	[38]
	Salivary cortisol	An LFIA for salivary cortisol (0.3–60 ng/mL) with an LOD of 0.3 ng/mL.	[37]
SPR detection	β2M	An angle-resolved SPR detection system for β2M with an LOD of 0.1 μg/mL.	[26]
Electrochemical detection	*Pf*HRP2 biomarker for malaria	Detects *Pf*HRP2 in 15 min with an LOD of 16 ng/mL using a compact embedded circuit, disposable microfluidic chip, and capillary flow.	[18]
Cytometry	Optofluidic fluorescent imaging–based flow cytometer	Compact, lightweight, and cost-effective SP-based flow cytometer with a fluorescent resolution of ~2 μm and capability to analyze samples >0.1 mL.	[28]
	RBCs, WBCs, and Hb	Determines the density of RBCs, WBCs, and Hb in blood using just 10 μL sample.	[15]
Other detection techniques	*Salmonella* and *E. coli*	A paper microfluidic device and a Mie scattering-based IA for *Salmonella* and *E. coli* (10^2–10^5 CFU/mL).	[53]
	TSH	An SP-based Mie scatter optimized LFA for TSH with an LOD of 0.4 mIU/L.	[8]
	AFB1	A reader and an LFIA using a competitive binding format for AFB1 with the linearity of 5–1000 μg/kg and an LOD of 5 μg/kg.	[54]

AuNP, gold nanoparticle; *CRP*, C-reactive protein; *IA*, immunoassay; *LFIA*, lateral flow IA; *LOD*, limit of detection; *QD*, quantum dot; *SBCR*, smartphone-based colorimetric reader; *SP*, smartphone; *TSH*, thyroid-stimulating hormone.

both functionalized with monoclonal mouse–antihuman β_2M. It detected a pathophysiological range of β_2M in humans with negligible nonspecific binding and an LOD of about 0.1 μg/mL.

An SP-based EC biosensor [18] (Fig. 16.3) was employed for the rapid, highly sensitive, and specific detection of *Pf*HRP2 with an LOD of 16 ng/mL. The procedure takes 15 min, uses two loading steps for the sample and reporter (*Pf*HRP2 detection antibodies conjugated to horse-radish peroxidase) solutions, and detects *Pf*HRP2 in human serum with an LOD of 16 ng/mL.

A ~28 g attachment-based SPIA, based on a QD-based sandwich IA in an anti-*Escherichia coli* O157: H7 antibody-functionalized glass capillary, was employed for fluorescent detection of *E. coli* O157 [24]. It detected *E. coli* with an LOD of ~5–10 CFU/mL in water and milk samples. The SP device has an imaging FOV of, e.g., 11 mm × 11 mm and can simultaneously monitor ~10 functionalized capillary tubes (length: ~11 mm; inner and outer diameters: ~100 and ~170 μm, respectively). Battery-powered UV LEDs directly butt-coupled to the capillary array acts as an optofluidic waveguide that leads to uniform excitation of QD-labeled *E. coli* particles. The emitted fluorescence light is imaged by the SP camera by placing an additional lens with a focal length of 15 mm between the capillary array and the SP camera. Recently, an SPIA based on paper microfluidics together with the Mie scattering was also used for detecting *Salmonella* and *E. coli* [53].

GENTAG, Inc. [55] together with MacroArray Technologies, LLC, developed a cell phone–based urine analysis platform and a disposable immunoassay, which detects a prostate cancer diagnostic marker 1 (PCADM-1), which is a marker for the diagnosis of prostate cancer. The developed SPIA for PCADM-1 would be a better test than the prostate-specific antigen (PSA)-based blood tests.

Mobile Assay Inc. employed their SP LFIA reader based on MIR for the screening of drugs (e.g., cocaine and benzoylecgonine), food pathogens (e.g., *Botrytis cinerea*) [12], water, and aflatoxin. It detected cocaine and benzoylecgonine in the concentration range of 0.1–300 and 0.003–0.1 ng/mL, respectively. The technology has the potential for the detection of food pathogens that might enable rapid and efficient tracking of the origin and severity of outbreaks. It may also help the food producers to take appropriate measures at an early stage to prevent the wide-scale distribution of contaminated food. The device also enables the early detection and tracking of *B. cinerea* (a fungus that causes significant damage to plants and flowers), which might increase agricultural productivity due to decreased crop losses.

3.2 Detection of Other Analytes

An SP colorimetric reader based on the use of a 40-g optomechanical SP attachment was employed as the food allergen testing platform, which detected varying amounts (1–25 part per million, ppm) of peanut concentrations in commercially available cookies within ~20 min with an LOD of ~1 ppm [23]. In another approach, an SP-based microchip ELISA was developed for the detection of HE4 cancer biomarker in urine [49]. Furthermore, an SP-based reader for one-dot LFIA using a competitive binding format was reported for aflatoxin B1 (AFB1) [54].

3.3 Detection of Blood Glucose

The SP-based blood glucose meters by iHealth Inc. are a striking development [56] (Fig. 16.4). The iHealth Align detects the entire diabetic blood glucose pathophysiological range (1.1–33.3 M) in 5 s using 0.7 μL of fresh capillary whole blood. It requires the users to perform the blood glucose measurement by pricking their fingers and uploading the sample onto the test strip plugged into the iHealth Align. The glucose measurement is performed by standard amperometric detection of blood glucose using glucose oxidase. The results are displayed on the SP's screen in a few seconds and stored securely inside the smart app, iHealth Gluco. In contrast, the Wireless Smart Gluco-Monitoring System is a stand-alone blood glucose meter, which once synced with the SP can be used for up to 500 glucose measurements without any need for SP as the results are stored in the device itself. The measurements are transferred to the SP via Bluetooth. These devices can be purchased online for US$ 16.95 and use only iHealth test strips (costs US$ 12.50 for a pack of 50 strips). The iHealth Gluco app enables users to view trends and statistics of their blood glucose readings for up to 90 days and automatically determines the remaining test strips in the vial and their expiry based on which it alerts the users to buy a new vial. The app acts as an automatic logbook where users can add notes to their stored readings and set up reminders for medication and insulin dosages, and recording of pre- and postmeal glucose. The users can share their results with family members, friends, caregivers, and doctors.

An SP-based painless transdermal glucose sensing and monitoring system has also been developed by GENTAG, Inc. in collaboration with Georgetown University and Science Applications International Corporation. The system comprises disposable skin patches with wireless sensors, where each patch can be used for more than 100 glucose measurements for prolonged duration at customized time intervals. Furthermore, high density patches of up to 40,000 test sites can be produced. The technology obviates the finger-pricking as it employs the withdrawn interstitial fluid (IF) by applying heat, which ablates the stratum corneum that delivers the IF to the detector layer within the skin patch through capillary action.

4. CONCLUSIONS, CHALLENGES, AND FUTURE TRENDS

The developments in the field have shown the excellent merger of various disciplines to create innovative SPIA. The SP-based devices with associated smart applications are being used for personalized general healthcare monitoring, notably analyses of blood glucose, weight, heart rate, etc., a major driving force in the creation of new enterprises such as iHealth LLC, Runtastic, AliveCor, GENTAG, and Holomic LLC. The success of SP-based devices and smart applications together with improved diagnostic procedures [57,58] will warrant further developments and applications of commercially viable SPIA in the coming years. The SP-based devices, employing the FDA-cleared and/or CE-marked products such as LFIA strips are expected to have a higher market penetration compared with innovative concepts or technologies that require stringent clinical validation.

One major challenge is to identify relevant algorithms for highly reproducible imaging in a particular SP model or a specific application. Cloud computing significantly reduces healthcare costs but is challenged by serious concern about the security and privacy of personal data. The national data security guidelines of most countries require confidential and sensitive data to be physically stored within their national boundaries. The cloud service providers, such as Amazon Web Services and Google, have started complying with these guidelines. Efforts are also under way to define an international cloud computing standard, e.g., the creation of EuroCloud and Google's Data Liberation Front [5]. In addition, confidential data in personal health records must be encrypted to prevent any abuse.

Considering the integration of various contributing technologies in SP, there will be continuous improvement in the SPIA's performance and features. The coming years will witness the emergence of new biosensing concepts as well as the use of recently developed concepts for important and routine SP-IVD applications.

REFERENCES

[1] Measuring the Information Society Report. http://www.itu.int/en/ITU-D/Statistics/Pages/publications/mis2016.aspx.

[2] S.K. Vashist, O. Mudanyali, E.M. Schneider, R. Zengerle, A. Ozcan, Cellphone-based devices for bioanalytical sciences, Anal. Bioanal. Chem. 406 (14) (2014) 3263–3277.

[3] S.K. Vashist, J.H.T. Luong, Trends in in vitro diagnostics and mobile healthcare, Biotechnol. Adv. 34 (3) (2016) 137–138.

[4] S.K. Vashist, P.B. Luppa, L.Y. Yeo, A. Ozcan, J.H.T. Luong, Emerging technologies for next-generation point-of-care testing, Trend Biotechnol. 33 (11) (2015) 692–705.

[5] S.K. Vashist, E.M. Schneider, J.H.T. Luong, Commercial smartphone-based devices and smart applications for personalized healthcare Monitoring and Management, Diagnostics 4 (3) (2014) 104–128.

[6] A. Ozcan, Mobile phones democratize and cultivate next-generation imaging, diagnostics and measurement tools, Lab Chip 14 (17) (2014) 3187–3194.

[7] D. Erickson, D. O'Dell, L. Jiang, V. Oncescu, A. Gumus, S. Lee, et al., Smartphone technology can be transformative to the deployment of lab-on-chip diagnostics, Lab Chip 14 (17) (2014) 3159–3164.

[8] D.J. You, T.S. Park, J.Y. Yoon, Cell-phone-based measurement of TSH using Mie scatter optimized lateral flow assays, Biosens. Bioelectron. 40 (1) (2013) 180–185.

[9] O. Mudanyali, S. Dimitrov, U. Sikora, S. Padmanabhan, I. Navruz, A. Ozcan, Integrated rapid-diagnostic-test reader platform on a cellphone, Lab Chip 12 (15) (2012) 2678–2686.

[10] D.C. Cooper, B. Callahan, P. Callahan, L. Burnett, Mobile image ratiometry: a new method for instantaneous analysis of rapid test strips, Nat. Preced. (2012). https://doi.org/10.1038/npre.2012.6827.1.

[11] B.A. Cadle, K.C. Rasmus, J.A. Varela, L.S. Leverich, C.E. O'Neill, R.K. Bachtell, et al., Cellular phone based image acquisition and quantitative ratiometric method for detecting cocaine and benzoylecgonine for biological and forensic applications, Subst. Abuse 4 (2010) 21–33.

[12] D.C. Cooper, Mobile image ratiometry for the detection of Botrytis cinerea (Gray Mold), Nat. Preced. (2012). https://doi.org/10.1038/npre.2012.6989.1.

[13] O. Mudanyali, D. Tseng, C. Oh, S.O. Isikman, I. Sencan, W. Bishara, et al., Compact, lightweight and cost-effective microscope based on lensless incoherent holography for telemedicine applications, Lab Chip 10 (11) (2010) 1417–1428.

[14] W. Bishara, U. Sikora, O. Mudanyali, T.W. Su, O. Yaglidere, S. Luckhart, et al., Holographic pixel super-resolution in portable lensless on-chip microscopy using a fiber-optic array, Lab Chip 11 (7) (2011) 1276–1279.

[15] H. Zhu, I. Sencan, J. Wong, S. Dimitrov, D. Tseng, K. Nagashima, et al., Cost-effective and rapid blood analysis on a cell-phone, Lab Chip 13 (7) (2013) 1282–1288.

[16] Z.J. Smith, K. Chu, A.R. Espenson, M. Rahimzadeh, A. Gryshuk, M. Molinaro, et al., Cell-phone-based platform for biomedical device development and education applications, PLoS One 6 (3) (2011) e17150.

[17] D.N. Breslauer, R.N. Maamari, N.A. Switz, W.A. Lam, D.A. Fletcher, Mobile phone based clinical microscopy for global health applications, PLoS One 4 (7) (2009) e6320.

[18] P.B. Lillehoj, M.C. Huang, N. Truong, C.M. Ho, Rapid electrochemical detection on a mobile phone, Lab Chip 13 (15) (2013) 2950–2955.

[19] G.E. Geiger, J.W. Oberding, R.N. Ward, K.D. White, Blood Glucose Meter/Modem Interface Arrangement, vol. 20, February 2007. U.S. Patent No. 7,181,350.

[20] J.P. Peeters, Diagnostic Radio Frequency Identification Sensors and Applications Thereof vol. 13, December 2011. U.S. Patent No. 8,077,042.

[21] X. Wang, M.R. Gartia, J. Jiang, T.-W. Chang, J. Qian, Y. Liu, et al., Audio jack based miniaturized mobile phone electrochemical sensing platform, Sens. Actuators B Chem. 209 (2015) 677–685.

[22] Y. Lu, W. Shi, J. Qin, B. Lin, Low cost, portable detection of gold nanoparticle-labeled microfluidic immunoassay with camera cellphone, Electrophoresis 30 (4) (2009) 579–582.

[23] A.F. Coskun, J. Wong, D. Khodadadi, R. Nagi, A. Tey, A. Ozcan, A personalized food allergen testing platform on a cellphone, Lab Chip 13 (4) (2013) 636–640.

[24] H. Zhu, U. Sikora, A. Ozcan, Quantum dot enabled detection of *Escherichia coli* using a cell-phone, Analyst 137 (11) (2012) 2541–2544.

[25] C.M. McGeough, S. O'Driscoll, Camera phone-based quantitative analysis of C-reactive protein ELISA, IEEE Trans. Biomed. Circuits Syst. 7 (5) (2013) 655–659.

[26] P. Preechaburana, M.C. Gonzalez, A. Suska, D. Filippini, Surface plasmon resonance chemical sensing on cell phones, Angew. Chem. 51 (46) (2012) 11585–11588.

[27] A.F. Coskun, A.E. Cetin, B.C. Galarreta, D.A. Alvarez, H. Altug, A. Ozcan, Lensfree optofluidic plasmonic sensor for real-time and label-free monitoring of molecular binding events over a wide field-of-view, Sci. Rep. 4 (2014) 6789.

[28] H. Zhu, S. Mavandadi, A.F. Coskun, O. Yaglidere, A. Ozcan, Optofluidic fluorescent imaging cytometry on a cell phone, Anal. Chem. 83 (17) (2011) 6641–6647.

[29] L. Shen, J.A. Hagen, I. Papautsky, Point-of-care colorimetric detection with a smartphone, Lab Chip 12 (21) (2012) 4240–4243.

[30] S.K. Vashist, E. Marion Schneider, R. Zengerle, F. von Stetten, J.H.T. Luong, Graphene-based rapid and highly-sensitive immunoassay for C-reactive protein using a smartphone-based colorimetric reader, Biosens. Bioelectron. 66 (0) (2015) 169–176.

[31] S.K. Vashist, T. van Oordt, E.M. Schneider, R. Zengerle, F. von Stetten, J.H.T. Luong, A smartphone-based colorimetric reader for bioanalytical applications using the screen-based bottom illumination provided by gadgets, Biosens. Bioelectron. 67 (2015) 248–255.

[32] V. Oncescu, M. Mancuso, D. Erickson, Cholesterol testing on a smartphone, Lab Chip 14 (4) (2014) 759–763.

[33] N. Lopez-Ruiz, V.F. Curto, M.M. Erenas, F. Benito-Lopez, D. Diamond, A.J. Palma, et al., Smartphone-based simultaneous pH and nitrite colorimetric determination for paper microfluidic devices, Anal. Chem. 86 (19) (2014) 9554–9562.

[34] Q. Wei, R. Nagi, K. Sadeghi, S. Feng, E. Yan, S.J. Ki, et al., Detection and spatial mapping of mercury contamination in water samples using a smart-phone, ACS Nano 8 (2) (2014) 1121–1129.

[35] V.K. Rajendran, P. Bakthavathsalam, B.M. Jaffar Ali, Smartphone based bacterial detection using biofunctionalized fluorescent nanoparticles, Microchim. Acta 181 (15–16) (2014) 1815–1821.

[36] F.M. Walker, K.M. Ahmad, M. Eisenstein, H.T. Soh, Transformation of personal computers and mobile phones into genetic diagnostic systems, Anal. Chem. 86 (18) (2014) 9236–9241.

[37] M. Zangheri, L. Cevenini, L. Anfossi, C. Baggiani, P. Simoni, F. Di Nardo, et al., A simple and compact smartphone accessory for quantitative chemiluminescence-based lateral flow immunoassay for salivary cortisol detection, Biosens. Bioelectron. 64 (2015) 63–68.

[38] A. Roda, E. Michelini, L. Cevenini, D. Calabria, M.M. Calabretta, P. Simoni, Integrating biochemiluminescence detection on smartphones: mobile chemistry platform for point-of-need analysis, Anal. Chem. 86 (15) (2014) 7299–7304.

[39] E. Petryayeva, W.R. Algar, Multiplexed homogeneous assays of proteolytic activity using a smartphone and quantum dots, Anal. Chem. 86 (6) (2014) 3195–3202.

[40] S. Lee, V. Oncescu, M. Mancuso, S. Mehta, D. Erickson, A smartphone platform for the quantification of vitamin D levels, Lab Chip 14 (8) (2014) 1437–1442.

[41] H. Yu, Y. Tan, B.T. Cunningham, Smartphone fluorescence spectroscopy, Anal. Chem. 86 (17) (2014) 8805–8813.

[42] http://www.cellmic.com/.

[43] H. Zhu, O. Yaglidere, T.W. Su, D. Tseng, A. Ozcan, Cost-effective and compact wide-field fluorescent imaging on a cell-phone, Lab Chip 11 (2) (2011) 315–322.

[44] Q. Wei, H. Qi, W. Luo, D. Tseng, S.J. Ki, Z. Wan, et al., Fluorescent imaging of single nanoparticles and viruses on a smart phone, ACS Nano 7 (10) (2013) 9147–9155.

[45] I. Navruz, A.F. Coskun, J. Wong, S. Mohammad, D. Tseng, R. Nagi, et al., Smart-phone based computational microscopy using multi-frame contact imaging on a fiber-optic array, Lab Chip 13 (20) (2013) 4015–4023.

[46] J.I. Hong, B.Y. Chang, Development of the smartphone-based colorimetry for multi-analyte sensing arrays, Lab Chip 14 (10) (2014) 1725–1732.

[47] M. Mancuso, E. Cesarman, D. Erickson, Detection of Kaposi's sarcoma associated herpesvirus nucleic acids using a smartphone accessory, Lab Chip 14 (19) (2014) 3809–3816.

[48] J.E. Smith, D.K. Griffin, J.K. Leny, J.A. Hagen, J.L. Chavez, N. Kelley-Loughnane, Colorimetric detection with aptamer-gold nanoparticle conjugates coupled to an android-based color analysis application for use in the field, Talanta 121 (2014) 247–255.

[49] S. Wang, X. Zhao, I. Khimji, R. Akbas, W. Qiu, D. Edwards, et al., Integration of cell phone imaging with microchip ELISA to detect ovarian cancer HE4 biomarker in urine at the point-of-care, Lab Chip 11 (20) (2011) 3411–3418.

[50] C.F. Fronczek, T.S. Park, D.K. Harshman, A.M. Nicolini, J.-Y. Yoon, Paper microfluidic extraction and direct smartphone-based identification of pathogenic nucleic acids from field and clinical samples, RSC Adv. 4 (22) (2014) 11103.

[51] S.K. Ludwig, H. Zhu, S. Phillips, A. Shiledar, S. Feng, D. Tseng, et al., Cellphone-based detection platform for rbST biomarker analysis in milk extracts using a microsphere fluorescence immunoassay, Anal. Bioanal. Chem. 406 (27) (2014) 6857–6866.

[52] N.K. Thom, G.G. Lewis, K. Yeung, S.T. Phillips, Quantitative fluorescence assays using a self-powered paper-based microfluidic device and a camera-equipped cellular phone, RSC Adv. 4 (3) (2014) 1334–1340.

[53] T.S. Park, W. Li, K.E. McCracken, J.Y. Yoon, Smartphone quantifies *Salmonella* from paper microfluidics, Lab Chip 13 (24) (2013) 4832–4840.

[54] S. Lee, G. Kim, J. Moon, Performance improvement of the one-dot lateral flow immunoassay for aflatoxin B1 by using a smartphone-based reading system, Sensors 13 (4) (2013) 5109–5116.

[55] http://www.gentag.com/.

[56] http://www.ihealthlabs.com/.

[57] S.K. Vashist, E. Lam, S. Hrapovic, K.B. Male, J.H.T. Luong, Immobilization of antibodies and enzymes on 3-aminopropyltriethoxysilane-functionalized bioanalytical platforms for biosensors and diagnostics, Chem. Rev. 114 (21) (2014) 11083–11130.

[58] V. Gubala, L.F. Harris, A.J. Ricco, M.X. Tan, D.E. Williams, Point of care diagnostics: status and future, Anal. Chem. 84 (2) (2012) 487–515.

Chapter 17

Immunoassays: Future Prospects and Possibilities

Sandeep K. Vashist[1], John H.T. Luong[2]
[1]*IDS Immunodiagnostic Systems Deutschland GmbH, Frankfurt am Main, Germany;* [2]*University College Cork, Cork, Ireland*

1. IMMUNOASSAYS: TRENDS AND PROSPECTS

Significant advances in immunoassay (IA) formats and technologies during the last few decades [1−3] have drastically shortened the duration of an IA procedure manyfold in comparison with conventional radioimmunoassay (RIA) and enzyme-linked immunosorbent assay (ELISA). ELISA is also difficult to develop for point-of-care (POC) use and requires significant technical expertise. Furthermore, the IA procedure is simplified as many process steps are integrated and automated using dedicated instruments. The evolution of microfluidics [4] and lab-on-a-chip (LOC) technologies [5] based on automated microfluidic protocols [6−8] has resulted in prospective semi- and fully automated immunoassays, enabling point-of-care testing (POCT) of analytes at remote settings. As the number of steps in the IA has also been critically reduced, a wide range of rapid IA chemistries have resulted in novel IA formats to detect analytes in less than 30 min [9−14]. The use of emerging micro- and nanomaterials has paved the way for IAs to achieve detectivity in the range of fg/mL. Such low or even lower detection limits are needed for early detection of cancer or the prompt detection of relapse after cancer treatment and therapy. To date, advances together with novel detection concepts such as chemiluminescent resonance energy transfer and fluorescence resonance energy transfer have further pushed down the limits of sensitivity to attogram levels. The use of nanomaterials and novel IA concepts results in the emergence of visualized IAs [15,16], which can read and analyze pertinent information based on the visual detection. This interesting trend in immunodiagnostics has instigated an interest toward the development of instrument-free IAs.

The surface plasmon resonance (SPR)-based BiaCore instruments for label-free IAs from GE Healthcare (previously Pharmacia Biosensor AB,

Handbook of Immunoassay Technologies. https://doi.org/10.1016/B978-0-12-811762-0.00017-7

Biacore AB Corporation) are another striking advance, leading to the emergence of real-time label-free IAs. These systems have become the global standard to probe biomolecular interactions, the development of rapid IAs for analytes, and screening of appropriate immunological components for an IA [17,18]. Besides GE Healthcare as the leader in the field of SPR, other competitors have also developed prospective SPR-based instruments for IAs. The main characteristic features of SPR-based IAs are critically reduced sample-to-answer time of only a few minutes; minimal samples and reagents requirement of only tens of microliters; a label-free microfluidic IA procedure; and a fully automated analysis with a high throughput. Moreover, a large number of surface-functionalized SPR chips, immobilization chemistries, regeneration buffers, and assay protocols further facilitate the rapid development of SPR-based IAs.

The low-cost IA formats including the disposable paper-based IAs [19–22] have been the long-cherished vision as they would be ideal for the developing nations and remote settings. The developed complementary technologies during the last decade have led to the emergence of several prospective paper-based IAs. However, they still require intensive efforts as their bioanalytical performance is not optimal for real sample testing and inferior compared with that of the established IA formats.

The most significant trend during the last decade is the emergence of smartphone (SP)-based readers, bioanalytical platforms, LOC formats, and IA formats [2,3]. Considering the enormous outreach and penetration of cellphones with over 7 billion users, they are seen as the ideal devices for POCT [3,23,24]. Indeed, several SP-based IAs have been developed with superior bioanalytical performances and capabilities far exceeding those performed using advanced central laboratory instruments. The SP-based IAs can be performed at any place and time without the need for a continuous power supply, highly skilled analysts, and healthcare professionals. The IA data and results with spatiotemporal tagging are transmitted securely by cloud computing and Internet/wireless connectivity to the central server. The results are analyzed by authorized healthcare professionals and their feedback is transmitted back to the users. A large number of Food and Drug Administration (FDA), the United States, and Conformité Europėene (CE) compliant SP-based devices for the monitoring of basic healthcare parameters have been used by millions of users [24]. Several SP-based IAs have been developed and demonstrated [3,23], such as the lateral flow IA, ELISA, colorimetric IA, chemiluminescent IA, fluorescent IA, spectroscopy, SPR, Mie scattering, paper-based IA, electrochemical IA, and LOC-based IA.

Doubtlessly, the future IA formats would be based on the multiplex detection of biomarkers [25–28] for probing a particular disease using the effective clinical scoring algorithm. Single biomarkers often have inadequate predictive value, only about 70% for prostate-specific antigen, which is known as one of the better single biomarkers. Thus, the highly reliable prediction of a specific cancer requires measuring a number of relevant biomarker proteins. The multiplex bead-based Luminex assays by Thermo Fisher Scientific [26] and multiplex IAs by Meso Scale Discovery [25] are the two critical advances in this field. Other prospective clinical scoring algorithms have already been demonstrated and used in the clinical diagnosis [29–33]. Such IA formats based on multiplex detection and clinical scoring would lead to better and sustained healthcare by enabling effective monitoring and management of disease, therapeutic regimen, and personalized or healthcare intervention.

The coming years would witness continuous innovation and breakthroughs in IA formats, bioanalytical platforms, biosensing concepts, detection devices, and complementary technologies. The next-generation IA formats would be based on highly simplified IA procedures, requiring minimal process steps and critically reduced sample volumes without any washing steps. They would be fully automated, cost-effective, and equipped with smart features such as mobile healthcare capabilities by interfacing or integration with SP and gadgets. They could be deployed and operated at centralized, decentralized, remote, and personalized settings with minimal or no requirement for power supply and skilled analysts.

2. EVOLVING HEALTHCARE

IAs provide the precise diagnosis of diseases based on the quantification of their specific biomarkers in the patient's sample. The early-stage detection of such biomarkers with high analytical precision, specificity, and sensitivity is an essential requirement to establish a proper treatment at a very early stage of disease more effectively with minimal costs. As the recent focus of healthcare has shifted to the rapid POCT of biomarkers [34], several POC analyzers are commercially available from Roche, Siemens, Abbott, and other companies. Such systems are being used in the intensive care units, neonatal stations, and emergency rooms for the rapid, precise, and accurate determination of disease biomarkers.

The recent years have witnessed an evolving trend toward the development of IA formats and technologies for the multiplex detection of biomarkers [25–28]. There has also been an extensive increase in efforts toward the development of clinical scoring algorithms [29–33]. The clinical score offers more effective disease diagnosis and management as each biomarker for a disease is determined and assigned a weightage based on its contribution toward the specific disease.

An international consensus and focus on developing healthcare technologies would empower the patients and the general population to effectively monitor and manage their own health by personalized lifestyle and nutritional interventions. Of notice is the emergence of mobile healthcare technologies with an exponential increase during the recent years [2,3]. A wide range of FDA- and CE-approved mobile healthcare devices for basic health parameters have been successfully commercialized and widely used [24]. Other potential IA formats and devices have been demonstrated for various healthcare applications. These advances have inculcated a tremendous interest toward the development of next-generation mobile healthcare technologies. SP- and gadget-based mobile healthcare devices and IAs offer adequate analytical performance, compared with that of an established IA using central laboratory instruments [12,14,35–39]. As equipped with easy-to-use intuitive software, superior biosensing capabilities, and advanced features, the SP-based devices provide better healthcare delivery by ensuring better and sustainable health outcomes. The ability to set up reminders for diagnostics, medications, and medical appointments would considerably improve the patience's compliance. Such resulting enriched data would be indispensable for healthcare, policy-making, and economics [40–42].

Currently, most governments focus on the self-empowerment of the general population, which can only be feasible via predictive diagnostics and personalized medicine, especially by the use of evolving mobile healthcare technologies. The commercial wearable healthcare gadgets by Apple, Samsung, Fitbit, iHealth, Garmin, and Runtastic are the trendsetters in this direction [23,24]. The collection of all basic health parameters of the users onto a dashboard via an intuitive smart application is a significant advance, enabling them to monitor their health and pursue an active lifestyle. The data obtained from these devices are stored on the gadgets, processed immediately, and transmitted securely to the cloud. iHealth has further designed the advanced software for healthcare management, i.e., the healthcare professionals and users can effectively monitor and manage the health of the users. Many essential IAs would be developed on such mobile healthcare technologies in the near future. This would come with an easy-to-use and easy-to-understand clinical score that is calculated in real time on the gadgets and enables the users to monitor and manage their health. There is an increase in international efforts for the development of predictive diagnostics for predominant diseases such as diabetes, cancer, cardiovascular diseases, obesity, hypertension, and psychological stress. The recently developed telecommunication and networking tools together with integrated smart applications for mobile healthcare have made the direct communications possible between the doctors and their patients. Therefore, the field of personalized medicine with telemedicine capabilities is coming up [43].

The striking changes in diagnostics, healthcare, smart health applications [44], and mobile healthcare technologies have led to the creation of new businesses, a prerequisite of future endeavors. The emergence of novel technologies will change the landscape of healthcare delivery by improved healthcare monitoring and management, leading to revolutionary advances in healthcare.

3. CHALLENGES

Among a wide range of developed IA formats, platforms, and devices, the most widely used IA formats used in clinical diagnostics are still conventional manual ELISA and automated analyzer–based enzyme-linked IA (EIA). The former is used in developing countries having limited healthcare resources, whereas the latter is the accepted standard for clinical testing in developed countries. There are several developed IA formats during the last two decades with superior analytical performance and appealing features, but they might not be clinically accredited. Unless this requirement is fulfilled, they cannot be employed for clinical testing and clinical decision-making. Most of the new IA formats, such as those based on LOC, point-of-care testing, and biosensors, have not been able to demonstrate the desired high throughput, high sensitivity, and specificity, precision, and robustness as shown by ELISA and automated EIA. Nevertheless, the mass spectroscopy–based IAs have shown considerable promise for selected clinical analytes and are being used routinely in the clinical diagnostics. Similarly, SPR-based IAs have become the global standard in industrial, healthcare, and bioanalytical settings for the real-time, label-free detection of analytes. The study of biomolecular detections via SPR has given rise to several breakthroughs and advances in immunodiagnostics. However, IAs based on mass spectroscopy and SPR require advanced and expensive instruments, costly consumables (such as SPR chips), and highly skilled analysts. Therefore, they are not prospective for large-scale testing of samples in clinical diagnostics.

Similarly, the lateral flow IA (LFIA) has been very useful for the rapid qualitative or semiquantitative analysis of various disease biomarkers in just a few minutes. Although they are mainly used for home pregnancy testing, they have also been employed by primary health workers, physicians in remote and decentralized settings, and emergency health professionals for the diagnosis of various infectious and highly prevalent diseases. The recent developments in the field, especially the SP-based integrated rapid diagnostic readers for LFIA [36], provide the quantitative determination of analytes and enhance the potential applications of LFIA in remote healthcare monitoring and management. However, the main limitation of this assay format is the low throughput and insufficient sensitivity in comparison with the highly sensitive EIA. It also requires manual handling and is thus not fully automated.

Several micro- and nanomaterial-based IA formats demonstrated recently have very high sensitivity. However, only the magnetic and latex beads have been successful in terms of commercial viability [45,46]. Magnetic beads–based IAs have become the global standard for many clinical analyzer–based IAs [45]. In contrast, latex beads are being used for the turbidimetric IA. Considering very low detection limits and high selectivity offered by labeled immunoassays, this approach will continue playing an important role in the detection of multiple biomarkers. Labels can be fluorophores, magnetic beads, and active enzymes with easily detectable products. The nanomaterials field, with unprecedented optical tunability, electrical and electrochemical qualities, will be responsible for significant developments in the design of novel immunoassays. Among several metallic nanoparticles (NPs), gold NPs are very popular because of their unique strong absorption in the UV-visible spectrum up to the near-infrared region and offer facile bioconjugation. They can be used to improve the sensitivity for protein detection in immunoassays based on SPR, quartz crystal microbalance (QCM), fluorescence spectroscopy, etc. Other metal-based NPs with redox properties could also trigger significant surface impedance and conductance changes once NPs are present as labeling tags or modifiers of transducer surfaces. Quantum dots, semiconductor NPs with spectral characteristics, will be exploited as unique fluorescent markers. However, their applications in more complex samples are facing problems related to nonspecific binding, reproducibility, and long-term operation. Several paper-based IAs have been attempted, but they still lack the high specificity and sensitivity desired for the testing of clinical samples [3].

The SP-based IAs, devices, and bioanalytical platforms are the most recent developments [2,3,23,24], which have tremendously extended the outreach of diagnostics and have made feasible the testing of an analyte in remote settings. The developed IA formats have demonstrated superior analytical performance and advanced features in comparison with central laboratory instruments–based IAs. Apart from the IA results, the devices provide enriched data, trends, communication and telemedicine tools, and advanced features. Doubtlessly, many SP-based IAs will be adopted by the healthcare, especially for the remote and personalized monitoring of patients and tackling of emergencies and epidemics. The ongoing potential concerns of the data safety, ownership, and transmission are posing a serious threat for the acceptance of web-based monitoring and mobile healthcare tools in hospitals and clinics. However, there is an international consensus and a strong demand for the mobile healthcare and POC technologies to foster more effective healthcare monitoring and management. Accordingly, most of the existing limitations would be counteracted effectively in the near future by the policy-makers, pertinent researchers, and developers, which would pave the way to the augmented use of smart IA technologies in the healthcare settings.

Label or label-free is another subject of debate and intensive research. To date, most immunoassays still require a label such as fluorophores, magnetic beads, and active enzymes, attached to the target. In general, labeling a recognition biomolecule can drastically change its binding properties and other bioanalytical performances [47]. Albeit this issue is a minor concern for DNA biosensors, it is very problematical for protein targets. No label is required for SPR or impedance sensing and this feature is, perhaps, the major motivation for the development of SPR and impedance immunoassays. Unlabeled immunoassays have also been attempted using atomic force cantilevers, QCM, field-effect transistors (FETs), or nanowire transistors. For label-free measurements with high detection sensitivity, the target must be substantially larger than the probe or exhibit significantly different properties from the probe. For instance, the SPR signal is extremely sensitive to the dielectric constant as well as the mass of the material in contact with the Au film. Thus, one key question is whether such label-free methods can offer sufficient selectivity for use in real-world applications. For example, the cancer biomarker prostate-specific antigen needs to be detected at c.2 ng/mL, whereas cytokines exist in the blood in only pg/mL levels. Apparently, the serum will have some biomarkers at very low concentrations, and other members at much higher concentrations. In general, conventional ELISAs often achieve pg/mL detection limits if the probe–target affinity is high. However, SPR and QCM still require labels to reach sensitivity levels needed for clinical cancer biomarker detection. Label-free impedance immunosensing has received considerable attention, but this simple and low-cost technique may not be able to provide ultrasensitive protein analyses [48]. Hence, metal NP labels or gold NP labels are still needed to achieve the required low detection limit. These solutions are contrary to the goal of creating a POC detection owing to the requirement of extra time, more sample preparation steps, and increasing complexity. To date, label-free impedance-based immunoassays have not gained significant commercial success. Label-free methods such as AFM and ultrasensitive nanowire transistors also require intensive development and validation together with significant cost reductions to compete with currently more accessible, lower-cost procedures. Another critical issue is nonspecific binding (NSB) of nonanalyte serum proteins in the aforementioned nonspecific label-free methods as any adsorbed molecule will contribute to the NSB background. Extensive testing and validation using serum or saliva will be necessary to establish detection sensitivity, selectivity, and reliability. As the biological samples consist of "nontarget" complex mixtures of proteins, lipids, and carbohydrates at high concentrations, the achievement of high specificity and low levels of NSB is always a daunting task. Cancer detection and diagnostics will require accurate detection of 4–10 biomarkers in such samples and the progress on this front is "somewhat" sluggish.

4. FUTURE DIRECTIONS

Among several IA formats developed during the last decade, only a select few have made their way successfully into the healthcare and bioanalytical settings. However, the significant developments made in the field of immunodiagnostics have critically improved the immunoanalytical systems and led to many additional features to significantly increase the instrument capabilities. The clinical analyzers have become highly advanced, automated, and equipped with high-end information technology (IT) features for high-throughput automated sample testing and data processing. The current generation of analyzers has advanced from barcode coding of sample reagents and quality controls to sample chip-based integrated coding. The automated IT capabilities provide fully automated solutions for data management, storage, and processing, which considerably reduce the workload for the healthcare professionals. Other complementary technologies would further be incorporated into the clinical analyzers to render them further advanced in terms of a fully automated immunoanalytical system with complete system integration.

The emergence of mobile healthcare and SP/smart gadget–based IA readers and formats has further provided the futuristic IA formats for POCT applications. They have considerably increased the outreach of diagnostics and healthcare by providing the much desired IA formats that mankind has been looking for. The coming years would witness a significant increase in such IA formats and devices together with the mobile healthcare capabilities that would change the landscape of healthcare delivery and clinical diagnostics. The current limitations in mobile healthcare [40], such as data security, encryption, transfer, storage, processing, and ownership, will be counteracted effectively together with the growing trend toward increasing use of cloud computing [49]. The continuing improvements in the camera, integrated sensors, smart applications, processors, and other features of SP would further increase the sensitivity and bioanalytical performance of the SP-based IA formats. Indeed, SP-based readers have a superior analytical performance in comparison with the advanced, bulky, and expensive reading instruments being used in the central laboratory [12,14,35–39]. Therefore, SP-based readers would be the ideal for IA applications in POCT. Several opto-mechanical SP attachments have enabled the SP to read several types of detection signals, such as colorimetry, spectroscopy, fluorescence, chemiluminescence, SPR, Mie scatter, and electrochemical. [2,3,23,24]. However, the availability of various SP models with varying specifications and features would make it really challenging to realize a universal SP-based IA that can be performed with same bioanalytical performance on all SP models. This is a challenge that could be tackled effectively via a stand-alone SP-interfaced device that has the same readout sensors and optomechanical attachments for a highly reproducible readout of IA detection signal. The development of such IA formats and devices is ongoing with intensive efforts.

Fully automated and integrated LOC IA formats for POC multiplex detection of disease biomarkers would be one of the major instruments to be developed. There is a need for novel strategies for the prolonged storage of dry, liquid, and freeze-dried reagents, which could be employed in LOC IAs for POCT [3]. In addition, the next-generation IAs would involve multiplex detection and rapid IA strategies. There is an evolving trend toward predictive diagnostics based on innovative clinical diagnostic algorithms, enabling the prediction of a particular disease much before its onset. This would allow the healthcare professionals and persons prevent the disease by medical or lifestyle intervention.

The advanced immunoanalytical systems such as BiaCore by GE Healthcare and mass spectrophotometers would assist the development of IAs by providing the critical information pertaining to biomolecular interactions and IA components. Similarly, a recently developed novel immunoanalytical system developed by Dynamic Biosensors GmbH, Germany, based on SwitchSENSE technology would enable the determination of the hydrodynamic radius of the molecule with an accuracy of 0.1 nm [50–52]. It will also act as a molecular ruler to determine the height of the bound molecule above the biochip substrate. These systems would critically improve the understanding of IAs by providing the desired information that would assist the development of IAs. Moreover, the continuous improvements in such systems based on the developments in technologies would further lead to improved immunodiagnostics.

The current trend is strongly inclined toward the empowerment of the general population with the desired technologies and tools, which would enable them to effectively monitor and manage their own health. The next-generation IAs will be rapid, cost-effective, easy-to-use, simplified, highly sensitive, automated, and integrated. They would be equipped with complementary technologies such as smart applications and information technology to provide better healthcare monitoring and management. More unknown biomarker proteins for a plethora of diseases with very low serum concentrations will be unraveled once immunoassays with very high sensitivity can be performed routinely.

REFERENCES

[1] P.B. Luppa, L.J. Sokoll, D.W. Chan, Immunosensors-principles and applications to clinical chemistry, Clin. Chim. Acta 314 (2001) 1–26.

[2] S.K. Vashist, J.H.T. Luong, Trends in *in vitro* diagnostics and mobile healthcare, Biotechnol. Adv. 34 (2016) 137–138.

[3] S.K. Vashist, P.B. Luppa, L.Y. Yeo, et al., Emerging technologies for next-generation point-of-care testing, Trends Biotechnol. 33 (2015) 692–705.

[4] G.M. Whitesides, The origins and the future of microfluidics, Nature 442 (2006) 368–373.

[5] M.I. Mohammed, M.P. Desmulliez, Lab-on-a-chip based immunosensor principles and technologies for the detection of cardiac biomarkers: a review, Lab Chip 11 (2011) 569–595.

[6] M. Karle, S.K. Vashist, R. Zengerle, et al., Microfluidic solutions enabling continuous processing and monitoring of biological samples: a review, Anal. Chim. Acta 929 (2016) 1–22.

[7] C.-C. Lin, J.-H. Wang, H.-W. Wu, et al., Microfluidic immunoassays, J. Lab. Autom. 15 (2010) 253–274.

[8] E. Livak-Dahl, I. Sinn, M. Burns, Microfluidic chemical analysis systems, Ann. Rev. Chem. Biomol. Eng. 2 (2011) 325–353.

[9] G.B. Stevens, D.A. Silver, A. Zgaga-Griesz, et al., Bioluminescence assay for the highly sensitive detection of botulinum neurotoxin A activity, Analyst 138 (2013) 6154–6162.

[10] S.K. Vashist, G. Czilwik, T. Van Oordt, et al., One-step kinetics-based immunoassay for the highly sensitive detection of C-reactive protein in less than 30 min, Anal. Biochem. 456 (2014) 32–37.

[11] S.K. Vashist, E. Marion Schneider, J.H.T. Luong, Rapid sandwich ELISA-based *in vitro* diagnostic procedure for the highly-sensitive detection of Human fetuin A, Biosens. Bioelectron. 67 (2015) 73–78.

[12] S.K. Vashist, E. Marion Schneider, R. Zengerle, et al., Graphene-based rapid and highly-sensitive immunoassay for C-reactive protein using a smartphone-based colorimetric reader, Biosens. Bioelectron. 66 (2015) 169–176.

[13] S.K. Vashist, E.M. Schneider, J.H.T. Luong, Surface plasmon resonance-based immunoassay for human fetuin A, Analyst 139 (2014) 2237–2242.

[14] S.K. Vashist, T. Van Oordt, E.M. Schneider, et al., A smartphone-based colorimetric reader for bioanalytical applications using the screen-based bottom illumination provided by gadgets, Biosens. Bioelectron. 67 (2015) 248–255.

[15] D. Cecchin, R. De La Rica, R. Bain, et al., Plasmonic ELISA for the detection of gp120 at ultralow concentrations with the naked eye, Nanoscale 6 (2014) 9559–9562.

[16] R. De La Rica, M.M. Stevens, Plasmonic ELISA for the ultrasensitive detection of disease biomarkers with the naked eye, Nat. Nanotechnol. 7 (2012) 821–824.

[17] L. Jason-Moller, M. Murphy, J. Bruno, Overview of Biacore systems and their applications, Curr. Protoc. Protein Sci. (2006) 19.13.11–19.13.14.

[18] P. Thillaivinayagalingam, J. Gommeaux, M. Mcloughlin, et al., Biopharmaceutical production: applications of surface plasmon resonance biosensors, J. Chromatogr. B Anal. Technol. Biomed. Life Sci. 878 (2010) 149–153.

[19] X. Ge, A.M. Asiri, D. Du, et al., Nanomaterial-enhanced paper-based biosensors, Trends Anal. Chem. 58 (2014) 31–39.

[20] J. Hu, S. Wang, L. Wang, et al., Advances in paper-based point-of-care diagnostics, Biosens. Bioelectron. 54 (2014) 585–597.

[21] A.W. Martinez, S.T. Phillips, E. Carrilho, et al., Simple telemedicine for developing regions: camera phones and paper-based microfluidic devices for real-time, off-site diagnosis, Anal. Chem. 80 (2008) 3699–3707.

[22] M. Santhiago, E.W. Nery, G.P. Santos, et al., Microfluidic paper-based devices for bioanalytical applications, Bioanalysis 6 (2014) 89–106.

[23] S.K. Vashist, O. Mudanyali, E.M. Schneider, et al., Cellphone-based devices for bioanalytical sciences, Anal. Bioanal. Chem. 406 (2014) 3263–3277.

[24] S.K. Vashist, E.M. Schneider, J.H.T. Luong, Commercial smartphone-based devices and smart applications for personalized healthcare monitoring and management, Diagnostics 4 (2014) 104–128.

[25] D. Dabitao, J.B. Margolick, J. Lopez, et al., Multiplex measurement of proinflammatory cytokines in human serum: comparison of the meso scale discovery electrochemiluminescence assay and the cytometric bead array, J. Immun. Methods 372 (2011) 71–77.

[26] S.A. Dunbar, M.R. Hoffmeyer, Microsphere-based multiplex immunoassays: development and applications using Luminex® xMAP® technology, in: The Immunoassay Handbook: Theory and Applications of Ligand Binding, ELISA and Related Techniques, vol. 157, 2013.

[27] S.F. Kingsmore, Multiplexed protein measurement: technologies and applications of protein and antibody arrays, Nat. Rev. Drug Discov. 5 (2006) 310–320.

[28] E. Petryayeva, W.R. Algar, Multiplexed homogeneous assays of proteolytic activity using a smartphone and quantum dots, Anal. Chem. 86 (2014) 3195–3202.

[29] D.J. Buysse, M.L. Hall, P.J. Strollo, et al., Relationships between the Pittsburgh sleep quality index (PSQI), Epworth sleepiness scale (ESS), and clinical/polysomnographic measures in a community sample, J. Clin. Sleep. Med. 4 (2008) 563–571.

[30] D.T. Krausman, R.P. Allen, Sleep Scoring Apparatus and Method, 2005. U.S. Patent No. 6,878,121.

[31] A.A. Van Der Heijden, M.M. Ortegon, L.W. Niessen, et al., Prediction of coronary heart disease risk in a general, pre-diabetic, and diabetic population during 10 years of follow-up: accuracy of the Framingham, SCORE, and UKPDS risk functions: the Hoorn Study, Diabetes Care 32 (2009) 2094–2098.

[32] J.E. Ware Jr., M. Kosinski, S.D. Keller, A 12-Item short-form health survey: construction of scales and preliminary tests of reliability and validity, Med. Care 34 (1996) 220–233.

[33] D.S. Weiss, The impact of event scale: revised, in: Cross-cultural Assessment of Psychological Trauma and PTSD, Springer, 2007, pp. 219–238.

[34] P.B. Luppa, C. Müller, A. Schlichtiger, et al., Point-of-care testing (POCT): current techniques and future perspectives, Trends Anal. Chem. 30 (2011) 887–898.

[35] B. Berg, B. Cortazar, D. Tseng, et al., Cellphone-based hand-held microplate reader for point-of-care testing of enzyme-linked immunosorbent assays, ACS Nano 9 (2015) 7857–7866.

[36] O. Mudanyali, S. Dimitrov, U. Sikora, et al., Integrated rapid-diagnostic-test reader platform on a cellphone, Lab Chip 12 (2012) 2678–2686.

[37] A. Venkatesh, F.V. Stetten, R. Zengerle, et al., Smartphone Easy ELISA-a Point-of-Care Platform for Cost-Effective and Easy-to-Use In Vitro Diagnostics, MikroSystemTechnik Kongress, 2015, in: https://www.vde-verlag.de/proceedings-en/454100124.html.

[38] Q. Wei, W. Luo, S. Chiang, et al., Imaging and sizing of single DNA molecules on a mobile phone, ACS Nano 8 (2014) 12725–12733.

[39] H. Zhu, I. Sencan, J. Wong, et al., Cost-effective and rapid blood analysis on a cell-phone, Lab Chip 13 (2013) 1282–1288.

[40] H. IaT, I. Yaqoob, N.B. Anuar, et al., The rise of "big data" on cloud computing: review and open research issues, Inf. Syst. 47 (2015) 98–115.

[41] T.B. Murdoch, A.S. Detsky, The inevitable application of big data to health care, JAMA 309 (2013) 1351–1352.

[42] W. Raghupathi, V. Raghupathi, Big data analytics in healthcare: promise and potential, Health Inf. Sci. Syst. 2 (2014) 3.

[43] C. Pattichis, E. Kyriacou, S. Voskarides, et al., Wireless telemedicine systems: an overview, IEEE Antennas Propag. Mag. 44 (2002) 143–153.

[44] A.S. Mosa, I. Yoo, L. Sheets, A systematic review of healthcare applications for smartphones, BMC Med. Inf. Decis. Making 12 (2012) 67.

[45] M.A. Gijs, Magnetic bead handling on-chip: new opportunities for analytical applications, Microfluid. Nanofluid. 1 (2004) 22–40.

[46] C.T. Lim, Y. Zhang, Bead-based microfluidic immunoassays: the next generation, Biosens. Bioelectron. 22 (2007) 1197–1204.

[47] S.K. Vashist, Effect of antibody modifications on its biomolecular binding as determined by surface plasmon resonance, Anal. Biochem. 421 (2012) 336–338.

[48] J.S. Daniels, N. Pourmand, Label-Free impedance biosensors: opportunities and challenges, Electroanalysis 19 (2007) 1239–1257.
[49] L. Griebel, H.U. Prokosch, F. Kopcke, et al., A scoping review of cloud computing in healthcare, BMC Med. Inf. Decis. Making 15 (2015) 17.
[50] J. Knezevic, A. Langer, P.A. Hampel, et al., Quantitation of affinity, avidity, and binding kinetics of protein analytes with a dynamically switchable biosurface, JACS 134 (2012) 15225–15228.
[51] U. Rant, W. Kaiser, P.A. Hampel, et al., Apparatus and Method for Evaluating Characteristics of Target Molecules, 2015. U.S. Patent No. 9,164,055.
[52] U. Rant, E. Pringsheim, W. Kaiser, et al., Detection and size analysis of proteins with switchable DNA layers, Nano Lett. 9 (2009) 1290–1295.

Index

'*Note*: Page numbers followed by "f" indicate figures and "t" indicate tables.'

A

Abaxis Inc., 10–11, 115–116, 121–122
Abbott Diagnostics, 5
Accuracy, bioanalytical requirement for immunoassay, 82
Acoustic wave–based immunoassays, 203–205
  clinical diagnostics, 205–213
    quartz crystal microbalance immunosensors, 205–212
      direct immunosensors, 205–211
      indirect immunosensors, 211–212
      sandwich-amplified immunosensors, 212
    surface acoustic wave immunosensors, 213
      direct immunosensors, 213
  commercial acoustic wave immunosensors, 227–228
  integrated acoustic wave immunosensors for point of care, 225–227
  market potential, 228
  microbial pathogens and toxins detection, 213–216
    quartz crystal microbalance immunosensors, 213–216
      direct immunosensors, 213–214
      indirect immunosensors, 214–215
      sandwich-amplified immunosensors, 215–216
    surface acoustic wave immunosensors, 216
      direct immunosensors, 216
      sandwich-amplified immunosensors, 216
  parasites detection, 216–217
    quartz crystal microbalance immunosensors, 216–217
      direct immunosensors, 216
      indirect immunosensors, 217
      sandwich-amplified immunosensors, 217
  quartz crystal microbalance and surface acoustic wave immunoassays
    electronic noses, 223
    in environmental monitoring and allergens detection, 224
  viruses detection, 217–223
    quartz crystal microbalance immunosensors, 217–222
      direct immunosensors, 217–221
      indirect immunosensors, 221
      sandwich-amplified immunosensors, 221–222
    surface acoustic wave immunosensors, 222–223
Alere, 119
Alere Afinion AS100 analyzer, 120–121, 120f
AliveCor, 449–450
Allergens detection
  quartz crystal microbalance and surface acoustic wave immunoassays, 224
AlphaLISA, 1–2, 8, 9f, 121–122, 425–426
Amino groups, 21–22
3-Aminopropyltriethoxysilane (APTES), 21, 334
Ammunition monitoring, 322–323
Amperometry, 362
Analytes detection, smartphone-based immunoassays, 448–449
Analytical measurement range, 88
Antibodies, 373–374
Antibody formats
  recombinant antibody formats, 50–51
  used for phage display, 48–50
Antibody immobilization chemistries, 23–37
  covalent, 23–32
  noncovalent, 34–36
    adsorption, 34
    affinity, 34–36
  oriented, 32–34
  peptide nucleic acid and deoxyribonucleic acid–directed, 37

Antibody immobilization chemistries (*Continued*)
  recombinant antibody, 37
  site-specific, 36–37
Antibody libraries, 58–63
  immune libraries, 59–60
  naïve natural libraries, 60
  naïve semisynthetic libraries, 61
  naïve synthetic libraries, 61–62
  special library designs, 62–63
  synthetic libraries from nonhuman species, 63
Antibody structure
  immunoassays, 3–5, 4f
Apple, 427
Approaches based on receptor–ligand–receptor binding, 367–368
Aptamer–antibodies hybrid, 375–376
Aptamer-based immunosensors, 399–401
ArcDia TPX, 113
Atomic force microscopy (AFM), 39t–40t, 305
AuNPs, 185
Automated clinical analyzer–based immunoassays, 6t–7t
Automatic detections, paper-based immunoassays, 192–194

B

Bacteria detection
  quartz crystal microbalance–based sensors, 343–345
Bacterial/cancer cells–based immunosensors, 381–388
BDBiosciences, 5
Bending behavior of microcantilevers, 312–314
BiaCore, 5–7, 11
Bias, bioanalytical performance parameter, 88
Bibliographic data, of lateral flow immunoassays, 174–175
Bimodal interferometers, 264–265
Bioanalytical performance parameters, 85–89
  analytical measurement range, 88
  bias, 88
  carryover, 87
  clinically reportable range, 88
  cross-reactivity, 89
  errors, 86–87
  Hook effect, 88
  interference, 87
  limit of blank, 85
  limit of detection, 85–86
  limit of quantification, 86
  linear interval limit, 86
  linear range, 87–88
  measuring interval limit, 86
  method comparison, 88–89
  quality controls, 87
Bioanalytical requirements, for immunoassay, 82–89
  accuracy, 82
  calibration curve, 84–85
  precision, 82
  recovery, 84
  reproducibility, 83
  selectivity, 82
  sensitivity, 82
  stability, 83
BioLegend, 5
BioMerieux SA, 5
Biomolecules detection
  microcantilever-based sensors, 317–319
    DNA, 317–319
    glucose, 318–319
    lipoproteins, 318
    myoglobin, 318
    prostate-specific antigen, 317–318
    tributyrin, 319
  quartz crystal microbalance–based sensors, 335–343
Biomolecules microorganisms detection, smartphone-based immunoassays, 444–448
Biopharmaceuticals, 5
Bio-Rad Laboratories, 5, 11
Biorecognition elements in immunosensors, 372–376
  antibodies, 373–374
  aptamer–antibodies hybrid, 375–376
  truncated (fragmented) antibodies, 374–375
Blood glucose detection, smartphone-based immunoassays, 449

C

Calibration curve, bioanalytical requirement for immunoassay, 84–85
Calixarene film–coated QCM sensors, 346
Campylobacter, 7–8
Campylobacter jejuni, 343–345
Cantilever mass sensors

dynamic mode, 307, 307f
static mode, 307, 307f
Capacitive detection methods, microcantilever-based sensors, 310–311
Carbon nanoparticles, 187
Carboxyl groups, 22
CardioCheck test strips, 444–445
Carryover, bioanalytical performance parameter, 87
Cellmic, 423–424, 439–441
Cellphone, lab-on-a-chip (LOC) immunoassays, 421–425
Cellulose, 20–21
Centrifugal immunoassay, 115–117
Charge-coupled device detection methods, microcantilever-based sensors, 312
Chemical activation of transducers, 255–256
Chemicals and metal ions detection, 320–321
quartz crystal microbalance–based sensors, 347–348
"Click" chemistry, 31–32
Clinical diagnostics, acoustic wave–based immunoassays, 205–213
quartz crystal microbalance immunosensors, 205–212
direct immunosensors, 205–211
indirect immunosensors, 211–212
sandwich-amplified immunosensors, 212
surface acoustic wave immunosensors, 213
direct immunosensors, 213
Clinical immunoassays, 5
Clinically reportable range, 88
Cloud Services and Test Developer, 423–424, 439–441
Colorimetric detection, smartphone-based immunoassays, 434–435
Colorimetric immunoassay, 279–285
electrochemical immunoassay, 285–290
lateral flow assay, 279–282
plate-based colorimetric immunoassay, 282–285
Colorimetric immunosensing approaches, 368–370
Colorimetric method, paper-based immunoassays, 185–188
AuNPs, 185
carbon nanoparticles, 187
enzymes, 185–187
magnetic nanoparticles, 187–188
Commercial acoustic wave immunosensors, 227–228
Commercial availability, of microcantilever-based sensors, 325
Commercial data, of lateral flow immunoassays, 174–175
Competitive enzyme-linked immunoassay, 101
Complex-specific antibodies, 64–65
Conjugation, site-specific, 67–69
CoPc/mesoporous silica–coated QCM, 348–349
Coupling, to detection principles range, 162–163
Covalent antibody immobilization chemistries, 23–32
Critiques, lab-on-a-chip (LOC) immunoassays, 425–427
Cross-reactivity, 89
CUBE analyzer, 119, 120f
Cyclic polyolefins, 20–21
Cysteine, 36–37
Cytometry, smartphone-based immunoassays, 443

D

Desulfotomaculum, 343–345
Diagnostic immunoassays, human controls in, 64–67
Diasorin, 119
Direct immunosensors
viruses detection, 217–221
quartz crystal microbalance immunosensors, 205–211, 213–214
parasites detection, 216
surface acoustic wave immunosensors, 213, 216
Direct immunosorbent assay, 98–99
Disease targets, surface plasmon resonance immunoassays, 144
DNA, microcantilever-based sensors, 317
Drug monitoring, 64–65
Dynabeads, 116–117
Dynamic Biosensors GmbH, 463

E

Electrochemical detection, smartphone-based immunoassays, 435–438
Electrochemical immunoassay, 285–290

Electrochemical immunosensors, 359–360
advanced approaches in, 366–372
approaches based on receptor–ligand–receptor binding, 367–368
colorimetric immunosensing approaches, 368–370
electrochemical lateral flow assay, 366–367
redox cycling–based approaches, 370–372
biorecognition elements in immunosensors, 372–376
antibodies, 373–374
aptamer–antibodies hybrid, 375–376
truncated (fragmented) antibodies, 374–375
electrochemical immunosensors, 376–401
aptamer-based immunosensors, 399–401
bacterial/cancer cells–based immunosensors, 381–388
enzyme-labeled immunosensors, 388–394
label-free protein-based immunosensors, 377–381
redox-labeled immunosensors, 397–399
electrochemical techniques in immunosensing, 360–366
amperometry, 362
electrochemical impedance spectroscopy, 364–365
electrochemiluminescence, 363–364
photoelectrochemical techniques, 365–366
potentiometry, 363
voltammetry, 361–362
outlook, 401–403
Electrochemical immunosensors, 376–401
aptamer-based immunosensors, 399–401
bacterial/cancer cells–based immunosensors, 381–388
enzyme-labeled immunosensors, 388–394
label-free protein-based immunosensors, 377–381
redox-labeled immunosensors, 397–399
Electrochemical impedance spectroscopy, 364–365
Electrochemical lateral flow assay, 366–367
Electrochemical method, paper-based immunoassays, 188–189
Electrochemical techniques in immunosensing, 360–366
amperometry, 362
electrochemical impedance spectroscopy, 364–365
electrochemiluminescence, 363–364
photoelectrochemical techniques, 365–366
potentiometry, 363
voltammetry, 361–362
Electrochemiluminescence, 363–364
Electronic nose, microcantilever-based sensors, 323
Elimination of nonspecific binding, 260
ELISA, 8
EMD Biosciences Inc., 5
Environmental monitoring
quartz crystal microbalance and surface acoustic wave immunoassays, 224
Enzyme-labeled immunosensors, 388–394
Enzyme-linked immunoassays, 97–98
competitive, 101
conventional types, 98–101
critiques, 121–122
emerging, 102–119
highly simplified rapid sandwich, 104–105
high-sensitivity sandwich, 102–104
microfluidics-based, 114–117
multiplex, 108–111
nano/micromaterial-based, 111–113
paper-based, 113–114
smartphone-based, 118–119
wash-free, 105–108
immunosorbent assay, 98–101
direct, 98–99
indirect, 99
sandwich, 99–101
outlook, 121–122
portable analyzer–based immunoassays, 119–121
Enzymes, and colorimetric method, 185–187
Epoxy groups, 23
Errors, bioanalytical performance parameter, 86–87
Escherichia coli O157: H7, 7–8
Ethyl chloroformate, 20–21
1-Ethyl-3-(3-dimethylaminopropyl) carbodiimide hydrochloride (EDC), 21
Ethylenediamine, 21
EuroCloud, 427
Eurolyser Diagnostica, 119
Explosives detection, 322–323

F

Fabrication of microcantilevers, 314–315
Field emission SEM (FESEM), 39t–40t
Fluorescent detection, smartphone-based immunoassays, 434–435
Fluorescent immunoassay, 290–299
  fluorescence resonance energy transfer assay, 293–299
  heterogeneous immunoassay, 290–293
Fluorescent polystyrene microsphere, 274–275
Food, and immunoassays, 7–8
Fragmented antibodies, 374–375
Future prospects and possibilities, immunoassays
  challenges, 459–461
  evolving healthcare, 457–458
  future directions, 461–463
  trends and prospects, 455–457

G

Garmin, 427
Gaseous analytes detection, 319–320
  quartz crystal microbalance–based sensors, 348–349
GE Healthcare, 2, 5–7, 11, 455–456
GENTAG, Inc., 424, 448–450
Glass, 20–21
Glucose, microcantilever-based sensors, 318–319
Google's Data Liberation Front, 427
Graphene nanoplatelet (GNP)-based IA, 111–113, 112t
Gyrolab Bioaffy CDs, 416
Gyrolab xPlore, 416
Gyros AB, 10–11
Gyros Protein Technologies, 115–116, 121–122, 416

H

Heavy chains, 4–5
Heterogeneous immunoassay, 290–293
Highly simplified rapid sandwich enzyme-linked immunoassay, 104–105
High-sensitivity sandwich enzyme-linked immunoassay, 102–104
Holomic LLC, 449–450
Holomic rapid diagnostic reader (HRDR), 10–11, 10f
Holomic Rapid Diagnostic Test Reader, 423–424, 439–441
Holomic Substance Test Assistant (HSTA), 439–441
Hook effect, 88
Humidity detection, 321–322
Hydroxyl groups, 20–21

I

IBIS Technologies, 11
iHealth Align, 449
iHealth Gluco, 449
iHealth Labs, 427
iHealth LLC, 449–450
Immobilization of biomolecules, 20
Immobilization of recognition molecules, 256–260
Immune libraries, 59–60
Immunoassays, 81
  antibody structure, 3–5, 4f
  bioanalytical performance parameters, 85–89
    analytical measurement range, 88
    bias, 88
    carryover, 87
    clinically reportable range, 88
    cross-reactivity, 89
    errors, 86–87
    Hook effect, 88
    interference, 87
    limit of blank, 85
    limit of detection, 85–86
    limit of quantification, 86
    linear interval limit, 86
    linear range, 87–88
    measuring interval limit, 86
    method comparison, 88–89
    quality controls, 87
  bioanalytical requirements for, 82–89
    accuracy, 82
    calibration curve, 84–85
    precision, 82
    recovery, 84
    reproducibility, 83
    selectivity, 82
    sensitivity, 82
    stability, 83
  critiques, 89–92
  formats, 8–12
  future trends, 12–14
  need for, 5–8
    clinical, 5
    environment and security, 7
    food, 7–8

Immunoassays (*Continued*)
- industrial, 5–7
- personalized healthcare, 8
- outlook, 89–92
- overview, 1–3

Immunodiagnostics, 19–20
- antibody immobilization chemistries, 23–37
  - covalent, 23–32
  - noncovalent, 34–36
    - adsorption, 34
    - affinity, 34–36
  - oriented, 32–34
  - peptide nucleic acid and deoxyribonucleic acid–directed, 37
  - recombinant antibody, 37
  - site-specific, 36–37
- challenges, 38–41
- future trends, 38–41
- surface characterization, 37
- surface functionalization chemistries, 20–23
  - amino groups, 21–22
  - carboxyl groups, 22
  - epoxy groups, 23
  - hydroxyl groups, 20–21
  - sulfhydryl groups, 22–23

Immunoglobulin A (IgA), 4–5

Immunoglobulin D (IgD), 4–5

Immunoglobulin E (IgE), 4–5

Immunoglobulin G (IgG), 3–4

Immunoglobulin M (IgM), 4–5

Immunohistochemistry antibodies and tissue panning, 64

Immunosorbent assay, 98–101
- direct, 98–99
- indirect, 99
- sandwich, 99–101

In vitro selection of antibodies, 63–69
- complex-specific antibodies, 64–65
- conjugation, site-specific, 67–69
- diagnostic immunoassays, human controls in, 64–67
- drug monitoring, 64–65
- future, 70–72
- immunohistochemistry antibodies and tissue panning, 64
- outlook, 69–72
- sandwich pair selection, 64–65

Indirect immunosensors
- quartz crystal microbalance immunosensors, 211–212, 214–215
  - parasites detection, 217
- viruses detection, 221

Indirect immunosorbent assay, 99

Industrial immunoassays, 5–7

Integrated acoustic wave immunosensors for point of care, 225–227

Interference, bioanalytical performance parameter, 87

Interferometric sensors, 247–253

Interferometry detection methods, microcantilever-based sensors, 311–312

Interferometry-based immunoassays
- application, 260–265
  - bimodal interferometers, 264–265
  - Mach–Zehnder interferometers, 261–263
  - Young Interferometers, 263–264
- future perspectives, 265–266
- general context, 241–243
- operation, principles of, 243–253
  - interferometric sensors, 247–253
  - label-free optical sensing, 243–247
- sensor surface functionalization, 253–260
  - chemical activation of transducers, 255–256
  - elimination of nonspecific binding, 260
  - immobilization of recognition molecules, 256–260

K

KOH, 20–21

L

LabDisk, 10–11, 115–116, 116f, 416

Label-free optical sensing, 243–247

Label-free protein-based immunosensors, 377–381

LABGEO IB10, 119

Lab-on-a-chip (LOC) immunoassays, 415–425
- cellphone, 421–425
- critiques, 425–427
- microfluidic, 416–419
- outlook, 425–427
- paper, 420

Langmuir–Blodgett films, 334

Lateral flow assay, 279–282

Lateral flow assays, smartphone-based immunoassays, 439–442

Lateral flow immunoassays, 157–162
advances in, 162–170
coupling to a range of detection principles, 162–163
multianalyte and quantitative lateral flow immunoassays, 163–166
reading multispot lateral flow assays, 166–170
lateral flow reader for microarrays, 166–167
reading arrays by a smartphone application, 170
real-time video reader, 167–170
basic technology, 158–159
bibliographic data, 174–175
challenges and future directions, 171–174
combination with amplification procedures, 172–173
commercial data, 174–175
history, 157
integration of lateral flow immunoassays with paper diagnostics, 173–174
recognition elements, 159
signal labels, 160–161
storage of lateral flow devices, 161–162
updated SWOT analysis, 171–172
opportunities, 171
threats, 171–172
weaknesses, 171
LFIA, 10–11
Life Technologies Corporation, 5
Limit of blank, 85
Limit of detection, 85–86
Limit of quantification, 86
Linear interval limit, 86
Linear range, 87–88
Lipoproteins, microcantilever-based sensors, 318
Lower limit of quantification (LLOQ), 82
Luminescent detection, smartphone-based immunoassays, 434–435
Luminex, 2
Luminex IA, 12
Luminex xMAP, 121

M

Mach–Zehnder interferometers, 261–263
MacroArray Technologies, LLC, 424, 448
Magnetic method, paper-based immunoassays, 189–190
Magnetic microbeads, 275–277
Magnetic nanoparticles, 187–188
Market potential, quartz crystal microbalance and surface acoustic wave immunoassays, 228
Mass spectroscopy–based IAs, 2–3
Measuring interval limit, 86
Meso Scale Diagnostics LLC, 425–426
Microbial pathogens and toxins detection, 213–216
quartz crystal microbalance immunosensors, 213–216
direct immunosensors, 213–214
indirect immunosensors, 214–215
sandwich amplified immunosensors, 215–216
surface acoustic wave immunosensors, 216
direct immunosensors, 216
sandwich-amplified immunosensors, 216
Microcantilever-based sensors, 305–306, 315–323
ammunition monitoring, 322–323
applications, 316t
bending behavior of microcantilevers, 312–314
biomolecules detection, 317–319
DNA, 317–319
glucose, 318–319
lipoproteins, 318
myoglobin, 318
prostate-specific antigen, 317–318
tributyrin, 319
chemicals and metal ions detection, 320–321
commercial availability, 325
detection methods, 308–312
capacitive, 310–311
charge-coupled device, 312
interferometry, 311–312
optical, 308–309
optical diffraction grating, 312
piezoelectric, 311
piezoresistive, 309–310
electronic nose, 323
explosives detection, 322–323
fabrication of microcantilevers, 314–315
future trends, 326
gaseous analytes detection, 319–320
humidity and pH detection, 321–322
nanocantilevers, 323–325

Microcantilevers
and their modes of operation, 306–308
for cantilever mass sensors, 307–308
Microfluidic, lab-on-a-chip (LOC) immunoassays, 416–419
Microfluidic ELISA, 2
Microfluidic immunoassays, 2
Microfluidics-based enzyme-linked immunoassay, 114–117
centrifugal, 115–117
optimiser, 114–115
Microgravimetry, 2–3
Micromaterial-based immunoassay, 274–279
fluorescent polystyrene microsphere, 274–275
magnetic microbeads, 275–277
nanomaterial-based immunoassay, 277–279
Microorganisms detection, smartphone-based immunoassays, 444–448
Microscopy, smartphone-based immunoassays, 442–443
Microtiter plate (MTP) readers, 1–2
Millipore Corporation, 5
Mobile Assay Inc., 441–442, 448
Mobile Image Ratiometry (MIR), 441–442
Monoclonal antibody generation, 47–51
antibody formats used for phage display, 48–50
antibody libraries, 58–63
immune libraries, 59–60
naïve natural libraries, 60
naïve semisynthetic libraries, 61
naïve synthetic libraries, 61–62
special library designs, 62–63
synthetic libraries from nonhuman species, 63
phage display development, 47–48
phage display selection, 51–58
advantages, 53–56
affinity improvement, 57–58
other selection technologies, 58
in vitro strategies, 56–57
recombinant antibody formats, 50–51
in vitro selection of antibodies, 63–69
complex-specific antibodies, 64–65
conclusion and outlook, 69–72
conjugation, site-specific, 67–69
diagnostic immunoassays, human controls in, 65–67
drug monitoring, 64–65
future, 70–72
immunohistochemistry antibodies and tissue panning, 64
sandwich pair selection, 64–65
Multianalyte lateral flow immunoassays, 163–166
Multiplex enzyme-linked immunoassay, 108–111
Multispot lateral flow assays readings, 166–170
lateral flow reader for microarrays, 166–167
reading arrays by a smartphone application, 170
real-time video reader, 167–170
Myoglobin, microcantilever-based sensors, 318

N

Naïve natural libraries, 60
Naïve semisynthetic libraries, 61
Naïve synthetic libraries, 61–62
Nanocantilevers, 323–325
Nanomaterial-based immunoassay, 277–279
Nano/micromaterial-based enzyme-linked immunoassay, 111–113
NanoSPR Devices, 11
NaOH, 20–21
Noncovalent antibody immobilization chemistries, 34–36, 35f
adsorption, 34
affinity, 34–36
NycoCard, 120–121, 120f

O

Odor sensing, 349–351
Optical detection methods, microcantilever-based sensors, 308–309
Optical diffraction grating detection methods, microcantilever-based sensors, 312
Optimiser ELISA, 2, 8, 419
Optimiser immunoassay, 114–115
Optimiser MTP–based IAs, 114–115
Oriented antibody immobilization chemistries, 32–34
Orion Diagnostica, 119–121
Ortho-Clinical Diagnostics Inc., 5

P

Paper, lab-on-a-chip (LOC) immunoassays, 420
Paper diagnostics, integration with lateral flow immunoassays, 173–174

Paper-based enzyme-linked immunoassay, 113–114
Paper-based immunoassays
  colorimetric method, 185–188
    AuNPs, 185
    carbon nanoparticles, 187
    enzymes, 185–187
    magnetic nanoparticles, 187–188
  development of the paper-based immunoassays devices, 190–195
    automatic detections, 192–194
    quantification detection, 194–195
    semiquantification detection, 194–195
    sensitivity improvement, 191–192
  electrochemical method, 188–189
  magnetic method, 189–190
  strategies and detection principles, 184–190
  thermal method, 188
Parasites detection, 216–217
  quartz crystal microbalance immunosensors, 216–217
    direct immunosensors, 216
    indirect immunosensors, 217
    sandwich-amplified immunosensors, 217
Pathogens and viruses, surface plasmon resonance immunoassays, 141–144
PDMS, 20–21
Peptide nucleic acid and deoxyribonucleic acid–directed Ab immobilization chemistries, 37
Peptostreptococcus magnus, 33–34
Perkin Elmer, 121–122, 425–426
Personalized healthcare, and immunoassays, 8
pH detection, 321–322
Phage display
  antibody formats, 48–50
  development of, 47–48
  selection, 51–58
    advantages, 53–56
    affinity improvement, 57–58
    other selection technologies, 58
    in vitro strategies, 56–57
1,4-Phenylene diisothiocyanate (PDITC), 31
Photoelectrochemical techniques, 365–366
Piccolo Xpress, 416
Piezoelectric detection methods, microcantilever-based sensors, 311
Piezoresistive detection methods, microcantilever-based sensors, 309–310
Piranha solution, 20–21
Plasma-enhanced CVD, 22
Plate-based colorimetric immunoassay, 282–285
PMMA, 20–21
Polyelectrolytes, 334
Polyethylene, 20–21
Polyhistidine, 36–37
Polymers, 334
Polystyrene, 20–21
Poly(vinylidene fluoride) (PVDF)-coated QCM sensor, 345
Portable analyzer–based immunoassays, 119–121
Potentiometry, 363
Precision, bioanalytical requirement for immunoassay, 82
Principal component analysis (PCA), 345
Prostate-specific antigen, microcantilever-based sensors, 317–318
Proteins, 32
Pullulan, 426

Q

Quality controls, bioanalytical performance parameter, 87
Quantification detection, paper-based immunoassays, 194–195
Quantitative lateral flow immunoassays, 163–166
Quantum dot (QD)-based sandwich IA, 118
Quartz crystal microbalance immunosensors, 205–222
  applications, 336t–338t
  commercially available, 335t
  direct immunosensors, 205–211, 213–214
    parasites detection, 216
    viruses detection, 217–221
  electronic noses, 223
  in environmental monitoring and allergens detection, 224
  indirect immunosensors, 211–212, 214–215
    parasites detection, 217
    viruses detection, 221
  sandwich-amplified immunosensors, 212, 215–216
    parasites detection, 217
    viruses detection, 221–222

Quartz crystal microbalance–based sensors, 333–334
analytical applications, 351–352
special applications, 349–351
bacteria detection, 343–345
biomolecules detection, 335–343
chemical analytes detection, 347–348
future trends, 352–353
gaseous analytes detection, 348–349
volatile organic compounds detection, 345–346
QuikRead go, 120–121

R

R&DSystems, 5
Reading arrays by a smartphone application, 170
Reading multispot lateral flow assays, 166–170
lateral flow reader for microarrays, 166–167
reading arrays by a smartphone application, 170
real-time video reader, 167–170
Real-time video reader, 167–170
Recombinant antibody, 37
Recombinant antibody formats, 50–51
Recovery, bioanalytical requirement for immunoassay, 84
Redox cycling–based approaches, 370–372
Redox-labeled immunosensors, 397–399
Reichert Technologies, 11
Reproducibility, bioanalytical requirement for immunoassay, 83
Roche, 5
Runtastic, 427, 449–450

S

Salmonella, 7–8
Salmonella typhimurium, 343–345
Samsung, 115–116, 119, 121–122, 427
Sandwich immunosorbent assay, 99–101
Sandwich pair selection, 64–65
Sandwich-amplified immunosensors
viruses detection, 221–222
Sandwich-amplified quartz crystal microbalance immunosensors, 212, 215–216
parasites detection, 217
Sandwich-amplified surface acoustic wave immunosensors, 216
Scanning electron microscopy (SEM), 39t–40t
Scanning tunneling microscopy (STM), 305
Security, and immunoassays, 7
Selectivity, bioanalytical requirement for immunoassay, 82
Semiquantification detection, paper-based immunoassays, 194–195
SensiQ, 11
Sensitivity, bioanalytical requirement for immunoassay, 82
Sensitivity improvement, paper-based immunoassays, 191–192
Sensor surface functionalization, 253–260
chemical activation of transducers, 255–256
elimination of nonspecific binding, 260
immobilization of recognition molecules, 256–260
Sepharose, 20–21
Siemens, 5
Signal detection, in smartphone-based immunoassays, 434–443
colorimetric detection, 434–435
cytometry, 443
electrochemical detection, 435–438
fluorescent detection, 434–435
lateral flow assays, 439–442
luminescent detection, 434–435
microscopy, 442–443
surface plasmon resonance detection, 439
Signal labels, lateral flow immunoassays, 160–161
Siloam Biosciences Inc., 419
Site-specific antibody immobilization chemistries, 36–37
Small molecular weight targets, surface plasmon resonance immunoassays, 141
SMART analyzer, 119
Smartphone application, reading arrays by, 170
Smartphone-based enzyme-linked immunoassay, 118–119
Smartphone-based immunoassays, 433–434
applications of, 444–449
analytes detection, 448–449
biomolecules and microorganisms detection, 444–448
blood glucose detection, 449
challenges, 449–450

future trends, 449–450
signal detection in, 434–443
colorimetric detection, 434–435
cytometry, 443
electrochemical detection, 435–438
fluorescent detection, 434–435
lateral flow assays, 439–442
luminescent detection, 434–435
microscopy, 442–443
surface plasmon resonance detection, 439
SP-based EasyELISA, 9
Special antibody library designs, 62–63
Stability, bioanalytical requirement for immunoassay, 83
Storage of lateral flow devices, 161–162
Sulfhydryl groups, 22–23
Surface acoustic wave immunoassays
electronic noses, 223
in environmental monitoring and allergens detection, 224
Surface acoustic wave immunosensors, 213, 216
direct immunosensors, 213, 216
sandwich-amplified immunosensors, 216
viruses detection, 222–223
Surface characterization, antibody immobilization, 37
Surface functionalization chemistries, 20–23
amino groups, 21–22
carboxyl groups, 22
epoxy groups, 23
hydroxyl groups, 20–21
sulfhydryl groups, 22–23
Surface plasmon resonance, 129–130
biosensor platforms, 132–135
principles, 130–131
Surface plasmon resonance (SPR)-based Bia-Core instruments, 2
Surface plasmon resonance detection, smartphone-based immunoassays, 439
Surface plasmon resonance–based immunoassays, 135–144
antibody
introduction, 135
screening, 138–140
future trends, 144–150
applications, 148–150
disease targets, 144
mass spectrometry, 144–148
pathogens and viruses, 141–144
small molecular weight targets, 141
point-of-care applications, 148–150
CD-based, 148–149
mobile phone based, 149–150
Surface plasmon resonanceebased BiaCore systems, 11f
SWOT analysis, for lateral flow immunoassays, 171–172
opportunities, 171
threats, 171–172
weaknesses, 171
Synthetic libraries from nonhuman species, 63

T

Thermal gravimetric analysis (TGA), 39t–40t
Thermal method, paper-based immunoassays, 188
Thermo Fisher Scientific Inc., 5, 12
Time-of-flight secondary ion mass spectroscopy (TOF-SIMS), 39t–40t
p-Toluenesulfonyl chloride, 20–21
Transmission electron microscopy (TEM), 39t–40t
Tributyrin, microcantilever-based sensors, 319
Truncated antibodies, 374–375

V

van der Waals forces, 34
Vibrio anguillarum, 343–345
Vibrio harveyi, 343–345
Vibrio parahaemolyticus, 343–345
Vibrio vulnificus, 343–345
Viruses detection, 217–223
quartz crystal microbalance immunosensors, 217–222
direct immunosensors, 217–221
indirect immunosensors, 221
sandwich-amplified immunosensors, 221–222
surface acoustic wave immunosensors, 222–223
Volatile organic compounds detection
quartz crystal microbalance–based sensors, 345–346
Voltammetry, 361–362

W

Wash-free enzyme-linked immunoassay, 105–108
Wash-free IAs, 121
Wireless Smart Gluco-Monitoring System, 449

X

X-ray photoelectron spectroscopy (XPS), 39t–40t

Y

Young Interferometers, 263–264

Z

ZnO nanorodsecoated QCM sensor, 349